PROGRESS IN FULLERENE RESEARCH

PROGRESS IN FULLERENE RESEARCH

MILTON LANG
Editor

Nova Science Publishers, Inc.
New York

For permission to use material from this book please contact us:
Telephone 631-231-7269; Fax 631-231-8175
Web Site: http://www.novapublishers.com

Library of Congress Cataloging-in-Publication Data

Progress in Fullerene research / Milton Lang, editor.
p. cm.
Includes index.
ISBN-13: 978-1-60021-841-5 (hardcover)
ISBN-10: 1-60021-841-5 (hardcover)
1. Fullerenes--Research. 2. Nanotubes--Research. I. Lang, Milton.
QD181.C1P76 2008
546'.681--dc22

2007030063

Published by Nova Science Publishers, Inc. ✢ New York

CONTENTS

PREFACE

Until 1985, the chemical element Carbon was only known to exist in two forms - diamond and graphite. This changed when Kroto and co-workers discovered an entirely new form of carbon, which became known as C_{60} or the fullerene molecule. (This discovery later led to their award of the 1996 Nobel Prize in Chemistry.) The original discovery of C_{60} was in the soot produced from the laser ablation of graphite. Since then, other methods of production have been developed. It is also thought that isolated C_{60} molecules may be found in stars and interstellar media. It was soon discovered that C_{60} is not the only ball-like carbon molecule possible (although it is the most stable and the most dominant). The rugby-ball shaped C70 molecule is another possibility. In nanotechnology, the potential applications of carbon nanotubes (formed by combining hexagonal rings of carbon atoms only, rather than hexagons and pentagons as in C_{60}) for very small electronic devices are currently the subject of much activity. This new book presents research advances from around the globe.

Chapter 1 - This chapter reviews a fundamental study on fullerene derivatives and fullerene composites for proton conducting membranes used in polymer electrolyte fuel cells. New fullerene derivatives have been synthesized and fullerene composites have been fabricated by using Nafion and sulfonated poly (ether ether ketone) as the host polymers. AC impedance measurements have been performed to evaluate the proton conductivity of both fullerene pellets and fullerene composites under low humidity conditions. Thorough characterizations have been conducted to examine the relevant properties of the composites: the water uptake, the morphology, the water characterization, the interactions between the fullerene and the host polymer. Computer simulations have been also utilized to gain an insight into the basic nature of the fullerenes and the fullerene composites. It has been found that the fullerenes reduce the MeOH crossover in the Nafion membrane without increasing the membrane resistance and also they improve the performance of aromatic hydrocarbon polymer-based membranes under low humidity conditions.

Chapter 2 - Boron nitride (BN) nanocage clusters, BN nanohorns and BN nanotubes and were synthesized by arc-melting and thermal annealing methods, and characterized by high-resolution electron microscopy and molecular orbital calculations, and their properties were discussed. The BN clusters consisted of 4-, 6-, 8- and 10-membered BN rings satisfying the isolated tetragonal rule, which was optimized by molecular orbital calculations. Total energy calculation showed that some elements stabilize and expand the $B_{36}N_{36}$ structure. Bandgap energies of the $B_{36}N_{36}$ clusters were found to be reduced by introducing a metal atom inside the cluster, which indicates controllability of the energy gap. BN nanohorns with tetragonal BN rings were synthesized and proposed, and multi-walled BN nanohorns would be stabilized by stacking of BN nanohorns. Chiralities of BN nanotubes with zigzag-type structures were directly determined from high-resolution images, and structure models are proposed. Total energies of BN nanotubes with a zigzag-type structure were lower than those of armchair-type structure, and these results agreed well with the experimental data. Fe-filled BN nanotubes were also produced, and the [110] direction of the Fe nanowires was parallel to the BN nanotube axis. A new process for Fe or Co nanoparticles coated with BN layers in large quantity was developed. These BN nanocapsules encaging Fe or Co nanoparticles exhibited a soft magnetic characteristic and good oxidation resistances. These unique structures would be suitable materials for nanoelectronics devices, magnetic recording media and biological sensors with excellent protection against oxidation and wear. Possibility of hydrogen gas storage in BN clusters was also investigated by molecular orbital calculations, which indicated possibility of hydrogen storage of ~5 wt. %. The present work indicates that the new BN nanocage fullerene materials with various atomic structures and properties can be produced, and a guideline for designing the BN fullerene materials is summarized.

Chapter 3 - Fullerenes - or cage compounds built exclusively from carbon atoms - and their metal-containing derivatives metallofullerenes were for the first time observed in the gas phase about twenty years ago, and prepared in crystalline form about fifteen years ago. Still, an enormous amount of observed and computed data has been obtained during the time period. In addition to spheroidal fullerene cages, fullerene science also deals with other objects like elongated cylindrical bodies known as nanotubes, prepared soon after mastering the fullerene synthesis, nanocones or peapods. All the species exhibit a substantial application potential, especially for molecular electronics. Fullerenes and metallofullerenes have attracted a considerable interest from both experiment as well as computations and theory. Their experimental characterization is mostly based on ^{13}C NMR spectroscopy and more recently also on ^{3}He NMR spectroscopy. Fullerenes are commonly defined as polyhedral cages containing only carbon atoms arranged into five- and six-membered rings. Quasi-fullerenes are cages containing also other types of cycles like (topological) squares and heptagons. The very early history of carbon clusters starts with a mass-spectrometric observation of clusters up to C_{15} by Hahn and his coworkers sixty years ago. In the fifties and sixties experiments could expand up to C_{33}. At this stage, first simple computations had also been performed on small carbon clusters. Moreover, various qualitative estimations for larger carbon cages including C_{60} were presented. In the time of the C_{60} discovery, computations could support fullerene and metallofullerene research with increasingly reliable data. It is of historic interest that the four IR bands computationally predicted for C_{60} were used in analysis of the carbon-arc C_{60} synthesis. Later on, fullerene research has indeed been characterized by a close theory-experiment cooperation and application of large-scale computations. Although numerous surveys are available, the sole computations of fullerenes and metallofullerenes are reviewed relatively rarely. Structural elucidations for the known higher fullerenes C_n have been based on so called isolated pentagon rule (IPR) that claims that especially stable fullerenes should have all pentagons surrounded just by hexagons. Connectivity patterns for such cages can be generated by various topological schemes and in fact the cages can be quite numerous. Beyond n=76 isomerism of the IPR structures should generally play a role in observations. Several such mixtures of empty fullerene isomers have indeed been computed and an agreement with experiments found for even-numbered cages from C_{76} till C_{96}, while there is still only a computational prediction for C_{98}. A similar isomeric interplay has also been described for smaller fullerene systems like C_{32} or C_{36}. Various metallofullerenes can coexist in several isomeric forms as well, for example Ca@C_{72}, Yb@C_{74}, Ca@C_{82}, La@C_{82}, Tm@C_{82}, Ti_2@C_{80}, Sc_2@C_{84}, Ti_2@C_{84}, or Sc_2@C_{76}. Still other interesting isomeric sets can be created by derivatives of fullerenes or nanotubes. Although the inter-isomeric separation energies are important, they alone cannot predict the relative stabilities of the isomers. Owing to very high temperatures, entropy contributions can even over-compensate the enthalpy terms. Hence, the enthalpy-entropy interplay represents an essential feature of fullerenes and metallofullerenes and is surveyed in this article.

Chapter 4 - A variety of investigations has been done for physical and chemical properties of C_{60}, and revealed unique potentialities of C_{60} crystals such as superconductivity and magnetism. C_{60} molecules are reported to crystallize into a face-centered cubic structure on crystalline Si and GaAs substrates, and the crystalline quality is confirmed to be surprisingly high. However, C_{60} crystals are very fragile and chemically unstable due to the fact that C_{60} crystals are formed by van der Waals force, which is very weak compared with other bonding structures such as covalent and ionic bindings. Because of these characteristics, conventional techniques such as etching and ion milling are not suitable for fabricating nanostructures with C_{60}. Area-selective epitaxy techniques make it possible to achieve damage-free fabrications. Therefore, the authors have carried out area-selective epitaxial growth of C_{60} crystals on SiO2-masked GaAs substrates by molecular beam epitaxy, and the fabrication of C_{60} nanoscale structures is successfully achieved. During C_{60} deposition, C_{60} molecules landing on SiO2 masks either evaporate or migrate to the open area enhancing growth rate of C_{60} layers. Mean diffusion length of C_{60} molecules on SiO2 mask at 200°C is evaluated to be 200-300 nm.

Mechanical strength and chemical stability are very important for practical device applications. To study feasibility of C_{60} layers, it is inevitable to obtain harder and more stable C_{60} layers. It has been reported that metal-C_{60} interactions are stronger than van der Waals interactions between C_{60} molecules. Therefore, metal doping in C_{60} layers may produce much harder and chemically stable C_{60} films. Multivalent impurity-doped C_{60} films (Al, Ga, Ge) are grown on GaAs and quartz glass substrates. All the impurity-doped C_{60} films show no X-ray diffraction peak, indicating that the films have an amorphous structure. Vickers hardness values of all the impurity-doped C_{60} films are considerably enhanced, and the films are confirmed to be undissolved in organic solvents. New photoluminescence peaks of electron transitions between HOMO and LUMO states of C_{60} molecules are found in Al-doped and Ga-doped C_{60} films. Optimized bonding structures of these impurity atoms to C_{60} molecules are determined by using ab initio calculations. Stable bonds between impurities and C_{60} molecules are verified to be formed.

Chapter 5 - The discovery of fullerenes encapsulated inside single-wall carbon nanotubes (SWCNTs) by Smith *et al.* [Nature 396, 323 (1998)], provides a compelling opportunity to study the nanotubes using fullerenes as local probes. It is reviewed herein, how fullerenes and magnetic fullerenes such as N@C_{60} and $C_{59}N$, and ^{13}C enriched fullerenes can be used to yield unprecedented insight into the nanotube electronic, vibrational, and optical properties and into the growth of SWCNTs. Encapsulated C_{60} fullerenes are transformed to inner tubes by high temperature annealing. The unique, low defect concentration and well shielded nature of inner tubes makes them ideal to study the effect of diameter dependent treatments such as opening and closing of the tubes. Electron-spin resonance studies on the encapsulated magnetic fullerenes indicates a metallic-like finite density of states for an ensemble SWCNT sample, which contradicts the current view of SWCNTs as containing separate metallic and semiconducting nanotubes. The linear spin chains of the encapsulated magnetic fullerenes is suggested as a building element for quantum information processing. The possibility to grow inner tubes from encapsulated fullerenes allows the isotope selective enrichment of the inner tubes if the source fullerene is ^{13}C enriched. Nuclear magnetic resonance on such samples evidences the presence of a low energy spin-gap which is thought to open due to a Peierls transition.

Chapter 6 - At the present time the "atomic park" of endohedral fullerenes, $A@C_{2n}$, counts approximately 50 elements of the Periodic Table. The purpose of this review is to highlight examples of progress in chemical physics of endohedral fullerenes and point out fertile grounds for future applications of ENF in engineering and medicine. In large part, the authors concentrate on paramagnetic metallofullerenes, $M@C_{2n}$, since the efforts of their group were aimed mainly at synthesis and study of these objects. Atoms entrapped inside the fullerene cage manifest unusual, even unique, structure and electron properties. In paramagnetic $La@C_{2n}$, for example, there is the partial runoff of electron spin density outside the fullerene cage ("spin leakage") testified by EPR, ENDOR and NMR data. In $Be@C_{60}$, $N@C_{60}$, $P@C_{60}$, on the other hand, there is the accumulation of electron spin density on the entrapped atoms as if the electronic shells of the trapped atoms are compressed ("spin compression"). There is a strong tendency of ENF to form nanoparticles in liquid and solid states. The MAS (magic angle spinning) NMR study of solid complexes of La-ENF and Y-ENF with hexamethylphosphoramide has revealed the vast range of chemical shifts for the 31P NMR, along with the enormous shifts for the 139La NMR. Being unchanged with temperature, these shifts suggest their Knight nature caused by the density of conduction electrons at the positions of the NMR observed nuclei. Furthermore, the fullerene cages can be modified in plenty of organic-functionalized derivatives. Yet, chemical and physical properties of ENF are still far from being clear since these compounds are not easily accessible. Paramagnetic ENF hold much promise as the basic elements for realizing nanometer scale devices, amongst them – molecular magnets, superconductors with low critical temperatures, molecular memory devices, field effect transistors, quantum dots, fullerene-based quantum computers. Furthermore, there are the fields in medicine where the employment of ENF appears to have considerable promise too. For example, the water proton relaxivity of some $Gd@C_{82}(OH)n$ is much higher than that of Gd(III)-DTPA, which is in current clinical use as the contrast agent for magnetic resonance imaging. $M@C_{2n}$ with the appropriate particle-emitting radionuclide inside, like β-emitting 90Y, 47Sc, 149Pr, 153Sm, 166Ho or 177Lu, can be synthesized in macroscopic quantities, while advances in the molecular biotechnology provide targeting vectors to deliver therapeutic doses of the ionizing radiation with high specificity for treatment of cancer cells. Yet, neither of these ideas has been realized up to date.

Chapter 7 - Both the eigenfunctions and the local density of states (DOS) near the pentagonal defects on the fullerenes was calculated analytically as a numerically. The results shows that the low-energy DOS has a cusp which drops to zero at the Fermi energy for any number of pentagons at the tip except three. For three pentagons, the nonzero DOS across the Fermi level is formed. Graphite is an example of a layered material that can be bent to form fullerenes which promise important applications in electronic nanodevices. The spheroidal geometry of a slightly elliptically deformed sphere was used as a possible approach to fullerenes. The authors assumed that for a small deformation the eccentricity of the spheroid is much more smaller then one. The authors are interested in the elliptically deformed fullerenes C_{70} as well as in C_{60} and its spherical generalizations like big C_{240} and C_{540} molecules. The low-lying electronic levels are described by the Dirac equation in (2+1) dimensions. The authors show how a small deformation of spherical geometry evokes a shift of the electronic spectra compared to the sphere and both the electronic spectrum of spherical and the shift of spheroidal fullerenes were derived. In the next study the expanded field-theory model was proposed to study the electronic states near the Fermi energy in spheroidal fullerenes. The low energy electronic wave functions obey a two-dimensional Dirac equation on a spheroid with two kinds of gauge fluxes taken into account. The first one is so-called K spin flux which describes the exchange of two different Dirac spinors in the presence of a conical singularity. The second flux (included in a form of the Dirac monopole field) is a variant of the effective field approximation for elastic flow due to twelve disclination defects through the surface of a spheroid. The authors consider the case of a slightly elliptically deformed sphere which allows us to apply the perturbation scheme. The authors show exactly how a small deformation of spherical fullerenes provokes an appearance of fine structure in the electronic energy spectrum as compared to the spherical case. In particular, two quasi-zero modes in addition to the true zero mode are predicted to emerge in spheroidal fullerenes. The effect of a weak uniform magnetic field on the electronic structure of slightly deformed fullerene molecules was also studied. It was shown how the existing due to spheroidal deformation fine structure of the electronic energy spectrum splitted in the presence of the magnetic field. The authors show that the fine structure of the electronic energy spectrum is very sensitive to the orientation of the magnetic field. They found that the magnetic field pointed in the x direction does not influence the first electronic level whereas it causes a splitting of the second energy level. Exact analytical solutions for zero-energy modes are found. The HOMO (highest occupied molecular orbital)-LUMO (lowest unoccupied molecular orbital) gap was calculated within the continuum field-theory model of fullerenes.

Chapter 8 - Three aspects of fullerene formation were studied. the influence of electron concentration in ionized carbon vapor on C_{60} the processes of fullerene and metallofullerene formation, influence of metal atom properties on metallofullerene formation, and influence of buffer gas on these processes. Total rates of formation of C_{60} fullerene and $Me@C_{84}$ (Me=Sc, Fe, Pt) endohedral metallofullerene were calculated using classical collision theory and quantum-chemical calculations of ionization potentials of reacting particles. It was shown that fullerene and metallofullerene formation rates essentially depend on electron concentration and temperature of carbon containing plasma. The temperature dependence of optimal electron concentration corresponded to maximum formation rate was determined. Also the estimations made showed that waves of electron concentration in plasma can increase the rate of C_{60} fullerene assembling. For metallofullerenes it was shown that the metal atom Me adheres to the growing carbon cluster at the last stage of $Me@C_{84}$ molecule formation and the probability of the formation is sensitive to the metal-cluster chemical bonding and metal potential ionization value. Further, the heating and cooling of carbon clusters by buffer gas during the fullerene formation process was considered. Reduced description scheme allowed us to calculate kinetics of a huge number of reactions: coagulation, isomerization and heating/cooling by buffer gas. The kinetic calculations gave the qualitatively correct correlation between the fullerene yields in helium and argon.

Chapter 9 - Polymer membrane efficiency depends on polymer structure and specific properties. Polymer modification by nano-size particles is a way for the development of new membrane materials. Polyphenylene oxide (PPO) modification by fullerene incorporation can be made by two modes. One of them is dispersing fullerene C_{60} and PPO that leads to composition with donor-acceptor binding. Another mode is chemical modification by a three-step synthesis that leads to obtaining covalently bonded PPO-C_{60}. Transport properties of dispersed and bonded PPO-C_{60} are compared exclusively in gas separation. PPO modified by dispersing fullerene is successfully used for preparation of homogeneous (dense) and asymmetric (porous) membranes. The effect of fullerene addition on structure and transport properties of these membranes is considered in gas separation, pervaporation, and ultrafiltration processes. Homogeneous membranes based on PPO-C_{60} compositions are effective in pervaporation of the reacting mixture: ethanol – acetic acid – water – ethyl acetate in binary and quaternary combination of components. The use of membranes modified by fullerene leads to shifting of the reaction equilibrium and maximizing ethyl acetate yield in the hybrid process. The asymmetric porous membranes containing fullerene can be applied for the removing estrogenic pollutants from water. The membrane efficiency to estrogens removal is estimated in adsorption and ultrafiltration processes. Transport properties of membranes under study are improved by fullerene incorporation.

Chapter 10 - A great interest has been taken, for a long time, in tetrahedral oxides like silicates and aluminates, due to their numerous applications in various fields ranging from catalysis, ion exchange, molecular sieves, to thermo and photoluminescence. These compounds often exhibit complex frameworks characterized by the presence of hexagonal and pentagonal windows, but no structural relationships had been established with the different forms of carbon and especially with fullerenes. For the first time, three years ago, has been reported a complex oxide, which exhibits a 3D-framework of AlO_4 tetrahedra forming huge "Al_{84}" spheres similar to those of the D2d isomer of the C_{84} fullerenes, labeled by the way *"fullerenoïd"*[1]. The detailed analysis of its structure and non-stoichiometry mechanisms has opened a promising route for the generation of a large family of fullerenoïd oxides.

Chapter 11 - In the present study, the molecular orbital properties of 1,3-dipolar cycloaddition products of certain substituted pyridinum betaines with C_{60} have been investigated theoretically at the level of PM3 (RHF) type semi empirical quantum chemical calculations. The betaines act as 4π-components across their 2,6-positions and C=C bond of C_{60} present either between the two hexagonal rings (66-type) or the pentagonal and hexagonal rings (56-type) acts as 2π-component in the 1,3-dipolar cycloadditions considered presently. The results indicate that in all the cases, the cycloadducts formed with 66-type dienophile are more stable and less endothermic than adducts formed with the 56-type. Some structural, spectral and physicochemical properties of these cycloadducts have been studied.

Chapter 12 - In this chapter, the results of structural modification of fullerene thin films induced by multiple charged ion bombardment have been presented. The fullerene thin films have been bombarded by different multiple charged ions: N^{5+}, B^{3+}, Fe^{6+}. The applied energies were in the range of 45 to 90 keV. As-deposited and irradiated films have been characterized by atomic force microscope, Raman and Fourier transform infrared spectroscopy and UV/Vis spectrophotometry.

The results of Raman spectroscopy indicate the formation of highly disordered structure with sp^3 bonds spread throughout sp^2 structures during irradiation by multiple charged ions. The morphology of bombarded films has been changed significantly. Results of FTIR spectroscopy indicate the formation of new C-N and B-C bonds while Fe-C bonds could not be identified. The optical band gap was found to be varied from 1.7 to 0.7 eV depending on the type of used ions.

After irradiation of fullerene thin films by multiple charged nitrogen ions, amorphous carbon nitride phase was formed inside the bombarded films. During bombardment by miltiple charged boron ions, amorphous boron carbide films could be observed. New Fe-C bonds could not be observed during bombardment of fullerene thin films by multiple charged iron ions. Results of Raman measurements indicate that the percentage of sp^3 bonds is raised after bombardment by multiple charged iron ions.

In: Progress in Fullerene Research
Editor: Milton Lang, pp. 1-39
ISBN: 978-1-60021-841-5

Chapter 1

FULLERENE-BASED PROTON CONDUCTING MEMBRANES FOR HYDROGEN FUEL CELLS AND DIRECT METHANOL FUEL CELLS

Ken Tasaki
MC Research & Innovation Center, Inc.*,
601 Pine Ave., Goleta, CA 93117

Abstract

This chapter reviews a fundamental study on fullerene derivatives and fullerene composites for proton conducting membranes used in polymer electrolyte fuel cells. New fullerene derivatives have been synthesized and fullerene composites have been fabricated by using Nafion and sulfonated poly(ether ether ketone) as the host polymers. AC impedance measurements have been performed to evaluate the proton conductivity of both fullerene pellets and fullerene composites under low humidity conditions. Thorough characterizations have been conducted to examine the relevant properties of the composites: the water uptake, the morphology, the water characterization, the interactions between the fullerene and the host polymer. Computer simulations have been also utilized to gain an insight into the basic nature of the fullerenes and the fullerene composites. It has been found that the fullerenes reduce the MeOH crossover in the Nafion membrane without increasing the membrane resistance and also they improve the performance of aromatic hydrocarbon polymer-based membranes under low humidity conditions.

1. Introduction

Fullerenes possess extremely low electron affinities, e.g., 2.65 eV for C_{60}, and have attracted intense attentions as semiconductors in the areas such as electronics, opt-electronics, photovoltaic devices, and others in the past decade.[1,2,3] The same unique property may be beneficial for another conductivity: ionic conductivity. Yet, their potential as ionic conductor has received little attention until recently.

---

* A subsidiary of Mitsubishi Chemical Corp.

In this chapter, we present a fundamental study on fullerene-based proton conducting membranes for polymer electrolyte fuel cells (PEFC) in which water management is one of the most difficult issues. A breakdown in water balance at the cathode side often results in water flood, while the anode interface with the membrane may suffer from water depletion due to water transportation by electro-osmotic drag, both of which increase the cell overpotential. Furthermore, the most commonly used proton exchange membranes (PEM) based on sulfonated perfluoropolymers need to be fully humidified during the operation, thus not only requiring a humidifier, but also an even distribution of water across the membrane becomes additional concern. Dry operation of PEFC may alleviate some of the water management problems.

To realize dry operation of PEFC, there have been intensive efforts underway. One approach is to replace water altogether. Kreuer et al. have reported PEM with low volatile solvents such as imidazole to replace water.[4] Though the proton conductivity of 10^{-2} S cm^{-1} has been achieved at high temperatures, imizadole is known to poison Pt and also subject to diffusing out of the membrane. The phosphonic acid group, more stable and benign to the catalyst, has also been used as the protogenic group, but with the lower conductivity, 10^{-3} S cm^{-1} due to the aggregation of the phosphonic groups.[5] the other hand, Savinell et al. have pioneered the work of polybenzimidazole membrane doped by H_3PO_4 (PBI/H_3PO_4).[6,7] Yet, H_3PO_4 leached out by water on the cathode side. Although Benicewicz et al. have significantly improved the performance of the PBI/H_3PO_4 membrane through the use of polyphosphoric acid,[8] the poor performance at low temperatures and leaching out of H_3PO_4 by water condensation still remain to be solved. In another approach to replace water, Haile and co-workers have used inorganic solid acids such as $CsHSO_4$.[9] However, such solid acids can be problematic: reduction of sulfur in the $CsHSO_4$ electrolyte may undergo over time, the reaction with hydrogen forms hydrogen sulfide, and also a poisoning to the Pt catalyst may occur. Although these problems can be mostly alleviated by replacing the S atom with the P atom in the solid acid, there still remain problems of the water solubility and the mechanical plasticity of the material. The decomposition temperatures of the solid acids too close to their operation temperatures is another concern.

An alternative approach may be rather to take advantage of water as the excellent proton carrier. However, conventional perfluoropolymer-based PEMs have limitations in that decreasing the equivalent weight to increase the performance under low humidity often leads to unacceptable swelling of the membrane. Composite membranes of conventional PEMs with inorganic/organic additives may avoid such a problem. Fabrication of Nafion and SiO_2 or heteropolyacids (HPA) have been tried.[10] Especially, HPA has been widely used to improve the performance of proton conducting membranes.[11] The problems of HPA, however, is that they are water-soluble and their proton conductivity can be sensitive to humidity as well.[12] Hence, immobilization of HPA in a membrane has become an important issue.[13]

Fullerenes may provide new tools to improve the performance of PEMs. Ionic conductivities of fullerenes have been first observed by Saab et al.;[14] yet the nature of their ionic conductivity was not thoroughly explored. Subsequently, Hinokuma et al. have developed proton conducting materials based on functionalized fullerenes. For example, a pellet of polyhydroxyl hydrogen sulfated fullerene (PHSF), $C_{60}(OH)_n(HSO_3H)_{12-n}$ (n~6), exhibited a conductivity of 10^{-2} S cm^{-1} under dry conditions, opening possibilities for dry proton conducting membranes.[15] Their work demonstrated a potential of functionalized fullerenes as proton conducting membranes for fuel cells.

Chemical functionarization of fullerenes is well-established; thus their performance (the conductivity, the thermal, the chemical, and the mechanical stability) can be controlled and even fine-tuned chemically. Furthermore, the high acidic nature of C_{60}, a prerequisite for a proton conducting membrane, is demonstrated by a very low pK_a value of $C_{60}H_2$, 4.7, in DMSO.[16] By comparison, pK_a of CCl_2HCO_2H is 6.4 in DMSO.[17] Moreover, functionalized fullerenes have a high surface and volumetric density of proton conductive functional groups. For example, PHSF has about twelve functional groups packed in a cavity of 1 nanometer in diameter. They may present a new type of tailored proton conductors.

Yet, there are several issues not addressed in the work by Hinokuma et al.[15] First, it was suspected that the proton conduction was due to the proton hopping between the functional groups attached to C_{60} since the measurements were performed under dry N_2 gas, though no detailed examination of the proton transport mechanism was reported. Yet, so far, it is not clear the conductivity of their fullerene membrane is due to the functional groups or the fullerene itself possesses a conductive nature. No proton conductivity of C_{60} has ever been reported. As the first subject of this chapter, we will examine the proton conductive nature of C_{60}.

Secondly, though PHSF achieved the high conductivity, it is prone to hydrolysis and decomposition around 100 °C. We address the chemical functionalization of C_{60} in light of designing chemically and thermally stable fullerene derivatives. This is important in fuel cell operations at higher temperatures and also under occasional high humidity conditions resulting in the flooding at the cathode side.

Thirdly, the membrane fabricated by Hinokuma et al. [15] was a pellet of fullerene derivatives pressed under high pressures which had little practical use for PEM. When mixed in a polymer, the conductivity was dropped to 10^{-5} S cm^{-1}.[18] We will describe the new fabrication of fullerene–polymer composites as the second topic. Improved fabrication methods for incorporation of fullerenes in polymeric materials will be also beneficial in general for the use of fullerenes in a variety of existing as well as new applications.

We start with the computational work on C_{60} to investigate its inherent proton conductivity, followed by the synthesis of fullerene derivatives, the fabrication of the fullerene composite membranes, the membrane characterization, and the electrochemical measurements.

2. Calculations

We have investigated the proton conductive nature of C_{60} by calculating its proton transfer activation barrier. The apparent activation energy, which can be obtained from the Arrhenius plots of the proton conductivity against the temperature, represents the activation energy averaged over the vehicular and the structural transport mechanism including those assisted by contaminants such as residual water, and it is often difficult to separate their contributions. Computational techniques can offer powerful complementary tools to experimental data and may shed a light to the proton conductive nature of fullerene itself. However, accurate calculations of activation energies for such large molecules as C_{60} are computationally burdensome. Thus, we have chosen a semi-empirical method, PM3, to make a qualitative assessment.

Due to the little known performance of the PM3 method in calculating the proton transfer activation barrier, we have conducted an extensive validation of this method, using a variety of levels of calculations for a series of polyaromatic hydrocarbons (PAH) wherever computationally possible. The proton transfer in these PAH's concerns the transfer of a proton going from one carbon to the adjacent carbon dividing two hexagons:

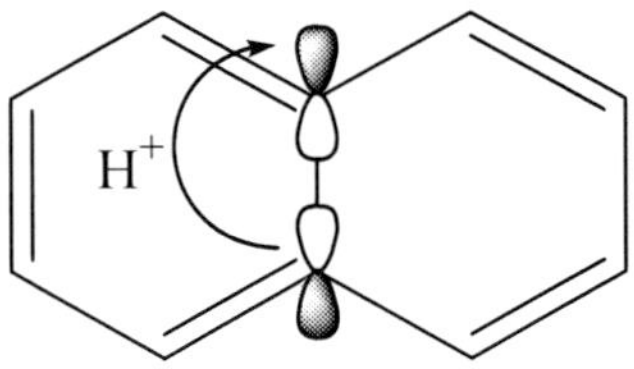

The activation barrier was obtained by determining the transition state for the proton transfer in a protonated PAH (PAH**H**$^+$) and then taking the energy difference between the transition state and the equilibrium structure, equivalent to "the reactant" or "the product" in the internal reaction coordinate calculations. We define the transition state as a PAH**H**$^+$ in which a proton is at the mid point between two carbon atoms, while the equilibrium state is a PAH**H**$^+$ in which a proton is attached to one of the carbons. All calculations were performed in the gas phase, and geometry optimization was carried out both for the equilibrium and transition state. Various levels of theory have been applied when computationally feasible: MP2/6-31+G(2d,p)//MP2/6-31+G(2d,p), B3LYP/6-31+G(2d,p)//B3LYP/6-31+G(2d,p), B3LYP/6-31+G(2d,p)//HF/6-31G(d), HF/6-31G(d)//HF/6-31G(d), and PM3.

Next, to examine the hygroscopic nature of the fullerenes, the water binding energy was calculated for C_{60} and polyhydroxy fullerene (PHF), $C_{60}(OH)_{12}$, at the PM3 level. As far as the relative energy is concerned, PM3 has been able to reproduce the relative energies of various isomers of fullerenes calculated by B3LYP calculations.[19] To calculate the water binding energy, a water molecule was first placed near C_{60} or one of the OH groups of PHF, respectively, and the resultant fullerene–water complex was subject to full optimization. For PHF, both neutral and anion forms were examined. For the anion form, a proton was removed from the OH group to which the water molecule was hydrogen-bonded. The water binding energy (ΔE_{bind}) was estimated by the following equation:

$$\Delta E_{bind} = E(C_{60}X\text{–}H_2O) - E(C_{60}X) - E(H_2O) \qquad (1)$$

where $E(C_{60}X\text{–}H_2O)$, $E(C_{60}X)$, and $E(H_2O)$ are the ground state SCF energy for the fullerene–water complex, the fullerene (C_{60} or PHF, with X being nothing or $(OH)_{12}$, respectively), and a water molecule in the gas phase. To compare with the fullerenes, the water binding energy of $C_2F_5SO_3H$ was also calculated similarly.

Thirdly, to study the interactions between the fullerene and the host polymer, Nafion, molecular dynamics (MD) simulations were performed in the gas phase for a C_{60} or PHF molecule together with a single Nafion oligomer having the following structure:

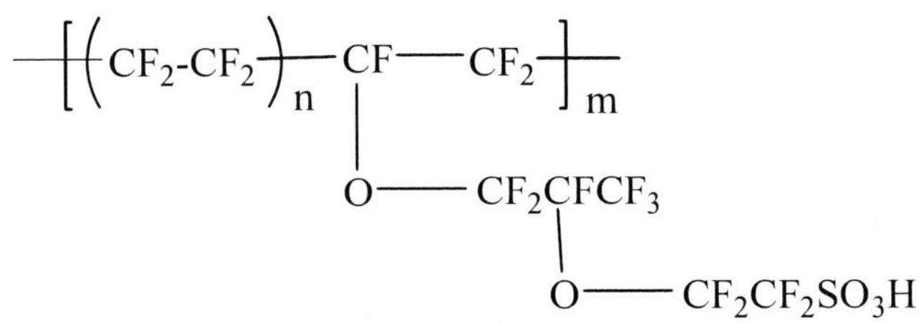

where n and m are 5 and 8, respectively. During the simulations, the hydrogen atoms were kept connected to the sulfonic acid groups of the Nafion chain, since no water was present to dissociate them. For MD simulations, all atoms were subject to integration of Newton's equation of motion with a time step of 1 fs and a periodic boundary condition was applied. The force field was COMPASS,[20] which has been parametrized for condensed-phase simulations of mainly organic molecules including water. The electrostatic interactions were treated by the Ewald summation.[21] Bonds with hydrogens were constrained using SHAKE algorithm.[22]

3. Experiments

3.1. Synthesis

(a) Polyhydroxyfullerene ($C_{60}(OH)_n$: PHF)

C_{60} was purchased from SES Research, Houston TX. $C_{60}(OH)_n$ (n~12) was synthesized according to Chiang et al.,[23] through sulfonation of C_{60} and the subsequent hydrolysis.

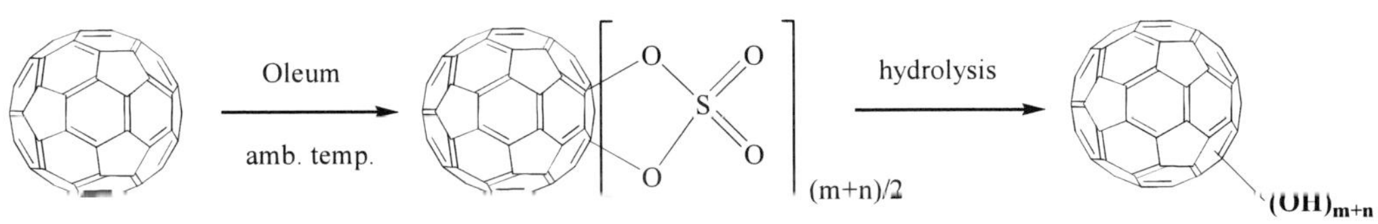

(b) Polyhydroxy hydrogen sulfated fullerene ($C_{60}(OH)_n(OSO_3H)_{12-n}$: PHSF)

$C_{60}(OH)_m(OSO_3H)_{12-n}$ (n~12) was synthesized according to Chiang et al.,[23] through sulfonation of $C_{60}(OH)_n$.

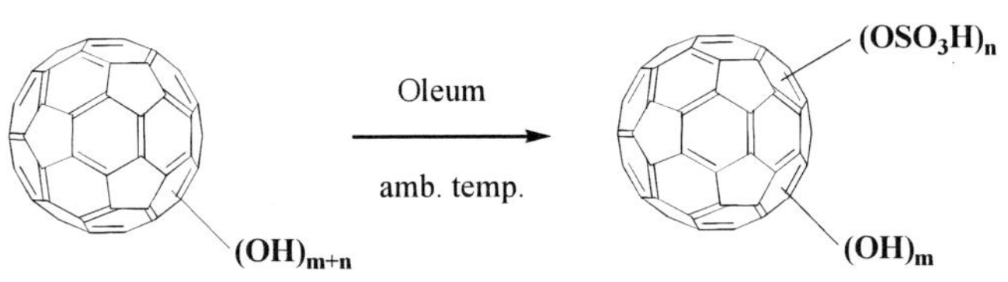

(c) Hydrogenated fullerene ($C_{60}H_n$)

The synthesis of $C_{60}(H_2)_n$ (n=1) was according to Craig et al.[24]

$C_{60}H_{18}$ was provided by MER Corp of Tucson, Arizona.

(d) Methanofullerene phosphonic acid ($C_{60}[C(PO_3H)_m]_n$: MFPA)

$C_{60}[C(PO_3H)_m]_n$ (m=2; n=2) was synthesized according to Andrey et al.,[25] as follows:

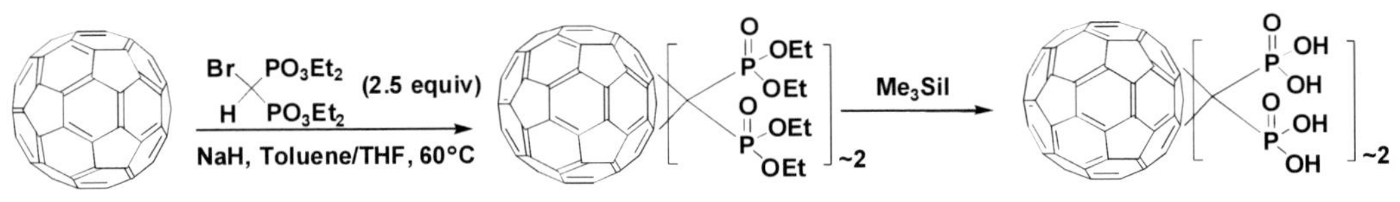

C_{60} $C_{60}\{C[PO(OEt)_2]_2\}_{\sim 2}$ $C_{60}\{C[PO(OH)_2]_{2\sim 2}$

(e) Fullerene phosphonic acid ($C_{60}[(PO_3H)_m]_n$: FPA)

A solution of $NaPO(OEt)_2$ (70ml, 0.1M, 10 equiv.) in anhydrous THF was added to a solution of C_{60} in ortho-dichlorobenzene (100ml, 0.007M) at room temperature (RT). The solution turned into blue immediately and was stirred at RT for 1 hr. Then, $ClPO(OEt)_2$ (1.2g, 0.007mol) was added to the solution and stirred for 2 hrs. The resulting brown solution was dried under vacuum and washed with water several times to give a brown solid of $C_{60}(PO(OEt)_2)_{\sim 5}$ (yield 0.75g, ~80%). This brown solid, $C_{60}(PO(OEt)_2)_{\sim 5}$, (200mg, ~0.15mmol) was dissolved in 100ml of $CHCl_3$, then Me_3SiI (200 μl) was added drop wise. The solution was refluxed for 1 hr, then cooled to RT and 0.5ml of water was added. The precipitated brown solid was filtrated and washed with $CHCl_3$ several times to give $C_{60}(PO(OH)_2)_{\sim 5}$ as a brown solid (yield 140mg, ~85%).

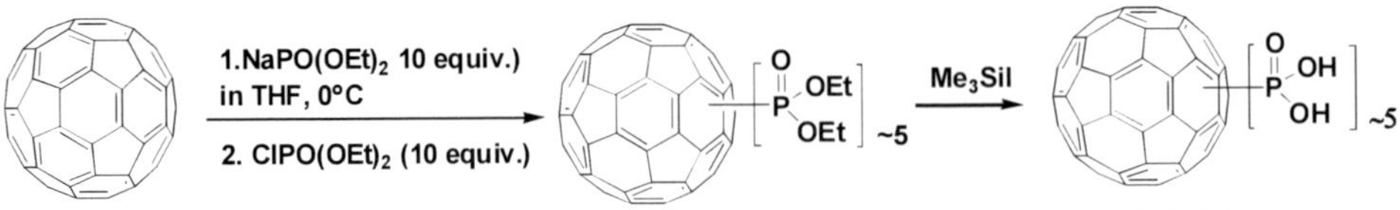

(In ODCB) $C_{60}[PO(OEt)_2]_{\sim 5}$ $C_{60}[PO(OH)_2]_{\sim 5}$

(f) Hydrogen tricyano fullerene ($HC_{60}(CN)_3$:HTCF)

The synthesis of cyano fullerenes is shown in Figure 1. First, $C_{60}(CN)_2$ was synthesized by reacting C_{60} with one equivalent of NaCN, then quenched with $CH_3C_6H_4SO_2CN$. A degassed solution of NaCN (76 mg, 1.2 eq.) in DMF (100 mL) was added to a degassed solution of $C_{60}(CN)_2$ (1g, 1.3 mmol.) in 500 ml of *ortho*-dichloro benzene (ODCB) via cannula at room temperature. After quenching with $HClO_4$, $HC_{60}(CN)_3$ was produced as a mixture of regio-isomers. A very strong acid such as $HClO_4$ was necessary to achieve a high yield, which suggested that the hydrogen on $HC_{60}(CN)_3$ was highly acidic. By repeating this sequence, it is possible to synthesize hydrofullerenes with more than three cyano groups. After 10 minutes of stirring, the mixture was concentrated under vacuum. The resulting brown solid was purified by chromatography on silica gel (eluent: CS_2 / Toluene (1:3)) then washed with methanol to give $HC_{60}(CN)_3$ as a brown solid (530 mg, yield : 51 %). $HC_{60}(CN)_3$ was a dark brown crystalline solid, which was moderately soluble in polar organic solvents such as ODCB and dimethyl acetamide (DMAc). The synthesis was performed at University of California, Los Angeles. The characterization has been published elsewhere.[17]

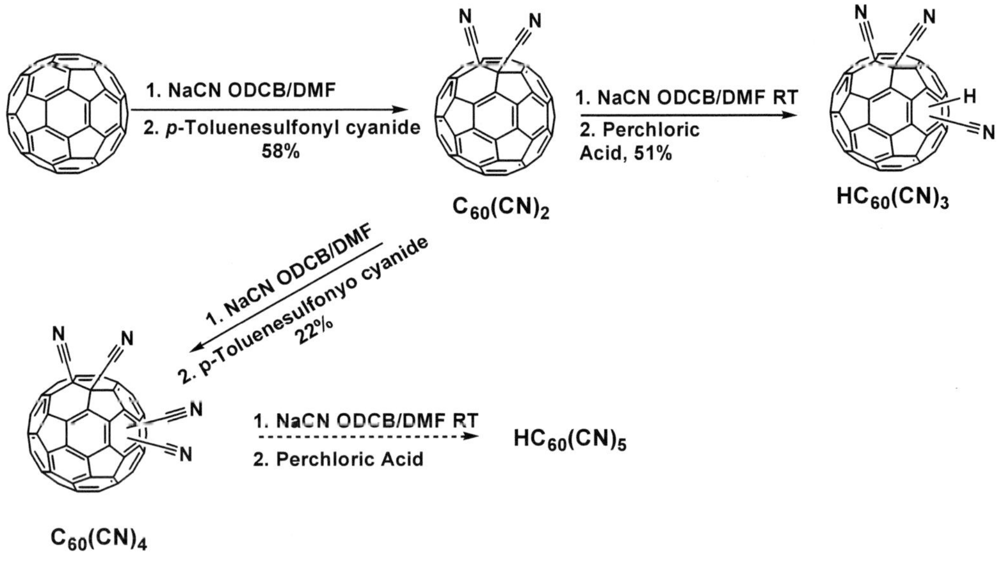

Figure 1. Synthetic scheme for hydrogen cyano fullerenes.

(g) poly[tri(ethylene oxide)benzyl]fullerene ($C_{60}[CH_2C_6H_4(OCH_2CH_2O)_3OCH_3]_n$ (n ~ 5):$C_{60}TEG_5$)

The synthesis of $C_{60}(TEO)_5$ was based on the atom transfer radical addition (ATRA) reaction in four steps to attach multiple ethylene oxide chains to C_{60}, as illustrated in Figure 2. In a typical ATRA step, C_{60} (720 mg, 1 mmol), tri-ethylene oxide benzyl bromide (8 mmol) and bipyridine (1.56g, 10 mmol) were dissolved in 100 ml ODCB. The solution was degassed for 10 minutes and CuBr (0.789g, 8 mmol) was added quickly. The vessel was sealed and heated at 110°C for 2 days until a green precipitation occurred. H_2S was bubbled through the solution to completely precipitate the Cu residue, then the solution was filtrated and ODCB was removed under vacuum. The detail of the characterization has been published elsewhere.[26]

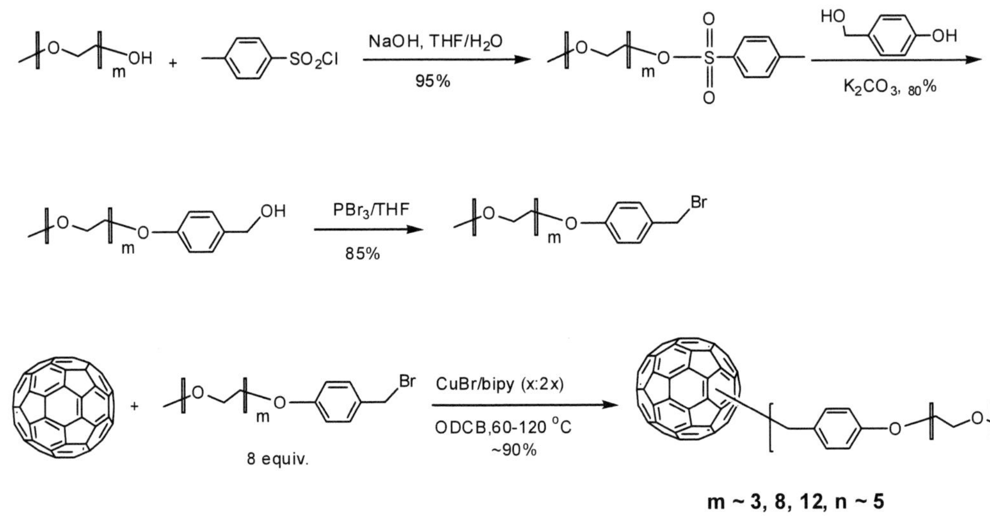

Figure 2. Synthetic scheme for poly[(ethylene oxide)benzyl]fullerenes

3.2. Pellet Fabrication

Pellets were pressed using the Carver press with a pressure of 15 MPa at room temperature for 1 hr. Spacers were used to control the thickness of the pellet, 0.5 mm.

3.3. Composite Membrane Fabrication

Two host membranes have been chosen to prepare the fullerene composite membranes: Nafion 1100 and sulfonated poly(ether ether ketone) (SPEEK). The preparation of the solution cast fullerene–Nafion composites was as follows: a 5% Nafion solution, purchased from Alpha Aesar, was dried at 80°C overnight. The dried Nafion film was dissolved in DMAc. This solution (to be referred to as the Nafion–DMAc solution) was stirred for several hrs and cast onto a Teflon substrate. The film was dried at 120°C overnight, then annealed at 150°C for 1 hr and later at 170°C for half an hour (to be referred to as the recast Nafion). To fabricate the C_{60}–Nafion composite, a small amount of ODCB was first added to the Nafion–DMAc solution at 80°C under vigorous stirring. Subsequently, a known amount of C_{60} was dissolved in chlorobenzene (CB), and the C_{60}–CB solution was added to the Nafion–DMAc solution while stirring. The resulting solution was cast onto a Teflon substrate and dried overnight. Then, the cast membrane was annealed at 150°C for 1 hr and then at 170°C for half an hour (to be referred to as C_{60}–Nafion). The amount of C_{60} in Nafion was 1 wt%. As to the PHF composite, PHF was first dissolved in DMAc, which was added to the separately prepared the Nafion–DMAc solution. The mixed solution was cast onto a glass substrate and dried in an oven at 120°C overnight, followed by annealing at 150°C for 1 hr and then at 170°C for half an hour (to be referred to as PHF–Nafion). The weight percentage of PHF ranged from 1% to 18%. Other fullerene composites, MFPA– and FPA–Nafion membranes were prepared in the same way as that for PHF–Nafion membrane. The weight percentage was 1 wt%, respectively.

SPEEK was obtained through sulfonation of PEEK as follows:

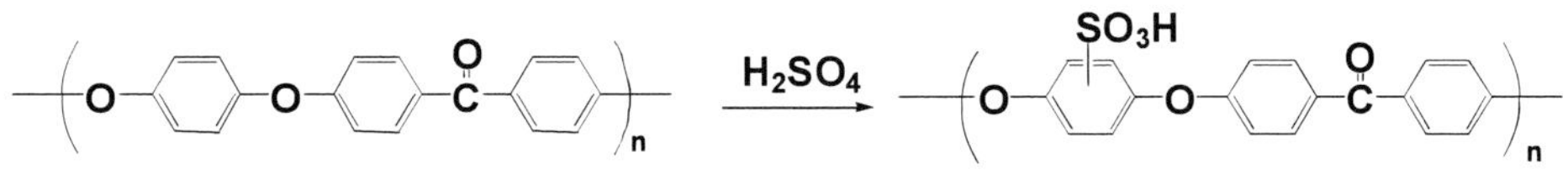

The degree of sulfonation was controlled by reaction temperature and time to yield a series of polymers with IEC of 1.12, 1.41, 1.68, and 1.92 meq g^{-1}. We denote the degree of sulfonation by the percentage following SPEEK such as SPEEK66 which has 66 % of sulfonation, equivalent to IEC of 1.92 meq g^{-1}. SPEEK and PHF were dissolved in DMAc separately at 80 °C. The PHF solution was mixed in the SPEEK–DMAc solution and stirred for 1 hr, then cast into a Teflon dish at 80 °C in an oven overnight. The composite membrane was then further dried in vacuum for 2 hrs at 80 °C; then washed in deionized water until neutral, dried in the air, to be referred to as the PHF–SPEEK membrane.

Finally, all the membranes were treated by boiling in 1 M H_2SO_4 aqueous solution for 1 hr, followed by washing by deionized water.

3.4. Water Uptake Measurements

The water uptake was measured by soaking the membrane in water at room temperature for 24 hrs. The membrane was removed from water and blotted gently to remove surface water, then transferred quickly into a sealed vial and weighed (W_{wet}). The membrane was then vacuum dried at 105°C overnight, transferred quickly into a sealed vial under dry N_2 protection and weighed (W_{dry}). The *wet* water uptake was determined by the following equation:

$$\textit{Wet}\text{ water uptake} = \frac{(W_{wet} - W_{dry})}{W_{dry}} \tag{2}$$

To measure the water uptake under 25% RH, the membranes were equilibrated under 25 % RH for two days and weighed afterward as $W_{25\%}$. *Dry* water uptake was estimated from the following equation:

$$\textit{Dry}\text{ water uptake} = \left(\frac{W_{25\%} - W_{dry}}{W_{dry}}\right) \tag{3}$$

3.5. Fullerene Extraction from Nafion Composite Membranes

The amount of PHF extracted from the Nafion composite membrane in water was determined to examine the physical stability of fullerene derivative in the Nafion polymer matrix. First, the PHF-doped Nafion 117 composite (to be referred to as PHF/Nafion) membrane was dried in a flask by heating at 105 °C under vacuum for 1 hr on a sand bath. Then, the flask was purged with nitrogen and the membrane's weight was determined. Subsequently, the membrane was immersed in water at ambient temperature for 120 hrs, after which the membrane was dried in the same manner as above and the weight was measured. From the weight difference between before and after the immersion in water, the extraction of PHF was determined. As to the extraction from the PHF–Nafion composite (solution cast) membrane, UV-vis spectroscopy was used to quantify the extraction by identifying the peaks characteristics to PHF in the extracted solution from the PHF–Nafion composite. Perkin Elmer Lambda 20, UV-vis Spectrometer was used.

3.6. AC Impedance Measurements

For AC impedance measurements of pellets, the pellet was first sandwiched between two E-TEK gas diffusion electrodes to form a pellet-electrode assembly (PEA). The PEA was placed in a Teflon fixture for AC impedance measurements which were performed in the frequency range of 1 to 10^5 Hz using Solartron SI 1260 Impedance Analyzer and SI 1287 Electrochemical Interface. The Pt loading for both electrodes was 0.5 mg cm^{-2}. The measurement was conducted through the plane. For the composite membranes, the membrane was directly mounted to the Teflon fixture and subjected to on-plane measurements.

Both dry N_2 and H_2 were used for the C_{60} pellet, while only dry N_2 was used for other pellets and the composite membranes. The relative humidity (RH) was controlled by adjusting the ratio of dry and wet N_2 gas flow, and the internal RH of the membrane was monitored using a humidity meter. For high temperature measurements, the RH was controlled by adjusting the bubbler temperature. The sample was equilibrated under a given RH for several hrs prior to the impedance measurements. The resistance associated with the pellet or the membrane at zero phase angle was used to estimate the proton conductivity of the pellet or the membrane using the equation, $\sigma = (1/R)(L/A)$, where R is the bulk resistance of the pellet or the membrane, L represents the pellet thickness or the membrane length, and A the pellet area or the cross section area of the membrane. The thickness of the composite membranes varied from 110 μm to 150 μm.

3.7. MeOH Crossover

The membrane-electrode assembly was fabricated by the decal method.[27] The anode catalyst for the methanol oxidation was Pt-Ru (RV 30-30, E-TEK, Inc.), while the cathode catalyst for the hydrogen reduction was Pt black (20 m^2 g^{-1}, Johnson-Matthey). The catalyst ink was prepared by adding 5% Nafion solution the catalyst. The ink composition was 85 wt% Pt-Ru and 15 wt% Nafion for the anode, while that for the cathode was 85 wt% Pt and 15 wt% Nafion. Both inks were uniformly painted onto two Teflon blanks (2.85 cm^2) to give the desired catalyst loading which was 5 and 0.5 mg cm^{-2} for the anode and the cathode, respectively. During the measurement of the limiting current density, the 1M methanol solution was fed to the cathode at the rate of 5 mL min^{-1}, while the humidified nitrogen gas to the anode at the rate of 35 mL min^{-1}. The limiting current density of the oxidation of methanol, at the anode, permeating through the membrane was determined voltammetarically.[28] The cell voltage sweeping rate was 10 mA min^{-1}.

3.8. Fuel Cell Measurements

The membrane-electrode-assembly (MEA) was prepared by the decal method. Pt was used as the catalyst for both electrodes with the loading of 0.5 mg cm^{-2}. The catalyst ink was prepared by adding 5 wt% Nafion solution to the catalyst. The composition of the ink was 85 wt% Pt and 15 wt% Nafion ionomer. Both inks were uniformly painted onto two Teflon blanks with the area of 2.85 cm^2 to give the desired catalyst loading. After drying at 100 °C, the catalyst layers were transferred to and bound to the composite membrane by hot pressing at 125 °C and 10.6 MPa for 120 s. After the Teflon blanks were peeled off the MEA, the cell was assembled for fuel cell measurements. At the room temperature measurements, the RH was controlled by adjusting the ratio of the dry and the wet N_2 gas flow, and the internal RH of the membrane was monitored using a humidity meter. For high temperature measurements, the RH was controlled by adjusting the bubbler temperature. The flow rate of H_2 gas and air was both 60 mL min^{-1}.

3.9. Thermogravimetirc Analysis (TGA)

After all samples were equilibrated in water for 24 hrs, they were blotted on the surface before the measurement. For constant temperature measurements, the weight of the sample was monitored under dry N_2 gas flow at 50 ml min^{-1} at 30°C for 40 min. For heating measurements, the sample was first subjected to dry N_2 gas for 30 min at 50 ml min^{-1} prior to heating cycle and gradually heated at the rate of 2°C min^{-1}. Mettler Toledo TGA/SDTA 851e was used for the TGA measurements.

3.10. ^{1}H pulse NMR T_2 Measurements

The experiments were performed on a Bruker Avance 300MHz solid state instrument. The membranes were cut to the desired size to fit into the solid state NMR sample holder. Prior to the measurements, the membranes were soaked for 24 hrs in water and then removed from the water and the surface was blotted. The sample was then placed in the sample holder, which was then placed in the NMR probe. T_2 experiments were run using the Carr-Purcell-Meiboom-Gill pulse sequence,[29,30] with a 90° pulse of 4.4 μs at room temperature. Using a 2τ of 100 μs, 92 points on the decay curve were sampled at various evolution times, for the following samples: the recast Nafion, 1% PHF–Nafion, 3% PHF–Nafion, and 1% C_{60}–Nafion composites. Sixteen scans were co-added to improve the signal to noise ratio.

3.11. Small Angle X-ray Scattering (SANS)

The source for SANS measurements is a Rigaku ultraX 18, which is an 18 kW, direct drive, rotating copper anode source. The actual operating condition used is 5 kW (54 kV x 90 mA). The white radiation that is emitted then goes through a double-graphite monochromator (OSMIC Confocal Max-Flux focusing multilayer mirror) that only reflects the Cu K-α line. The beam is defined by a 4-pinhole collimation system in an evacuated flight path. The detector is an 18 cm diameter image plate detector which was made by Mar USA. The sample to detector distance used was 758 mm. The sample was placed in the beam path.

3.12. ATR Fourier Transform Infrared (ATR FT-IR) Spectroscopy

ATR FT-IR spectroscopy was employed to investigate the interactions between the fullerene and the Nafion polymer. Measurements were recorded using Nicolet Impact 400 FT-IR spectrometer. After saturation in water overnight, the fullerene composites were dried in an oven at 80°C for 6 hrs, followed by further drying under a vacuum at room temperature for 18 hrs, prior to the measurements.

4. Results and Discussion

4.1. Thermal Stability

Before the calculations of the proton conductive nature of fullerenes, we briefly describe the thermal stability of the fullerene derivatives. Figure 3 shows the TGA curves for two representative fullerene derivatives synthesized in this work: PHF and HTCF. The initial weight loss for PHF below 100 °C is due to the loss of water, PHF being hygroscopic. The second weight reduction occurs around 200 °C, due to the dehydroxylation. On the other hand, HTCF showed little weight loss up to near 300 °C. In any case, both compounds are thermally stable up to 200 °C, which is well above the ideal operating temperature suggested by Gasteiger et al., 120 °C.[31] Other fullerene derivatives with the phosphonic acid group, not shown in the figure, are stable as well, except for PHSF whose instability has been mentioned above.

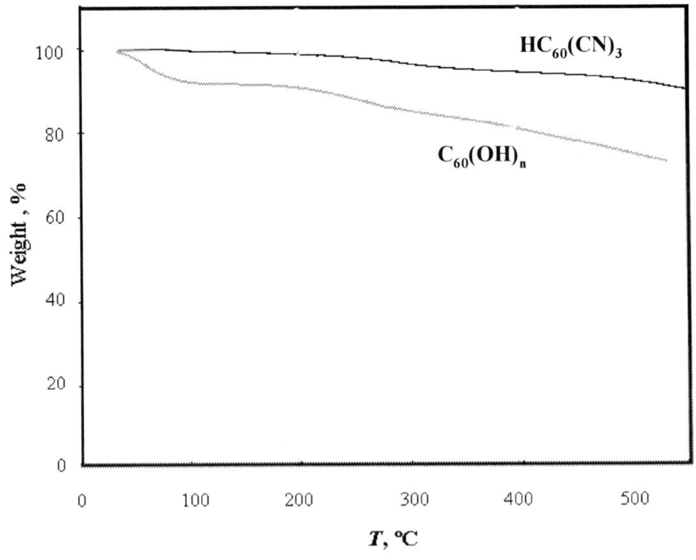

Figure 3. TGA of HTCF and PHF.

4.2. Proton Conductive Nature of C_{60}

Recently, carbon-based solid acids have been prepared from aromatic compounds and have shown the high proton conductivity.[32] The process to prepare the carbon acids was simple and the materials were low cost, and the proton conductivity of some of the acids reached as high as 0.11 S cm^{-1} at 323 K under 100 % relative humidity. The conductivity under dry conditions was not reported; thus, it is not clear whether the carbon acids can be conductive in the absence of water. In this section, we used a series of PAHs with a mixture of hexagon and pentagon rings in order to first investigate their proton conductive nature as a function of the size and second to examine the performance of the PM3 level through comparison with higher level calculations. Obviously, many carbonaceous materials are electrically conductive; yet, their ionic conductive nature has been little explored. Among PAHs, C_{60} has a very high electric resistance of 10^{14} Ω cm.[33]

Figure 4 illustrates the activation barriers of a series of protonated polyaromatic PAH compounds as a function of the number of π electrons in the system calculated at several levels of theory displayed in the figure. The proton transfer path in question was along the central C-C bond dividing two hexagons. The results indicate that the difference in the results among the various levels of calculations tended to become smaller as the number of π electrons increased, and above 20 π electrons the difference remained almost constant, ~ 0.2 eV. From a computational point of view, several points can be derived from Figure 4. First, the B3LYP/6-31+G(2d,p)//HF/6-31G(d) calculations seem to give about 0.2 eV higher barriers than the MP2/6-31+G(2d,p)//MP2/6-31+G(2d,p) calculations, though only two results are presented by the latter calculations which were used as the reference point calculations. This is encouraging since not only are MP2 calculations limited to systems of small size and secondly, but also optimization at the DFT level costs much more than the HF level. Secondly, the PM3 method always gives the highest barriers among the levels of calculations examined. Thus, the PM3 activation barrier can be taken as the upper limit of the calculated barrier, at least for these aromatic compounds. Furthermore, the difference between the PM3 values and the values calculated by other higher levels of calculations seems to become smaller, for the systems above 20 π electrons, down to as small as about 0.2 eV from the DFT calculations.

As to C_{60}, we have examined two proton activation barriers for the fullerene: the intra and the intermolecular barriers. For the intramolecular H^+ transportation in C_{60}, there are two possible transfer paths: along the C-C bond dividing two hexagons (C_h-C_h) and along the C-C bond dividing a hexagon and a pentagon (C_p-C_h). The activation barriers for the two paths were 1.00 eV and 1.16 eV, respectively, by PM3 calculations. The reported activation barrier for H_3PO_4 liquid is also included by a straight line across Figure 4 for comparison.[6] As to the intermolecular activation barrier of H^+ transportation between two C_{60} molecules, the calculated activation barrier was only 0.13 eV. These numbers suggest a fast proton transfer in C_{60}. Our calculations are qualitatively consistent with the experimental observation. Reed et al. has reported that a single ^{13}C-NMR spectrum was observed for the $C_{60}H^+$ solid at ambient temperature, suggesting a rapid movement of a proton around a C_{60} molecule.[34]

Table.1 Proton Activation Barrier of C_{60}.[a]

	ΔE, eV
Intramolecular Barrier	1.00,[b] 1.16[c]
Intermolecular Barrier	0.13

[a] Calculated by PM3.

[b] Along the C-C bond dividing two hexagons.

[c] Along the C-C bond dividing a hexagon and a pentagon.

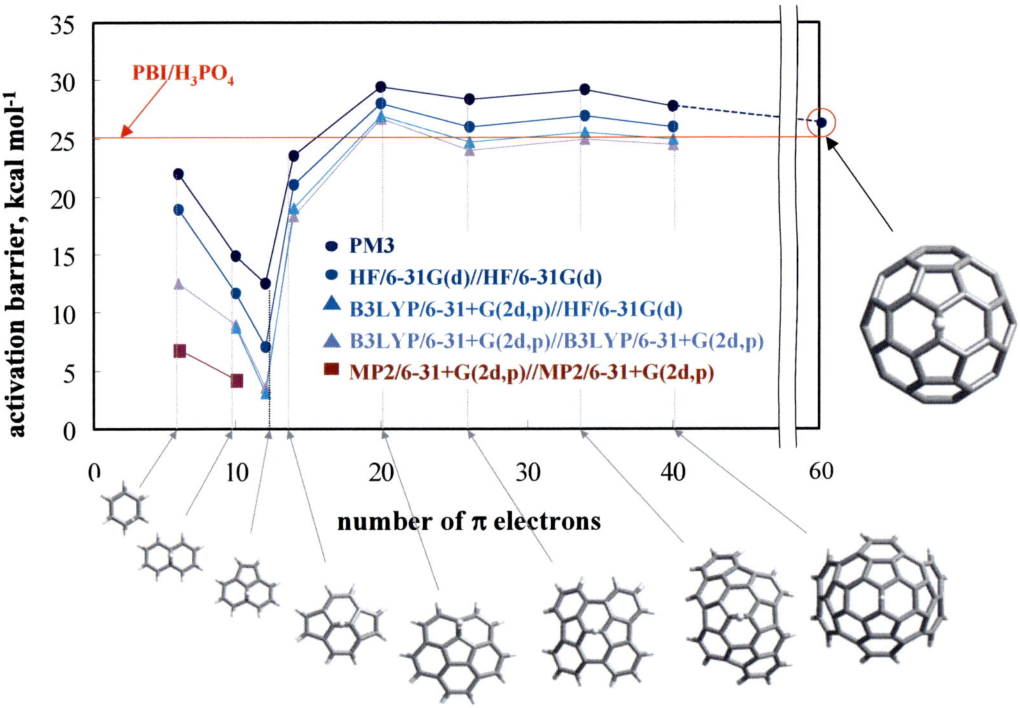

Figure 4. The proton transfer activation barriers of a series of protonated polyaromatic hydrocarbons as a function of the number of π electrons calculated by a variety of computational models listed in the plot.

C_{60} can be functionalized as proton conductor as well. We also examined $C_{60}(OH)_2$ as a model compound for fullerenol in order to compare the proton hopping between the functional groups of C_{60} and the hopping on C_{60}. Two hydroxy groups were attached to two carbons of C_{60}, followed by geometry optimization of $C_{60}(OH)_2$, and then a proton was attached to one of the OH groups. The transition state for the intramolecular transfer was determined by placing a proton in between the two OH groups. As to the intermolecular transfer, several orientations of the two $C_{60}(OH)_2$ molecules with respect to one another were tried to locate the lowest activation barrier.

Figure 5 illustrates the potential energy profiles for the intra- and intermolecular proton hopping between the OH groups in $C_{60}(OH)_2$ obtained at the PM3 level. The internal coordinate calculations gave the barriers of 0.64 eV and 0.47 eV, respectively. The equilibrium and the transition state structure for each transfer are also inserted in the figure.

Experimentally, the activation energy for $C_{60}(OH)_{12}$ in a pellet has been reported to be 0.35 eV.[35] While C_{60} has a higher intramolecular barrier than $C_{60}(OH)_2$, C_{60} possesses a much lower intermolecular barrier than $C_{60}(OH)_2$. Our calculations suggest that functionalization may counteract the inherently conductive nature of fullerene, though to what extent it is counterproductive depends on the chemical functionalization.

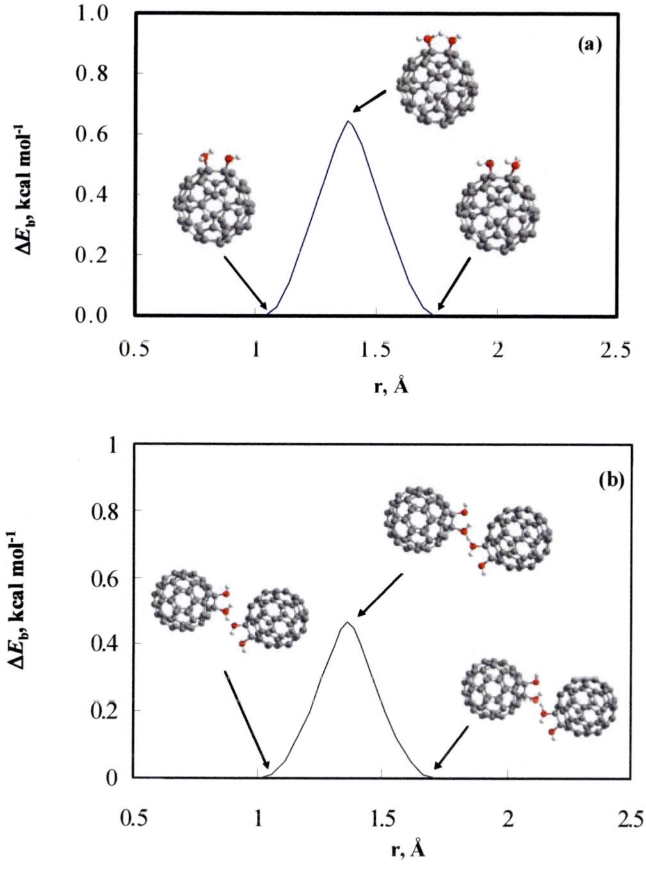

Figure 5. (a) The potential energy profile for a proton hopping between two OH groups of $C_{60}(OH)_2$ and (b) between two OH groups of different $C_{60}(OH)_2$ molecules. The abscissa is the distance between the proton and the carbon to which the proton was originally attached before the transfer.

4.3. Evaluation of Acidity of Fullerene Derivatives

The pK_a value of a proton conducting material in water is a relevant property for fuel cell applications. However, many fullerene derivatives are water insoluble, posing a challenge in determining the pK_a values in water for a direct comparison to other proton conducting membranes. We have developed a simple method to obtain a rough estimate of the acidity of various acids.

In a proton dissociation process, $AH \leftrightarrow A^- + H^+$, the pK_a may be written by eq 4 where K_a is the equilibrium constant for the process, G, H, and S are the free energy, the enthalpy, and the entropy, respectively, associated with the process and A the constant. A variation in ΔH values is larger than that in ΔS values in general since the proton association is dominated by the stability of ions, thus primarily the heat, and also the process always releases a proton. Assuming ΔS and T are constant, there should be a linear relationship between pK_a and ΔH, as is in eq 4:

$$\begin{aligned} pKa &= -\log Ka \\ &= (1/2.303RT)\Delta G + A \\ &= (1/2.303R)(\Delta H/T - \Delta S) + A \end{aligned} \tag{4}$$

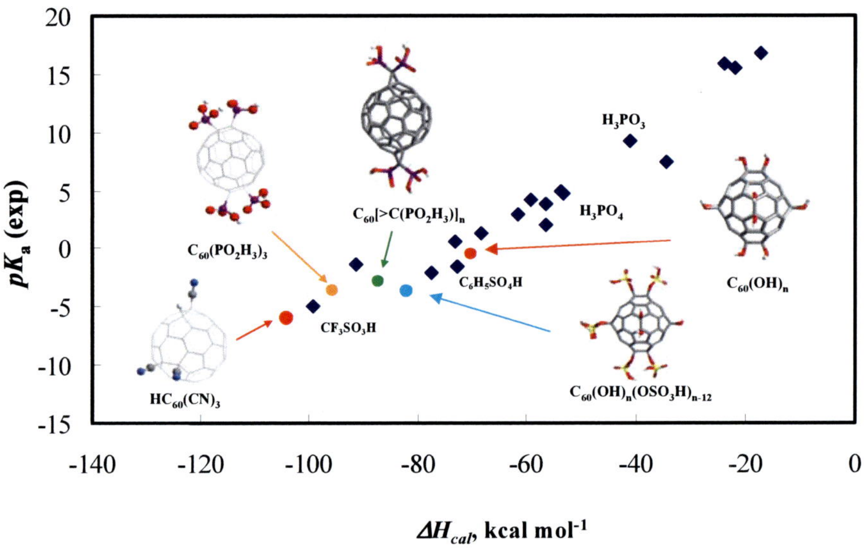

Figure 6. pK_a vs. proton dissociation energies for various acids including fullerene derivatives.

Figure 6 plots a relationship between the experimental pK_a values and the calculated ΔH for known acids, shown by a black rhombus, some of which are captioned in the figure such as $C_6H_5SO_3H$. The calculations were performed by the PM3 method in which only the electronic energy difference (ΔE) between ($A^- + H^+$) and AH in the dissociation process above was taken and used as ΔH with an assumption that ΔE is nearly equal to ΔH, while ΔS is constant. There seems, though possibly fortuitous, given the approximations, a linear correlation-like relationship between pK_a values and ΔH in Figure 6. Using this rough relationship, the calculated values of ΔH for several fullerene derivatives were plotted against pK_a, thus these pK_a values are only roughly estimated values. Our calculations suggest that $C_{60}(OH)_n$ (n=12) is somewhat weaker acid than $C_6H_5SO_3H$, whereas the other fullerene derivatives are all stronger than $C_6H_5SO_3H$, but weaker than CF_3SO_3H. Through chemical modification, fullerene derivatives having a higher acidity are possible to synthesize.

4.4. Proton Conductivity of Fullerene Pellets

Hinokuma et al. have reported that the proton conductivity of the $C_{60}(OH)_n(OSO_3H)_{12-n}$ pellet was 10^{-2} S cm^{-1} under dry conditions and the conduction mechanism was through the proton hopping transfer between the functional groups.[15] Either C_{60} or $C_{60}(H_2)_n$ (n=1,9) differs from their fullerene derivative in that there is no proton acceptor or donor for the proton transfer unlike $C_{60}(OH)_n(OSO_3H)_{12-n}$ in which both hydroxyl and hydrogen sulfated groups work as either the proton acceptor or donor. Thus, it is of interest to examine the conductivity of C_{60} and hydrogenated C_{60} from a scientific point of view. It is expected that the proton transfer, if any, should occur between the carbon atoms of the C_{60} molecules, given the high acidity of the protons attached to C_{60} and the low proton activation barriers predicted by the above calculations.

Based on our AC impedance measurements, it was found, however, that neither C_{60} nor $C_{60}(H_2)_n$ (n=1,9) showed any significant conductivity under dry conditions, contrary to our calculations presented above. Figure 7a exhibits the Nyquist plots of the $C_{60}H_{18}$ pellet under two humidity conditions, 25 %RH and 80 %RH, to illustrate the humidity dependence of its conductivity. The $C_{60}H_{18}$ pellet exhibits little conductivity under low humidity conditions, but showed a meaningful conductivity only under higher RH under which water works as a proton carrier. This finding suggests that $C_{60}H_{18}$ is only conductive in the presence of water, but not conductive all by itself. However, it is possible that the distance between the fullerene molecules may be too large for the protons to hop. The proton hopping distance between the hopping sites has been suggested to be within 2.5 Å.[36] The crystallographic study, on the other hand, has reported that the nearest-neighbor distance of C_{60} molecules, the distance between the center of gravity of two adjacent C_{60} molecules, is 10 Å which gives rise to the distance of 3Å between the nearest carbon atoms belonging to the two neighboring C_{60} molecules.[37] The proton transfer potential barrier quickly increases with the hopping distance. The distance of 3Å between the hopping sites may be too large for a proton to hop. On the other hand, the functional groups of $C_{60}(OH)_n(OSO_3H)_{12-n}$ extend themselves to the neighboring fullerenes, bringing the hopping sites close to one another, well within 2.5 Å.

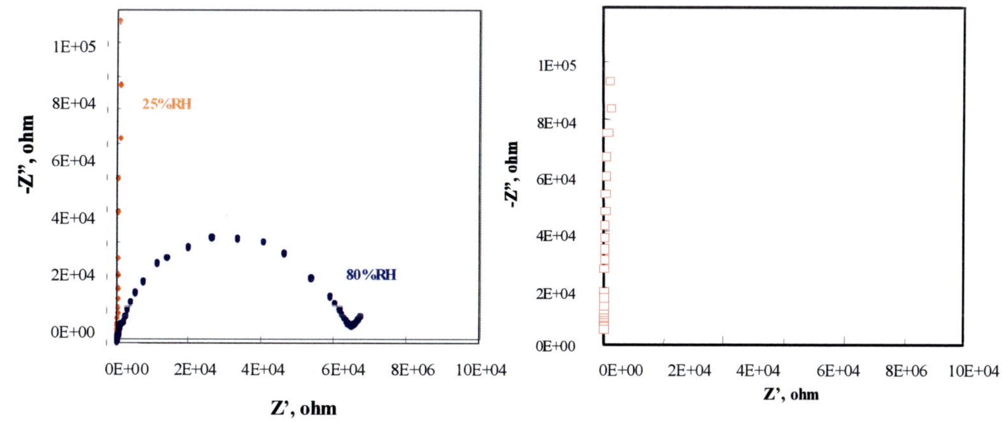

(a) (b)

Figure 7. The Nyquist plots of (a) the $C_{60}H_{18}$ pellet under 25 and 80%RH and (b) poly(ethylene oxide) under 80%RH.

The ^{13}C-NMR study only suggests the proton moving around the C_{60} molecule, not necessarily between the C_{60} molecules.[34]

Separately, the conductivity of poly(ethylene oxide) (PEO) was also measured under 80 %RH, the Nyquist plots of which are shown in Figure 7b. PEO is known to be conductive and has exhibited the conductivity under dry conditions in the presence of acid.[38] Figure 7b shows little conductivity even under high humidity conditions in the absence of acid. This is an interesting comparison to the results in Figure 7a showing some conductivity of $C_{60}H_{18}$ under the same relative humidity, 80 %RH. We believe that this is due to the acidic nature of $C_{60}H_{18}$. Under 80 %RH, water serves as the carrier of the proton which is dissociated from the fullerene.

Table 2 summarizes the proton conductivity of various fullerene pellets. Neither C_{60} nor the hydrogenated C_{60} showed much conductivity under various conditions (the conductivity immeasurable indicated by a dash line "–"). Under low humidity conditions, they behave more as an insulator than a conductor. It is important to note that C_{60} has a high electric resistivity, ~10^{14} Ω cm, as is demonstrated by the results in Table 2. Once chemically functionalized, fullerenes lose π electrons, thus becoming even more insulating. The conductivity of PHSF is similar to the literature value.[15] The order in the magnitude among PHSF, FPA, MFPA, PHF, and HTCF does not necessarily correlate with their acidity shown in Figure 6. It is believed that the proton conduction in these functionalized fullerenes is due to the proton hopping between the functional groups except for HTCF. The fullerene derivatives such as FPA and MFPA have smaller numbers of functional groups, thus less available sites for proton hopping, than PHSF or PHF may have.

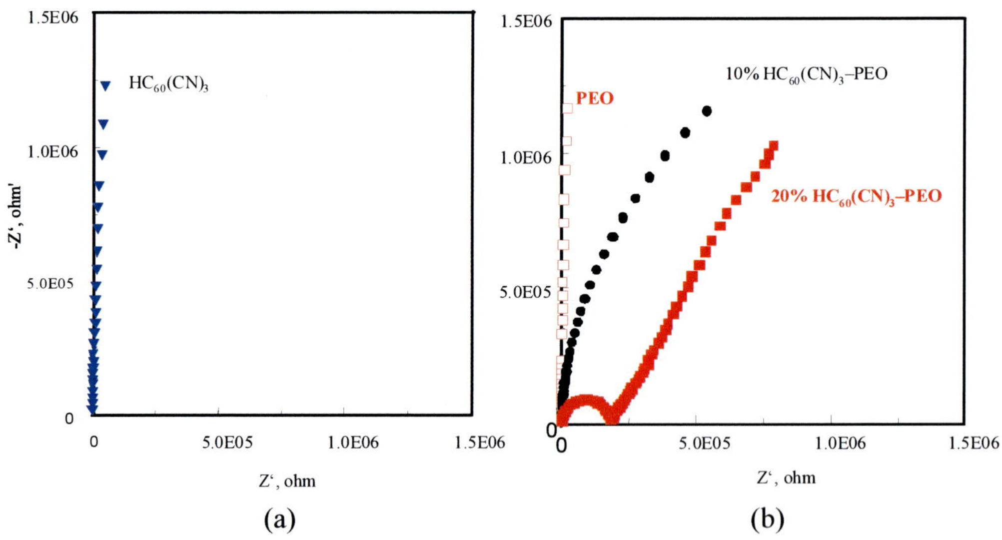

Figure 8. The Nyquist plots of (a) the $HC_{60}CN_3$ pellet and (b) poly(oxyethylene) with $HC_{60}CN_3$ under 80%RH.

Here, we take HTCF to further examine its conductivity. While the HTCF pellet exhibited little conductivity (Fig. 8a), when mixed with PEO (captioned as $HC_{60}(CN)_3$–PEO), however, the conductivity was recovered as the concentration of HTCF was increased, as is demonstrated in Figure 8b.

Table 2. The Proton Conductivity of Fullerene Pellets.

	C_{60} [a]		$C_{60}H_2$		$C_{60}H_{18}$		PHF	PHSF	MFPA	PFA	$HC_{60}(CN)_3$	
Temp., °C	20~100	20	20~100	20	20	20	20	20	20	20	20	20
RT, %	25	80	25	80	25	80	25	25	25	25	25	80
σ, S cm^{-1}	–	–	–	10^{-7}	–	$4x10^{-7}$	$6x10^{-7}$	10^{-2}	$3x10^{-6}$	$2x10^{-4}$	10^{-8}	10^{-6}

[a]The hydrogen gas was fed to the electrode. The dry N_2 gas was used as the feed gas for the rest of the samples.

The $HC_{60}(CN)_3$–PEO composite was prepared by dissolving PEO (MW=4M) in chrolobenzene (CB), folloed by drop-wise addition of HTCF to the PEO–CB solution, and then dried under vacuum. The HTCF's proton, though acidic, may trot around the molecule, but unlikely to be transferred between the molecules, due to the bulky cyano groups which keep the molecular apart (see the illustration below). MD simulations of a single HTCF molecule in the gas phase revealed oscillating motions of the cyano groups which prevent the neighboring molecules from coming close to one another. Thus, the proton of HTCF does not contribute to the conductivity.

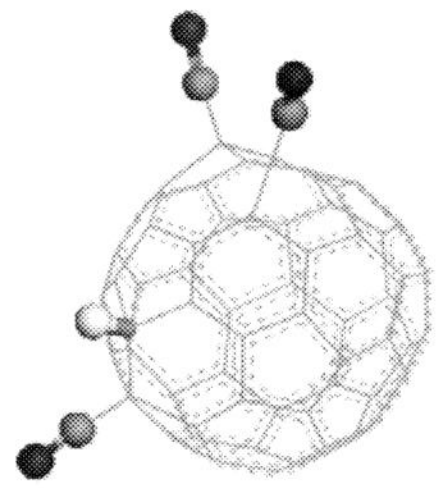

Once mixed with PEO, known as a conductive medium, however, the acidic proton of HTCF is carried away along the PEO chains through the membrane for conductivity, as is illustrated in Figure 9.

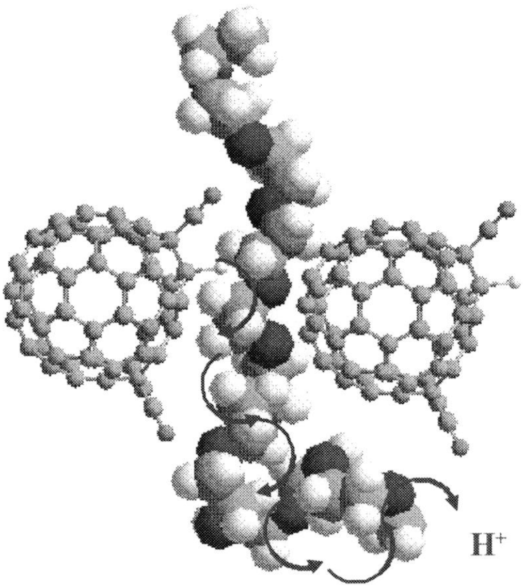

Figure 9. Schematic illustration of the proton transfer in $HC_{60}(CN)_3$–PEO composite.

4.5. Fullerene Composite Characterization

It was found that a successful fabrication of solution-cast fullerene–Nafion composites requires a careful selection of the solvents. Simple mixing of fullerene with commercially available Nafion solution, consisting primarily of alcohol and water, only gives a composite membrane with poor dispersion of fullerene (see Fig. 10a). A difficulty of choosing the solvent for solution casting of fullerene–Nafion composites stems from sharply different solvents available for fullerenes from those for Nafion in general. For example, good solvents for C_{60} are primarily aromatic hydrocarbons such as benzene, chlorobenzene, and dichlorobenzene, all of which are poor solvents for Nafion. On the other hand, even PHF does not dissolve well in alcohols, good solvents for Nafion. These circumstances make the fullerene–Nafion composite difficult to prepare, especially when the fullerene is hydrophobic. This is the difference from other fullerene–polymer systems, some of which have shown homogeneous dispersion of fullerene in polymers such as polydimethylsiloxane[39] and polyaniline.[40] So far, no co-solvent for hydrophilic fullerenes such as C_{60} and Nafion has been found.

Our search for the solvents was based on the thermodynamic compatibility of two solvents, of which at least one solvent was a good solvent for the fullerene and another for Nafion. The rational was that though each of the two solvents was a good solvent for only one component (the fullerene or Nafion) of the composite, if the two solvents are miscible, the two immiscible components, the fullerene and Nafion, may be brought together, carried by their good solvents. The other consideration was the boiling points of the two solvents which are preferably close to one another for a simpler membrane drying process. The solvents thus chosen were ODCB and DMAc for the solution cast of the C_{60}–Nafion composite. ODCB has a high solubility of C_{60}, while DMAc is often used to recast Nafion membranes. Moreover, ODCB and DMAc are completely miscible with one another, having their b.p. at 180°C and at 165°C, respectively. As to the PHF–Nafion composite, since alcohols are in general poor solvents for PHF and also DMAc happens to be a good solvent for both PHF and Nafion, DMAc was used as the co-solvent for the solution casting.

(a) (b) (c)

Figure 10. Photographs of (a) the PHF–Nafion composite film cast from iso-propanol solution, (b) the PHF–Nafion composite film cast from DMAc solution, and (c) the C_{60}–Nafion composite film cast from ODCB and DMAc. See the text for the detail. The loading was all 1wt%.

Figure 10 displays the photographs of the solution cast fullerene–Nafion composite films using the solvents mentioned above, along with the PHF–Nafion composite film (Fig. 10a) prepared by using iso-propanol as the main co-solvent for both PHF and Nafion. In the preparation of the film shown in Figure 10a, PHF was first mixed in iso-propanol, then the PHF–iso-propanol solution was added to the Nafion solution obtained from Alfa Aesar, and the solution was cast and dried. The loading of the fullerene in the Nafion composite was all 1wt % in the photographs. The PHF–Nafion film cast out of the iso-propanol solution shows a poor dispersion of PHF in the film with a visibly extensive aggregation of PHF. On a sharp contrast, the PHF–Nafion composite prepared by DMAc as the co-solvent (Fig. 10b) displays a much more improved dispersion of the fullerene in the film, as is the case for the C_{60}–Nafion composite (Fig. 10c). No example for the C_{60}–Nafion film cast out of iso-propanol solution is shown due to a complete immiscibility of C_{60} in iso-propanol. The color of the C_{60}–Nafion composite film is dark brown, while that of the PHF–Nafion composite film is reddish brown which is a characteristic of PHF's red shift in the UV-vis spectrum from that of C_{60}.

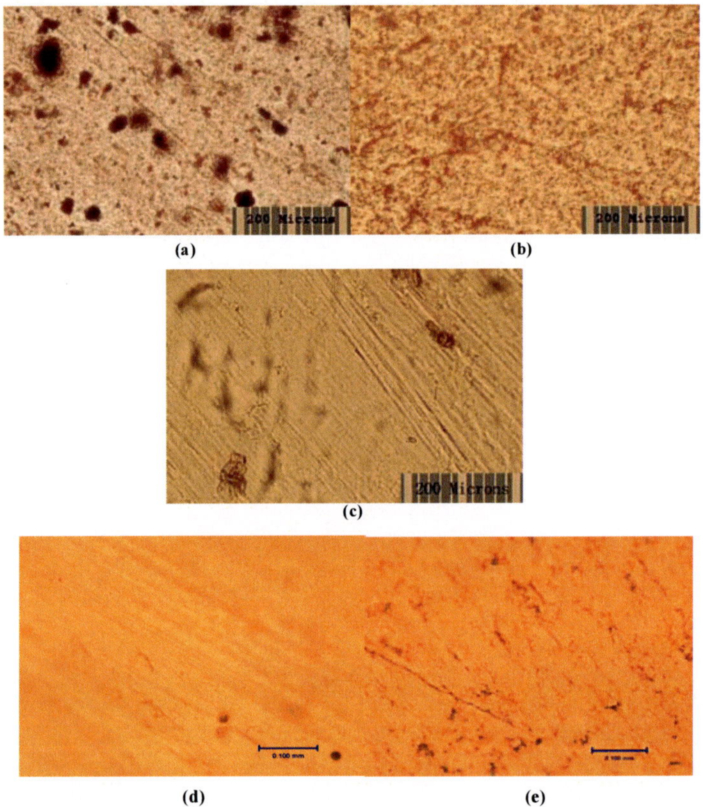

Figure 11. Optical micrographs of (a) C_{60}(1%)–Nafion, (b) PHF(1%)–Nafion, (c) FPA (1%)–Nafion, (d) PHF(35%)–SPEEK, and (e) FPA(3%)–SPEEK (% is by weight).

Figure 11 illustrates the optical micrograms (OM) of various fullerene composites, revealing the dispersion of the fullerenes in more detail. The OMs were chosen according to the size of the fullerene aggregates in the composite membranes. The contrasts among the composites mainly reflect the differences in the solubility of the fullerene in the host polymer. Obviously, C_{60}, highly hydrophobic, has the lowest solubility in Nafion among the fullerene derivatives studied here, exhibiting a significant agglomeration (Fig. 11a) with the size of agglomerates ranging up to as much as 50 μm. On the other hand, PHF has a better solubility to Nafion with much finer agglomerates than those in the C_{60}-Nafion film. FPA dissolves even better in Nafion due to the more polar nature of the functional group, the PO_3H group compared to the OH group. The FPA composite films were essentially transparent. As to the SPEEK composites, fullerenes have a favorable miscibility with aromatic hydrocarbon polymers in general. Thus, even at 3 wt%, both PHF and FPA composite films are transparent. Figure 12 illustrates the SEM images of the 3% PHF–Nafion composite at different scales. The average particle size is approximately 5 μm in diameter, much smaller than that of the 0.5wt% C_{60}–Nafion composite. PHF is easy to synthesize and has a reasonable solubility in Nafion; thus, was extensively studied in this work.

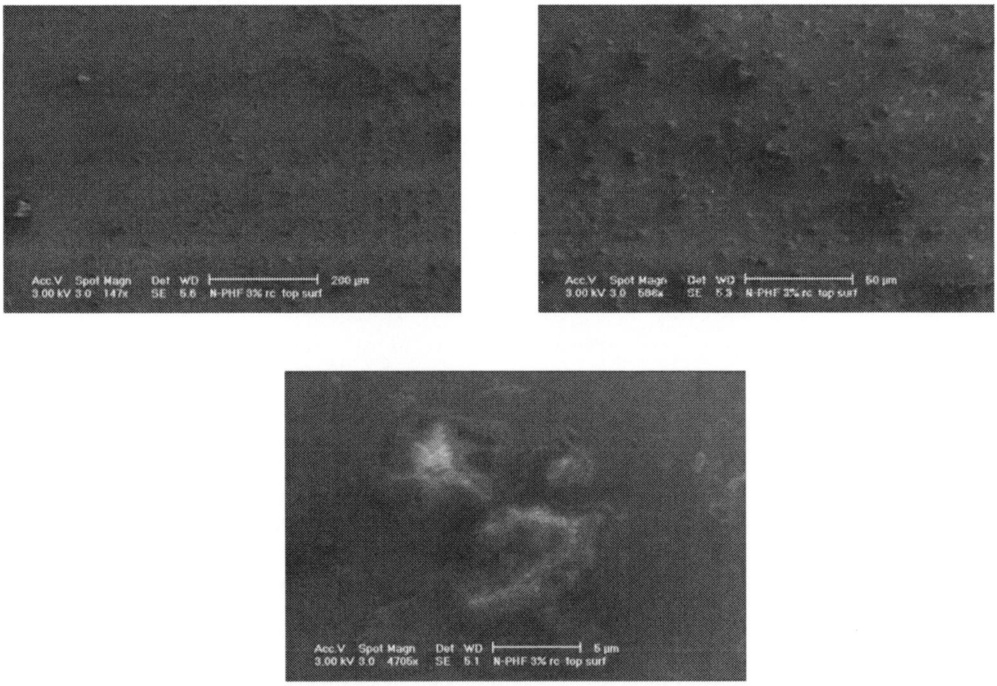

Figure 12. SEM images of 3% PHF–Nafion at different scales.

The physical stability of the fullerene in a Nafion composite film is a concern since it is not chemically bonded to the Nafion polymer. The fullerene's physical stability depends primarily on the integration of the fullerene in the Nafion matrix, subject to the robustness of the fabrication method used for a given fullerene. Since C_{60} is water insoluble, extraction of the fullerene from the Nafion composite membrane into water was carried out only for the PHF/Nafion 117 composite and the solution cast PHF–Nafion composites. The doped composite was prepared by soaking Nafion 117 in iso-propanol overnight and mixed with PHF dissolved in THF. The extraction was measured by the membrane weight loss between the dried membrane and the membrane which had been soaked in water at 20 °C for 120 hrs. Table 3 lists the results from the extraction test. For PHF/Nafion 117, the extraction of PHF out of the composite was 65 % of the initial amount in the composite membrane. The large loss of PHF is no surprise. In the doped composite, PHF was introduced into the hydrophilic pores of the Nafion membrane which had been swollen by alcohol. Once the composite was soaked in water, the pores can be opened again, and thus PHF was prone to diffuse out into the water.

As to the PHF–Nafion composite, the amount of extracted PHF was not detected by the weight measurements. Accordingly, UV-vis spectroscopy was used for more sensitive determination of PHF in the extracted solution. Separately, a calibration curve was developed to measure the PHF concentration in water from the absorbance in UV-vis spectroscopy. Using the calibration curve, the amount of extracted PHF out of the Nafion composite was determined and listed in Table 3. As Table 3 indicates, no more than 2.5 wt% of the original PHF was extracted from the composite film, demonstrating a significantly improved stability of PHF in the composite, compared to the doped film. In comparison, the extraction of phosphotungstic acid from the Nafion recast composite membrane has been also reported, and the loss of HPA after immersing the composite membrane in water for 48 hrs at 30 °C was 24 %.[11] Additionally, another PHF–Nafion composite was soaked in boiling water for two hrs and the extracted amount was measured to be less than 1 % of the original PHF, showing the excellent stability even in a more drastic environment. Apparently, the solution cast composite membrane holds the fullerene more tightly than the doped membrane, demonstrating the sharply increased integration of PHF into the Nafion polymer network. The improved stability in the solution cast composite is consistent with the better dispersion of PHF in the PHF–Nafion film by the optical microgram above. Optimization of composite membrane fabrication should improve further the retention of the fullerene in the composite membrane.

Table 3. Extraction of PHF out of Nafion Composites.

	1% PHF/Nafion 117[a]	**1% PHF–Nafion[b]**	
Temperature, °C	20[c]	20[c]	100[d]
Extracted PHF, [e] wt%	65[f]	2.5 [g]	<1 [g]

[a]The doped composite. [b]The solution-cast composite. [c]Soaked in water for 120 hrs. [d]Soaked in boiling water for 2 hrs. [e]The weight percentage of the extracted PHF relative to the original amount in the composite. [f]Determined by weight measurements. [g]Determined by UV-vis spectroscopy.

Titlation experiments were performed to determine the values of equivalent weight (*E.W.*) for both recast Nafion and 3% PHF–Nafion composite, listed in Table 4. The values of *E.W.* for the PHF composites with higher loadings are also included. The results show that though PHF is an acid, it dilutes the Nafion's *E.W.*; hence, the loading should be maintained low.

Table 4. The Equivalent Weight for the PHF–Nafion Composites and the Recast Nafion.

	Recast Nafion	**3% PHF–Nafion**	**9% PHF–Nafion**	**18% PHF–Nafion**
E.W. g meq^{-1}	1095	1087	1124	1183

We also performed AC impedance measurements for the PHF–SPEEK composite. Figures 13 displays the humidity dependence of 15% PHF–SPEEK66 (IEC=1.92 meq g^{-1}) and SPEEK66 at 20°C, respectively. Though the conductivity of SPEEK66 alone was too low to measure under 25%RH, it seems to be increased by the addition of PHF by almost an order of magnitude under higher humidity conditions. Mecheri et al. have reported the conductivity of SPEEK80 mixed with 23% of $WO_{32}H_2O$ to be approximately 5 x10^{-3} S cm^{-1} under more than 80 %RH at ambient temperature.[41] Obviously, PHF can achieve the higher conductivity at a lower loading with SPEEK having a lower degree of sulfonation. Yet, it is not clear the mechanism by which the improvement is achieved by PHF.

Table 5 summarizes the conductivity of various fullerene composites at 20 and 120 °C under 25 %RH except for the SPEEK membranes which were measured under 50 %RH since the conductivity under 25 %RH was too low to measure. First, the fullerenes do not increase the membrane resistance of either Nafion or SPEEK composites, at least up to the loading tested, 3% for Nafion and 15% for SPEEK. At 18% of loading, the PHF–Nafion composite loses the conductivity, but this is due to the increased *E.W.*, as shown in Table 4. Secondly, the fullerenes do not improve the conductivity of Nafion at either temperature tested. We have observed that this is true under a wide range of relative humidity.

The conductivity of SPEEK66, on the other hand, is clearly improved by the addition of PHF. At the higher temperature, 120°C, however, the improvement seems to become small. This may be due to the low activation energy of the PHF–SPEEK composite. The contrast between the Nafion and SPEEK composites seems to be due to the difference in the miscibility of the fullerene in the host polymer. Different degrees of miscibility give rise to different morphologies, thus, various distributions of the fullerene in a host polymer matrix which should affect the proton transfer through the membrane, hence the different effect on the conductivity. Further work is needed to understand the difference in the effect of the fullerene on the two host polymers.

Table 5. The Proton Conductivity of Fullerene Composites at 120 °C under 25% RH (mS cm^{-1}).

Host Polymer	Nafion[a]								SPEEK66[b,c]	
Temp. °C	117	recast	C_{60} 1%	PHF 1%	PHF 3%	PHF 18%	MFPA 1%	FPA 1%	recast	PHF 15%
20	4.2	3.9	3.7	3.5	6.1	1.5	1.3	1.2	0.09	0.82
120	4.5	4.3	4.1	3.9	5.5	1.9	3.9	4.1	1.3	1.5

[a]*E.W.*=520. [b]*E.W.*=520. [c]under 50%RH.

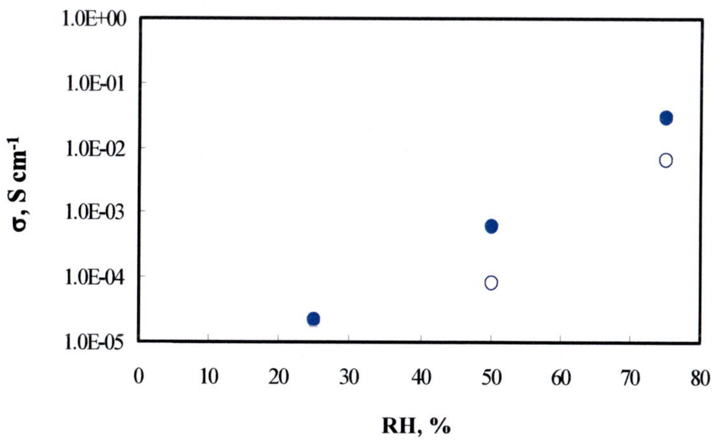

Figure 13. The conductivity of 15%PHF–SPEEK66 composite (the blue circle) and SPEEK66 (the white circle) as a function of humidity at 20 °C.

Figure 14 presents the *I-V* curves for 3%PHF–Nafion, as a representative of the fullerene composites, and the recast Nafion at 45 and 80 °C under 25 %RH. Although the differences between the fullerene composite and the recast Nafion are small, the 3% PHF–Nafion composite does perform better. It is not clear why the 3% PHF–Nafion composite is somewhat improved from the recast Nafion since no detailed analysis on the polarization data was performed. The work is still in progress.

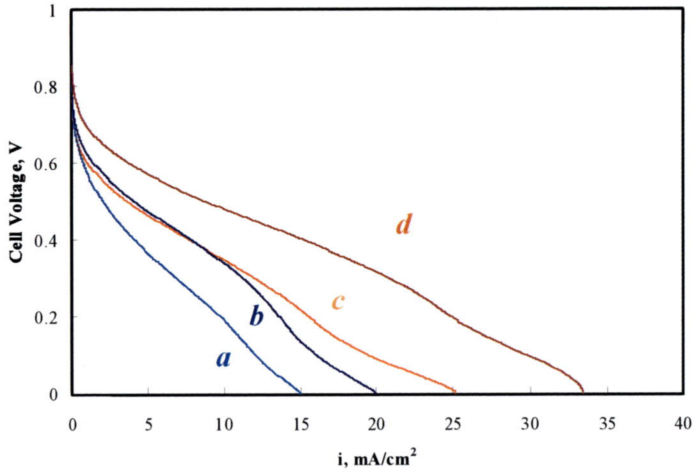

Figure 14. I-V curves of (a) the recast Nafion at 45 °C and (b) at 80 °C, respectively, and (c) 3% PHF–Nafion at 45 and (d) at 80 °C, respectively, all under 25 %RH.

The MeOH crossover of the fullerene–Nafion composites was measured voltammetarically. Figure 15 displays the inverse of the limiting current density of various fullerene composites as a function of the membrane thickness.

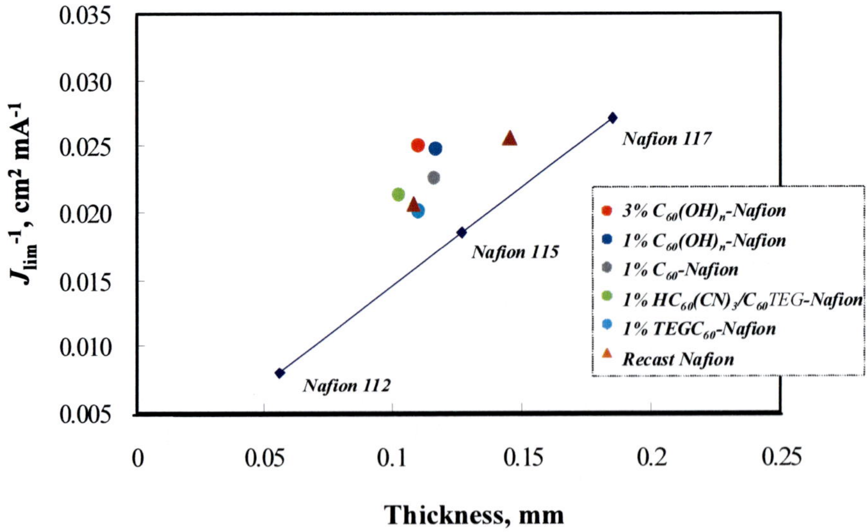

Figure 15. Inverse of the limiting current density of the fullerene–Nafion composites as a function of the membrane thickness.

The plots for the commercially available Nafion membranes follow a straight line, as is expected. All the fullerene–Nafion composites show higher values of the inverse of the limiting current density than those for the Nafion membranes with the comparable thickness. This suggests a lower MeOH crossover for the fullerene–Nafion composites than the corresponding Nafion membrane, though some fullerene composites have lower crossover than others. The lowest crossover was achieved by the 3% PHF–Nafion composite, about 30% lower than that of Nafion 115. Even the recast Nafion shows a lower crossover than Nafion 117. Yet, the limiting current density of the fullerene composites are lower than that of the recast Nafion; thus, the fullerene derivatives seem to have favorable effects on reducing the MeOH crossover.

The water characteristics play an important role in reducing the MeOH crossover. We have extensively characterized the fullerene–Nafion composites and focus primarily on them below. First, the water uptake was measured both from soaking in water (wet) and also under 25% RH (dry).

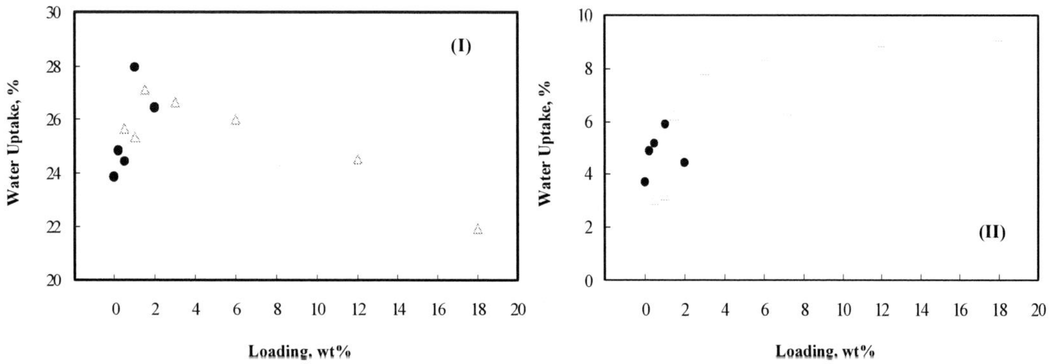

Figure 16. (I) The wet water uptake of the C_{60}–Nafion (the filled circle) and the PHF–Nafion composites (the red triangle) as a function of the fullerene loading and (II) the dry water uptake of the C_{60}–Nafion and the PHF–Nafion composites as a function of the fullerene loading.

Figure 16 displays both wet and dry water uptake of the C_{60} and PHF composites as a function of the loading. Both composites show a higher wet and dry water uptake than the recast Nafion in general. The C_{60} composite has the peak in the wet water uptake around 1%, a 4% increase in the water uptake from that of the recast Nafion (at loading = 0%). As to the PHF composite, the peak resides between 1 and 3% of loading, an approximately 3% increase from the recast Nafion. As to the dry water uptake, it increased asymptotically to around 9% with the loading after which it became plateau for the PHF composite, while it peaked for the C_{60} composite at around 1wt% and then decreased. The results are summarized in Table 6 for 1% C_{60}–Nafion, 3%PHF–Nafion, and the recast Nafion. Apparently, both fullerene composites held more water than the recast Nafion, especially under 25% RH.

Table 6. Water Uptake for Fullerene–Nafion Composite Membranes (%).

Water uptake	Recast Nafion	1 % C_{60}–Nafion	3% PHF–Nafion
Wet	23.87	27.96	26.62
Dry	3.71	5.89	7.96

Our previous TGA measurements indicated that PHF lost four water molecules per PHF molecule, while Nafion 117 lost 1.8 water molecules up to the temperature of 120 °C.[42] It was suspected that approximately twelve hydroxy groups on C60 helped bind water molecules tightly, a manifestation of its extremely high volumetric density of the functional groups. Our molecular dynamics simulations have also revealed the water dynamics and the retention around the hydroxyl groups of C60 in water.[42] The higher wet and dry water uptakes of the PHF composites, relative to those of the recast Nafion, are believed to be due to the highly hygroscopic nature of PHF. As to the C60 composite, the cause for the higher water uptakes is not clear at this point. There are several possibilities for this observation such as the water trapped in the interface between the C60 aggregates and the Nafion domain or the morphological change of Nafion caused by the introduction of C_{60} into the Nafion matrix. Due to the small amount of C_{60}, ~1%, no experimental evidence has been found to support either possibility. Only the morphological change by PHF is discussed later.

Aromatic hydrocarbon polymer-based membranes have been extensively studied as alternative membranes to perfluoro polymers such as Nafion. Though they are promising in terms of the cost and the high temperature performance, the swelling of the membranes having low *E.W.*s poses a serious problem. Figure 17 shows the wet water uptake of 15%PHF–SPEEK and SPEEK as a function of IEC.

As the IEC increases, the water uptake is dramatically suppressed by the presence of PHF. The reduction of the water uptake by PHF is as much as 2/5 for SPEEK having IEC of 1.92 meq g^{-1}. Mecheri et al. have reported that the water uptake of SPEEK in the presence of 23% of $WO_{32}H_2O$ was even higher than that of SPEEK in the absence of $WO_{32}H_2O$.[41] It is possible that PHF acts as a cross-linker between the sulfonic acid groups of SPEEK to reduce the membrane swelling, as is illustrated in Figure 18.

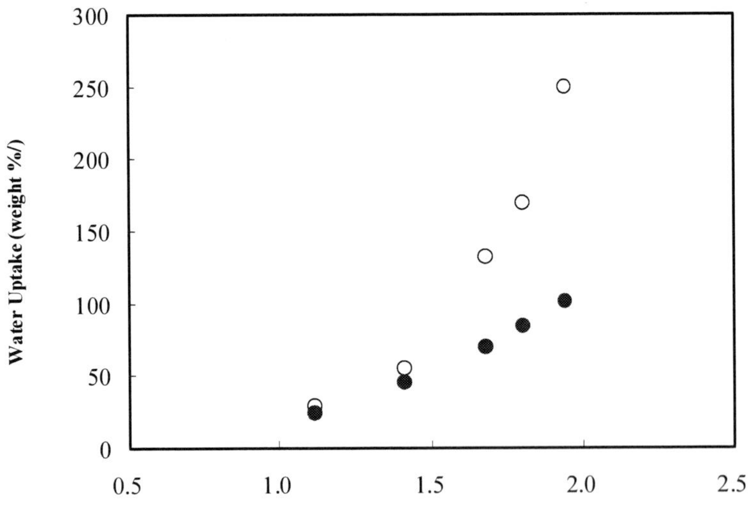

Figure 17. Water uptake of (a) 15 wt% PHF–SPEEK (the blue circle) and (b) SPEEK (the white circle) as a function of IEC.

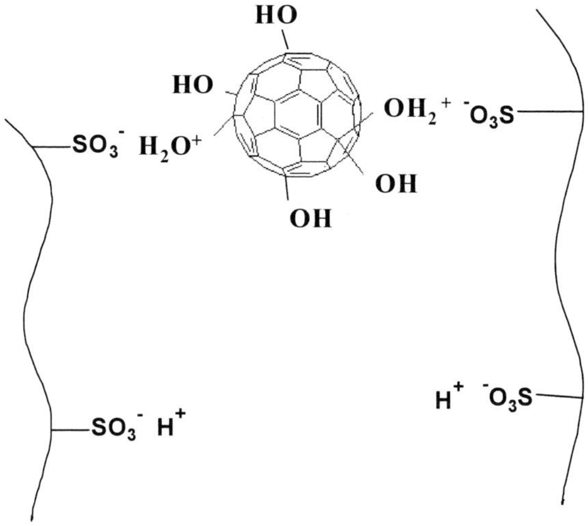

Figure 18. A schematic drawing of possible cross-linking between PHF and SPEEK.

Next, the water in the composites was further characterized by two techniques: TGA and NMR spectroscopy. First, the water retention in the composites was examined by TGA in which the weight loss of the composites was monitored at 30°C for 40 min, assuming the weight loss under this condition is primarily due to the water loss. Figure 19 (I) displays the weight loss decays of various composite films including the recast Nafion at 30°C. The weight losses of the same samples during heating up to 250°C are shown in Figure 19 (II). Both figures indicate considerable differences among the membranes. The 3% PHF–Nafion composite exhibits the highest water retention under both conditions. The recast Nafion also shows the lowest water losses in both TGA measurements; however, it had the smallest water uptake to begin with, as shown in Table 6. Table 7 summarizes the water characteristics in each film. The data is consistent to the results of the dry water uptake. Under both conditions, the 3% PHF–Nafion composite maintained more water in the membrane than any other membranes studied, suggesting extremely high water retention. On the other hand, the recast Nafion shows a smaller water retention. The 1% C_{60}–Nafion composite lost more water than the 3% PHF–Nafion composite, though it had more wet water uptake than the latter. It is clear that it does not hold water as tightly as the 3% PHF–Nafion composite. Our previous MD simulations have revealed strong interactions between the OH groups of $C_{60}(OH)_{12}$ and the SO_3H groups of Nafion.[42] The sharp difference in the water loss of the 3% PHF composite from any other membranes warrants further study.

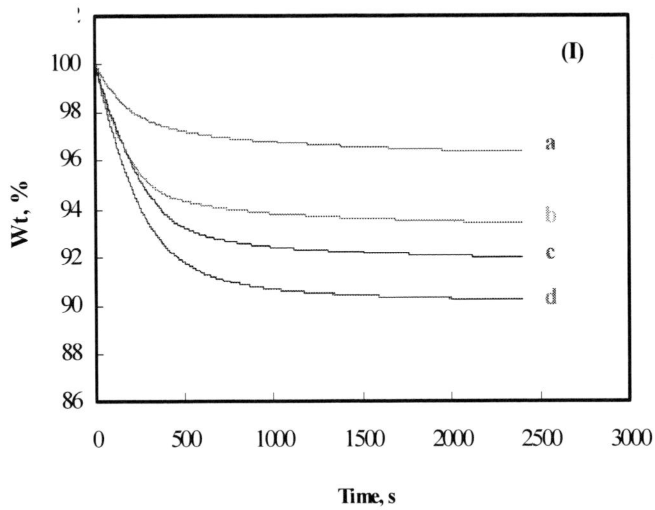

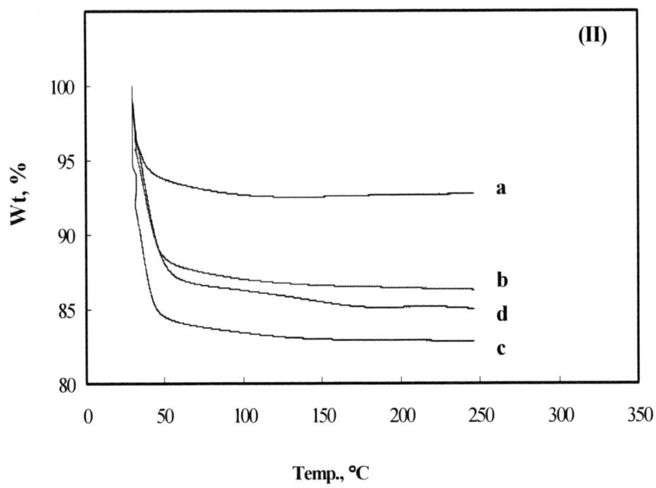

Figure 19. The weight loss decays of (I) the membranes under dry N_2 gas flow at 30°C and (II) under dry N_2 gas flow at the heating rate of 2°C min^{-1} measured by TGA : (a) 3% PHF–Nafion, (b) the recast Nafion, (c) 1% PHF–Nafion, and (d) 1% C_{60}–Nafion.

Table 7. Water Characteristics in the Membranes.[a]

at 30°C after 40 min[b]

	Recast Nafion	1% C_{60}–Nafion	1% PHF–Nafion	3% PHF–Nafion
Water loss, %	6.58	9.79	8.04	3.66
Water remained, %	17.29	18.35	18.35	22.96

[a]Under dry N_2 gas flow.

[b]Obtained from Figure 19 (I).

at 250°C with a heating rate of 2°C min^{-1} [c]

	Recast Nafion	1% C_{60}–Nafion	1% PHF–Nafion	3% PHF–Nafion
Water loss, %	13.72	14.12	17.15	7.28
Water remained, %	10.15	13.84	9.24	19.34

[c]Obtained from Figure 19 (II).

Table 8 lists the water binding energies for C_{60} and PHF for both neutral and anion forms. The anion form of PHF was $C_{60}(OH)_{11}O^{-}$. It is of interest to note that even C_{60}, strongly hydrophobic, has a negative binding energy. Recently, it has been reported that there was an attractive interaction between water and C_{60} through the lone pair orbital of water and the anti-bonding π orbital of C_{60}.[43] The results for $C_2F_5SO_3H$ are also included as a model for a sulfonated perfluoropolymer. The binding energies of PHF are larger than those of $C_2F_5SO_3H$ for both neutral and anon forms. In general, the stronger the acid is, the less the water binding energy becomes since the water binding site of a stronger acid is more delocalized which leads to smaller electrostatic interactions with water. Our calculations of proton dissociation energies at the PM3 level (see Fig.6) suggested that $C_2F_5SO_3H$ is stronger in acidity than PHF. Furthermore, the optimized structures of both PHF neutral and anion forms showed a bridge hydrogen bond of a water molecule between the adjacent OH groups, as is depicted below.

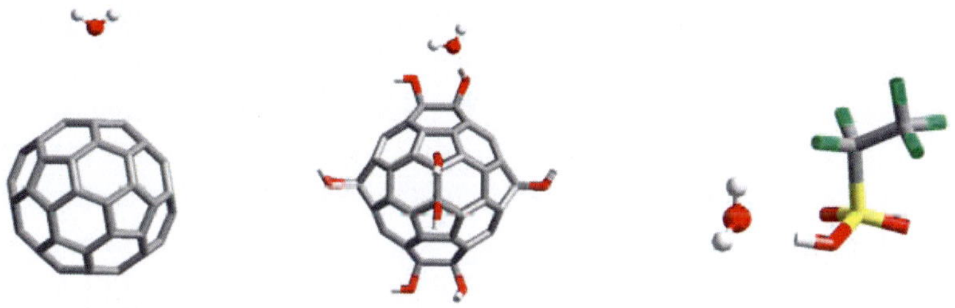

$C_{60}\cdots H_2O$ $\quad$ $C_{60}(OH)_{12}\cdots H_2O$ $\quad$ $C_2F_5SO_3H\cdots H_2O$

Table 8. The Water Binding Energy (ΔE_{bind}, kcal mol^{-1}) for Fullerene Derivatives.

		C_{60}	PHF	$C_2F_5SO_3H$
ΔE_{bind}	neutral	-0.7	-3.56	-7.24
	anion	–	-10.75	-8.87

T_2 relaxation time determined from ^1H pulse NMR spectroscopy can provide valuable information on the water mobility in the membrane.[44,45] Figure 20 displays T_2 decays of the fullerene composites including the recast Nafion.

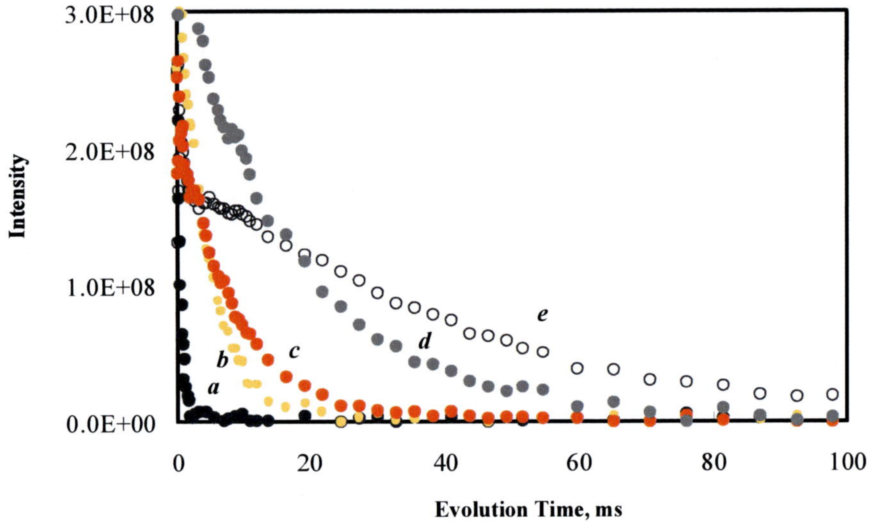

Figure 20. T_2 decays of (a) the recast Nafion, (b) 3% PHF-Nafion, (c) 1% PHF-Nafion, and (d) 1% C_{60}-Nadion, and (e) Nafion 117.

Sharp differences were observed among the membranes. The gradual, slow decay for Nafion 117 is typical of having a large number of free water, similar to the observation made by other researchers.[45] The less gradual decay of the 1% C_{60}–Nafion composite suggests less mobile water in the membrane, a manifestation of weak interactions between water and its surrounding in the C_{60}–Nafion composite. On the other hand, the recast Nafion shows a very quick decay, indicative of the immobile water, rather unexpected. Our separate DSC measurements confirmed that the recast Nafion had little free water, showing the endothermic heat flow peak due to the melting of ice well below 0 °C. With little free water in the membrane, the water in the recast Nafion can be mostly characterized as bound water, thus the slow decay. This may be a result of the annealing at 170 °C. Lee et al. have found that a recast Nafion annealed at high temperatures had smaller ionic clusters and lower water uptake than those annealed at lower temperatures.[46] Kim et. al. have reported similar T_2 results for their membranes having little free water.[45] The water characteristics of the recast Nafion needs further study. The PHF composites exhibit the decays in between 1% C_{60}–Nafion and the recast Nafion, suggesting the state of the water in those membranes to range from somewhat tightly bound water to loosely bound water.

The location of the fullerene in the Nafion matrix is critical in understanding the composite properties. Yet, the probing of location of C_{60} in Nafion was difficult due to the small loading and a rapid growth of aggregation with increasing loading made any change in spectroscopy obscure. Accordingly, we focused on the examination of the location of PHF in the Nafion matrix. Figure 21 displays the ATR FT-IR spectra for the dry samples of the recast Nafion (a), 1% PHF–Nafion (b), and 3% PHF–Nafion (c) in the region for the SO_3^- symmetry stretching where the largest change was observed upon the addition of PHF to Nafion. Though the 1% PHF composite shows little change, the 3% PHF composite exhibits a clear shift to a lower frequency which suggests interactions between PHF and the SO_3^- group of Nafion. This change is consistent to the hydrophilic nature of PHF which is likely to reside more in the ionic cluster of the Nafion matrix.

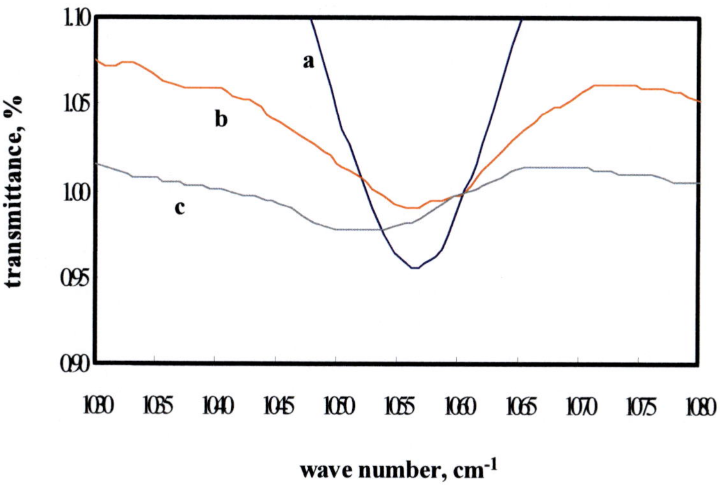

Figure 21. ATR-IR spectra of (a) the recast Nafion, (b) 1% PHF–Nafion and (c) 3% PHF–Nafion. The spectra are expanded for the region due to the SO_3^- symmetry stretching.

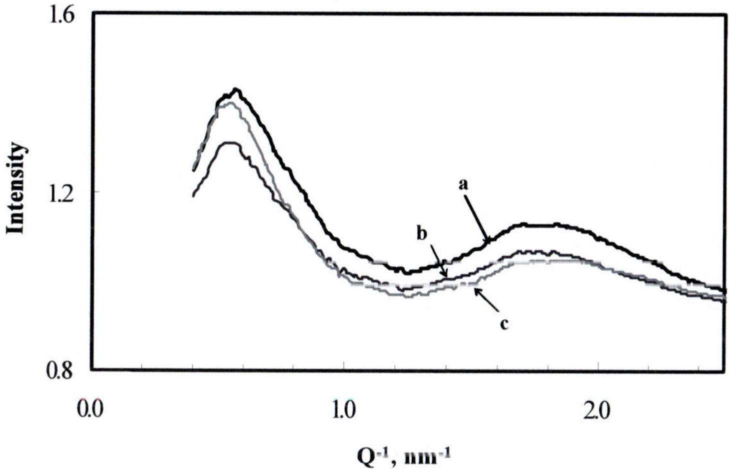

Figure 22. SAXS profiles of (a) the recast Nafion, (b) 1% PHF–Nafion, and (c) 3% PHF–Nafion.

Small angle x-ray scattering (SAXS) is a powerful tool to examine the structure and the morphology of polycrystalline polymers, but can be also used to locate additives in a polymer matrix. Figure 22 shows the intensity of SAXS plotted against Q^{-1} for the dry samples of the recast Nafion, 1% PHF–Nafion, and 3% PHF–Nafion. The SAXS profile of the recast Nafion is similar to that of perfluorosulfonated ionomer reported elsewhere.[47,48] Previous studies have identified that the largest peak around Q^{-1} = 0.6 nm^{-1} is due to the hydrophobic domain of Nafion and the second largest peak less than Q^{-1} = 1.5 nm^{-1} the hydrophilic domain.[46,49] The increased scattering vector for the hydrophilic domain of the recast Nafion, ~1.8 nm^{-1}, suggests a reduced ionic cluster size which may be consistent with the fast decay of the T_2 NMR data above. Lee et al. have reported a relationship between the annealing temperature, the ionic cluster size, and the water uptake.[46] The reduced peaks for both domains by the PHF composites suggest the increased amorphous regions in the composite by the addition of

PHF to Nafion, a sign of a morphological change by PHF in the Nafion composite. Though C_{60} is hydrophobic, a similar effect on the morphology can be expected by the addition of C_{60} to Nafion, suggested by the above simulations. This morphology change can be likely caused by rearrangement of the Nafion chains and entanglements of PHF by Nafion through the interactions between PHF and Nafion, most likely involving the SO_3^- groups, as is suggested by the ATR FT-IR results and MD simulations. Interactions of PHF with the ionic cluster of Nafion can certainly bring about a disruption in the hydrophobic region as well. As to the shift in the peaks, any change upon the mixing of the fullerene to the Nafion matrix is inconclusive from the current SAXS data. Further study is warranted. Our earlier computational study suggested that a PHF molecule can be accommodated in the ionic channel of Nafion without much enthalpic penalty.[42]

Computer simulations may give a valuable insight into the interactions between the fullerene and the Nafion polymer. Although detailed analysis requires many samplings including the fullerene located in various domains of the Nafion polymer after adequate simulation time, demanding serious computer resources, simple MD simulations may provide the information on the essential aspects of the fullerene–Nafion interactions.

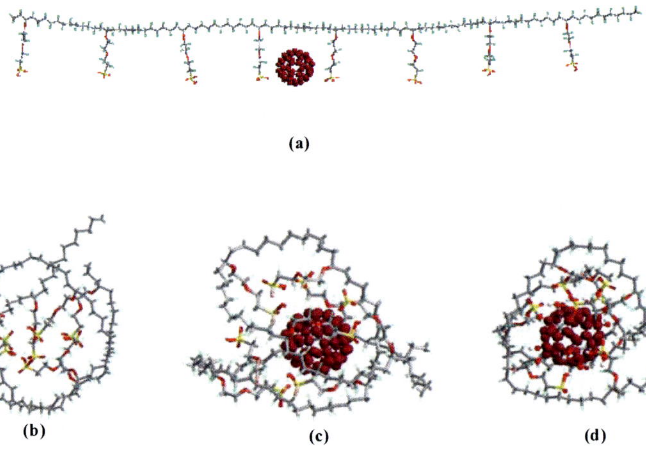

Figure 23. (a) The initial structure of C_{60} and the Nafion oligomer, and the snapshots of (b) the Nafion oligomer, (c) C_{60} and the Nafion oligomer, and (d) PHF and the Naifon oligomer, the last three taken after 1 ns from MD simulations.

Realistic modeling of the fullerene–Nafion amorphous system is out of the scope in this study. Here, we performed MD simulations of a C_{60} or PHF molecule together with a single Nafion oligomer in the gas phase. During the simulations, the hydrogen atoms were kept connected to the sulfonic acid groups of the Nafion chain, since no water was present to dissociate them. Figure 23a depicts the initial structure where the Nafion oligomer chain had both backbone and side chains stretched in all trans conformation adjacent to which the fullerene was placed.

First, the Nafion oligomer chain alone was subjected to MD simulations. After 200 ps, the chain transformed itself to a coil with the side chains pointing inside and the backbone outside and largely remained the same, illustrated in Figure 23b. This conformation ensures the maximum interactions among the hydrophilic side chains. Then, MD simulations of the fullerene–Nafion complex were performed. The initial structure, shown in Figure 23a, gradually underwent a transition to the fullerene completely wrapped around by the Nafion oligomer after approximately 200 ps for both complexes (Figs. 23c and 23d) and stayed unchanged afterwards. Figure 24 illustrates the history of the radius of gyration for each system. It is of particular interest to note that the radius of gyrations of all three systems converge to more or less the same value, ~ 10Å, after 400 ps. This observation suggests that the Nafion oligomer can accommodate either fullerene in the ionic cluster without much perturbation of its inherent chain dimension. Independent simulations with the fullerene originally placed on the other side of the Nafion backbone, thus, the fullerene facing the hydrophobic domain of the Nafion oligomer, eventually resulted in configurations similar to those in Figure 23, but the simulations took longer.

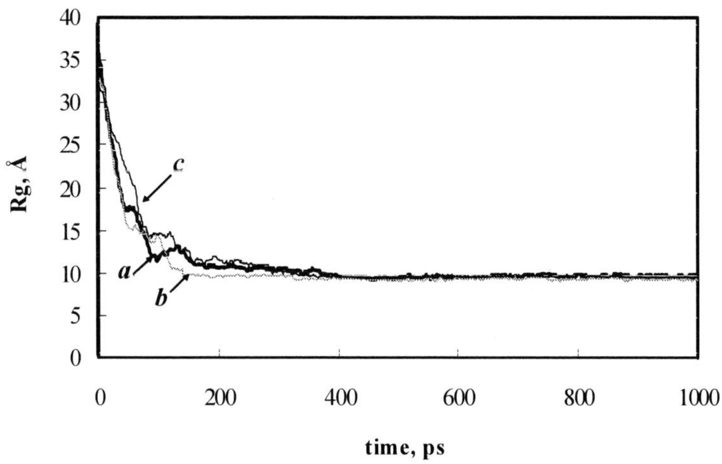

Figure 24. The radius of gyration of (a) the Nafion oligomer, (b) C_{60} and the Nafion oligomer, and (c) PHF and the Nafion oligomer as a function of time.

The detailed inspection in the case of the PHF–Nafion oligomer indicated that the hydroxy groups of PHF and the SO_3H groups of Nafion had close contact within the distances of hydrogen bonds. This is consistent with the shift observed in the above ATR FT-IR data. After several independent simulations, it seemed that the Nafion chain tried to accommodate the PHF molecule by having the maximum interactions through the SO_3H groups where all SO_3H groups were more or less occupied by the hydroxyl groups of PHF. It is of interest to note that C_{60}, highly hydrophobic, is still surrounded by the Nafion chain with its side chains pointing toward C_{60}. C_{60} is electronically neutral with zero charge on each atom in the force field used, thus creating no electrostatic repulsions between C_{60} and the Nafion oligomer during the simulations and the interactions were strictly controlled by the van der Waals interactions. Furthermore, the configuration shown in Figure 23c not only ensures the maximum contact between C_{60} and the oligomer, but also preserves the Nafion's stable conformation (shown in Fig.23b). The C_{60}–Nafion side chain interactions are essentially hydrophobic in origin between C_{60} and the CF_2 groups of the side chain. These interactions as well as the attractive interactions among the side chains pulled the side chains toward C_{60}. Though the interactions between C_{60} and the oligomer side chains may not be as strong as the PHF–Nafion side chain interactions, the smaller radius of a C_{60} molecule, approximately 2Å less than that of a PHF molecule, gave rise to the comparable average radius of gyration for both complexes: 9.6 and 9.8 Å. Yet, the C_{60}-side chain interactions created packets between C_{60} and the backbone chain of the Nafion oligomer. These voids may accommodate some water molecules which are trapped in the C_{60}–Nafion complex to increase water uptake.

The stabilization energy was calculated from the following equation:

$$\Delta E(\text{fullerene–Nafion}) = E(\text{fullerene}) - E(\text{Nafion}) - E(\text{fullerene–Nafion}) \qquad (5)$$

where E(fullerene) and E(Nafion) are the potential energies for the fullerene and the Nafion oligomer, respectively, and E(fullerene–Nafion) is the potential energy for the fullerene–Nafion oligomer complex, all in the gas phase. The stabilization energy should be distinguished from the interaction energy between the fullerene and the Nafion oligomer since the conformation of the Nafion oligomer alone is different from that in the fullerene–complex; thus, E(Nafion) is always lower than the energy of the Nafion oligomer in the fullerene complex. The average was taken over the period in which no significant change in the radius of gyration and in the conformation was observed. While the stabilization energy for the PHF–Nafion oligomer complex was 58.94 kcal mol^{-1}, that for the C_{60}–Nafion oligomer was only 0.2 kcal mol^{-1}. Clearly, Nafion favorably interacts with PHF due to their hydrophilic nature. As to C_{60}, the value suggests that Nafion can accommodate C_{60} in the hydrophilic domain without much penalty. Since no calculations for the condensed phase were performed, such as the sublimation energy of the fullerene, no discussion on the favorable location of the fullerene in the Nafion matrix can be made.

5. Conclusion

Several fullerene derivatives have been synthesized to examine their effects on proton conducting membranes, some of which are novel fullerene derivatives such as $C_{60}[(PO_3H)_m]_n$, $C_{60}H_{18}$, and $HC_{60}CN_3$. The AC impedance measurements on the fullerene pellets have shown that the fullerene directly attached by hydrogens such as $C_{60}H_{18}$ or $HC_{60}CN_3$ do not exhibit much conductivity, at least under the conditions used, despite the suggestions made by our calculations. It is suspected that the intermolecular distance between the carbon atoms of C_{60} is too large for a proton hopping to take place. It is known that interstitial voids exist between C_{60} molecules.[33] On the other hand, the fullerenes such as PHF, PHSF, MFPA, and FPA have the functional groups extending to the neighboring fullerene molecules which help a proton to transfer between the functional groups. Still, the conductivity among them seems to depend on the acidity of the functional group and the volumetric density of the functional groups. In fact, PHSF had the highest acidity and density among the fullerene derivatives studied except for HTCF; thus, its conductivity the highest.

Fullerene composites have been successfully fabricated through our new solution-cast technique, the first time for the Nafion and SPEEK host polymers. Though the fullerene derivatives did not seem to improve the conductivity of Nafion under low humidity conditions, they did not increase the membrane impedance, either. As to SPEEK, PHF did exhibit a significant improvement under low humidity conditions, almost by an order of magnitude. The difference seems to be due to the different miscibility of the fullerene in the host polymer. The conductivity may depend on the distribution of the fullerene in a host polymer matrix, thus, the morphology of the composite which depends on the miscibility of the fullerene.

PHF also showed another benefit which is to reduce the water uptake of highly sulfonated SPEEK which could dissolve in water, otherwise. Membrane swelling is a serious problem in fuel cell operations and this type of fullerene derivatives may find a new way to improve membrane swelling while maintaining the high proton conductivity through high IECs.

The fullerene–Nafion composites have been subjected to thorough characterization such as the water uptake, the morphology, the physical stability, the water characterization, and the interactions between the fullerene and Nafion. The PHF composite has shown a significantly improved physical stability from the previously prepared doped membrane, suggesting PHF mostly integrated into the Nafion matrix at a microscopic level. Somewhat large aggregations were observed in the C_{60}–Nafion composite due to the C_{60}'s poor miscibility in Nafion, while the PHF–Nafion membrane displayed a uniform dispersion throughout the composite. The TGA and T_2-NMR measurements have shown that the fullerene composites have the increased water retention and the decreased water mobility, compared to Nafion 117. The implication is that the fullerene composites have less free water than Nafion 117, which may be beneficial to alleviate the flooding in the cathode and also to reduce the MeOH crossover. Yet, despite the significant difference observed for the fullerene composites in the water characteristics, relative to Nafion 117, little improvement in the proton conductivity over Nafion 117 was observed, though the MeOH crossover of Nafion seems to be reduced in the presence of the fullerenes. The increased water retention of the PHF composite is a result of the highly hygroscopic nature of PHF, a direct consequence of the high volumetric density of the hydroxy groups, approximately twelve OH groups in a cavity of 1 nm in diameter.

Based on the results from the ATR FT-IR, the MD simulations, and the SAXS studies, it is postulated that PHF resides largely in the hydrophobic domain of Nafion while interacting with the SO_3^- groups when the loading is small. As a result, Nafion expands its ionic cluster, possibly the ionic channel as well, to accommodate PHF. PHF, in turn, being highly hygroscopic, brings extra water into the ionic channel of Nafion and holds it tightly under low humidity conditions, increasing the water retention in the channels, but not entirely eliminating the mobility of water, all consistent to the data from the water uptake, TGA and T_2-NMR measurements.

The results of the fullerene–SPEEK composite were encouraging in improving the performance of SPEEK under dry conditions and also alleviating the membrane swelling and further study is needed to understand the mechanism and to improve the performance.

Acknowledgement

The author wishes to thank the work by my collaborators: Drs. Arun Venkatesan, Ryan DeSousa, Hengbin Wang, Jeff Gasa, all of MC Research & Innovation Center, and Drs. Perumal Pugazhendhi and Raouf Loutfy of MER Corp., and Dr. Bruno Jousselme and Prof. Fred Wudl of University of California, Los Angeles, and Prof. Galen Stucky of University of California, Santa Barbara. They have contributed through synthesis, characterization, and valuable discussions. We received a financial support from DOE (the contract number 21724-001-06)).

References

[1] Dodabalapur, A.; Katz, H. E.; Torsi, L.; Haddon, R. C. *Science (Washington, D. C.)* 1995, 269(5230), 1560.

[2] Kubozono, Y.; Takabayashi, Y.; Shibata, K.; Kanbara, T.; Fujiki, S.; Kashino, S.; Fujiwara, A.; Emura, S. *Physical Review B: Condensed Matter and Materials Physics* 2003, 67(11), 115410/1.

[3] Mizuseki, H.; Igarashi, N.; Belosludov, R.V.; Farajian, A. A.; Kawazoe, Y. S*ynthetic Metals* 2003, 138(1-2), 281.

[4] Kreuer, K.D.; Fuchs, A.; Ise, M.; Spaeth, M.; Maier, M.J. *Electrochim. Acta* 1998, 43 (10-11), 1281.

[5] Steininger, H.; Schuster, M.; Kreuer, K. D.; Kaltbeitzel, A.; Bingoel, B.; Meyer, W. H.; Schauff, S.; Brunklaus, G.; Maier, J.; Spiess, H. W. *Phys. Chem. Chem. Phys.* 2007, 9(15), 1764.

[6] Fontanella, J.J.; Wintersgill, M.C.;Wainright, J.S.; Savinell, R.F.; Litt, M. *Electrochimica Acta* 1998, 43 (10-11), 1289.

[7] Ma, Y.-L.; Wainright, J.S.; Litt, M.H.; Savinell, R.F. *J. Electrochem. Soc.*2004, 151(1), A8.

[8] Jayakody, J. R. P.; Chung, S. H.; Durantino, L.; Zhang, H.; Xiao, L.; Benicewicz, B. C.; Greenbaum, S. G. *J. Electrochem. Soc.* 2007, 154(2), B242.

[9] Haile, S.M.;Boysen, D.A.;Chisholm, C.R.I.;Merle, R.B.; *Nature (London, United Kingdom)*, 2001, 410 (6831), 910.

[10] Shao, Z.-G.;Joghee, P.; Hsing, I.-M. *J. Membr. Sci.* 2004, 229, 43.

[11] Kim, Y.S.; Wang, F.; Hickner, M.; Zawodzinski, T.A.; McGrath, J.E. *J. Membr Sci.* 2003, 212, 263.

[12] Katsoulis, D.E. *Chem. Rev.* 1998, 98, 359.

[13] Herring, A.M.; Turner, J.A.; Dec, S.F.; Sweikart, M.A.; Malers, J.L.;Meng, F.; Pern, J.;Horan, J.;Vernon, D. *Abs. 228th Am. Chem. Soc.* National Meeting, Philadelphia, PA, August 22-26, 2004, FUEL-053.

[14] Saab, A. P.; Stucky, G. D. *Fullerene Science and Technology*, 1998, 6, 227.

[15] Hinokuma, K. *J. Electrochem. Soc.* 2003, 150, A112.

[16] Niyazymbetov, M. E.; Evans, D. H.; Lerke, S. A.; Cahill, P. A.;Henderson, C. C. *J. Phys. Chem.* **98**(49), 1994, 13093.

[17] Jousselme, B.; Sonmez, G.; Wudl, F, *J. Mat. Chem.* 2006, 16(34), 3478

[18] Hinokuma, K.; Pietzak, B.; Rost, C. G.; Ata, M. US Patent 6,495,290 B1.

[19] Chen, Z.; Thiel, W. *Chem. Phys. Lett.* 2003, 367, 15.

[20] Sun, H. *J. Phys. Chem. B*, 1998, 102, 7338.

[21] Ewald, P. *Ann. Phys.* 1918, 54, 519.

[22] van Gunsteren, W.F.; Berendsen, H.J. *Mol. Phys.*,1977, 34, 1311.

[23] Chiang, L.Y.;Wang, L.-Y.; Swirczewski, J.W.; Soled, S.; Cameron, S. *J. Org. Chem.* 1994, 59, 3960.

[24] Craig, C.; Henderson, P.;Cahill, A. *Science*, 1993, 259, 1885.

[25] Andrey, L.; Mirakyan, J.;Wilson, L.J.*J. Chem. Soc., Perkin Trans.*, 2002, 2, 1173.

[26] Wang, H.; DeSousa, R.; Gasa, J.; Tasaki, K.; Stucky, G..; Jousselme, B.; Wudl, F., *J. Memb. Sci.* 2007, 289(1+2), 277.

[27] Ren, X. Wilson, M. S. Gottesfeld, S. *J. Electrochem. Soc.* 1996; 143, L12.
[28] Ren X, Zawodzinski TA, Springer TE, Gottesfeld S. *J. Electrochem. Soc.* 2000, 147, 466.
[29] Carr, H.Y.; Purcell, E. M. *Physical Review* 1954, 94, 630.
[30] Meiboom, S.; Gill, D. *The Review of Scientific Instruments* 1958, 29, 688.
[31] Gasteiger, H. ; Makharia, R. ; Mathias, M. F. *Interface*, 2005, 14, 24.
[32] Hara, M.; Yoshida, T.; Takagaki, A.; Takata, T.; Kondo, J.; Hayashi, S.; Domen, K. *Angew. Chem. Int. Ed.* 2004, 43, 2955.
[33] Dresselhaus, M.S.; Dresselhaus, G.; Eklund, P.C. *Science of Fullerenes and Carbon Nanotubes*; Academic Press: New York, 1996; 301.
[34] Reed, C.; Kim, K.-C.; Bolskar, R.D.; Mueller, L. J. *Science*, 2000, 289, 101.
[35] Hinokuma, K.; Ata, M. *Chem. Phys. Lett.* 2001, 341, 442.
[36] Kreuer, K.D. *Chem. Mater.* 1996, 8, 610.
[37] Stephens, P. W.; Mihaly, L.; Lee, P. L.; Whetten, R. L.; Huang, S. M.; Kaner, R.; Diederich, F.; Holczer, K. *Nature (London)* 1991, 351, 632.
[38] Miyatake, K.; Fukushima, K.; Takeoka, S.; Tsuchida, E. *Chemistry of Materials*, 1999, 11, 1171.
[39] Ouyang, J.; Pan, Y.; Zhou, S.; Goh, S.H.; *Polymer*, 2006; 47, 6140.
[40] Lu, J. ; Dai, L. ; Mau, A.W.H. *Acta Polym.* 1998, 49, 371.
[41] Mecheri, B.; D'Epifanio, A.; Vona, M. L. D.; Traversa, E.; Licoccia, S.; Miyayama, M. *J. Electrochem. Soc.* 2006, 153, A463.
[42] Tasaki, K; DeSousa, R; Wang, H; Gasa, J.; Venkatesan, A; Pugazhendhi, P.; Loutfy, R. O.", *J. Memb. Sci.* 2006, 281, 570.
[43] Fomina, L.; Reyes, A.; Guadarrama, P.; Fomine, S. *Int. J. Quantum. Chem.* 2004, 97, 679.
[44] Kanekiyo, M.; Kobayashi, M.; Ando, I.; Kurosu, H.; Ishii, T.; Amiya, S. *J. Mol. Struct.* 1998, 447, 49.
[45] Kim, Y. S.; Dong, L.; Hickner, M. A.; Glass, T. E.; V. Webb, V.; McGrath, J. E. *Macromolecules* 2003, 36, 6281.
[46] Lee, K.; Ishihara, A.; Mitsushima, S.; Kamiya, N.; Ota, K. *J. Electrochem. Soc.* 2004, 151, A639.
[47] Gebel, G.; Moore, R. B. *Macromolecules* 2000, 33, 4850.
[48] Page, K. A.; Landis, F. A.; Phillips, A. K.; Moore, R. B. *Macromolecules* 2006, 39, 3939.
[49] Fujimura, M.; Hashimoto, T, Kawai, H *Macromolecules* 1990, 14, 1309.

In: Progress in Fullerene Research
Editor: Milton Lang, pp. 41-83
ISBN: 978-1-60021-841-5

Chapter 2

BORON NITRIDE NANOCAGE CLUSTERS, NANOTUBES AND NANOHORNS

Takeo Oku*

Department of Materials Science, The University of Shiga Prefecture
Hassaka 2500, Hikone, Shiga 522-8533, Japan

Abstract

Boron nitride (BN) nanocage clusters, BN nanohorns and BN nanotubes and were synthesized by arc-melting and thermal annealing methods, and characterized by high-resolution electron microscopy and molecular orbital calculations, and their properties were discussed. The BN clusters consisted of 4-, 6-, 8- and 10-membered BN rings satisfying the isolated tetragonal rule, which was optimized by molecular orbital calculations. Total energy calculation showed that some elements stabilize and expand the $B_{36}N_{36}$ structure. Bandgap energies of the $B_{36}N_{36}$ clusters were found to be reduced by introducing a metal atom inside the cluster, which indicates controllability of the energy gap. BN nanohorns with tetragonal BN rings were synthesized and proposed, and multi-walled BN nanohorns would be stabilized by stacking of BN nanohorns. Chiralities of BN nanotubes with zigzag-type structures were directly determined from high-resolution images, and structure models are proposed. Total energies of BN nanotubes with a zigzag-type structure were lower than those of armchair-type structure, and these results agreed well with the experimental data. Fe-filled BN nanotubes were also produced, and the [110] direction of the Fe nanowires was parallel to the BN nanotube axis. A new process for Fe or Co nanoparticles coated with BN layers in large quantity was developed. These BN nanocapsules encaging Fe or Co nanoparticles exhibited a soft magnetic characteristic and good oxidation resistances. These unique structures would be suitable materials for nanoelectronics devices, magnetic recording media and biological sensors with excellent protection against oxidation and wear. Possibility of hydrogen gas storage in BN clusters was also investigated by molecular orbital calculations, which indicated possibility of hydrogen storage of ~5 wt. %. The present work indicates that the new BN nanocage fullerene materials with various atomic structures and properties can be produced, and a guideline for designing the BN fullerene materials is summarized.

---

* E-mail address: oku@mat.usp.ac.jp

Keywords: boron nitride, nanocage, cluster, nanohorns, nanotube, nanoparticles, fullerene, hydrogen storage, structure, electron microscopy, mass spectrometry, molecular orbital calculation, molecular mechanics calculation, endohedral cluster, chirality, total energy, nanowire, electronic state

1. Introduction

Since the discovery of C_{60} fullerene [1] and carbon nanotubes [2], various carbon-based nanocage structures, such as fullerene clusters, nanotubes, nanocapsules, nanopolyhedra, cones, cubes and onions, have been studied because of the great potential for using materials with low dimensions in an isolated environment [1-5]. Boron nitride (BN) nanostructured materials with a bandgap energy of ~6 eV and non-magnetism are also expected to show several unique properties [4]. Recently, many studies have been reported on BN nanomaterials such as nanotubes [6-23], bundled tubes [24,25], nanocorns [26-28], nanocapsules [4,29-33], nanoparticles [34-36], clusters [3,4,37-40] and BN metallofullerenes [4,41,42], which are expected to be useful as electronic devices [43], field-effect transistors [44], high heat-resistant semiconductors, insulator lubricants, nanowires [45-47], magnetic nanoparticles [48,49] and gas storage materials [4,50-52]. Theoretical calculations on BN nanotubes [53-58], cluster-included BN nanotubes [59], BN clusters [60-69], BN metallofullerenes [70] and BN cluster solids [71-73] have also been carried out for prediction of the properties. By controlling the size, layer numbers, helicity, compositions and included clusters, these cluster-included BN/C nanocage structures with bandgap energy of 0-6 eV and

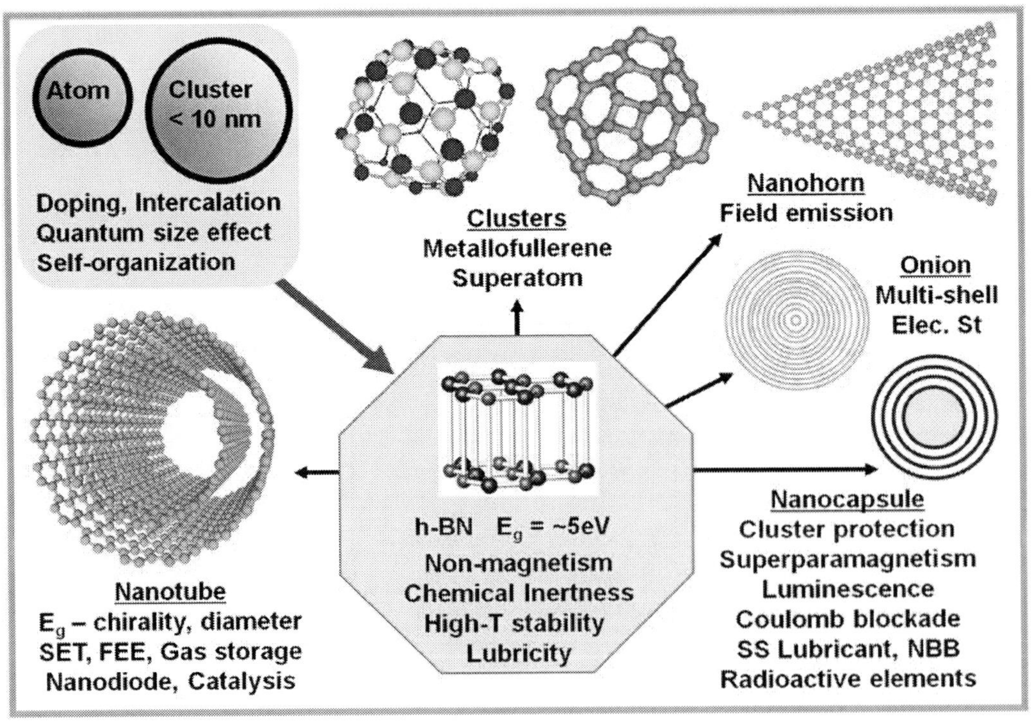

Figure 1. Structures and properties of BN fullerene materials.

non-magnetism are expected to show various electronic, optical and magnetic properties such as Coulomb blockade, photoluminescence and superparamagnetism, as shown in Fig. 1(a), and the differences between BN and carbon nanomaterials are summarized as listed in Table 1.

Table 1. Differences between BN and carbon nanomaterials

	C	BN
Structure	5-, 6-, 7-membered rings	4-, 6-, 8-membered rings
Oxidation resistance	~600 °C	~900 °C
Electronic property (E_g)	Metal-semiconductor (0∼1.7eV)	Insulator (∼6eV)
Band structrue	Indirect transition	Direct transition

The purpose of the present work is threefold. The first is to prepare the new BN nanocage clusters, nanotubes and nanohorns. In the present work, an arc-melting method was selected for the formation of the BN nanomaterials. An yttrium (Y) element was selected as catalysis for the BN nanomaterial formation. Yttrium had been reported to show excellent catalytic properties for producing single-walled carbon nanotubes [74]. The second purpose is to understand the atomic structures of these BN nanocage materials from high-resolution electron microscopy (HREM), which is a powerful method for direct structured analysis on an atomic scale [75-77]. The third purpose is to investigate hydrogen storage in the BN clusters by calculations. In order to confirm the atomic structures and to investigate stabilities, electronic states and hydrogen storage, total energy calculations by molecular mechanics and molecular orbital calculations were carried out. These studies will give us a guideline for designing the size of the tube, wall, thickness, chirality etc., and synthesis of the BN fullerene materials, which are expected for the future nanoscale devices.

2. Experimental Procedures

The YB_6 powder (4.0 g, 99.6%, Kojundo Chemical Lab. Co., Ltd) was used for the formation of BN clusters [78]. BN nanotubes were also synthesized from YB_6 with Ni powder [79]. The YB_6 powder (4.0 g, 99.6%, Kojundo Chemical Lab. Co., Ltd) and Ni powder (0.8g, 99.9%, Kojundo Chemical Lab. Co., Ltd) with the atomic ratio of 1:1 were set on a copper mold in an electric-arc furnace, which was evacuated down to 1×10^{-3} Pa. After introducing a mixed gas of Ar (0.025 MPa) and N_2 (0.025 MPa), arc-melting was applied to the samples at an accelerating voltage of 200 V and an arc current of 125 A for 10 s. Arc-melting was performed with a vacuum arc-melting furnace (NEV-AD03, Nisshin Engineering Co., Ltd), and gray to white powder were obtained around the copper mold.

An AXIMA-CFR (Shimadzu/Kratos, Manchester, UK) instrument was used to obtain laser desorption time-of-flight (LD-TOF) mass spectra [80]. The operating conditions were as follows: nitrogen laser (337 nm); linear mode; accelerating voltage at 20 kV; detection of positive ions. The sample powder (50 mg) was suspended in pyridine (C_5H_5N, 200 μl) with ultrasonination. The aliquot (1-2 μl) was spotted on the sample plate and dried at room temperature. The mass spectra were corrected by using C_{60} and C_{70} clusters [81].

Samples for HREM observation were prepared by dispersing the materials on carbon grids. HREM observation was performed with 300 kV electron microscope (JEM-3000F) having a point-to-point resolution of 0.17 nm. To compare observed images with calculated ones, HREM images were calculated by the multi-slice method using the MacTempas software (Total Resolution, CA, USA). The parameters used in the calculations are as follows: accelerating voltage = 300 kV, radius of the objective aperture = 5.9 nm^{-1}, spherical aberration Cs = 0.6 mm, spread of focus Δ = 8 nm, semi-angle of divergence α = 0.55 mrad, under defocus values Δf = -10 ~ 90 nm, unit cell (one cluster) = 2×2×2 nm, and crystal thickness (unit-cell thickness, 7 slices) t = 2 nm. space group P1, and assumed temperature factors [82,83] 0.02 nm^2 (B and N) and 0.14 nm^2 (Y).

Theoretical calculations for structural stability of the clusters and hydrogen storage were carried out by molecular mechanics calculations (MM2) and semi-empirical molecular orbital calculations (Hamiltonian : Parameterized Model Revision 5: PM5) using CS Chem3D Ultra (CambridgeSoft, MA, USA) and WinMOPAC Professional (Fujitsu Corp., Chiba. Japan). The energy levels and density of states were also calculated by the first principles calculation with discrete variational (DV)-Xα method.

3. BN Clusters

Figure 2(a) shows a mass spectrum for BN clusters in pyridine solution, and demonstrates the existence of $(BN)_n$ (n = 12~60) clusters. In addition, $(BN)_n$ (n = ~80) clusters were detected in the present work. Further, endohedral BN clusters $[Y_x@(BN)_n]$ with yttrium atoms encapsulated inside the BN clusters were detected. A mass spectrum for pyridine solution, which was used for matrix in LD-TOF mass spectrometry, is also shown for reference in Fig. 2(b). The mass spectrum peaks have a somewhat broad distribution because of the two isotopes of ^{10}B and ^{11}B. For localized structures of BN clusters, it should be noted that the isotopic ratio of boron atoms might be different from the natural averaged ratio because of minimization of the clusters' structural energy. Mass spectrum peaks for $(BN)_n$ clusters are observed in the range of m/z=600~1600, and the distribution peak is around n~40, which supports previously reported $B_{36}N_{36}$ clusters with high symmetry [4,40,63]. Figure 2(c) shows an enlarged mass spectrum for $B_{24}N_{24}$ clusters, and demonstrates the existence of $^{10}B_{24-x}{}^{11}B_x{}^{14}N_{24}$ (x = 1~6) clusters. The mass spectrum peaks have a somewhat broad distribution because of the two isotopes of ^{10}B and ^{11}B. Peaks of the cluster distribution are observed around x values of 3~5. For localized structures of BN clusters, it should be noted that the isotopic ratio of boron atoms might be different from the natural averaged ratio (^{10}B : ^{11}B = 19.9 : 80.1) because of the minimization of the clusters structural energy.

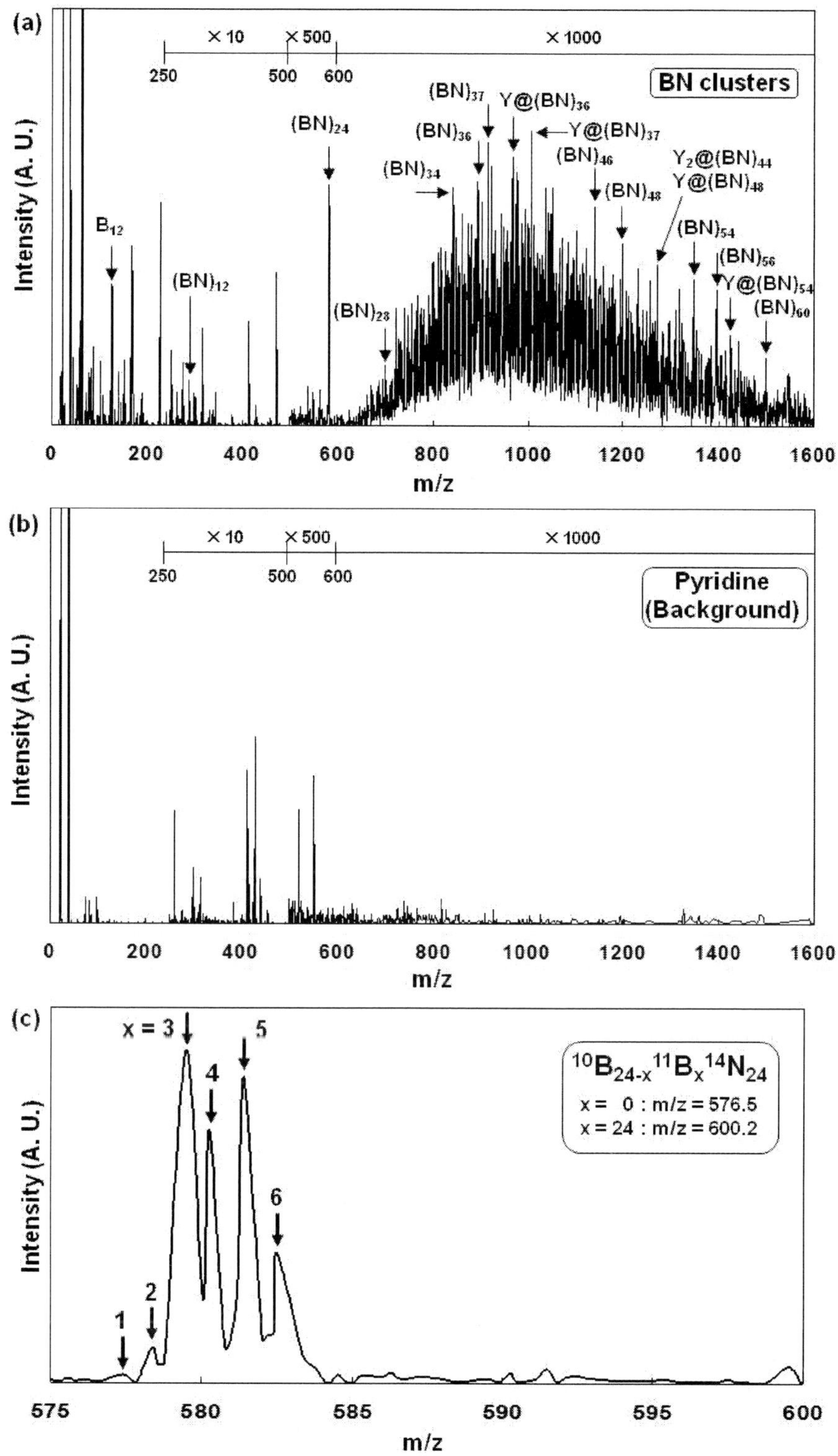

Figure 2. LD-TOF mass spectrum for (a) BN clusters and (b) pyridine (background). (c) Enlarged LD-TOF mass spectra for $B_{24}N_{24}$ cluster.

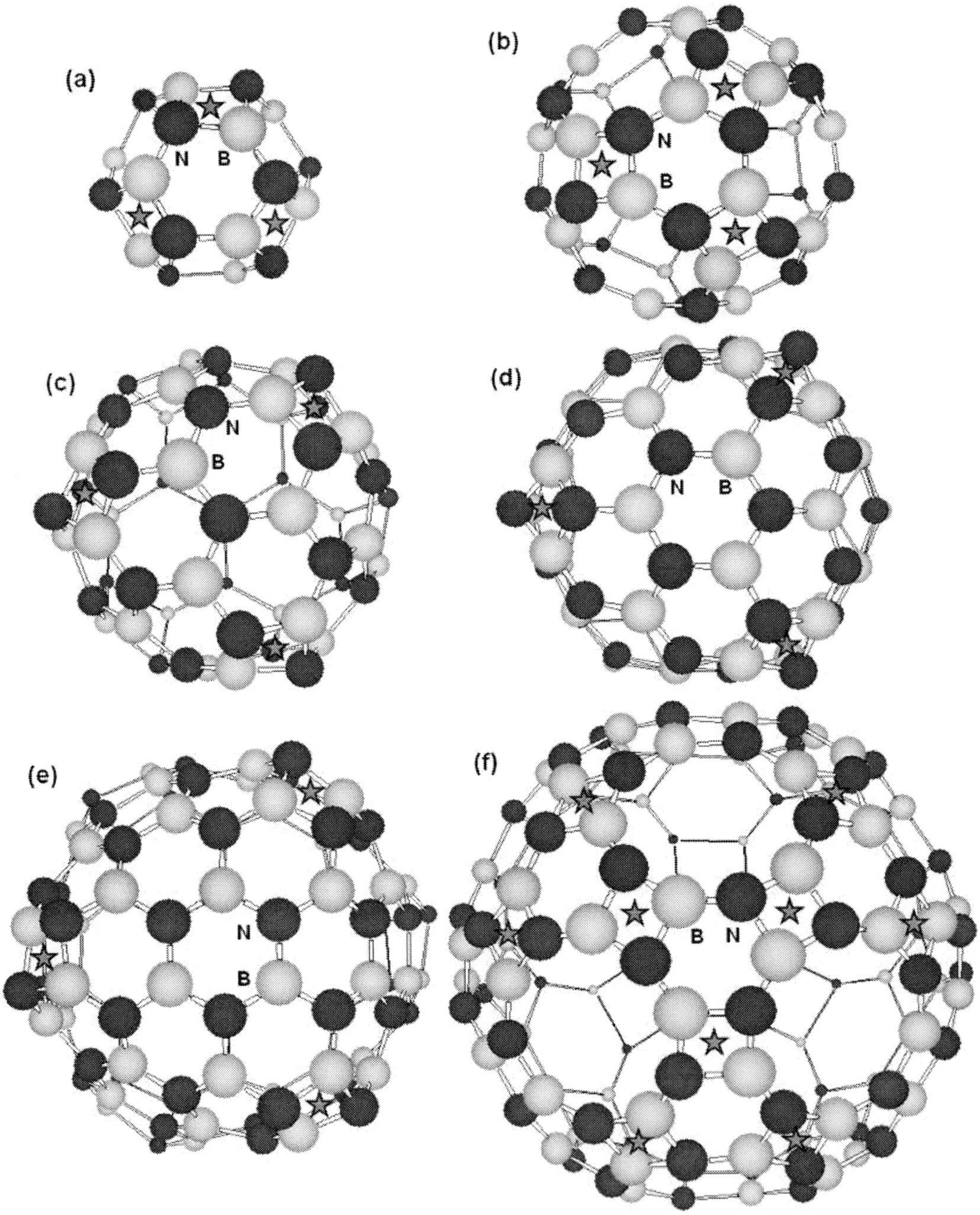

Figure 3. Atomic structure models of (a) $B_{12}N_{12}$, (b) $B_{24}N_{24}$, (c) $B_{28}N_{28}$, (d) $B_{36}N_{36}$, (e) $B_{48}N_{48}$ and (f) $B_{60}N_{60}$ clusters viewed along hexagonal BN rings. Tetragonal BN rings are indicated by star marks.

Although fullerenes satisfy the isolated pentagon rule, the present BN cage clusters satisfy the isolated tetragonal rule, as proposed for structural models of $(BN)_n$ (n = 12, 24, 28, 36, 48 and 60) in Fig. 3. All BN clusters have tetragonal BN rings isolated by hexagonal BN rings. In addition to the tetragonal and hexagonal rings, octagonal and decagonal BN rings [71] were introduced for $B_{24}N_{24}$ and $B_{60}N_{60}$ clusters in the present work.

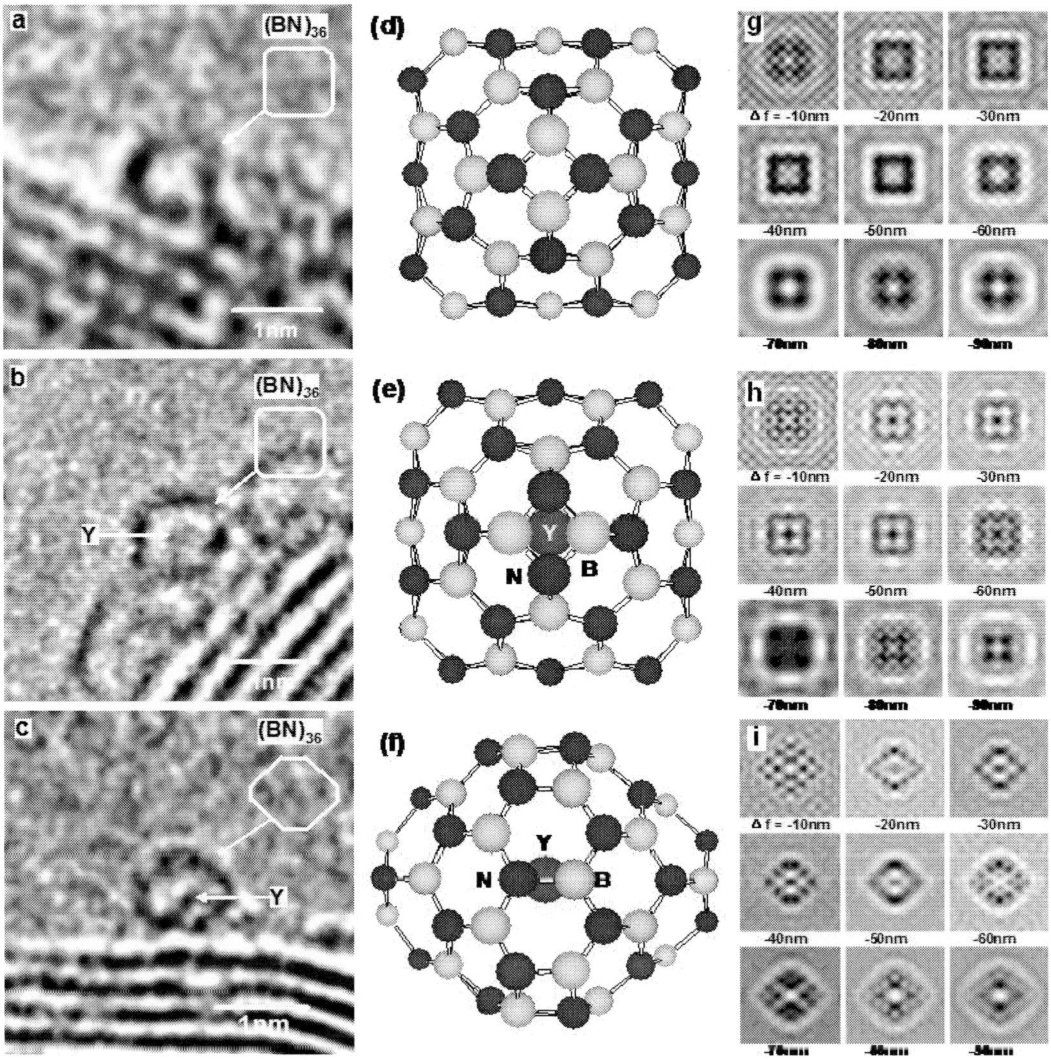

Figure 4. HREM images of (a) $B_{36}N_{36}$ and (b),(c) $Y@B_{36}N_{36}$ clusters. Structure models of (d) $B_{36}N_{36}$ and (e),(f) $Y@B_{36}N_{36}$ clusters corresponding to (a) and (b),(c), respectively. The structure model (e) is viewed perpendicular to (f). Calculated HREM images of (g) $B_{36}N_{36}$ and (h),(i) $Y@B_{36}N_{36}$ clusters as a function of defocus values along the same directions of (d-f), respectively.

In order to confirm the structure of the BN clusters, HREM observations were carried out. Hollow BN clusters were often observed on and inside the BN nanotubes, and the BN fullerene clusters had a single BN sheet, as shown in Fig. 4(a). Sizes of the BN clusters were in the range of 0.7-1.0 nm, which corresponds to the size of $B_{36}N_{36}$ clusters [4,42,65]. The $B_{36}N_{36}$ cluster consists of six 4-membered rings and thirty-two 6-membered rings, as shown in Fig. 4(d). In addition to the empty BN cage clusters, BN clusters containing doped atoms were also observed in HREM images. A typical HREM image of the BN clusters is shown in Fig. 4(b) and 4(c), and a ring contrast indicates $B_{36}N_{36}$ cluster. A dark contrast is observed inside the BN clusters, which indicates the existence of an yttrium atom inside the $B_{36}N_{36}$

clusters. During arc-melting of YB_6 powder, yttrium atoms could be introduced inside the $B_{36}N_{36}$ clusters [84]. Structure models for $Y@B_{36}N_{36}$ metallofullerene are proposed as shown in Fig. 4(e) and 4(f). The structures are optimized by molecular mechanics calculations (MM2), and the yttrium atom is closed to the center of the $B_{36}N_{36}$ cluster. Dark contrast is also observed outside of the BN cluster, which would be due to the quantum noise of electrons. However, inside the BN cluster, the contrast is strongly dependent on the diffraction conditions [4,42,75]. Therefore, the dark contrast inside the BN clusters is believed to be due to the existence of an yttrium atom.

Based on the projected structure models of Figs. 4(d-f), image calculations on the $B_{36}N_{36}$ and $Y@B_{36}N_{36}$ clusters were carried out to investigate the cluster structures. Figures 4(g-i) are HREM images of the $B_{36}N_{36}$ and $Y@B_{36}N_{36}$ clusters calculated along the [100] direction of the projected unit cell as a function of defocus values. The experimental images of Fig. 4(a-c) agree with calculated HREM images (Fig. 4g-i), especially at defocus value of -40 nm, which is nearly Scherzer defocus (Δf_s = -41.2 nm). This indicates the yttrium atom is included inside the BN cluster. The dark contrast corresponding to the yttrium atom inside the $B_{36}N_{36}$ cluster is smeared a little compared to the simulated image, which would be due to the vibration of the yttrium atom by electron beam irradiation. The formation of BN endohedral fullerene clusters including lanthanum and iron atoms has been previously reported [4]. In the present work, the endohedral BN metallofullerenes of $Y@B_nN_n$ (n = 36, 37 and 48) were produced and characterized from HREM and LD-TOF mass spectrometry. Various lanthanide elements could be introduced into the BN clusters, as reported in the carbon-based metallofullerenes [85].

Table 2. Calculated values for B_nN_n (n = 12-60) clusters. —: Data were not available.

	$B_{12}N_{12}$	$B_{24}N_{24}$	$B_{28}N_{28}$	$B_{36}N_{36}$	$B_{48}N_{48}$	$B_{60}N_{60}$
Heat of formation [kcal/mol]	-298.3	-832.4	-1162.6	-1597.8	-2312.3	-2311.1
Heat of formation [kcal/mol·atom]	-12.4	-17.3	-20.8	-22.2	-24.1	-19.3
Tetragonal BN rings	6	12	6	6	6	30
Hexagonal BN rings	8	8	24	32	52	20
Octagonal BN rings	0	6	0	0	0	0
Decagonal BN rings	0	0	0	0	0	12
$BN_{10\text{-}6}$ [nm]	—	—	—	—	—	0.1399
$BN_{10\text{-}4}$ [nm]	—	—	—	—	—	0.1492
$BN_{8\text{-}6}$ [nm]	—	0.1425	—	—	—	—
$BN_{8\text{-}4}$ [nm]	—	0.1503	—	—	—	—
$BN_{6\text{-}4}$ [nm]	0.1527	0.1529	0.1512	0.1509	0.1511	0.1532
$BN_{6\text{-}6}$ [nm]	0.1462	—	0.1493	0.1487	0.1472	—
d_{max} [nm]	0.488	0.683	0.780	0.743	1.063	1.102
d_{min} [nm]	0.488	0.682	0.647	0.882	0.862	1.102
ε_{HOMO} [eV]	-3.123	-3.127	-2.993	-3.051	-2.983	-2.966
ε_{LUMO} [eV]	1.957	1.818	2.230	2.316	2.221	1.944
Energy gap E_g [eV]	5.080	4.945	5.222	5.367	5.205	4.910

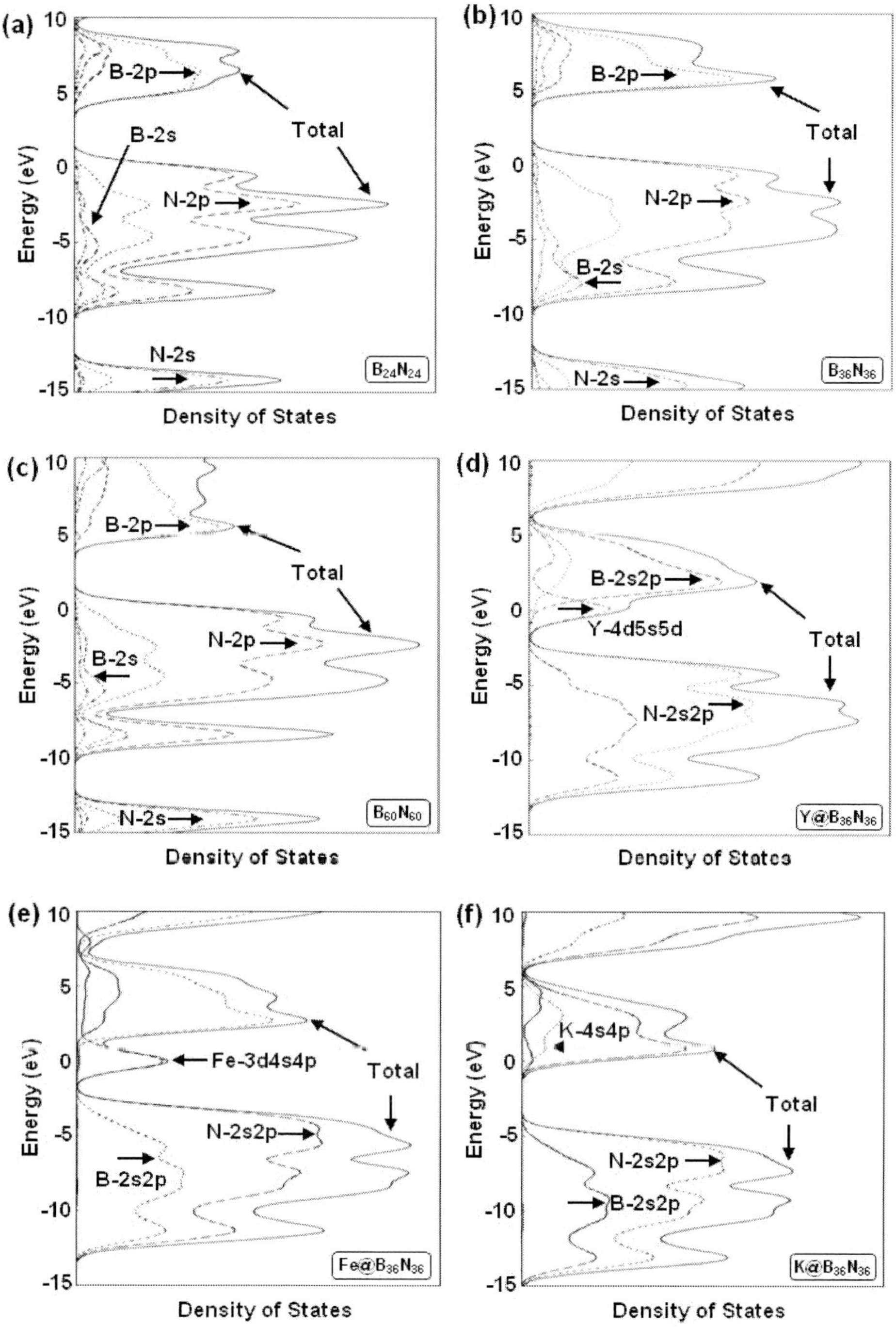

Figure 5. Densities of states for (a) $B_{24}N_{24}$, (b) $B_{36}N_{36}$, (c) $B_{60}N_{60}$, (d) Y@$B_{36}N_{36}$, (e) Fe@$B_{36}N_{36}$ and (f) K@$B_{36}N_{36}$ clusters.

Densities of states (DOS) for $B_{24}N_{24}$, $B_{36}N_{36}$ and $B_{60}N_{60}$ clusters are calculated and shown in Fig. 5(a-d), respectively. Strong effects of B-2p and N-2p orbitals on the densities of states are observed. The structural optimization and electronic structures of the BN clusters performed by molecular orbital calculations (PM5) are summarized in Table 2. The $B_{36}N_{36}$ cluster shows the largest energy gap of 5.367 eV between the highest occupied molecular orbital (HOMO) and the lowest unoccupied molecular orbital (LUMO), and the $B_{24}N_{24}$ and $B_{60}N_{60}$ clusters with octagonal and decagonal BN rings shows the smallest energy gap of 4.9 eV. The $B_{36}N_{36}$ and $B_{48}N_{48}$ clusters show the lowest total energies per atom as listed in Table 2, which agreed with the results of LD-TOF mass spectroscopy in the range of m/z = 600-1600.

Figure 6 is heat of formation (total energies) per atom of $B_{36}N_{36}$ clusters with doping element M (M@$B_{36}N_{36}$) clusters presented in periodic table. A total energy of $B_{36}N_{36}$ was calculated to be -19.68 kcal/atom. Comparing this value to other total energies in Fig. 6, energies of K@$B_{36}N_{36}$ and Ga@$B_{36}N_{36}$ show lower energies than that of $B_{36}N_{36}$ clusters. This indicates that potassium and gallium elements introduced inside the BN cluster stabilize the $B_{36}N_{36}$ structure, and the $B_{36}N_{36}$ clusters were found to be expanded by an atomic doping. Several other elements also have higher energies than that of $B_{36}N_{36}$ clusters. As indicated by asterisks, when molybdenum and barium elements were introduced, the structures of $B_{36}N_{36}$ were highly distorted or broken.

Electronic structures of Y@$B_{36}N_{36}$, Fe@$B_{36}N_{36}$ and K@$B_{36}N_{36}$ clusters were investigated as shown in Fig. 5(d-f), respectively. DOS of endohedral M@$B_{36}N_{36}$ clusters show effect of doping elements in BN clusters. Energy gaps of $B_{36}N_{36}$, Y@$B_{36}N_{36}$, Fe@$B_{36}N_{36}$ and K@$B_{36}N_{36}$ clusters were calculated to be 5.367, 0.114, 0.107 and 0.522 eV, respectively. Reduction of bandgap energies for Y@$B_{36}N_{36}$, Fe@$B_{36}N_{36}$ and K@$B_{36}N_{36}$ are due to the Y4d, Fe3d and K4s orbitals. The present results indicate that the energy gap E_g of the $B_{36}N_{36}$ can be controlled by introducing an atom inside the $B_{36}N_{36}$ cluster, and that some atoms such as potassium or gallium in the $B_{36}N_{36}$ cluster might stabilize the BN clusters by doping [86].

	1/I	2/II	3	4	5	6	7	8	9	10	11	12	13/III	14/IV	15/V	16/VI	17/VII	18/VIII
1	H -18.68																	He
2	Li -18.75	Be 16.38											B -17.47	C -16.42	N	O -17.79	F	Ne
3	Na -19.41	Mg -19.67											Al -18.49	Si -17.53	P -17.81	S -18.74	Cl -19.00	Ar
4	K -20.27	Ca -11.61	Sc	Ti	V	Cr	Mn	Fe -16.71	Co	Ni	Cu -17.93	Zn -17.81	Ga -19.74	Ge -18.97	As -18.48	Se -19.00	Br	Kr
5	Rb	Sr	Y	Zr	Nb	Mo *	Tc	Ru	Rh	Pd -18.84	Ag -17.78	Cd -17.40	In	Sn	Sb	Te	I	Xe
6	Cs	Ba *	La-Lu	Hf	Ta	W	Re	Os	Ir	Pt -15.17	Au	Hg -17.84	Tl	Pb -18.04	Bi	Po	At	Rn
7	Fr	Ra	Ac-Lr	Rf	Db	Sg	Bh	Hs	Mt									

Figure 6. Heat of formation (kcal/mol) of endohedral M@$B_{36}N_{36}$ clusters with doping element M. Total Energy of $B_{36}N_{36}$ per atom is -19.68 kcal/atom. Asterisks indicate $B_{36}N_{36}$ structure was highly distorted or broken.

The BN cluster is a molecule with polarity because of a positive charge at boron atom positions and a negative charge at nitrogen atom positions; so an electrophilic or nucleophilic reagent would work as a solution for BN clusters. Since C_{60} fullerene clusters have no polarity and are soluble in nonpolar solvents such as toluene [87] and benzene, they have difficulty in dissolving in a polar solvent. In the present work, pyridine (C_5H_5N) did work well for the extraction of BN clusters because of an electrophilic reaction; pyrrole (C_4H_4NH) would also work as a nucleophilic reagent. In order to investigate these BN nanocage clusters further, separation technique using high performance liquid chromatography should also be developed.

4. BN Nanotubes

A low magnification image of BN nanotubes produced from YB_6/Ni powder is shown in Fig. 7(a). The length and width of BN nanotubes is ~5 μm and 3-50 nm, respectively. A HREM image of a $B_{36}N_{36}$ cluster inside a BN nanotube is shown in Fig. 7(b). The BN nanotube has a

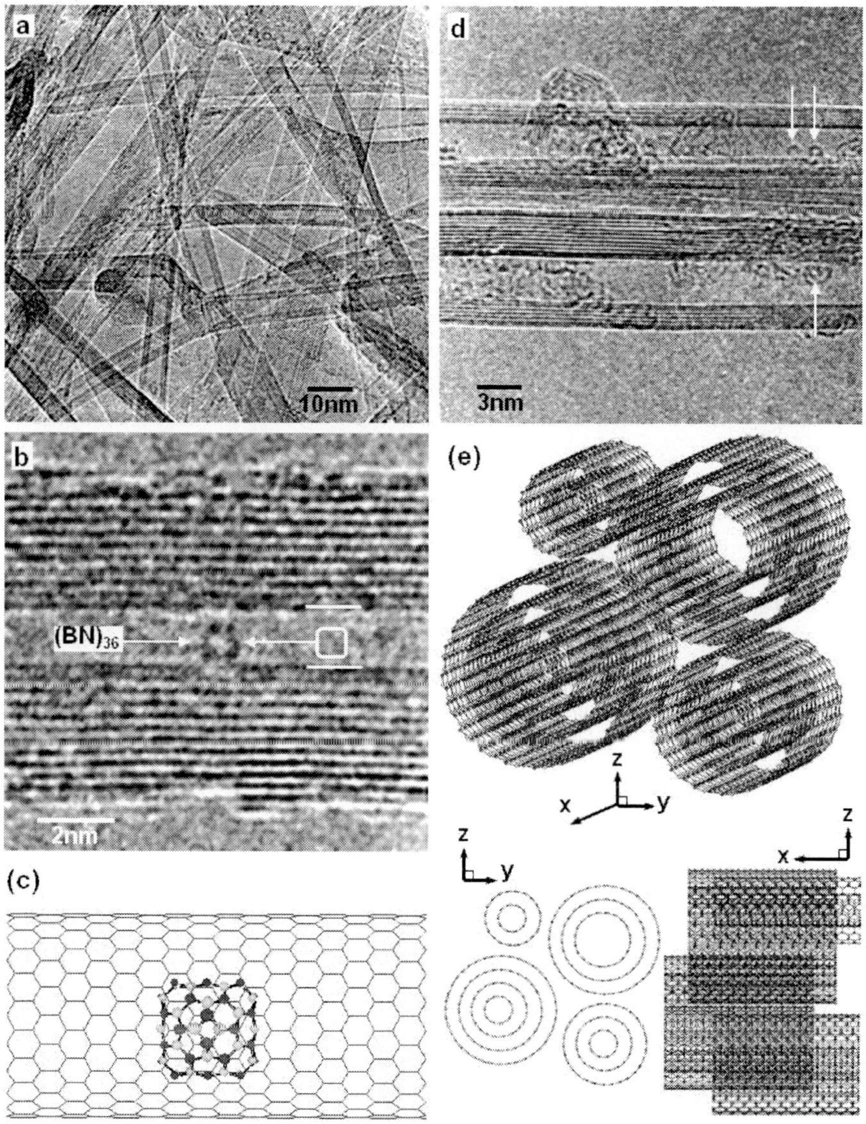

Figure 7. (a) Low magnification image of BN nanotubes. (b) HREM image of $B_{36}N_{36}$ cluster in BN nanotube. (c) Structure model of the center of (b). (d) HREM image of bundled BN nanotubes. BN clusters are indicated by arrows. (e) Atomic structure model from three different directions for bundled BN nanotubes.

multi-walled structure, and a diameter of the most inner tube is 1.75 nm. An atomic structure model of the center of Fig. 7(b) is shown in Fig. 7(c). Diameter and chirality of the BN nanotube are 1.747 nm and (22,0), respectively. This kind of peapod-type self-organized structure would be useful for the future nanoscale devices [23]. Another HREM image of BN nanotubes with a bundled structure is shown in Fig. 7(d), and an atomic structure model observed from three different directions is shown in Fig. 7(e). There are some spaces among the BN nanotubes, and the space would be useful for gas storage such as hydrogen.

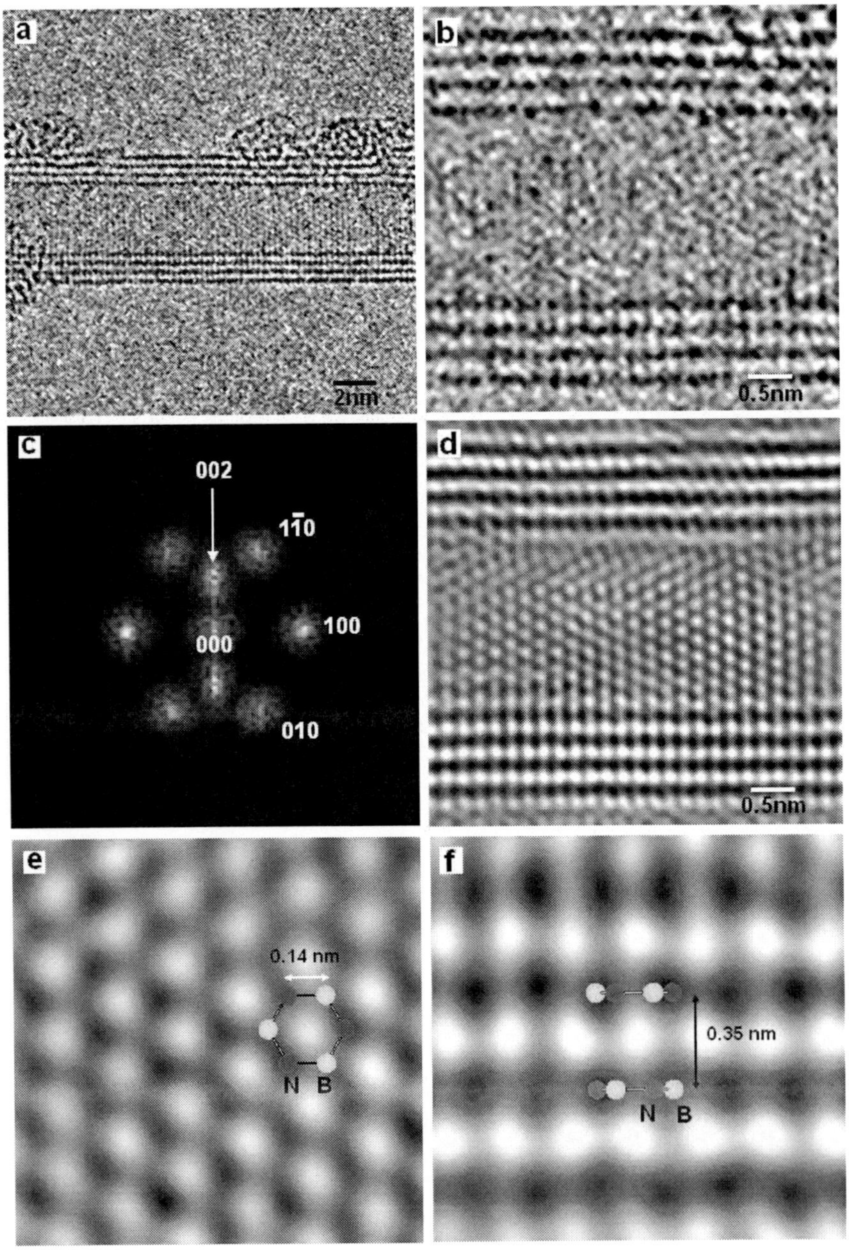

Figure 8. (a) HREM image of zigzag-type BN nanotube. (b) Enlarged HREM image of (a). (c) Filtered Fourier transform of (b). (d) Inverse Fourier transform of (c). (e), (f) Enlarged images of (d).

Figure 8(a) is a HREM image of a quadruple-walled BN nanotube. In the present work, all HREM images were taken close to the Scherzer defocus (Δf_s = -41.2 nm) in order to investigate atomic structures in detail. HREM observations and electron diffraction analysis on BN nanotubes have been reported [88,89], and direct observations of nanotube chirality were tried in the present work. An enlarged HREM image is shown in Fig. 8(b), which indicates lattice fringes in the BN nanotubes [90,91]. A filtered Fourier transform of Fig. 8(b) showed that this nanotube had a zigzag-type structure as shown in Fig. 8(c). A HREM image with clear contrast processed after Fourier noise filtering as shown in Fig. 8(d). The intervals of the bright and dark dots are 0.14 nm, which corresponds to the structure of hexagonal BN rings, as shown in Fig. 8(e). Layer intervals of each tube are 0.35 nm, as shown in Fig. 8(f). Diameters of each nanotube are 2.8 nm, 3.5 nm, 4.2 nm and 4.9 nm from the inside to outside.

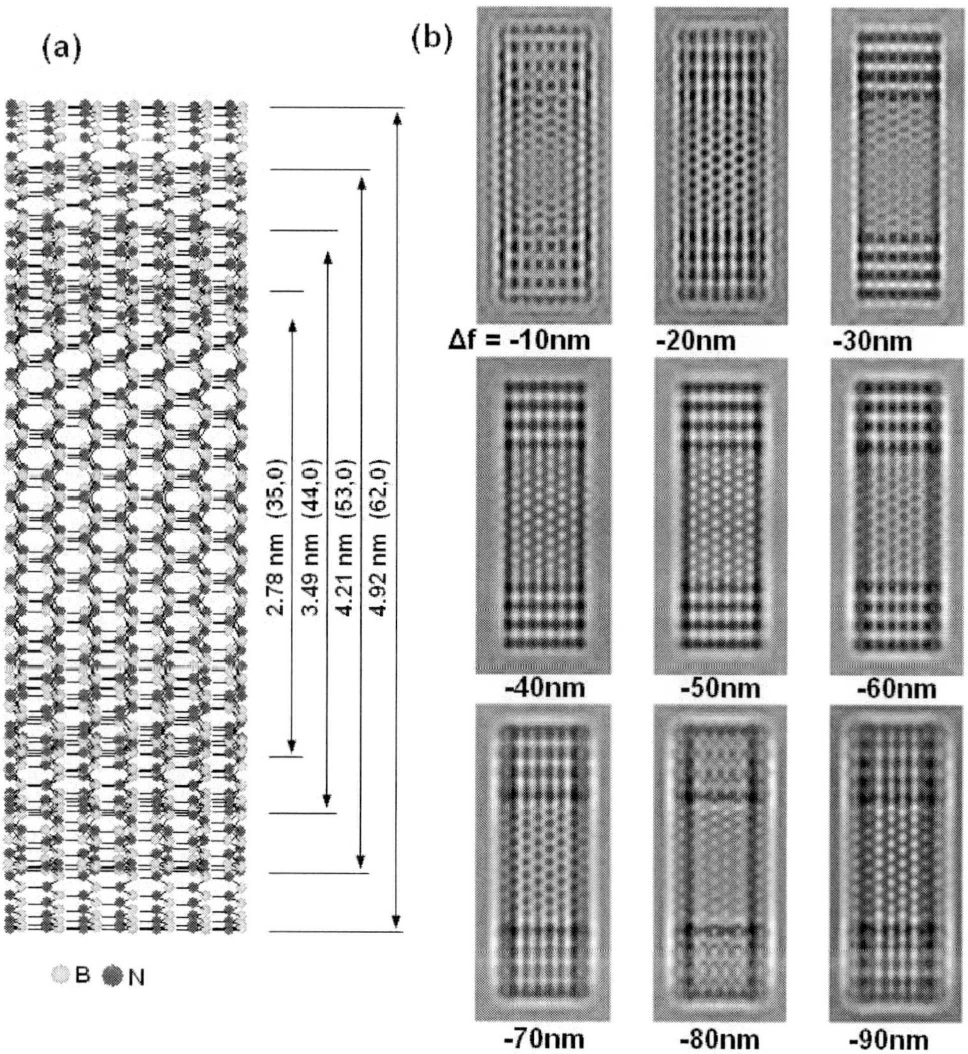

Figure 9. (a) Proposed structure model of quadruple-walled BN nanotube. Chiralities of zigzag BN nanotubes are (35,0), (44,0), (53,0) and (62,0) from inside to outside. (b) Calculated images of the proposed model (a) as a function of defocus values.

Atomic structure models were proposed from observed diameters of BN nanotubes, which were based on layer intervals of 0.34-0.35 nm. The chirality of (n, m) is derived from the equation

$$d_t = \frac{\sqrt{3}\, a_{B-N} \sqrt{n^2 + nm + m^2}}{\pi} \tag{1}$$

The d_t means a diameter of BN nanotube with nm scale, and the $a_{B\text{-}N}$ corresponds to the nearest distance of boron and nitrogen atoms. For the BN nanotubes, the value of $a_{B\text{-}N}$ is 0.144 nm. When a BN nanotube has a zigzag structure, the value of m is zero.

Figure 9(a) shows a proposed structure model of the quadruple-walled BN nanotube. Chiralities of each zigzag BN nanotube are (35, 0), (44, 0), (53, 0), and (62, 0) from the inside to outside. These chiralities were derived from the equation (1). The arrangement of boron and nitrogen atoms was reversed at each layer, as boron atoms exist just above the nitrogen atoms while maintaining the layer intervals of 0.35 nm. Calculated images of the proposed model as a function of defocus values are shown in Fig. 9(b). Contrast of hexagonal rings was clearly imaged at the defocus values in the range of -40 to -50 nm, and these simulated images agree well with the observed HREM image of Fig. 8(d).

When the zigzag-type BN nanotubes are characterized using HREM, the BN atom pairs at sides of the nanotubes are sometimes imaged as dots, as observed in Fig. 8(f). Taking such dot contrast would be difficult for armchair-type BN nanotubes because of high density distribution of atoms along nanotube axis, and HREM image contrast would show straight lines at the sides of nanotubes. Multi-helix BN nanotubes would also show unclear dot-contrast at the side of the nanotube, which indicates that the contrast of side edges of BN nanotubes would also give us useful information on the chirality. If clear dot-contrast is observed at the sides of BN nanotubes, possibility of zigzag-type BN nanotube is high. Image contrast also could be changed by rotation of nanotubes, and the further study on the rotation of BN nanotubes has been presented [92].

Structural stabilities of BN nanotubes were investigated and summarized as reported [92]. Since an effect of nanotube edges should be considered, nanotubes with short and long lengths were investigated. Total energies of zigzag-type structures showed lower values than those of armchair-type structures, which indicates the zigzag-type is more stable compared to the armchair-type structures. This also agreed with distorted nanotube structures in experimental data [91]. A similar experimental result was also reported [93], and the present calculations also confirmed the stability of the zigzag-type BN nanotubes. Encapsulation of a BN cluster in a BN nanotube showed reduction of total energy, which indicates that the encapsulation of the BN cluster in BN nanotube would stabilize the BN cluster [94].

5. BN Nanohorns

HREM images of BN nanohorns with various tip structures, which were synthesized by an arc-melting method, are shown in Fig. 10(a-e). The BN nanohorns have a hollow structure, and tip angles of BN nanohorns were 22°, 35°, 60°, 90° and 102°, respectively [96-98]. {002} layers of BN are clearly observed in these images. Since the distortion of apexes of BN nanohorns is very large, the structure may not be completely perfect, which results in the vague contrast at the apexes. Figure 10(f) is a HREM image of a BN nanohorn with an amorphous phase. As observed in the HREM images, the number of BN layers are in the range of 2~8.

EDX spectra of BN nanohorns with a hollow and an amorphous phase are shown in Fig. 11(a) and 11(b), respectively. Both graphs are normalized by nitrogen peaks. Figure 11(a) shows boron and nitrogen peaks, and Fig. 11(b) shows a peak of yttrium in addition to peaks of boron, nitrogen and oxygen. The composition ratio of B : N in Fig. 11(a) and 11(b) are calculated to be 45.8 : 54.2 and 49.2 : 50.8, respectively. Since the composition ratios of B : N in Fig. 2 are almost ~1.0, these B and N peaks are believed to be detected from BN layers. If the composition ratio of BN in Fig. 11(b) is assumed to be 1.0, composition of the amorphous phases is calculated as Y : O = 42.6 : 57.4, which indicates that the amorphous phase might be Y_2O_3 phase. Amorphous phases observed at the root of BN nanohorns would be starting points of BN growth.

Figures 12(a-e) are atomic structure models of triple walled $B_{1638}N_{1563}$, $B_{1635}N_{1635}$, $B_{1671}N_{1623}$, $B_{1605}N_{1605}$ and $B_{1608}N_{1575}$ nanohorns with tip angles of 20°, 40°, 60°, 84° and 112°, respectively, which agree with the HREM image in Fig. 1. Some different tip angles were also observed, which would be due to the lattice defects. Two and one tetragonal BN rings are introduced for the models in Fig. 12(b) and 12(d), respectively. Since a trigonal BN ring is introduced for structure models in Fig. 12(a) and 12(c), they have some B-B single bonds. Pentagonal BN rings are introduced for the model of Fig. 12(e), which is different from the model of Fig. 12(a) with trigonal BN rings. Total energies of the BN nanohorns calculated by molecular mechanics calculation are listed in Table 1. Distance between BN layers of nanohorn in HREM images were measured to be ~0.35 nm, and the basic structure models were constructed based on them. After molecular mechanics calculation, the layer distances were optimized as ~0.37 nm.

Figures 13(a) and 13(b) are HREM images of triple- and double-walled BN nanohorns synthesized by the present method, respectively [99]. The minimum numbers of BN layers in nanohorns synthesized by the present work is two, which would be due to stacking stability with two elements of boron and nitrogen. Tip angles of both BN nanohorns are ~60°. In Fig. 13(b), a dark contrast is observed inside the double-walled BN nanohorn, and this would be due to existence of a BN cluster with metal atom. Since $Y@B_{36}N_{36}$ clusters had already been detected by mass spectroscopy [10,19], it is believed that the dark contrast would correspond to a $Y@B_{36}N_{36}$ cluster. Atomic structure models for triple-walled BN nanohorn ($B_{1671}N_{1623}$) and double-walled BN nanohorn ($B_{1114}N_{1082}$) encaging $Y@B_{36}N_{36}$ are proposed as shown in Fig. 13(c) and 13(d), respectively. Since a trigonal BN ring is introduced for a model at the top in both models, the structure models have some B-B bonds.

Energy level diagrams of BN nanohorns show narrow energy gaps between the highest occupied molecular orbital (HOMO) and the lowest unoccupied molecular (LUMO), as shown in Table 1. The energy level diagram for BN nanohorn ($B_{24}N_{21}$) with a tip angle of 40° is shown in Fig. 14(a). Densities of states of $B_{39}N_{31}$, $B_{31}N_{31}$, $B_{27}N_{21}$, $B_{32}N_{32}$ and $B_{24}N_{21}$ nanohorns were calculated, as shown in Fig. 14(b-f), respectively. The HOMO-LUMO gaps of $B_{39}N_{31}$, $B_{31}N_{31}$, $B_{27}N_{21}$, $B_{32}N_{32}$ and $B_{24}N_{21}$were calculated to be 0.29, 0.80, 0.20, 0.85 and 0.22 eV, respectively [97,99].

Energy gaps calculated based on the atomic structure models of Fig. 3 show lower values (0.2 ~ 0.85 eV) compared to those of BN nanotubes or BN fullerenes (~ 6 eV). The low energy gaps would be due to trigonal and pentragonal BN rings and nanohorn structures. The band gaps of BN nanohorns were reduced, and various new applicable fields are expected by combining the BN nanohorns with BN nanotubes or BN fullerenes. Total energies of BN nanohorns were reduced by stacking of BN nanohorns, and it is believed that the structure of BN multi-walled nanohorns would be stabilized by multiplying hexagonal BN ring networks.

BN nanohorns with various tip angles were synthesized by the present arc-melting method [100]. Since the total energy of BN nanohorns with tip angles of 84° is fairly low in the present work as listed in Table 1, which agree well with experimental observation. The lowest total energy of BN nanohorns with tip angles of 112° would be strongly related with the nanoparticle growth of this structure up to 300 nm [36].

Figure 16(a) is a TEM image of chemically deposited BN synthesized from the BCl_3-NH_3-H_2 gas systems system at a deposition temperature of 1800°C and a total gas pressure of 5 Torr. Many nanoparticles with five-fold symmetry were observed in the sample [36]. Each region of the particle grows outwards in spite of even the shortness of twin boundaries in the particle. A HREM image of the center of a fivefold multiplytwinned h-BN nanoparticle is shown in Fig. 16(b). Twin boundaries are indicated by arrows. In the image, two dimensional lattice fringes are visible with separations of 0.18 and 0.22 nm, which correspond to the lattice spacings of {1 0 2} and {1 0 0}, respectively. The image clearly shows that the five parts have twin relations at their boundaries. The center of the fivefold axis indicates some distortion.

Figure 16(c) is an electron diffraction pattern of the fivefold multiply-twinned BN in Fig. 16(a). All diffraction spots in the pattern can be interpreted by the superposition of five twin-related electron diffraction patterns. A selected area electron diffraction pattern of one of the five parts can be indexed by the h-BN structure with lattice constants a = 0.25044 nm and c = 0.66562 nm [101]. The twin boundaries are {1 1 2}, and the fivefold axis (equal to incident beam direction) is [2 0 -1].

A projection of the atomic arrangement of the fivefold multiply-twinned h-BN nanoparticle is illustrated schematically in Fig. 16(d), where open circles correspond to rows of an alternate sequence of boron and nitrogen atoms lying parallel to the fivefold symmetry axis [2 0 -1]. The twin planes {1 1 2} are indicated by broken lines, and the h-BN cell is shown by solid lines.The fivefold axis, normal to the figure at the star mark, is tilted 37° from the c-axis of the hexagonal cells. The atomic cluster at the center of the particle is indicated by a pentagon. The central cluster consists of $B_{12}N_{13}$, and the structure was optimized by semi-empirical molecular orbital calculation (PM3), to give a calculative total energy of 6.7 kcal/mol, which is fairly small compared to that of diamond and Au clusters [34]. Taking account of the lattice parameters of h-BN, the misfit of the multiplytwinned h-BN particle is 1.6°, which is much smaller than the 7.3° of multiply-twinned fcc nanoparticles. This small amount of misfit allows the growth of relatively large sized particles with little inner distortion. In the $B_{12}N_{13}$ cluster, there are B-B and N-N bondings both inside and between the h-BN rings. The results of the calculations describe the distortions and defects in the experimental data of Fig. 15(b) rather well. BN clusters such as $B_{12}N_{12}$, $B_{16}N_{16}$ and $B_{36}N_{36}$ have been reported to have octahedral structure [37-39,102]. However, in the present work, the BN nanoparticles have fivefold symmetry. The total energy of $B_{12}N_{12}$ clusters was calculated to be 13.6 kcal/mol, which is similar in value to that calculated for the $B_{12}N_{13}$ cluster with fivefold symmetry. These results may indicate that the growth kinetics cause a change in the structure of BN clusters into decagonal shapes from octahedral shapes during cluster growth.

Table. 3. Total energies and energy gaps of BN nanohorns by molecular mechanics/orbital calculations.

	$B_{547}N_{520}$		$B_{545}N_{545}$		$B_{557}N_{541}$		$B_{535}N_{535}$		$B_{536}N_{525}$	
Nanohorn angle	20°		40°		60°		84°		112°	
Number of layers	1	3	1	3	1	3	1	3	1	3
Total energy [kcal/mol·atom]	1.03	0.64	0.98	0.47	1.08	0.53	0.99	0.38	0.03	-0.44
Energy gap [eV]	0.29 ($B_{39}N_{31}$)		0.80 ($B_{31}N_{31}$)		0.20 ($B_{27}N_{21}$)		0.85 ($B_{32}N_{32}$)		0.22 ($B_{24}N_{21}$)	

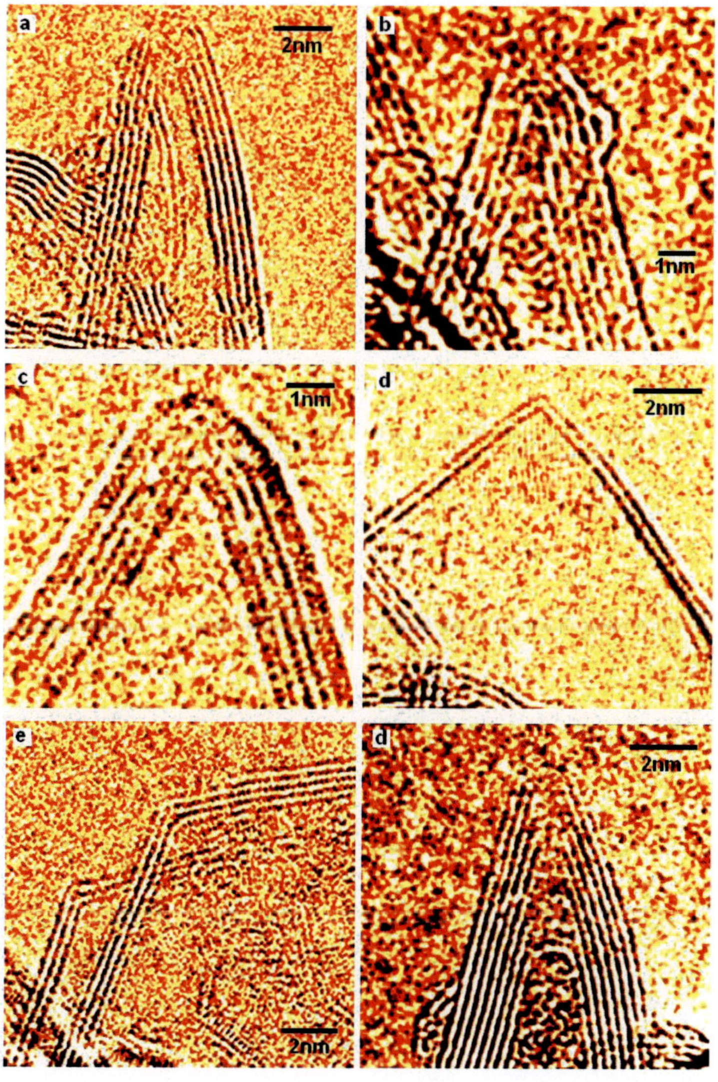

Figure 10. HREM images of (a-e) BN nanohorns with hollow structures, and (f) BN nanohorn with amorphous phase.

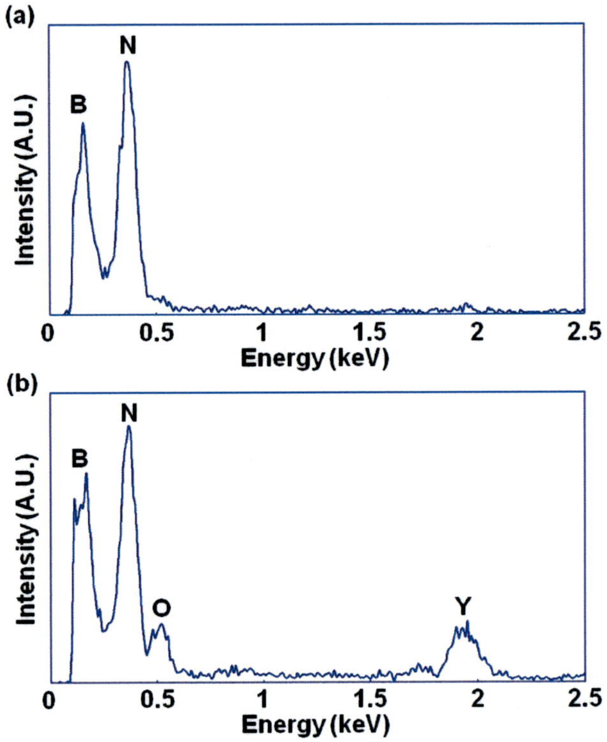

Figure 11. EDX spectrum of BN nanohorns with (a) a hollow structure and (b) amorphous phase.

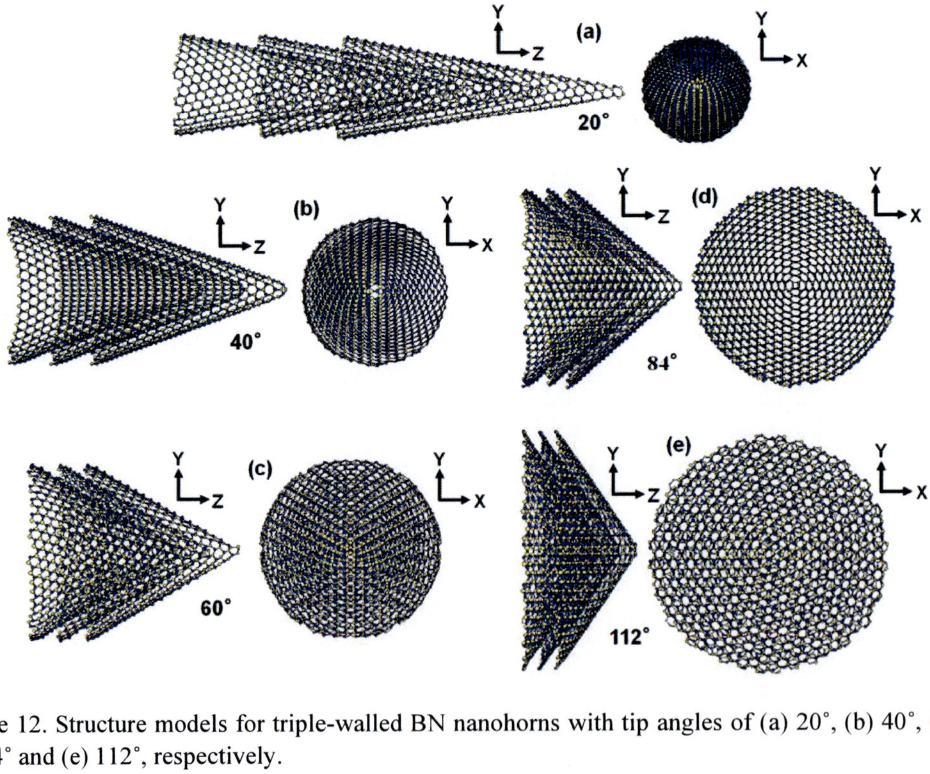

Figure 12. Structure models for triple-walled BN nanohorns with tip angles of (a) 20°, (b) 40°, (c) 60°, (d) 84° and (e) 112°, respectively.

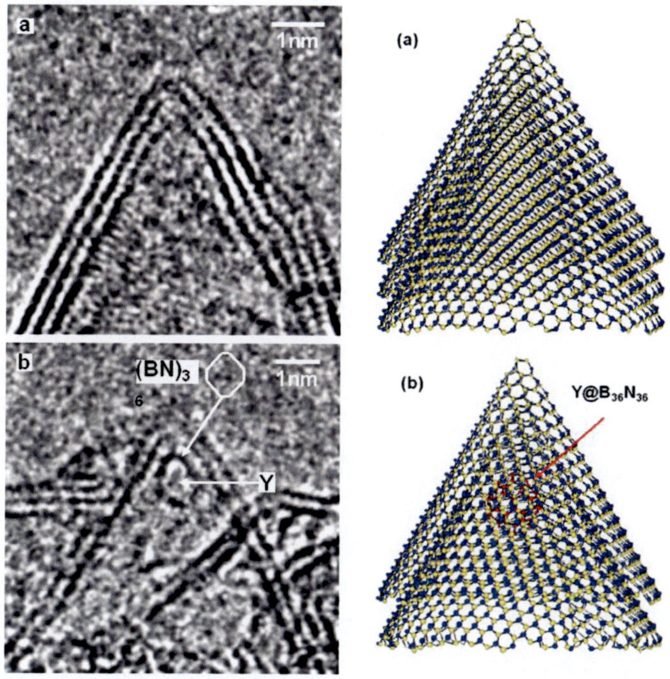

Figure 13. HREM images of (a) triple-walled BN nanohorn with tip angle of 60° and (b) double-walled BN nanohorn including Y@$B_{36}N_{36}$. Atomic structure models for (a) triple-walled BN nanohorn ($B_{1671}N_{1623}$) and (b) double-walled BN nanohorn ($B_{1114}N_{1082}$) encaging Y@$B_{36}N_{36}$.

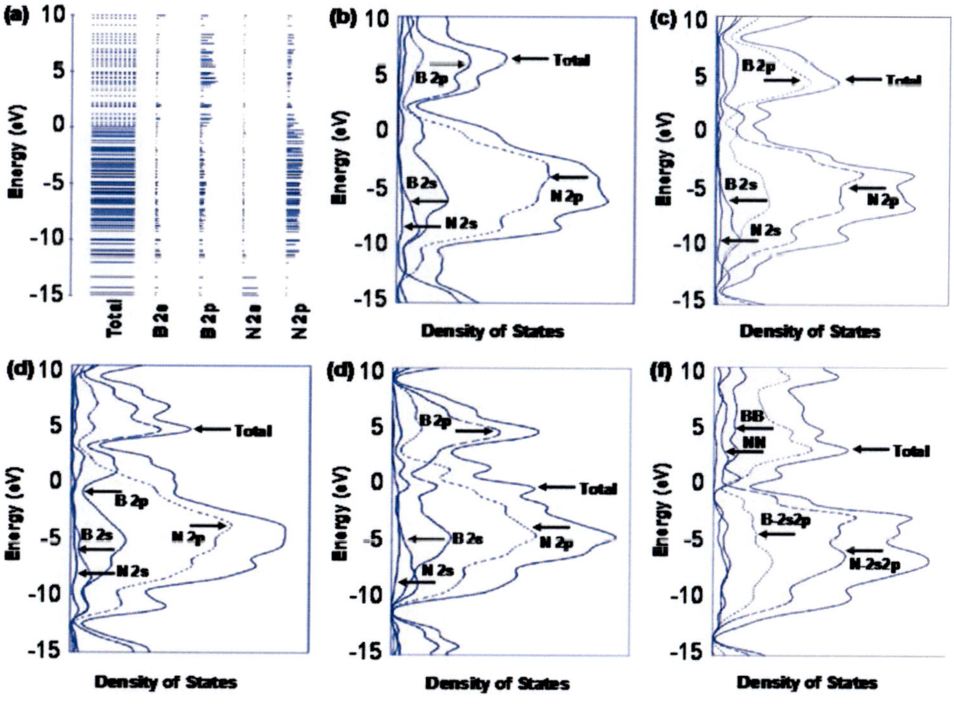

Figure 14. (a) Energy level diagram for BN nanohorn with tip angle of 40°. Density of states for BN nanohorns with tip angles of (b) 20°, (c) 40°, (d) 60°, (e) 84° and (f) 112°, respectively.

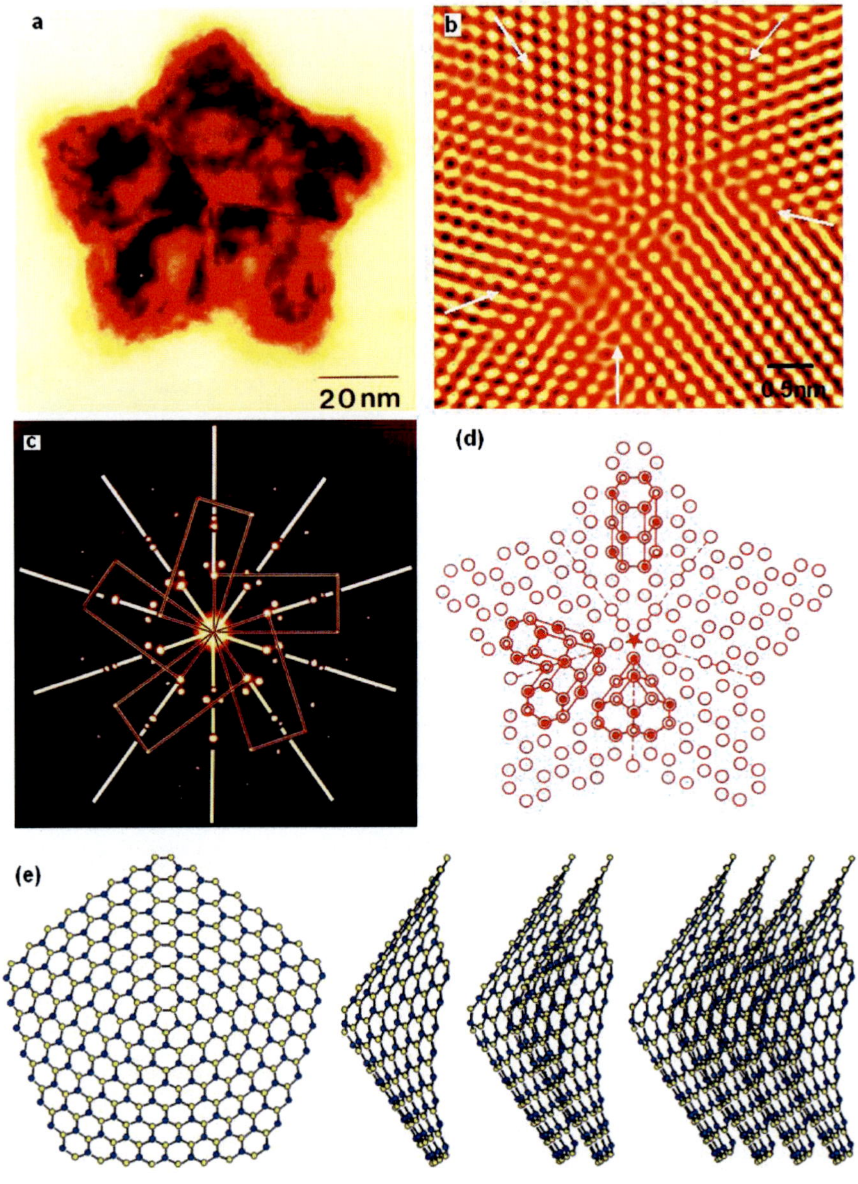

Figure 15. (a) TEM image of chemically vapor deposited BN nanoparticle. (b) Filtered HREM image of the center of the BN nanoparticle. (c) Electron diffraction pattern of the BN nanoparticle. (d) Atomic structure model. (e) Atomic structure models of $B_{164}N_{156}$, $B_{328}N_{312}$ and $B_{656}N_{624}$.

In the present work, yttrium worked as a good catalytic element to produce BN nanotubes. Catalytic metals for formation of BN nanotubes, nanocapsules and nanocages, which were confirmed by experiments on arc-method, are summarized in Fig. 15(a) as periodic table. It has been reported that Zr, Hf, Ta, W, Nb and La can be good catalytic metals for synthesis of BN nanotubes [6-23]. On the other hand, other metals could not form BN nanotubes, although BN nanocapsules or nanocages were formed. The relationship between catalytic metals and structures of BN fullerene materials would be summarized by formation enthalpy with nitrogen and boron. About some metals, formation enthalpies with boron ($H^{for}B$) and nitrogen

($H^{for}N$) were theoretically calculated [103]. Difference of formation enthalpy ($H^{for}N$ - $H^{for}B$) is also shown in Fig. 15(b). From our previous results [4,79], BN nanotubes and BN nanocapsules (or nanocages) were formed when $H^{for}N$ - $H^{for}B$ was negative and positive, respectively. From Fig. 10(b), BN nanotubes are formed when rare earth metals are used as catalytic metals. Y, Zr, Nb, La, Hf, Ta, Sc, Ti and V, would work as good catalytic elements, which agreed well with the experimental results.

In the case of YB_6/Ni powder, the yttrium and nickel worked as good catalytic elements to produce bundled BN nanotubes. From the results of only YB_6 powder, yttrium atoms would work as core element to produce BN nanotubes, and Ni atoms would have a role for combination of each BN nanotube. Therefore, existence of nickel atoms would have an effect on formation of bundled BN nanotubes, and the nickel atoms might exist among BN nanotubes. Further studies will be needed to understand the role of nickel atoms in bundled BN nanotubes.

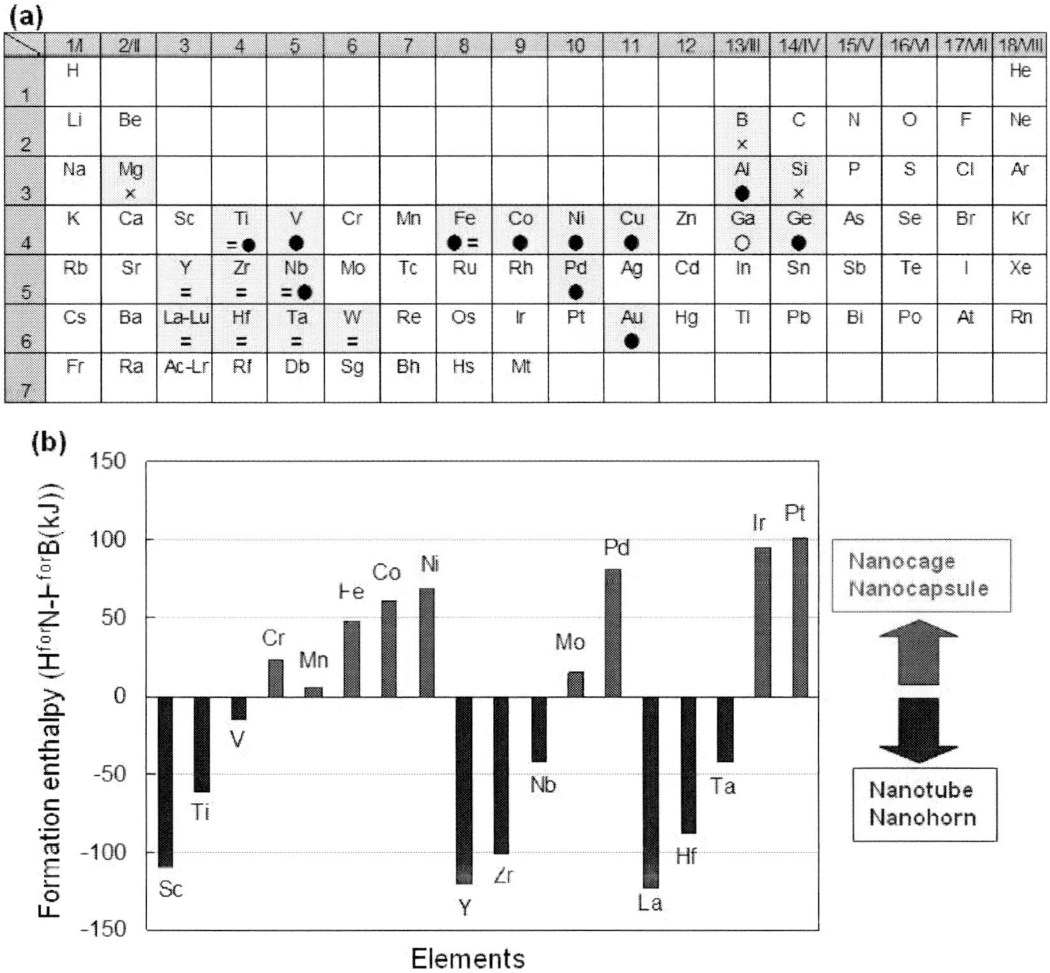

Figure 16. (a) Catalysis metals for BN fullerene nanomaterials confirmed by experiments on arc-method (=, BN nanotube; ●, BN nanocapsule; ○, BN nanocage; ×, non BN fullerene nanomaterials). (b) Difference formation enthalpy ($H^{for}N$ - $H^{for}B$) of nitrides and borides.

6. BN Nanotubes with Cup-Stacked Structures

Although the arc-melting method is good technique for produce BN nanocage materials, the amount is not so much, and a method for mass production should be developed. There are several methods for it, and one of the method is described here.

Fe_4N (99 %, Kojundo Chemical Laboratory (KCL) Co. Ltd., Saitama, Japan) and boron (B) powders (99 %, KCL) were used as raw materials [104,105]. Their particle sizes were about 50 and 45 μm, respectively. After the Fe_4N and B (weight ratio WR = 1:1) were mixed by a triturator, the samples were set on an alumina boat and annealed in the furnace. The furnace was programmed to heat at 6 °C/min from ambient to 1000 °C and hold for 1 h and then cooled at 3 °C/min to ambient temperature. Nitrogen pressure was 0.10 MPa, and its gas flow was 100 sccm (standard cubic centimeter per minute).

As-produced soot synthesized from Fe_4N/B via the above method was purified by the following steps. The as-produced soot were poured in 4 M HCl solution and stirred for 4 h at a room temperature. The green color of the solution provides an indication of the dissolution of Fe ions. After HCl treatment, the samples were poured in 1 M HNO_3 solution and stirred for 30 h at 50 °C. The yellow color of the solution provides an indication of the dissolution of boron. After both acid treatment, the solution was filtered and rinsed with deionized water until the pH of the filtrate became neutral and dried. Then, the samples were poured in pyridine to eliminate bulk BN, and high purity BN nanotubes with a cup-stacked structure were obtained by collecting supernatant.

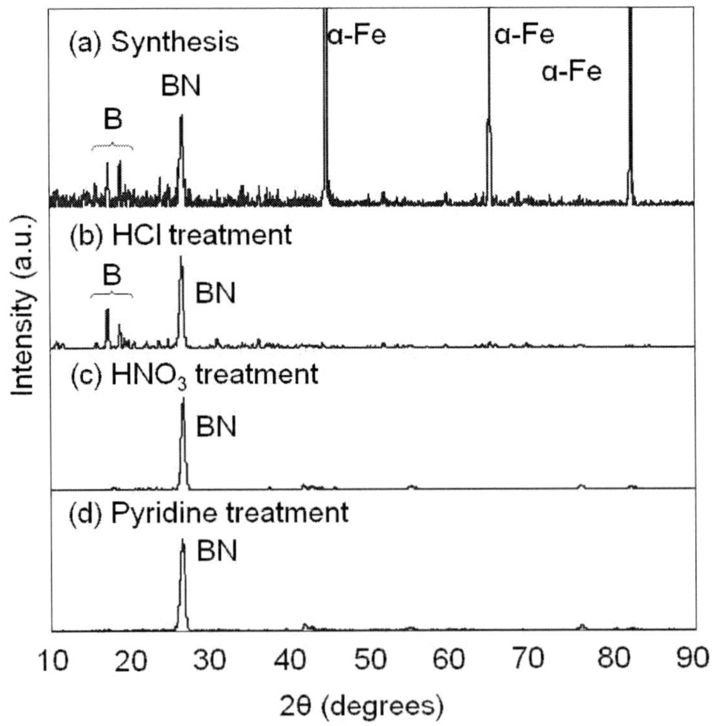

Figure 17. X-ray diffraction patterns of samples after (a) synthesis, (b) HCl treatment, (c) HNO3 treatment, and (d) pyridine treatment.

X-ray diffraction patterns in a purification process are shown in Fig. 17. Diffraction peaks of hexagonal BN, boron and α-Fe are observed for the sample at annealed at 1000 °C for 1 h as shown in Fig. 17(a). It is found that Fe was removed after HCl treatment as shown in Fig. 17(b), and boron was removed after HNO3 treatment as shown in Fig. 17(c). After pyridine treatment, a strong peak of BN was obtained as shown Fig. 17(d).

Figure 18(a) shows TEM image of BN nanotubes with a cup-stacked structure after purification process. Diameters and lengths of the BN nanotubes are in the range of 40-100 nm and 5-10 μm, respectively. Fe nanoparticles and bulk BN was eliminated during the process. An enlarged image of one of the BN nanotubes is shown in Fig 18(b), which shows a cup-stacked structure as indicated by lines of BN {002}. Figure 18(c) is an electron diffraction pattern of Fig. 18(b). 002 reflections of BN are splitting in Fig. 18(c), which indicates that the BN nanotube has a cup-stacked structure, and the cone angle between the BN layers at both nanotube walls is ~20°. Most of BN nanotubes (~90 %) have this cup-stacked structure with cone angle of ~20°, and normal structures with a cone angle of 0° were sometimes observed (~10 %).

An optical absorption spectrum of BN nanotubes is shown in Fig. 18(d). In Fig. 3, a strong peak is observed at 4.8 eV, which corresponds to the energy gap of BN nanotubes, and this value is almost same as the STM measurements. A broad, weak peak is also observed around 3.4 eV, which is considered to be impurity level (oxygen or hydrogen) of the BN layers as reported previously [4,16].

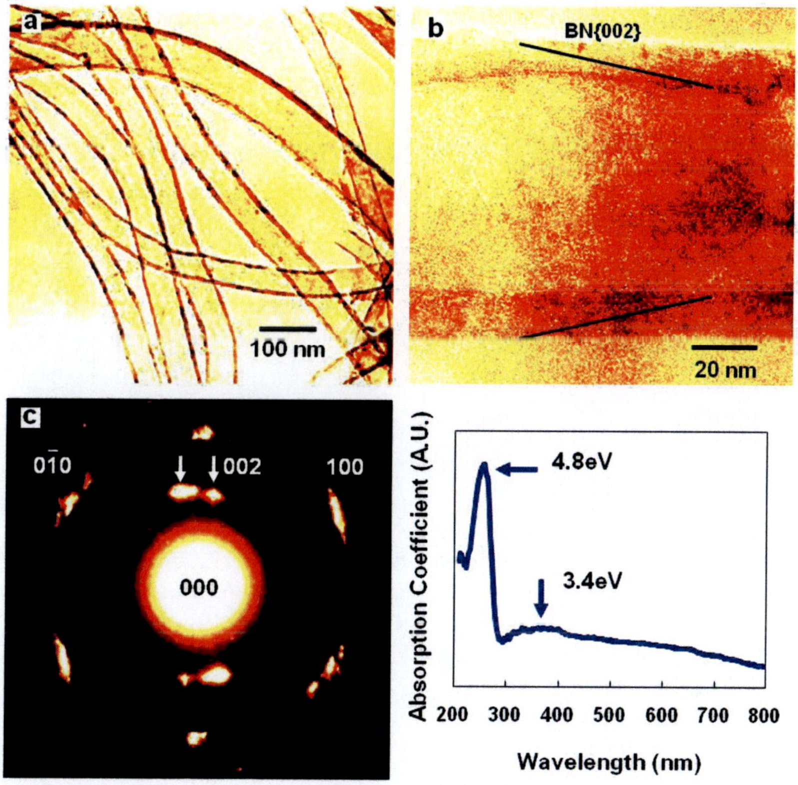

Figure 18. (a) TEM image of BN nanotubes. (b) Enlarged image of BN nanotube. (c) Electron diffraction pattern of (b). (d) Optical absorption spectrum of BN nanotubes.

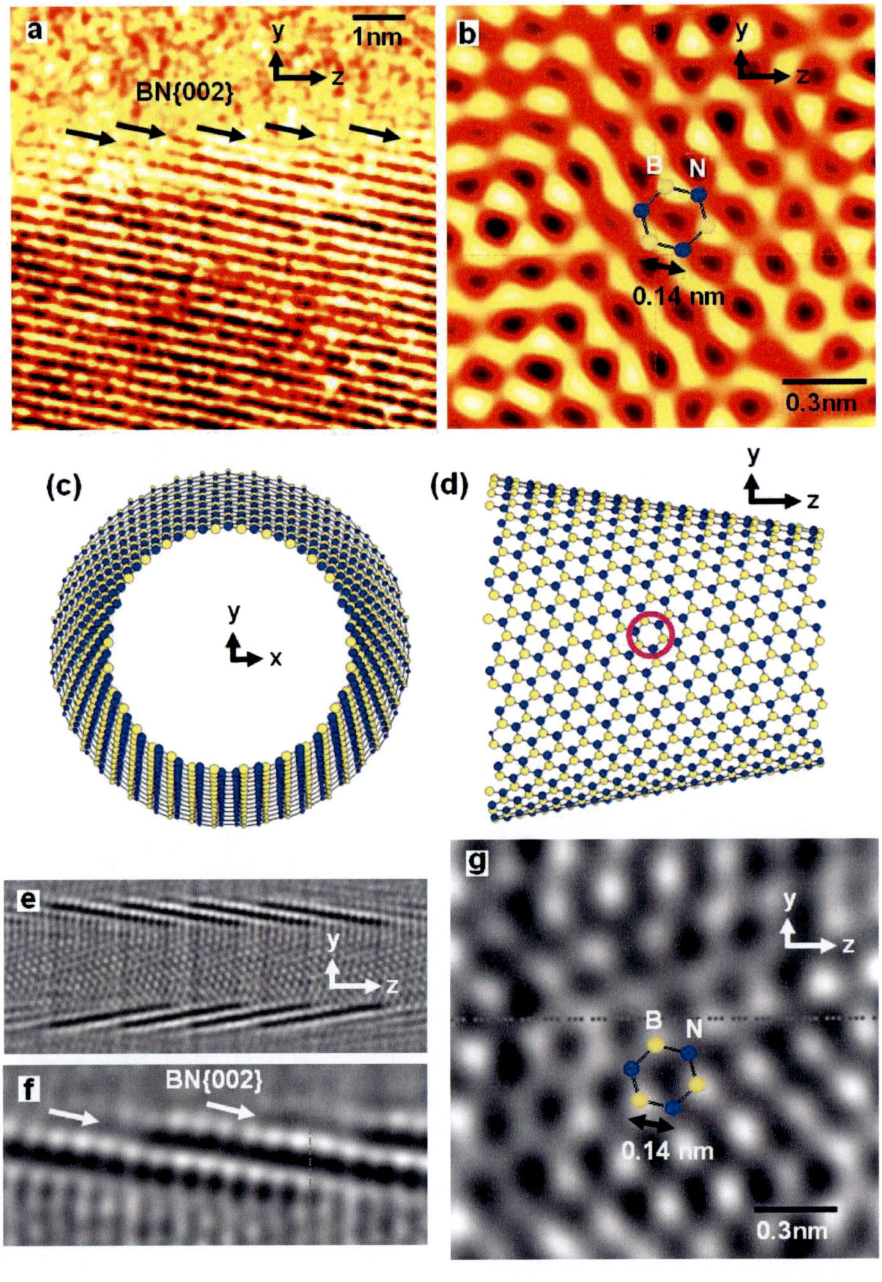

Figure 19. (a) HREM image of edge of BN nanotube wall in Fig. 17(b). (b) Processed HREM image after Fourier filtering of nanotube center of Fig. 17(b). Proposed model of BN cup-structure projected along (c) z-axis (nanotube axis) and (d) x-axis. (e) Calculated HREM image of 4-layered, cup-stacked BN nanotube at defocus values of -40 nm. Enlarged image of (f) edge and (g) center of BN nanotube in (e).

A HREM image of edge of the nanotube side wall in Fig. 18(b) is shown in Fig. 19(a), and a cup-stacked structure was observed. Edge structures are observed as indicated by arrows, and the BN {002} planes are inclined compared to nanotube axis (z-axis). Figure 19(b) is a processed HREM image after Fourier filtering of nanotube center of Fig. 19(b), and hexagonal arrangements of white dots are observed, which would correspond to BN six-membered rings. From these observations, a structure model for BN cup-structure was proposed, which consists only of hexagonal BN rings, as shown in Fig. 19(c) and 19(d).

In order to investigate the stability of the cup-stacked structure, four types of nanotubes are considered, as shown in Fig. 20. Figures 20(a) and 20(b) are atomic structure models of double-walled BN nanotubes with zigzag-type and armchair-type structures, respectively. Atomic structure models of 4-layered, cup-stacked BN nanotubes with different cone angles are shown in Fig. 20(c) and 20(d). These structures were summarized as Table 4 and Table 5. Total energies of these 4-type structures indicates that BN multi-layered nanotubes with and without a cup-stacked structure would be stabilized by stacking hexagonal BN networks. Distance between BN layers of nanotubes with a cup-stacked structure in a HREM image was

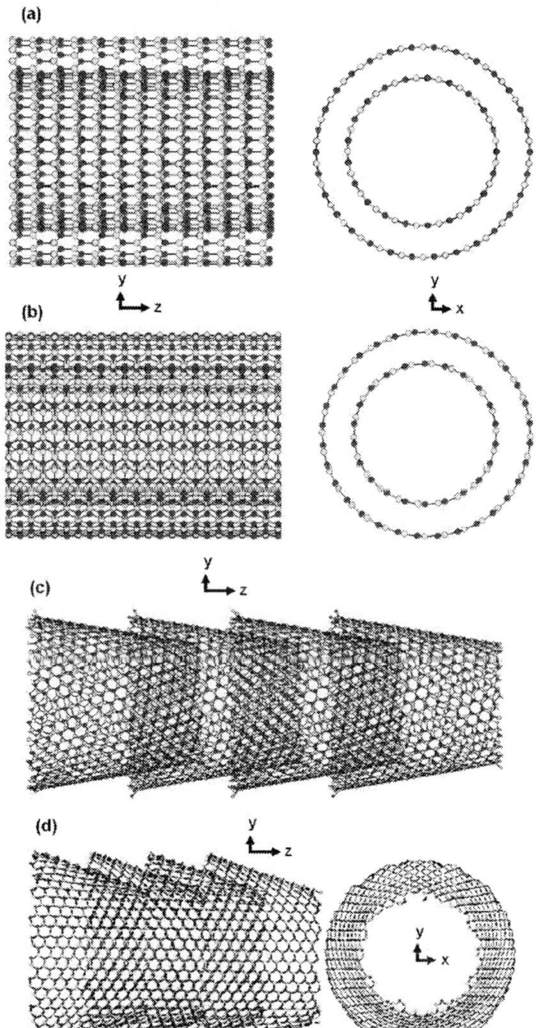

Figure 20. Atomic structure models of double-walled BN nanotubes with (a) zigzag-type and (b) armchair-type structures. Atomic structure models of 4-layered, cup-stacked BN nanotubes with cone angles of (c) 20° and (d) 36°.

found to be ~0.35 nm, and the basic structure model was constructed based on this observation. Geometry optimizations at molecular mechanics level result in the interlayer distances of ~0.38 nm. Comparing empirical total energies all considered structures, a cup-stacked structure ($B_{2240}N_{2240}$) with cone angle of 20° was found to be the lowest in energy, which indicates the high stability of this structure.

Based on the structure model of a cup-stacked $B_{2240}N_{2240}$ nanotube of Fig. 20(c), image calculation was carried out, as shown in Fig. 19(e). Enlarged calculated HREM images of edge and center of the BN nanotube in Fig. 19(e) is shown in Fig. 19(f) and 19(g), respectively. These calculated images agree with the experimental data of Fig. 19(a) and 19(b), respectively.

Cup-stacked carbon nanotubes with Pt nanoparticles in inner surface of the hollow core had been reported [106]. The BN nanotubes with cup-stacked structures in the present work would also be one of the candidates for atomic and gas storage, as well as carbon nanotubes. Cone angles of BN cup-stacks were measured to be ~20°, which agreed well with that of the model in Fig. 20(c). Although atomic structure models for BN nanotubes with a cup-stacked structure have been proposed from HREM observation and molecular mechanics calculations [100], the cone angles were ~36°, and they had bamboo-type structures. In the present work, there is no bamboo-type structure, which was removed during purification process, and the more stable cup-stacked structure with cone angle of 20° was formed.

Cone angles of carbon nanotubes with a cup-stacked structure were reported to be in the range of 45-80° [106]. The cause of the different cone angles of the present cup-stacked BN nanotubes would be due to the different stacking of BN layers along c-axis (B-N-B-N...) from carbon layers [107].

Table 4. Calculated values for various BN nanotubes.

	$B_{273}N_{273}$	$B_{390}N_{390}$	$B_{273}N_{273}$ @$B_{390}N_{390}$	$B_{264}N_{264}$	$B_{384}N_{384}$	$B_{264}N_{264}$ @$B_{384}N_{384}$
Structure type	**Zigzag-type**			**Armchair-type**		
Outer diameter (nm)		2.3	2.3		2.2	2.2
Inner diameter (nm)	1.6		1.6	1.5		1.5
Number of layers	1	1	2	1	1	2
Total energy [kcal/mol]	459.2	701.5	556.0	466.6	693.2	779.3
Total energy [kcal/mol·atom]	0.841	0.899	0.419	0.883	0.902	0.601

Table 5. Calculated values for various BN nanotubes with a cup-stacked structure.

	$B_{560}N_{560}$	$B_{1120}N_{1120}$	$B_{2240}N_{2240}$	$B_{494}N_{494}$	$B_{988}N_{988}$	$B_{1976}N_{1976}$
Corn angle [°]	20			36		
Outer diameter (nm)	3.4			4.2		
Inner diameter (nm)	2.4			2.4		
Number of layers	1	2	4	1	2	4
Total energy [kcal/mol]	31.456	-287.924	-936.415	895.1	1269	2062
Total energy [kcal/mol·atom]	0.028	-0.129	-0.209	0.906	0.642	0.522

7. BN Nanotubes Encaging Fe Nanowires

Figure 21(a) and 21(b) are TEM and high-angle annular dark-field scanning transmission electron microscopy (HAADF-STEM) images of Fe-filled BN nanotubes (WR = 9:1), which were remaining sediment after centrifugation [108]. The contrast in the TEM image is weak and direct observation of Fe-filled BN nanotubes is difficult. The same area imaged by HAADF-STEM shows excellent contrast and the morphology of Fe-filled BN nanotubes can be observed in detail. A great number of Fe-filled BN nanotubes were observed by HAADF-STEM. High WR of Fe_4N would be necessary for synthesis of Fe nanowires. TEM image of one of Fe-filled BN nanotubes is shown in Fig. 21(c). Figure 21(d) is an EDX spectrum of the Fe-filled BN nanotube. In Fig. 21(d), two peaks of boron, nitrogen are observed. This shows the atomic ratio of B:N = 46.5:53.5, which indicates formation of BN. A strong peak of Fe (0.70 keV) is also observed, while a Cu peak arises from the HREM grid.

Figure 21(e) is an enlarged image of Fig. 21(c). Fig. 21(f) is an electron diffraction pattern of the Fe-filled BN nanotube. Strong peaks of BN nanotubes correspond to the planes of (002) of BN. Strong peaks are also indexed as metallic Fe with a bcc structure, and the incident beam is parallel to the [111] zone axis of α-Fe.

Figure 22(a) is an enlarged HREM image of Fig. 21(e), and Fig. 22(b) is filtered Fourier transform of Fig. 22(a) [109]. Figure 22(c) is inverse Fourier transform of 22(b), and Fig. 22(d) is an enlarged image of 22(c). Figure 22(d) shows a lattice image of the bcc Fe-filled BN nanotube. The nanotube axis is parallel to the [110] direction of Fe, which indicates the bcc Fe is epitaxially grown to the [110] zone axis. The tubular layers around the nanowire have an average interlayer spacing of 0.34 nm, which corresponds to the (002) spacing of BN.

Figure 22(e) is inverse Fourier transform of Fig. 22(b) using 000, Fe (01-1 and 0-11) reflections, and Fig. 22(f) is an enlarged image of Fig. 22(e). Several edge-on dislocations are observed as indicated by arrows, which would be due to lattice distortion produced during Fe-filled nanotube growth. This lattice distortion is also observed as expansion in the electron diffraction pattern of Fig. 21(f), as indicated by arrows.

A small amount of nanocrystalline Fe_2B compounds were observed at the tip of the BN nanotube. Chemical formulas that Fe_4N reacts with B, and generates Fe and BN in the experiments can be proposed as follows:

$$Fe_4N + 3B \rightarrow BN + 2Fe_2B \quad (2)$$

Fe_2B and dissolution of boron were obtained, and BN was produced in the reaction expressed as eq. (2) because Fe2B is thermodynamically more stable than Fe_4N. Although the Fe_2B is stable to 1389°C, the Gibbs-Thompson effect shown that the melting occurs at a significantly lower temperature compared to values in the standard phase diagram. Therefore, fluid-like Fe2B can be attained more easily. In the next process, the reaction expressed as eq. (3) would take place.

$$2Fe_2B + N2(g) \rightarrow 2BN + 4Fe \quad (3)$$

Boron in liquid-like Fe_2B started to segregate on the surface of the particle. The boron would react with N_2 gas, and BN was produced. α-Fe in liquid-like Fe_2B is epitaxially grown to the [110] direction, and Fe nanowires were produced in the reaction of Eq. (3). In adeition, high WR would be mandatory for the formation of Fe-filled BN nanotubes. As the results of these reactions, the [110] of Fe is parallel to the BN nanotube axis.

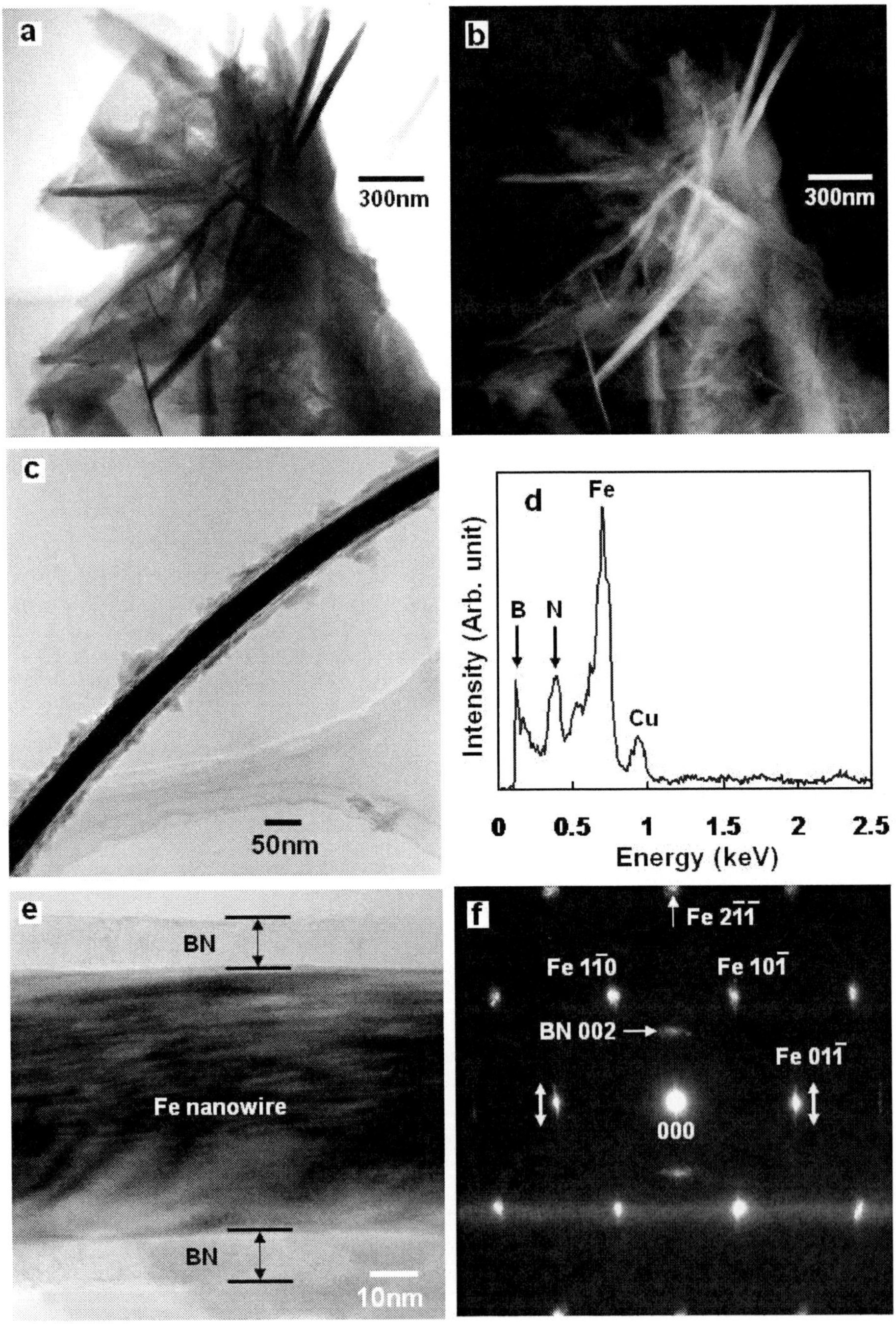

Figure 21. (a) TEM and (b) HAADF images of Fe-filled BN nanotubes. (c) TEM image of Fe-filled BN nanotube. (d) EDX spectrum of Fe-filled BN nanotube. (e) Enlarged image of (c). (f) Electron-diffraction pattern obtained from (e).

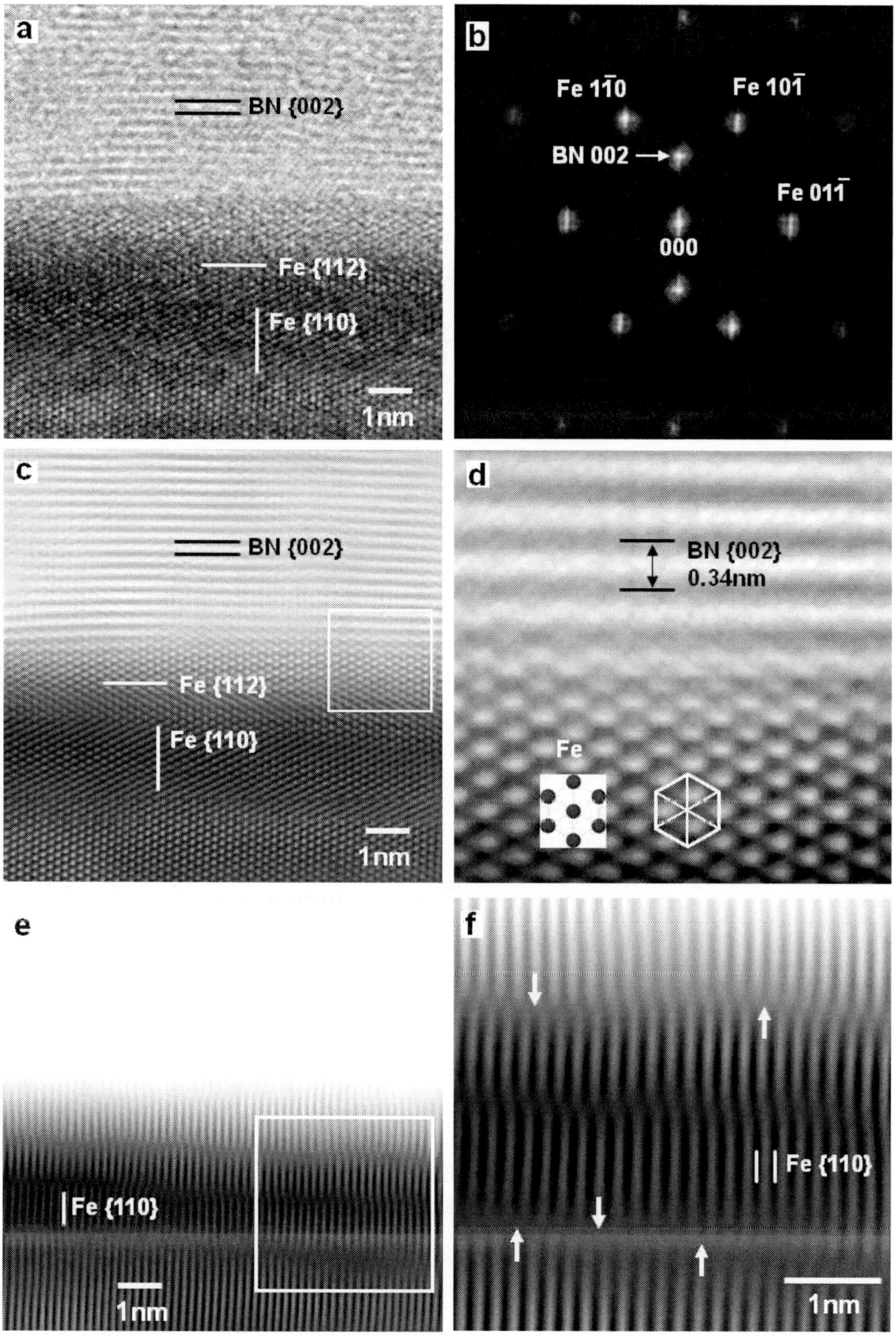

Figure 22. (a) HREM image of Fe-filled BN nanotube. (b) Filtered Fourier transform of (a). (c) Inverse Fourier transform of (b). (d) Enlarge image of square in (c). (e) Inverse Fourier transform of (b) using 000, Fe 01-1 and Fe 0-11 reflections. (f) Enlarged image of square in (e).

Gibb's energy on each formula is calculated as -89.4 and -23.2 kcal for the formulas (1) and (2) at 1000 °C, respectively. These negative values would stand for correctness of the proposed formulas. It is considered that a formation of Fe-B compounds might plays an important role for growth of the BN nanotubes, and that amorphous boron might change to BN and Fe_2B on the surface of the Fe_4N nanoparticles.

When magnetic materials are used as catalysis metals for BN nanotube formation, the magnetic nanoparticles would move around by magnetic field of a coil heater during the reaction process. Then, segments of BN {002} layers were produced in the tubes, which results in formation of bamboo structures. The interval of the BN layer segments might be related to the amount of iron nanoparticles, and further studies are expected on the control of the bamboo structure.

8. BN Nanocapsules

Nanosized particles have been extensively studied not only for their fundamental interests but also for their unique magnetic and electronic transport properties, and they have many potential for the future applications [110,111]. Especially, metal nanoparticles of Fe, Co, and some Fe-based alloys have an advantage of showing high magnetization. However, oxidation- and wear-resistances of the surface are weak points of these nanoparticles. In addition, most of magnetic materials are accompanied by exothermic heat due to an eddy current loss in high frequency. As a solution, it is well known that increasing of the resistance by using insulators is good.

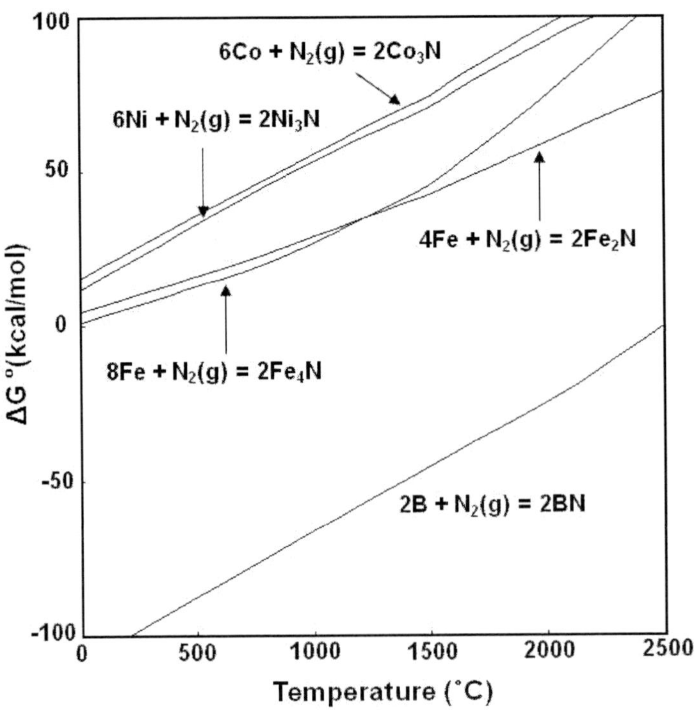

Figure 23. Ellingham diagram of Fe, Ni and Co nitrides for a N_2 molecule.

BN nanomaterials provide excellent protection against oxidation and wear, and its electrical conductance is insulator (~5eV). Therefore, magnetic nanoparticles coated with BN layers would have significant advantages for technological applications. Fe and Co nanocapsules coated with BN layers have been synthesized by an arc-discharge method [4,112]. However, the arc-discharge method is not suitable for mass production because of the limitation of the plasma area, and it is difficult to control the particle size and the number of BN layers. Fe nanocapsules coated with BN layers have been synthesized by annealing of boron and α-Fe_2O_3 powders [49,113]. Since they contained oxygen in the starting materials, a high annealing temperature was needed to remove the oxygen. Hu et al. also reported that nanocrystalline boron nitride had been synthesized with a whisker-like morphology by a reaction of KBH_4 and NH_4Cl at a temperature of 650 °C and a pressure of about 22 MPa [114]. This synthesis method does not contain oxides in the starting materials.

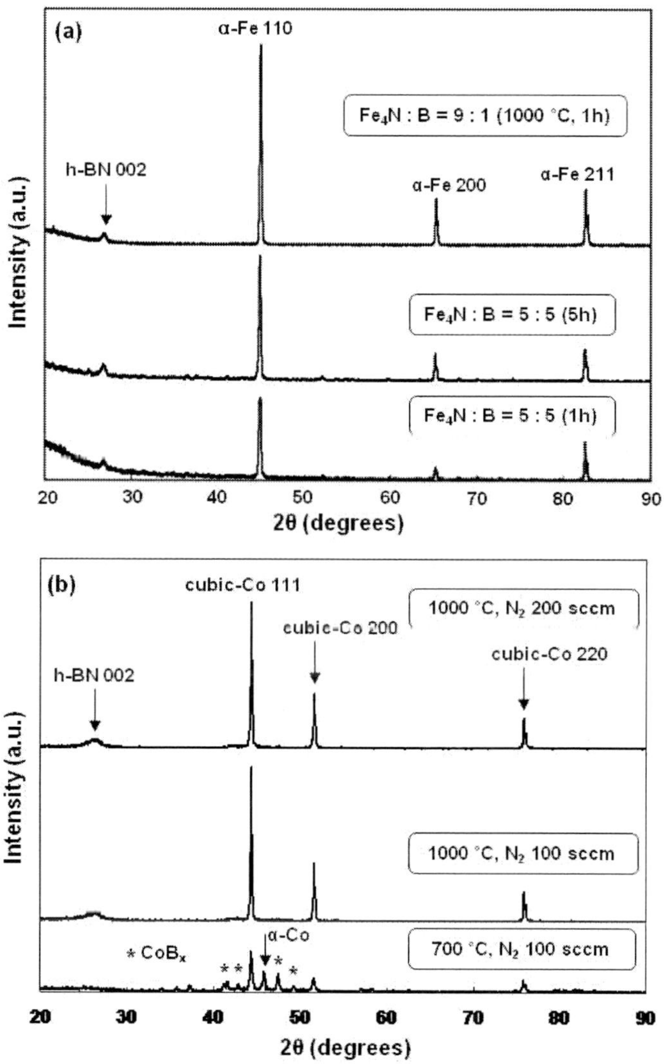

Figure 24. X-ray diffraction patterns of (a) Fe_4N/B samples annealed at 1000 °C and (b) $Co(NH_3)_6Cl_3/KBH_4$ samples annealed at 700-1000 °C for 2 h.

The purpose is to synthesize magnetic nanoparticles coated with BN layers, and to investigate the nanostructures and magnetic properties. An Ellingham diagram of nitride metals for N_2 gas per mol was calculated by HSC software as shown in Fig. 23. Fe_4N particles would be reduced to α-Fe completely by annealing with boron, because boron reacted with nitrogen more easily compared to Fe. Similarly, several nitrides would be reduced to pure metals by reaction with boron. In the present work, Fe and Co were selected for the magnetic nanoparticles, and mixture powders of Fe_4N/B and $Co(NH_3)_6Cl_3/KBH_4$ were used for the synthesis, respectively.

X-ray diffraction patterns of annealed samples of Fe_4N/B with various WR of Fe_4N: B annealed at 1000 °C are shown in Fig. 24(a). Peaks of hexagonal BN and α-Fe were confirmed for all samples, and no peak of Fe_4N and B was observed. Average diameters of Fe particles are summarized in Table 6, which were calculated from half-widths of α-Fe (110) by using the Scherrer equation. The sizes of Fe nanoparticles are below ~30 nm.

Figure 24(b) shows X-ray diffraction patterns of $Co(NH_3)_6Cl_3/KBH_4$ samples annealed at 700-1000 °C for 2 h. Although BN nanocapsules were formed, boron did not react with nitrogen perfectly yet for a sample annealed at 700 °C. Diffraction peaks of hexagonal BN and fcc-cobalt were confirmed for samples annealed at 1000 °C. Average particle sizes were calculated from half-widths of fcc-Co (111) by using Scherrer equation as listed in Table 7. The particle size depends on annealing temperatures rather than a flowing rate of N_2 gas.

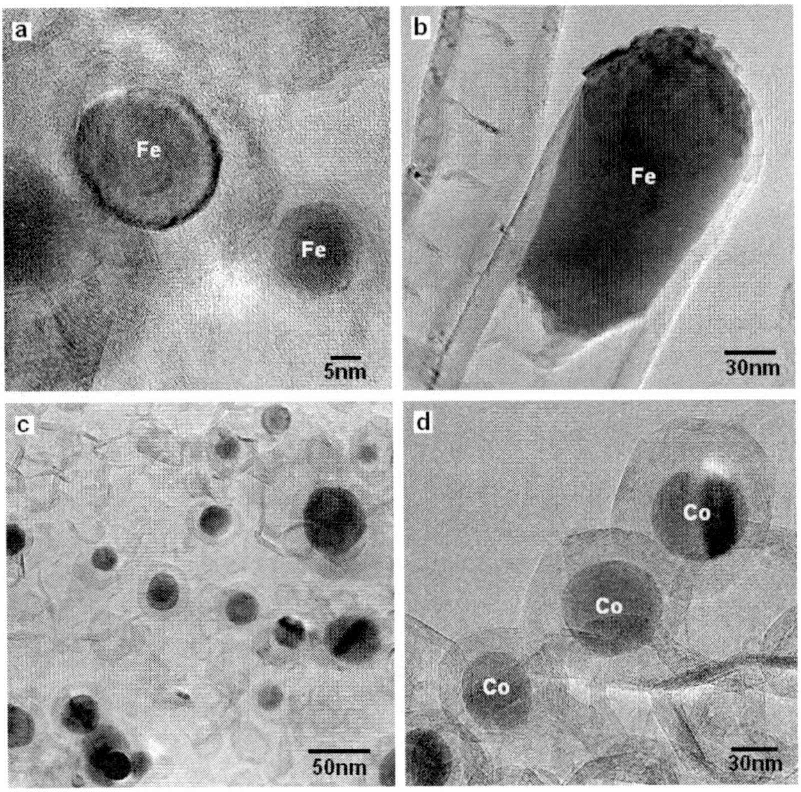

Figure 25. TEM images of (a) Fe nanocapsules coated with BN layers, (b) bamboo-type BN nanotubes with Fe nanoparticles, (c) BN nanocapsules encaging fcc-Co nanoparticles and (d) enlarged image of the BN nanocapsules with Co.

Figure 25(a) is a TEM image of Fe nanocapsules coated with BN layers produced by annealing a powder of Fe_4N/B (1: 1) at 1000°C for 1 h. Bamboo-type BN nanotubes with interval cells of 30 nm were also produced by annealing the mixture at 1000 °C for 5 h as shown in Fig 25(b). Length and width of bamboo-type BN nanotubes are approximately 10 μm and 40 nm, respectively. Fe nanoparticles were often observed at tips of bamboo-type BN nanotubes. Figure 25(c) shows a TEM image of fcc-Co nanoparticles coated with BN layers produced by annealing a mixture powder of $Co(NH_3)_6Cl_3/KBH_4$ at 1000 °C for 2 h with flowing 100 sccm N_2 gas. An enlarged images of BN nanocapsules encaging fcc-Co nanoparticles is shown in Fig. 25(d). Numbers of BN layers covering fcc-Co nanoparticles are in the range of 15~30 layers, and they increased more compared to sample annealed at 700 °C (5~10 layers).

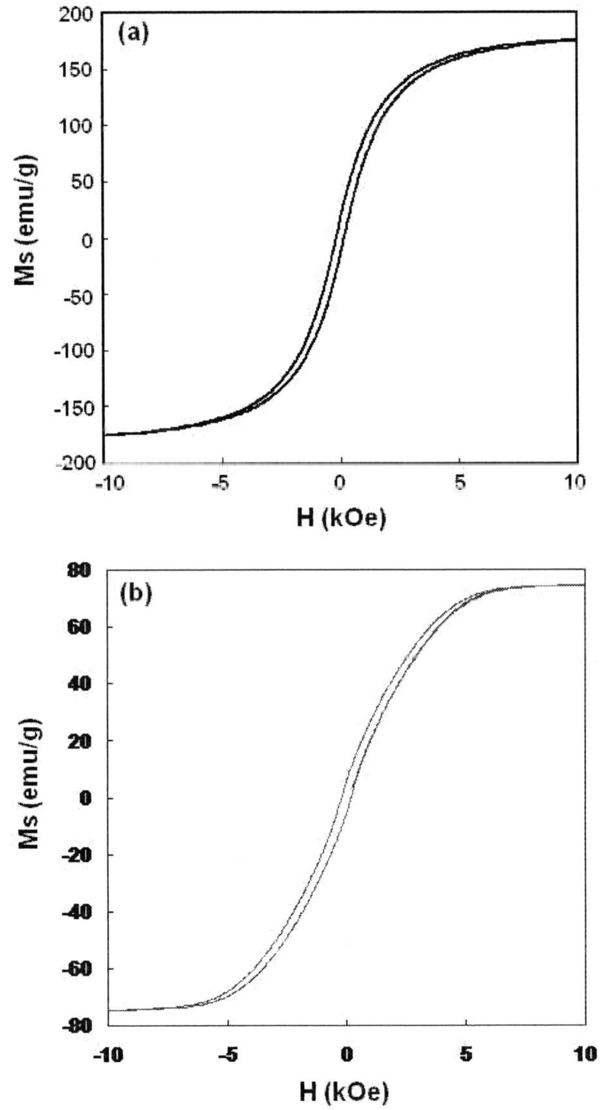

Figure 26. Hysteresis loops of BN nanocapsules encaging (a) Fe and (b) Co nanoparticles.

A magnetic hysteresis loop of BN nanocapsules encaging Fe nanoparticles is shown in Fig. 26(a), which was produced from Fe_4N/B (9: 1) sample annealed at 1000 °C for 1 h with flowing 100 sccm N_2 gas. The sample exhibits a soft magnetic characteristic, and saturation magnetization (M_s) and coercivity values (H_c) were 174.9 emu/g and 19.0 Oe, and the M_s and H_c values are 80% and 88% of bulk Fe (217.6 emu/g and 21.5 Oe), respectively. Results of VSM measurements are summarized in Table 6. Fe powders with only oxidized surface gives M_s value of 130 emu/g, and M_s value of magnetite (Fe_3O_4) is 92 emu/g. For the present Fe nanoparticles coated with BN layers, the M_s values were reduced due to the weight loss of BN layers, and are as high as pure Fe metal.

Figure 26(b) is a magnetic hysteresis loop of BN nanocapsules encaging Co nanoparticles synthesized at 1000 °C with flowing 100 sccm N_2 gas for 2 h. M_s and H_c values of the sample were 74.5 emu/g and 88 Oe, respectively, as summarized in Table 7. M_s values of samples annealed at 1000°C are larger than that of 700°C. On the other hand, H_c values decreased at elevated temperatures, which would be due to an effect of particle size. Herzer showed that H_c of soft magnetic metal particles depended on the particle size, and that the M_s values were inversely proportional to variance of the H_c values [15]. Therefore, decrease of the H_c values of the present samples annealed at 1000°C would be due to large particle sizes.

In order to investigate oxidation- and wear-resistances, M_s* and H_c* values of samples were measured by VSM after PCT, and the results are summarized in Table 6 and 7. Values of degauss coefficient were calculated according to the following equation: $(M_s^* - M_s)/M_s \times 100$ %. A sample of Fe_4N/B (7: 3) was more stable for oxidation- and wear-resistances than other samples for the BN nanocapsules encaging Fe nanoparticles. Although M_s value increased as increase of Fe_4N powder in the composition, appropriate amount of B powder is needed to keep the high M_s value for oxidation- and wear-resistances by covering the Fe nanoparticles with BN layers. BN nanocapsules with Co nanoparticles synthesized at 1000 °C were more stable than that of 700 °C, which would be due to that strong connection of boron and nitrogen atoms for the samples annealed at 1000 °C.

Table 6. Average diameters of Fe nanoparticles, and results of VSM measurements of Fe nanoparticles coated with BN layers at room temperature. Saturation magnetization (M_s*) and coercivity (H_c*) values of Fe nanoparticles coated with BN layers were measured after PC tests (120 °C ×12 h, humidity 100 %, 1 atm). Values of degauss coefficient were calculated according to the following equation: $(M_s^* - M_s)/M_s \times 100$ %.

Composition of Fe_4N: B	Annealing time at 1000 °C	Particle diameter (nm)	M_s (emu/g)	H_c (Oe)	M_s* (emu/g)	H_c* (Oe)	Degauss coefficient (%)
5: 5	1 h	24	95.0	24.4	78.3	42.5	-17.6
5: 5	5 h	28	92.6	22.5	78.9	39.8	-14.8
7: 3	1 h	-	134.2	20.9	117.4	33.8	-12.5
9: 1	1 h	30	174.9	19.0	149.9	37.5	-14.3

Table 7. Average diameters of Co nanoparticles, and VSM measurement of Co@BN cage nanomaterials at room temperature. M_s* and H_c* after PC tests. Values of degauss coefficient were calculated according to the following equation: (M_s*— M_s)/ Ms × 100 %.

Temperature (°C)	Flow rate (sccm)	Particle diameter (nm)	M_s (emu/g)	H_c (Oe)	M_s* (emu/g)	H_c* (Oe)	Degauss coefficient (%)
700	100	27	48.6	342.9	22.1	396.1	-54.6
1000	100	40	74.5	88.0	54.0	106.9	-27.5
1000	200	37	72.3	134.1	61.7	143.2	-14.7

9. Hydrogen Storage in BN and C Clusters

Many works have been reported on hydrogen storage ability of carbon (C) nanotubes, fullerenes and nanomaterials, as listed in Table 8. These results indicate the excellent storage ability of carbon nanotubes although the evaluation of hydrogen storage measurements is necessary. Recently, several studies on H_2 gas storage in boron nitride (BN) nanomaterials have also been reported, as listed in Table 8. BN nanomaterials are expected in prospective application because BN nanomaterials provide good stability at high temperatures with high electronic insulation in air. In the present work, H_2 gas storage ability was investigated for C/BN fullerene-like materials by theoretical calculations. Although huge amount of calculation is required to calculate nanotubes, it is considered that H_2 molecules enter from the cap of nanotubes. Barrier energy of ring structures, energy of chemisorption and stable hydrogen position inside the clusters were investigated in the minimum structure of nanotubes.

Table 8. Reported data for hydrogen storage in C/BN nanomaterials.

Structure	Hydrogen storage [wt. %]	References
Multi-walled carbon nanotube	2.5~4.6	115, 116
Alkali-doped carbon nanotube	2.61	117
Amorphous carbon nanotube	1.29~3.98	118
BN nanotube	2.6	119
BN nanotube and nanocapsule	~3	120

C_{60}, $B_{24}N_{24}$ and $B_{36}N_{36}$ clusters have energy barriers for H_2 molecules to pass through 4-, 5-, 6- and 8-membered rings. Figure 27(a) is a structural model of a H_2 molecule passing from the hexagonal rings of clusters. Single point energies for C_{60}, $B_{24}N_{24}$ and $B_{36}N_{36}$ clusters are summarized in Table 9. When H_2 molecules are set at the center and outside of the cage, the single point energy is equal. ΔE is regarded as the energy barrier for H_2 molecules to pass through 4-, 5-, 6- and 8-membered rings. ΔE of $B_{24}N_{24}$ octagonal rings showed the smallest value in the present calculation. This result was reached because the octagonal rings of $B_{24}N_{24}$ have a large curvature and ring size. Since single point energy is equal when H_2 molecules are set at the center and outside of the cage, the energy for H_2 discharge from fullerene-like materials is similar to that of H_2 storage.

Energies for hydrogen chemisorption on each position are summarized as Table 10. Figure 27(b) is schematic illustration for hydrogen chemisorption on $B_{36}N_{36}$. Hydrogen bonding with nitrogen is more stable than that with boron. In addition, hydrogen bonding on tetragonal ring is more stable than that of hexagonl ring. Chemisoption of hydrogen with C_{60} reduced the energy. When two hydrogen atoms were chemisorbed on carbon clusters, energies of carbon clusters increased.

To investigate stability of a H_2 molecule in clusters, energies were calculated for a H_2 molecule introduced inside clusters. One of the structural models for $B_{60}N_{60}$ is shown in Fig. 27(c). Energies of C_{60}, $B_{24}N_{24}$, $B_{36}N_{36}$ and $B_{60}N_{60}$ clusters with a H_2 molecule were calculated to be 0.58, -0.71, -0.93, -0.82 eV/mol·atom. This result indicates that $B_{24}N_{24}$, $B_{36}N_{36}$ and $B_{60}N_{60}$ with a H_2 molecule are more stable than C_{60} with H_2 molecule, and that $B_{36}N_{36}$ is the most stable in BN clusters.

Energies of H_2 absorption on $B_{36}N_{36}$ cluster were calculated, and structural models for $B_{60}N_{60}$ with 38 H_2 molecules are shown in Fig. 27(d). When more than 22 H_2 molecules were introduced in C_{60}, hydrogen atoms chemisorbed inside the cluster. When more than 26 H_2 molecules were introduced in the C_{60} clusters, 22 H_2 molecules were remained in the cluster and 8 hydrogen atoms chemisorbed inside the clusters, and a C-C bond was broken. Absorption of such large amount of hydrogen indicates rather unstable energy state. Maximum storages are 9, 20 and 38 H2 molecules for $B_{24}N_{24}$, $B_{36}N_{36}$ and $B_{60}N_{60}$, as summarized in Table 11. From Table 11, stability of H_2 molecules in $B_{24}N_{24}$, $B_{36}N_{36}$ and $B_{60}N_{60}$ seem to be higher than that of C_{60}. C and BN cluster showed possibility of hydrogen storage of ~6.5 and ~4.9 wt. %, respectively [121].

Carbon nanotubes are oxidized at 600°C in air. On the other hand, BN are stable up to 1000°C in air, which indicates BN fullerenes have higher thermal and chemical stability than those of carbon fullerenes. BN fullerenes with good thermal and chemical stability can store H_2 molecules with less energy, and they have the same chemisorption energy and higher stability, compared to carbon clusters. BN fullerene materials would be better candidates for H_2 storage materials.

Table 9. Single point energy for C_{60}, $B_{24}N_{24}$ and $B_{36}N_{36}$.

	Ring	Center of cage	Ring	Outside of cage	*ΔE (eV)
C_{60}	Pentagonal	-61891	-61871	-61891	20
	Hexagonal	-61891	-61874	-61891	17
$B_{24}N_{24}$	Tetragonal	-51806	-51780	-51806	26
	Hexagonal	-51806	-51791	-51806	15
	Octagonal	-51806	-51801	-51806	5
$B_{36}N_{36}$	Tetragonal	-77707	-77680	-77707	26
	Hexagonal	-77707	-77693	-77707	13

*ΔE(eV) = (Single point energy at ring)-(Single point energy at outside of cage).

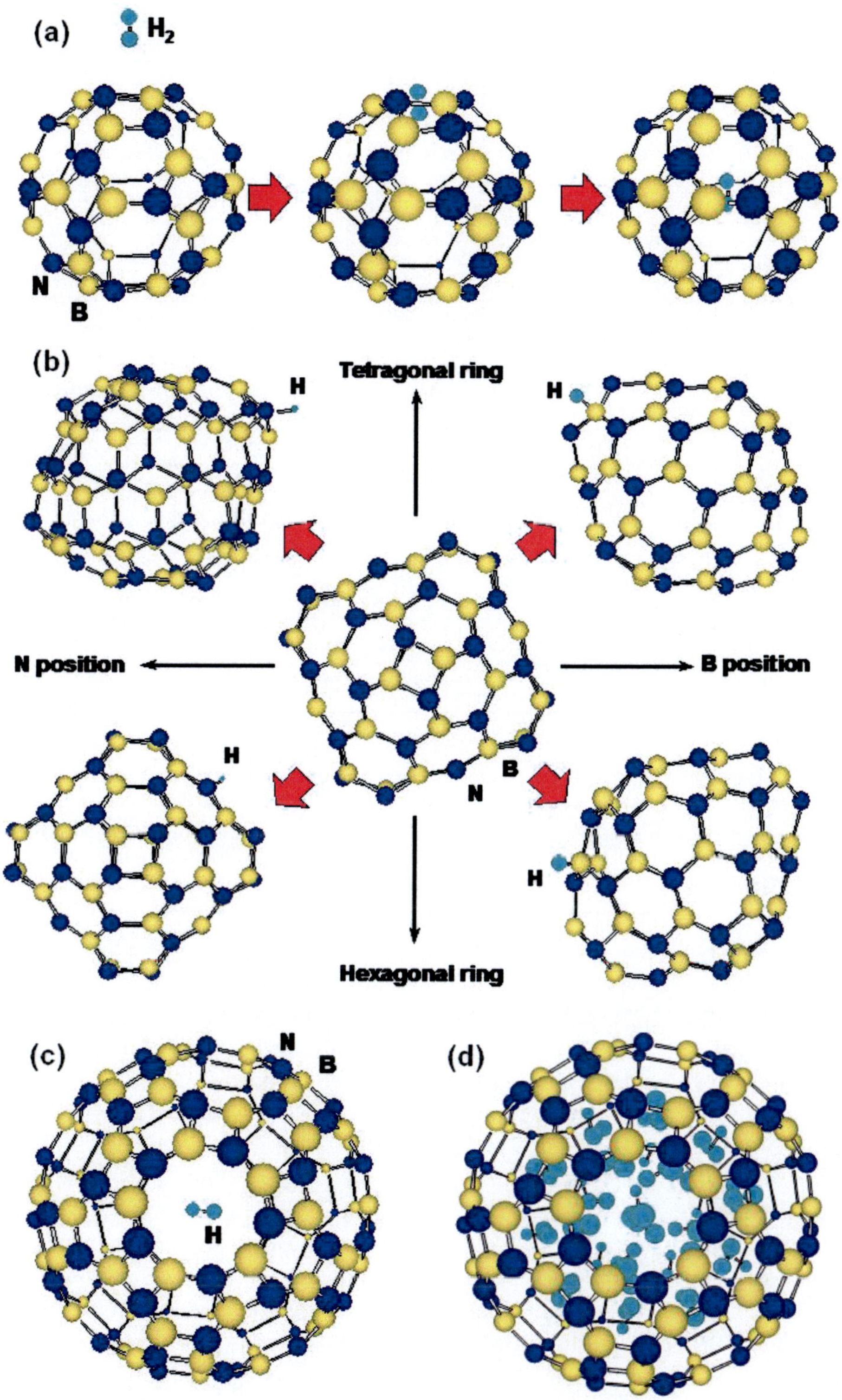

Figure 27. (a) Structural model of H_2 molecule passing from a hexagonal ring of $B_{24}N_{24}$. (b) Structure models for hydrogen chemisorption on $B_{36}N_{36}$. Optimized structural models of (c) one and (d) 38 H_2 molecules in $B_{60}N_{60}$ clusters.

Table 10. Chemisorption energy of hydrogen atoms on C_{60} and $B_{24}N_{24}$.

<table>
<tr><th rowspan="3"></th><th colspan="6">Heat of formation [eV/mol·atom]</th><th colspan="4" rowspan="2">*ΔE [eV/mol·atom]</th></tr>
<tr><th colspan="2" rowspan="2">Before addition</th><th colspan="4">Additional position of hydrogen</th></tr>
<tr><th>C H</th><th>C 2H</th><th>B</th><th>N</th><th>C H</th><th>C 2H</th><th>B</th><th>N</th></tr>
<tr><td>C_{60}</td><td colspan="2">35.21</td><td>35.03</td><td>35.81</td><td></td><td></td><td>-0.18</td><td>0.60</td><td></td><td></td></tr>
<tr><td>$B_{24}N_{24}$</td><td colspan="2">-36.12</td><td></td><td></td><td>-34.66</td><td>-35.67</td><td></td><td></td><td>1.46</td><td>0.45</td></tr>
<tr><td rowspan="2">$B_{36}N_{36}$</td><td>Tetragon</td><td rowspan="2">-69.33</td><td></td><td></td><td>-67.83</td><td>-69.16</td><td></td><td></td><td>1.50</td><td>0.17</td></tr>
<tr><td>Hexagon</td><td></td><td></td><td>-67.51</td><td>-68.83</td><td></td><td></td><td>1.83</td><td>0.51</td></tr>
</table>

*ΔE = (Heat of formation after hydrogen addition)-(Heat of formation before hydrogen addition).

Table 11. Energy of clusters with hydrogen. *A C-C bond was broken.

	Introduced H_2	Heat of formation [eV/mol·atom]	Chemisorbed H inside cluster	Hydrogen storage [wt. %]
C_{60}	0	35.21	0	5.8-6.5
	22	143.01	0	
	25	164.87	4	
	26	169.63	8*	
$B_{24}N_{24}$	0	-36.12	0	2.9
	9	-9.44	0	
$B_{36}N_{36}$	0	-69.33	0	4.3
	20	-6.66	0	
$B_{60}N_{60}$	0	-100.28	0	4.9
	38	-61.74	0	

10. Conclusions

B_nN_n (n = 12~60) nanocage clusters and endohedral Y@B_nN_n were synthesized by an arc-melting method from YB_6 powder in N_2/Ar mixture gas, and detected by LD-TOF mass spectrometry and HREM. The BN clusters consisted of tetragonal, hexagonal, octagonal and decagonal BN rings satisfying the isolated tetragonal rule, which was optimized by molecular orbital calculations. Atomic and electronic structures of endohedral M@$B_{36}N_{36}$ clusters with doping elements were also studied by molecular orbital calculations. Total energy calculation showed that potassium and gallium stabilize the $B_{36}N_{36}$ structure. Bandgap energies of the $B_{36}N_{36}$ clusters were found to be reduced by introducing a metal atom inside the cluster, which indicates controllability of the energy gap.

BN nanotubes with zigzag-type structures were synthesized and investigated by HREM, image simulation and total energy calculation. Hexagonal networks of BN nanotubes were directly observed by HREM in atomic scale, and chiralities of the BN nanotubes were directly determined from HREM images. An atomic structure models for quadruple-walled nanotubes was proposed, and simulated images based on these models agreed well with experimental HREM images. Molecular mechanics calculations showed good stability of a zigzag-type structure compared to the armchair-type structure, which agreed well with the experimental data of disordered armchair-type BN nanotubes.

In addition, BN nanohorns were synthesized, and an atomic strcture model with tetragonal BN rings was proposed. Multi-walled BN nanohorns would be stabilized by stacking of BN nanohorns, and the energy gap of BN nanohorn was calculated to be 0.2~0.8 eV. Possibility of hydrogen gas storage in BN clusters was also investigated by molecular orbital calculations, which indicated possibility of hydrogen storage of ~5 wt. %. Chemisorption calculation of hydrogen for BN clusters showed that hydrogen bondings with nitrogen atoms and tetragonal rings were the most stable.

Fe-filled BN nanotubes were also produced, and the [110] direction of the Fe nanowires was parallel to the BN nanotube axis. These unique structures would be suitable materials for nanoelectronics devices, magnetic recording media and biological sensors with excellent protection against oxidation and wear.

A new process for Fe or Co nanoparticles coated with BN layers in large quantity was developed. An Ellingham diagram drawn in the present work is useful for the design of BN nanomaterial formation. Fe_4N particles were reduced to α-Fe completely by the reaction. KBH_4 and $Co(NH_3)_6Cl_3$ powders were also used as starting materials for BN nanocapsules, and annealed at 1000°C for 2 h with flowing N_2 gas. These BN nanocapsules encaging Fe or Co nanoparticles exhibited a soft magnetic characteristic and good oxidation resistances. Gibbs energy calculation based on the reaction model agreed with the present results.

These new BN clusters, metallofullerenes, nanotubes and nanohorns can be expected to have applications in a wide variety of future nanodevices in combination with carbon-based fullerenes and nanotubes.

Acknowledgements

The authors would like to acknowledge I. Narita, N. Koi, A. Nishiwaki, H. Tokoro, K. Suganuma, M. Gonda, M. Nishijima, K. Hiraga, M. Hirabayashi, T. Matsuda and T. Hirai for experimental help, useful discussion and warm encouragements.

References

[1] H.W. Kroto, J.R. Heath, S.C. O'Brien, R.F. Curl and R.E. Smalley, *Nature* 318 (1985) 162.

[2] S. Iijima: *Nature* 354 (1991) 56.

[3] T. Oku, T. Hirano, M. Kuno, T. Kusunose, K. Niihara and K. Suganuma, *Mater. Sci. Eng. B* 74 (2000) 206.

[4] T. Oku, M. Kuno, H. Kitahara and I. Narita, *Int. J. Inorg. Mater.* 3 (2001) 597.

[5] T. Oku, I. Narita, A. Nishiwaki, N. Koi, K. Suganuma, R. Hatakeyama, T. Hirata, H. Tokoro, S. Fujii, *Topics in Appl. Phys.* 100 (2006) 187.

[6] N.G. Chopra, R.J. Luyken, K. Cherrey, V.H. Crespi, M.L. Cohen, S.G. Louie and A. Zettl, *Science* 269 (1995) 966.

[7] Loiseau, F. Willaime, N. Demoncy, G. Hug and H. Pascard, *Phys. Rev. Lett.* 76 (1996) 4737.

[8] M. Terrones, W.K. Hsu, H. Terrones, J.P. Zhang, S. Ramos, J.P. Hare, R. Castillo, K. Prassides, A.K. Cheetham, H.W. Kroto and D.R.M. Walton, *Chem. Phys. Lett.* 259 (1996) 568.
[9] Loiseau, F. Willaime, N. Demoncy, N. Schramchenko, G. Hug, C. Colliex and H. Pascard, *Carbon* 36 (1998) 743-752.
[10] D. Golberg, Y. Bando, K. Kurashima and T. Sato, *Chem. Phys. Lett.* 323 (2000) 185.
[11] J. Cumings and A. Zettl, *Chem. Phys. Lett.* 316 (2000) 211.
[12] T. Hirano, T. Oku and K. Suganuma, *Diamond Relat. Mater.* 9 (2000) 625.
[13] C.C. Tang, S.S. Fan, P. Li, Y.M. Liu and H.Y. Dang, *Mater. Lett.* 51 (2001) 315.
[14] R. Ma, Y. Bando and T. Sato, *Chem. Phys. Lett.* 337 (2001) 61-64.
[15] C.C. Tang, M.L. de la Chapell, P. Li, Y.M. Liu, H.Y. Dang and S.S. Fan, *Chem. Phys. Lett.* 342 (2001) 492.
[16] R. Ma, Y. Bando, T. Sato and K. Kurashima, *Chem. Phys. Lett.* 350 (2001) 434.
[17] M. Kuno, T. Oku and K. Suganuma, *Diamond Relat. Mater.* 10 (2001) 1231.
[18] Tang, Y. Bando and T. Sato: *Chem. Phys. Lett.* 362 (2002) 185.
[19] T. Oku, *Physica B* 323 (2002) 357.
[20] Goldberg, F.-F. Xu and Y. Bando, *Appl. Phys. A* 76 (2003) 479.
[21] Golberg, A. Rode, Y. Bando, M. Mitome, E. Gamaly, B. Luther-Davies, *Diamond Relat. Mater.* 12 (2003) 1269.
[22] W. Mickelson, S. Aloni, W.-Q. Han, J. Cumings and A. Zettl, *Science* 300 (2003) 467.
[23] Bengu and L.D. Marks, *Phys. Rev. Lett.* 86 (2001) 2385.
[24] D. Goldberg, Y. Bando, M. Mitome, K. Kurashima, T. Sato, N. Grobert, M. Reyes-Reyes, H. Terrones and M. Terrones, *Physica B* 323 (2002) 60.
[25] T. Oku, I. Narita and A. Nishiwaki, *Mater. Manuf. Process*. 19 (2004) 1215.
[26] L. Bourgeois, Y. Bando, W.Q. Han and T. Sato, *Phys. Rev. B* 61 (2000) 7686.
[27] M. Terauchi, M. Tanaka, K. Suzuki, A. Ogino and K. Kimura, *Chem. Phys. Lett.* 324 (2000) 359.
[28] M. Machado, R. Mota and P. Piquini, Electronic properties of BN nanocones under electric fields, *Microelectron. J. 34* (2003) 545.
[29] T. Oku, T. Kusunose, K. Niihara and K. Suganuma, *J. Mater. Chem*. 10 (2000) 255.
[30] J.F. Li, L.Z. Yao, C.H. Ye, C.M. Mo, W.L. Cai, Y. Zhang and L.D. Zhang, *J. Cryst. Growth* 223 (2001) 535.
[31] H. Kitahara, T. Oku, T. Hirano and K. Suganuma, *Diamond Relat. Mater.* 10 (2001) 1210.
[32] Narita and T. Oku, *Diamond Relat. Mater.* 11 (2002) 949.
[33] Xing, G. Chen, X. Song, X. Yuan, W. Yao and H. Yan, *Microelectron. Eng.* 66 (2003) 70.
[34] T. Oku and K. Hiraga, *Diamond Relat. Mater.* 10 (2001) 1398.
[35] T. Oku, K. Hiraga, T. Matsuda, T. Hirai and M. Hirabayashi, *Diamond Relat. Mater.* 12 (2003) 1138.
[36] T. Oku, K. Hiraga, T. Matsuda, T. Hirai and M. Hirabayashi, *Diamond Relat. Mater.* 12 (2003) 1918.
[37] Banhart, M. Zwanger and H.-J. Muhr, *Chem. Phys. Lett.* 231 (1994) 98.
[38] D. Golberg, Y. Bando, O. Stéphan, and K. Kurashima, *Appl. Phys. Lett.* 73 (1998) 2441.
[39] O. Stéphan, Y. Bando, A. Loiseau, F. Willaime, N. Shramchenko, T. Tamiya and T. Sato, *Appl. Phys. A* 67 (1998) 107.

[40] T. Oku, A. Nishiwaki, I. Narita and M. Gonda, *Chem. Phys. Lett.* 380 (2003) 620.
[41] T. Oku and K. Suganuma, *Diamond Relat. Mater.* 190 (2001) 1205.
[42] T. Oku, M. Kuno and I. Narita, *Diamond Relat. Mater.* 11 (2002) 940.
[43] S. Kokado and K. Harigaya, *Synthetic Metals* 135-136 (2003) 745.
[44] M. Radosavljević, J. Appenzeller, V. Derycke, R. Martel, Ph. Avouris, A. Loiseau, J.-L. Cochon and D. Pigache, *Appl. Phys. Lett.* 82 (2003) 4131.
[45] Y. Bando, K. Ogawa and D. Gorberg, *Chem. Phys. Lett.* 347 (2001) 349.
[46] C.C. Tang, Y. Bando and T. Sato, *Appl. Phys. A* 75 (2002) 681.
[47] R. Ma, Y. Bando and T. Sato, *Chem. Phys. Lett.* 350 (2001) 1.
[48] Tokoro, S. Fujii and T. Oku, *IEEE Trans. Mag.* 39 (2003) 2761.
[49] Tokoro, S. Fujii and T. Oku, *J. Mater. Chem.* 14 (2004) 253.
[50] T. Oku and I. Narita, *Physica B* 323 (2002) 216.
[51] Narita and T. Oku, *Diamond Relat. Mater.* 11 (2002) 945.
[52] T. Oku and M. Kuno, *Diamond Relat. Mater.* 12 (2003) 840.
[53] Rubio, J.L. Corkill and M.L. Cohen, *Phys. Rev. B 49* (1994) 5081.
[54] J.-Ch. Charlier, X. Blase, A. De Vita and R. Car, *Appl. Phys. A* 68 (1999) 267.
[55] Y.-H. Kim, K.J. Chang and S.G. Louie, *Phys. Rev. B* 63 (2001) 205408-1.
[56] Ş. Erkoç, *J. Molecular Structure (Theochem)* 542 (2001) 89.
[57] S. Okada, S. Saito and A. Oshiyama, *Physica B* 323 (2002) 224.
[58] Z. Peralta-Inga, P. Lane, J.S. Murray, S. Boyd, M.E. Grice, C.J. O'Connor and P. Politzer, *Nano Lett.* 3 (2003) 21.
[59] V.V. Ivanovskaya, A.A. Sofronov and A.L. Ivanovskii, *Phys. Lett. A* Vol. 297 (2002), p. 436.
[60] F. Jensen and H. Toflund, *Chem. Phys. Lett.* 201 (1993) 89.
[61] M.E. Zandler, E.C. Behrman, M.B. Arrasmith, J.R. Myers and T.V. Smith, *J. Molecular Structure (Theochem)* 362 (1996) 215.
[62] G. Seifert, R.W. Fowler, D. Mitchell, D. Porezag and Th. Frauenheim, *Chem. Phys. Lett.* 268 (1997) 352-358.
[63] Z. Slanina, M.-L. Sun and S.-L. Lee: *Nanostruc. Mater.* 8 (1997) 623.
[64] H.-Y. Zhu, T.G. Schmalz and D.J. Klein, *Int. J. Quantum Chem.* 63 (1997) 393.
[65] S.S. Alexandre, M.S.C. Mazzoni and H. Chacham, *Appl. Phys. Lett.* 75 (1999) 61.
[66] P.W. Fowler, K.M. Rogers, G. Seifert, M. Terrones and H. Terrones, *Chem. Phys. Lett.* 299 (1999) 359.
[67] K.M. Rogers, P.W. Fowler and G. Seifert, Chem. Phys. Lett. 332 (2000) 43.
[68] G. Will and P.G. Perkins, *Diamond Relat. Mater.* 10 (2001) 2010.
[69] S.S. Alexandre, H. Chacham and R.W. Nunes, *Phys. Rev. B* 63 (2001) 085406-1.
[70] Q. Wang, Q. Sun, T. Oku and Y. Kawazoe, *Physica B* 339 (2003) 105.
[71] V.V. Pokropivny, V.V. Skorokhod, G.S. Oleinik, A.V. Kurdyumov, T.S. Bartnitskaya, A.V. Pokropivny, A.G. Sisonyuk, D.M. Sheichenko, *J. Solid State Chem.* 154 (2000) 214.
[72] D.L. Strout, *J. Phys. Chem. A* Vol. 104 (2000), p. 3364.
[73] S.S. Alexandre, R.W. Nunes and H. Chacham, *Phys. Rev. B* 66 (2002) 085406-1.
[74] Y. Saito, M. Okuda, M. Tomita and T. Hayashi, *Chem. Phys. Lett.* 236 (1995) 419.
[75] T. Oku, *J. Ceram. Soc. Jpn.* 109 (2001) S17.
[76] T. Oku, *Chem. Comm.* (2002) 302.
[77] T. Oku, *J. Phys. Chem. Solids* 65 (2004) 363.

[78] Narita and T. Oku, *Solid State Commun.* 122 (2002), p. 465.
[79] Narita and T. Oku, *Diamond Relat. Mater.* 12 (2003) 1146.
[80] Tanaka, H. Waki, Y. Ido, S. Akita, Y. Yoshida and T. Yoshida, *Rapid Commun. Mass Spectrom.* 2 (1988) 151.
[81] H. Ajie, M.M. Alvarez, S.J. Anz, R.D. Beck, F. Diederich, K. Fostiropoulos, D.R. Huffman, W. Krätschmer, Y. Rubin, K.E. Schriver, D. Sensharma and R.L. Whetten, *J. Phys. Chem.* 94 (1990) 8630.
[82] P.W. Stephens, G. Bortel, G. Faigel, M. Tegze, A. Jánossy, S. Pekker, G. Oszlanyi and L. Forró, *Nature* 370 (1994) 636.
[83] Takata, B. Umeda, E. Nishibori, M. Sakata, Y. Saito, M. Ohno and H. Shinohara, *Nature* 377 (1995) 46.
[84] T. Oku, I. Narita and A. Nishiwaki, *J. Phys. Chem. Solids* 65 (2004) 369.
[85] H. Shinohara, *Rep. Prog. Phys.* 63 (2000), p. 843.
[86] Nishiwaki, T. Oku and K. Suganuma, *Physica B* 349 (2004) 254.
[87] W. Krätschmer, L.D. Lamb, K. Fostiropoulos and D.R. Huffman, *Nature* 347 (1990) 354.
[88] D. Golberg, Y. Bando, L. Bourgeois, K. Kurashima and T. Sato, *Appl. Phys. Lett.* 77 (2000) 1979.
[89] B.G. Demczyk, J. Cumings, A. Zettl and R.O. Ritchie, *Appl. Phys. Lett.* 78 (2001) 2772.
[90] Narita and T. Oku, *Chem. Phys. Lett.* 377 (2003) 354.
[91] Narita and T. Oku, *Sol. State Comm.* 129 (2004), p. 415.
[92] T. Oku and I. Narita, *Diamond Relat. Mater.* 13 (2004) 1254.
[93] D. Golberg, Y. Bando, K. Kurashima and T. Sato: *Sol. State Commun.* 116 (2000) 1.
[94] T. Oku, I. Narita, A. Nishiwaki, N. Koi, *Defects and Diffusion Forum* 226-228 (2004) 113.
[95] T. Oku and I. Narita, *J. Phys. Chem. Solids* 65 (2004) 359.
[96] Nishiwaki, T. Oku and I. Narita, *Sci. Tech. Adv. Mater.* 5 (2004) 629.
[97] Nishiwaki and T. Oku, *Diamond Relat. Mater.* 14 (2005) 1183.
[98] Nishiwaki and T. Oku, *J. Electron Microsc.* 54 (2005) i9.
[99] T. Oku and A. Nishiwaki, *Physica E* 29 (2005) 712.
[100] Nishiwaki and T. Oku, *J. Euro. Ceram. Soc.* 26 (2006) 435.
[101] Joint Committee on Powder Diffraction Standards (1984) 34-421.
[102] T. Oku, A. Nishiwaki and I. Narita, *Physica B* 351 (2004) 184.
[103] F.R. de Boer, R. Boom, W.C.M. Mattenes, A.R. Miedema and A.K. Niessen, *Cohesion in Metals - Transition Metal Alloys Vol. 1* (North-Holland, Amsterdam 1989).
[104] Narita, T. Oku, H. Tokoro and K. Suganuma, *J. Electron Microsc.* 55 (2006) 123.
[105] T. Oku, I. Narita and H.Tokoro, *J. Phys. Chem. Solids* 67 (2006) 1152.
[106] Endo, Y. A. Kim, T. Hayashi, T. Yanagisawa, H. Muramatsu, M. Ezaka, H. Terrones, M. Terrones and M. S. Dresselhaus, *Carbon* 41 (2003) 1941.
[107] T. Oku, N. Koi, K. Suganuma, R. V. Belosludov and Y. Kawazoe, *Sol. State Comm.* 143 (2007) 331.
[108] Koi, T. Oku and M. Nishijima, *Sol. State Comm.* 136 (2005) 342.
[109] T. Oku, N. Koi, I. Narita, K. Suganuma and M. Nishijima, *Mater. Trans.* 48(2007) 722.
[110] F. E. Kruis, H. Fissan, A. Peled, J. Aerosol Sci. 29 (1998) 511.
[111] S. Sun, C. B. Murray, D. Weller, L. Folks, A. Moser, *Science* 287 (2000) 1989.
[112] M. Kuno, T. Oku, and K. Suganuma, *Scripta Mater.* 44 (2001) 1583.

[113] H. Tokoro, S. Fujii, T. Oku, *Solid State Commun.* 133 (2005) 681.
[114] J. Q. Hu, Q. Y. Lu, K. B. Tang, S. H. Yu, Y. T. Qian, G. E. Zhou, X. M. Liu, J. X. Wu, *J. Solid State Chem.* 148 (1999) 325.
[115] P.-X. Hou, S.-T. Xu, Z. Ying, Q.-H. Yang, C. Liu and H.-M. Cheng, *Carbon* 41 (2003) 2471.
[116] W.Z. Huang, X.B. Zhang, J.P. Tu, F.Z. Kong, J.X. Ma, F. Liu, H.M. Lu and C.P. Chen, *Mater. Chem. Phys.* 78 (2002) 144.
[117] S. Challet, P. Azais, R.J.-M. Pellenq, O. Isnard, J.-L. Soubeyroux and L. Duclau, *J. Phys. Chem. Solids* 65 (2004) 541
[118] L. Ci, H. Zhu, B. Wei, C. Xu and D. Wu, *Appl. Surf. Sci.* 205 (2003) 39.
[119] R. Ma, Y. Bando, H. Zhu, T. Sato, C. Xu and D. Wu, *J. Am. Chem. Soc.* 124 (2002) 7672.
[120] T. Oku, M. Kuno and I. Narita, *J. Phys. Chem. Solids* 65 (2004) 549.
[121] Koi, T. Oku, K. Suganuma, *Physica E* 29 (2005) 541.

In: Progress in Fullerene Research
Editor: Milton Lang, pp. 85-121

ISBN 978-1-60021-841-5

Chapter 3

COMPUTED STABILITIES OF FULLERENES AND METALLOFULLERENES

Zdeněk Slanina[a,*], Filip Uhlík[b], Shyi-Long Lee[c], Ludwik Adamowicz[d] and Shigeru Nagase[a]

[a]Department of Theoretical Molecular Science
Institute for Molecular Science, Myodaiji, Okazaki 444-8585, Japan
[b]Department of Physical and Macromolecular Chemistry
School of Science, Charles University, 128 43 Prague 2, Czech Republic
[c] Department of Chemistry and Biochemistry
National Chung-Cheng University, Chia-Yi 62117, Taiwan
[d]Department of Chemistry, University of Arizona
Tucson, AZ 85721-0041, USA

Abstract

Fullerenes - or cage compounds built exclusively from carbon atoms - and their metal-containing derivatives metallofullerenes were for the first time observed in the gas phase about twenty years ago, and prepared in crystalline form about fifteen years ago. Still, an enormous amount of observed and computed data has been obtained during the time period. In addition to spheroidal fullerene cages, fullerene science also deals with other objects like elongated cylindrical bodies known as nanotubes, prepared soon after mastering the fullerene synthesis, nanocones or peapods. All the species exhibit a substantial application potential, especially for molecular electronics. Fullerenes and metallofullerenes have attracted a considerable interest from both experiment as well as computations and theory. Their experimental characterization is mostly based on ^{13}C NMR spectroscopy and more recently also on ^{3}He NMR spectroscopy. Fullerenes are commonly defined as polyhedral cages containing only carbon atoms arranged into five- and six-membered rings. Quasi-fullerenes are cages containing also other types of cycles like (topological) squares and heptagons. The very early history of carbon clusters starts with a mass-spectrometric observation of clusters up to C_{15} by Hahn and his coworkers sixty years ago. In the fifties and sixties experiments could expand up to C_{33}. At this stage, first simple computations had also been performed on small carbon clusters. Moreover, various qualitative estimations for

*E-mail address: zdenek@ims.ac.jp. (corresponding author)

larger carbon cages including C_{60} were presented. In the time of the C_{60} discovery, computations could support fullerene and metallofullerene research with increasingly reliable data. It is of historic interest that the four IR bands computationally predicted for C_{60} were used in analysis of the carbon-arc C_{60} synthesis. Later on, fullerene research has indeed been characterized by a close theory-experiment cooperation and application of large-scale computations. Although numerous surveys are available, the sole computations of fullerenes and metallofullerenes are reviewed relatively rarely. Structural elucidations for the known higher fullerenes C_n have been based on so called isolated pentagon rule (IPR) that claims that especially stable fullerenes should have all pentagons surrounded just by hexagons. Connectivity patterns for such cages can be generated by various topological schemes and in fact the cages can be quite numerous. Beyond n=76 isomerism of the IPR structures should generally play a role in observations. Several such mixtures of empty fullerene isomers have indeed been computed and an agreement with experiments found for even-numbered cages from C_{76} till C_{96}, while there is still only a computational prediction for C_{98}. A similar isomeric interplay has also been described for smaller fullerene systems like C_{32} or C_{36}. Various metallofullerenes can coexist in several isomeric forms as well, for example $Ca@C_{72}$, $Yb@C_{74}$, $Ca@C_{82}$, $La@C_{82}$, $Tm@C_{82}$, $Ti_2@C_{80}$, $Sc_2@C_{84}$, $Ti_2@C_{84}$, or $Sc_2@C_{76}$. Still other interesting isomeric sets can be created by derivatives of fullerenes or nanotubes. Although the inter-isomeric separation energies are important, they alone cannot predict the relative stabilities of the isomers. Owing to very high temperatures, entropy contributions can even over-compensate the enthalpy terms. Hence, the enthalpy-entropy interplay represents an essential feature of fullerenes and metallofullerenes and is surveyed in this article.

Introduction

Fullerenes, or cage compounds built exclusively from carbon atoms, and their metal-containing forms known as metallofullerenes were for the first time observed in the gas phase by Kroto *et al.* [1,2] more than twenty years ago and prepared in crystalline form by Krätschmer *et al.* [3] more than fifteen years ago. During the period, a considerable volume of observed and computed data has been accumulated (see, e.g., [4-10]). In addition to the spheroidal fullerene cages, fullerene science also deals with other objects like elongated cylindrical bodies known as nanotubes, prepared by Iijima [11] soon after the fullerene synthesis [3], nanocones [12] or peapods [13]. All the species exhibit a possible application potential, especially for quantum computing and molecular electronics [14].

Fullerenes and metallofullerenes have also been studied by theoretical techniques [15-17]. As their experimental characterization is mostly based [18-20] on ^{13}C NMR spectroscopy, computational support is essential in the structure elucidations. Fullerenes are polyhedral cages containing only carbon atoms arranged into five- and six-membered rings (*quasi*-fullerenes then contain [21] also other types of cycles) and this allows for an ample cage isomerism.

Historically, fullerenes are rooted [22-26] in mass-spectrometric observations of carbon clusters up to C_{15} by Hahn *et al.* [27] more than six decades ago, later on expanded [28-30] up to C_{33}. In the sixties first simple computations could also be done [31-34]. Moreover, even qualitative estimations for larger carbon cages including C_{60} were presented [35-39], especially Schultz's treatment [35] of fully hydrogenated fullerenes and *quasi*-fullerenes based on some regular (Platonic) and semi-regular (Archimedean) polyhedra which actually agrees even with present calculations [40]. When C_{60} was discovered [1], computations could readily supply useful supporting data [16,17,41], especially regarding its IR spectrum [42-48]. Fullerene research [49,50] has developed in a close theory-experiment cooperation, including large-scale computations [51,52]. Numerous surveys are available [53-80], however, the computations themselves of fullerenes and metallofullerenes are reviewed rather rarely [16,17,41,81-90].

Studies of C_{60} and higher fullerenes [19,91-93] have been based on so called isolated pentagon rule (IPR) [94,95] - especially stable fullerenes should have all pentagons surrounded just by hexagons. Such IPR cages can be quite numerous [96-99] and their mixtures have been computed in agreement with experiments for C_{76} - C_{96} [100-142]. A similar isomeric interplay has been described for smaller fullerene systems like [143-146] C_{32} or C_{36} (though the IPR pattern does not exist below C_{60}). Metallofullerenes can also coexist [41] in several isomeric forms as well, for example [147-171] $Ca@C_{72}$, $Mg@C_{72}$, $Ca@C_{74}$, $Ca@C_{82}$, $La@C_{82}$, $Tm@C_{82}$, $Ti_2@C_{80}$, $Sc_2@C_{84}$, $Ti_2@C_{84}$ or $Sc_2@C_{76}$. Isomeric sets are produced [172-176] by derivatives of fullerenes or nanotubes.

Although the inter-isomeric separation energies in such systems are important, they alone cannot predict the relative stabilities of the isomers. Owing to high temperatures, entropy contributions can in fact even over-compensate the enthalpy terms. Hence, the enthalpy-entropy interplay represents an essential issue in fullerene science.

Energetics and Thermodynamics of Clusters

The present quantum-chemical calculations of fullerenes start from the optimized geometries [177-181] obtained at semiempirical like [182-188] MNDO, AM1, PM3 or SAM1, *ab initio* Hartree-Fock Self-Consistent-Field (HF SCF), or density-functional theory (DFT) levels. Various *ab initio* procedures are mostly applied using Gaussian [189,190] and Spartan [191] program packages. Stability of the SCF wavefunction [192,193] can be an issue for some fullerenic structures. The geometry optimizations are frequently followed by harmonic vibrational analysis with an optional frequency scaling [194].

The relative concentrations of m isomers can be expressed as their mole fractions, w_i, using the isomeric partition functions q_i. In the terms of q_i and the ground-state energy changes $\Delta H^o_{0,i}$ the mole fractions are given [195-197]:

$$w_i = \frac{q_i exp[-\Delta H^o_{0,i}/(RT)]}{\sum_{j=1}^{m} q_j exp[-\Delta H^o_{0,j}/(RT)]}, \tag{1}$$

where R stands for the gas constant and T for the absolute temperature. Eq. (1) is a rigorous formula that can be directly derived [198] from the standard Gibbs energies of the isomers, supposing the conditions of the inter-isomeric thermodynamic equilibrium. Although the partition functions are to be constructed within the rigid-rotor and harmonic-oscillator (RRHO) approximation, eq. (1) itself is essentially exact. The partition functions q_i reflect the rotational, vibrational, electronic, symmetry, and chirality [199] contributions.

If the partition functions are neglected and the vibrational zero-point energy is extracted from $\Delta H^o_{0,i}$, the relative potential energies $\Delta E_{r,i}$ remain, giving the simple Boltzmann factors:

$$w_i^{,} = \frac{exp[-\Delta E^o_{r,i}/(RT)]}{\sum_{j=1}^{m} exp[-\Delta E^o_{r,j}/(RT)]}, \tag{2}$$

with no reference to entropy contributions. Clearly enough, the simple Boltzmann factors can never cross with a temperature change.

Fullerenes are cages built from three-coordinated (sp^2) carbon atoms arranged into two types of rings, pentagons and hexagons. Any polyhedra have to obey Euler's polyhedral theorem [200,201] (more precisely, convex polyhedra):

$$V + F = E + 2 \tag{3}$$

where V denotes number of vertexes (atoms), F number of faces (rings), and E number of edges (bonds). As only pentagons and hexagons are allowed, their numbers n_5 and n_6 give the total count of faces:

$$F = n_5 + n_6 \,. \tag{4}$$

As all atoms are three-coordinated, it must hold for the number of edges:

$$E = \frac{3V}{2} \tag{5}$$

as each bond is accounted twice. One can also count the edges through rings:

$$E = \frac{5n_5 + 6n_6}{2} \,. \tag{6}$$

Hence:

$$V = \frac{5n_5 + 6n_6}{3} \,. \tag{7}$$

Euler's theorem gives:

$$V + F = \frac{3V}{2} + 2 \tag{8}$$

One can thus get for the ring counts:

$$n_5 + n_6 = \frac{5n_5 + 6n_6}{6} + 2 \tag{9}$$

which can immediately be reduced to:

$$n_5 = 12 \,. \tag{10}$$

In any conventional fullerene C_n the number of five-membered rings must be equal to 12. The number of six-membered rings is variable, however, eq. (7) gives a relationship to the number of carbon atoms:

$$n_6 = \frac{n - 20}{2} \,. \tag{11}$$

According to eq. (11), the smallest possible fullerene has the stoichiometry C_{20}. As there are always 12 pentagons, then the smallest IPR cage should have 12 $\times$5 carbon atoms - known as truncated icosahedron or buckminsterfullerene C_{60}.

There is a smooth dependency [34,51,86,202-206] of the relative heats of formation $\Delta H^o_{f,298}/n$ on the number of atoms n in carbon clusters. The finding can readily be rationalized. Let us consider just IPR fullerenes. We deal with two types of bonds - 5/6 (between pentagons and hexagons) and 6/6 (shared by two hexagons). If these two types of bonds can be represented [207] by some uniform dissociation energies, $H_{5/6}$ and $H_{6/6}$, we can readily write for the atomization heat:

$$\Delta H_{at} = 60H_{5/6} + (\frac{3n}{2} - 60)H_{6/6} \tag{12}$$

as an IPR fullerene C_n always has sixty 5/6 bonds, while the number of the 6/6 bonds is $\frac{3n}{2} - 60$. The atomization and formation heats for carbon aggregates are related by the heat of vaporization of carbon, ΔH^o_{vap}:

$$\frac{\Delta H^o_f}{n} = -\frac{\Delta H_{at}}{n} + \Delta H^o_{vap} \,. \tag{13}$$

Thus:

$$\frac{\Delta H^o_f}{n} = -\frac{3}{2}H_{6/6} + \Delta H^o_{vap} + \frac{60}{n}(H_{6/6} - H_{5/6}). \tag{14}$$

This functional dependency can be expressed as:

$$\frac{\Delta H^o_f}{n} = A + \frac{B}{n}, \tag{15}$$

where B is a positive constant and A is a relatively small number, and eq. (15) indeed gives a smoothly decreasing curve.

From such a curve, however, one cannot see a particular stability of C_{60} or C_{70}. Just the opposite - it might suggest a rule that the stability would increase with the dimension which is of course not observed. Hence, a still more complex description of the fullerene synthesis is needed. If we treat the problem from the thermodynamic point of view, we should realize that the conventional standard pressure of 1 atm is considerably different from the actual synthetic conditions and this feature must be reflected [208-213]. For example, the equilibrium constant $K_{60/70}$ for an interconversion between the two clusters, expressed in partial pressures p, offers a deeper insight into the problem [208-212]:

$$K_{60/70} = \frac{p_{70}^{6/7}}{p_{60}} = \frac{(1 - x_{60})^{6/7}}{x_{60}} P^{-1/7} \,, \tag{16}$$

where P stands for the total pressure of the two clusters and x_{60} is the mole fraction of buckminsterfullerene. At higher pressures, C_{70} is more populated than C_{60}, but at the conditions of a saturated carbon vapor the stability order is reversed in favor of C_{60} so that an agreement with experiment is obtained. Although the isomeric stability problem is much better understood [198,214-216] at present than the relative stabilities of non-isomeric carbon clusters, there are some interesting results available also for the non-isomeric situation [208-212], especially a temperature increase of the clustering degree under the saturation conditions [217]. While the equilibrium constants for cluster formation decrease with temperature, the saturated pressure increases. It is just the competition between these two terms which decides the final temperature behavior [218-221].

Stabilities of Empty Fullerenes

The relative stabilities of small carbon clusters have frequently been studied by both theoretical [222-240] and experimental [241-264] techniques. Ion-chromatography observations by von Helden *et al.* [261,263,264] showed that starting from about n=7 cyclic rings always exist in addition to the linear forms. Cyclic structures were for example computed [236,237,240,265-271] for C_7 - C_{13}. C_{11} should be the first species for which the cyclic structure becomes dominant. Such transition from cyclic to polyhedral species was expected [272] around n=45. The linear and cyclic (rhombic, bicyclic) C_4 isomers were especially frequently computed. Newer estimates [273-277] suggest only a small separation between the forms. Then, the entropy contributions lead to the linear species at higher temperatures [245,278-280] in agreement with ion chromatography [261]. A similar effect was computed [214,222,240,281-283] for C_6.

Exhaustive topological generations of fullerene cages represent a crucial step for computational studies, systematically treated by Fowler *et al.* [96,97,284-302] and other groups [95,98,99,303-331], employing topological concepts like ring spiral [96,97,308], Goldberg polyhedra [286], leapfrog transformation [296], topological duals [306] or Stone-Wales transformations [304,332-335]. It was found [308-313] that, e.g., the number of C_{60} isomers is 1812. Odd-numbered cages are usually not considered in the enumerations though they are known [314,336,337]. The numbers of the IPR isomers are [96,97], for example: 2, 5, 7, 9, 24, 19, 35, and 46 for n = 76, 78, 80, 82, 84, 86, 88, and 90, respectively. There are also interesting applications of Pólya's enumeration theorem [198,338] to various substituted fullerenes [321,324,330,331].

Smaller fullerenes have also been studied. Von Helden *et al.* [261,272] observed gas-phase fullerene-like structures around C_{30}. Even the smallest fullerene C_{20} is available [339-342] and thus computed [216,343-346]. C_{36} fullerene was also isolated [347], computed [144,145,348,349], and linked to narrow nanotubes [350-353]. Another computed smaller fullerene is [143] C_{32}.

Studies of isomerism of the IPR cages from C_{76} till C_{98} are presently available [100-142]. In fact, C_{72} and C_{74} are the only species [19,92] among C_{60} - C_{96} not isolated yet (though some access to C_{74} is already possible [354,355]). A low solubility [356] in conventional solvents may be one reason for the difficulties. Hence, as C_{72} could only be recorded in gas phase [357,358], its structure is unknown. There is just one IPR satisfying structure for C_{72}, namely with D_{6d} symmetry [97]. However, a non-IPR (i.e., IPR-violating) structure with one pentagon-pentagon junction [148] is by a few kcal/mol lower in energy than the IPR cage. Moreover, it was demonstrated [359] on the Si_{60} case that the IPR/non-IPR stability order can be reversed by the entropy factors. Hence, the C_{72} system was investigated at semiempirical [360] and DFT level [361]. At the DFT level, the geometry optimizations were carried out [361] using the B3LYP/3-21G treatment. In the optimized B3LYP/3-21G geometries the harmonic vibrational analysis was carried out and also the B3LYP/6-31G* separation energies were evaluated. The electronic excitation energies were evaluated by means of the ZINDO method [362]. The following structures were considered [361]: the IPR cage (a), two non-IPR cages [148] with one pentagon-pentagon junction (b) and (c), a structure [148] with one heptagon (d), a cage [149] with two heptagons (e), and two structures [356] each with two pentagon-pentagon junctions (f), (g). Fig. 1 presents the B3LYP/3-21G optimized structures of the seven C_{72} isomers. The (c) structure of C_{2v} symmetry with just one pentagon/pentagon fusion represents the lowest energy isomer. Fig. 2 presents the DFT computed temperature development of the relative concentrations of the seven C_{72} isomers in a high temperature region. The lowest energy structure (c) is the most populated species at any temperature. On the other hand, the IPR structure (a) is always negligible.

The C_{74} case is however different. Recently, Shinohara *et al.* [355] recorded electronic spectrum of C_{74} anion and suggested that the cage could have D_{3h} symmetry (i.e., the only one available IPR structure [97]). This interesting finding prompted DFT calculations [363] that treated a set of six isomers, five of them being non-IPR species. The computational evaluation however shows that he IPR structure prevails at any relevant temperature in agreement with the experimental finding [355].

Stabilities of Metallofullerenes

The stability computations have been applied also to a few metallofullerenes like Ca@C_{72}, Ca@C_{74}, Ca@C_{82} or La@C_{82}. Ca@C_{72} was isolated [147], and recently even its second isomer [364], however structures are not yet known. It follows from its very first computations [148,149] that there are four isomers especially low in potential energy. The endohedral Ca@C_{72} species created by putting Ca inside the sole IPR cage has been labeled [148] by (a). The other three Ca@C_{72} isomers considered in ref. [148] are related to two non-IPR C_{72} cages (b) and (c), and to a C_{72} structure with one heptagon (d) - see Fig. 3. The extended computations [150] were performed in the 3-21G basis set for C atoms and a dz basis [365] and the effective core potential (ECP) on Ca with the B3LYP functional (B3LYP/3-21G~dz). The electronic excitation energies were evaluated by means of time-dependent DFT response theory [366] at the B3LYP/3-21G~dz level. Fig. 4 presents [150] the temperature development of the relative concentrations of the four Ca@C_{72} isomers in a high temperature region. At a temperature of 226 K (though such a low temperature is of no importance for the fullerene synthesis) the relative concentrations of the (c) and (b) structures are interchanged and beyond the point the (b) structure is always somewhat more populated.

In contrast to Ca@C_{72}, Ca@C_{74} was not only isolated [147] but even its structure was determined [153]. According to the ^{13}C NMR spectra recorded by Achiba *et al.* [153], Ca@C_{74} exhibits D_{3h} symmetry of its cage. There is only one IPR structure possible [97] for C_{74} and the sole C_{74} IPR cage has D_{3h} symmetry. A set of altogether six Ca@C_{74} isomers was subjected to the stability computations [154]. It turns out, in agreement with the experiment, that the encapsulate with the IPR cage has not only favorable enthalpy but

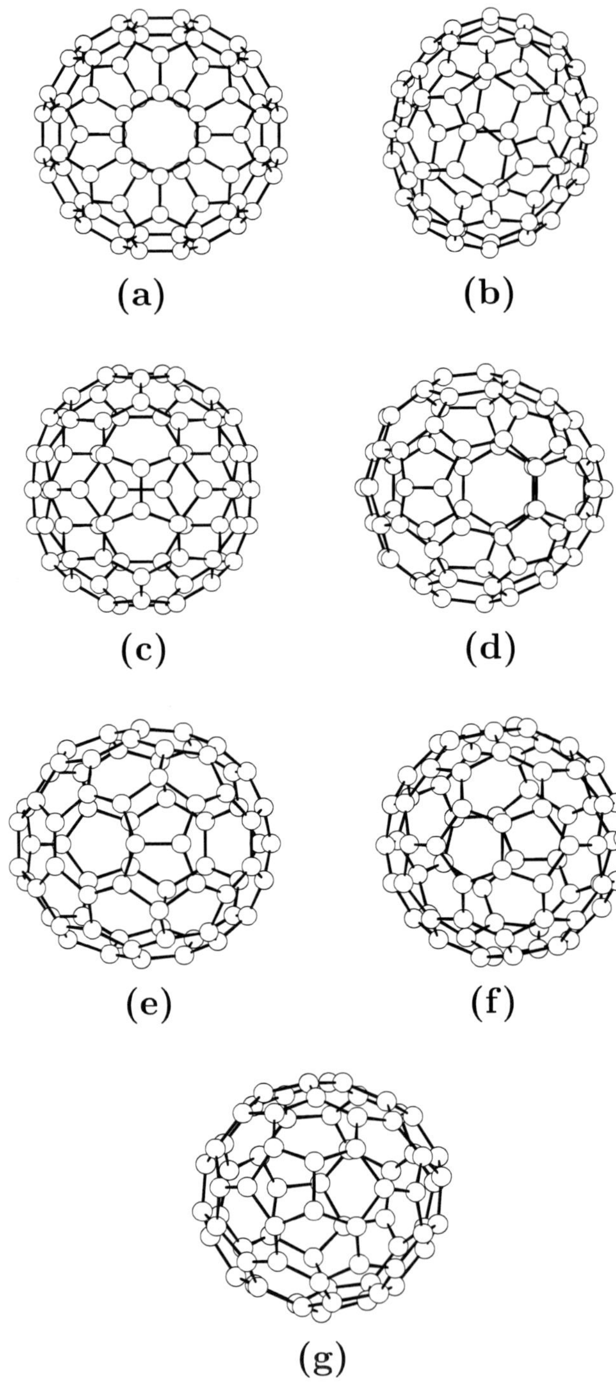

Figure 1. B3LYP/3-21G optimized structures [361] of C_{72} isomers: (a) IPR, (b) 5/5 pair, (c) 5/5 pair, (d) 7-membered ring, (e) two 7-membered rings, (f) two 5/5 pairs, (g) two 5/5 pairs.

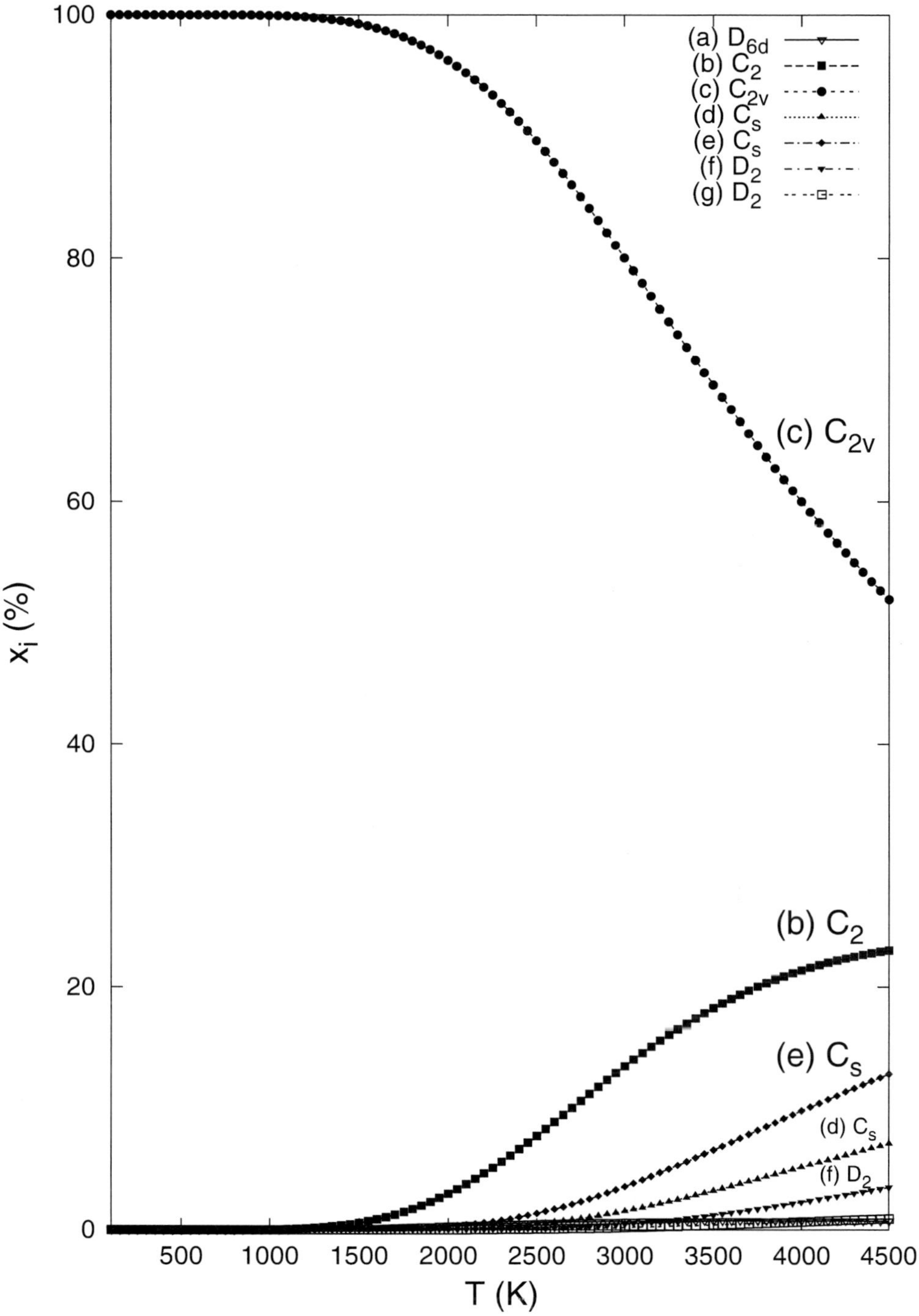

Figure 2. Relative concentrations [361] of the C_{72} isomers based on the B3LYP/6-31G* energetics and the B3LYP/3-21G & ZINDO entropy.

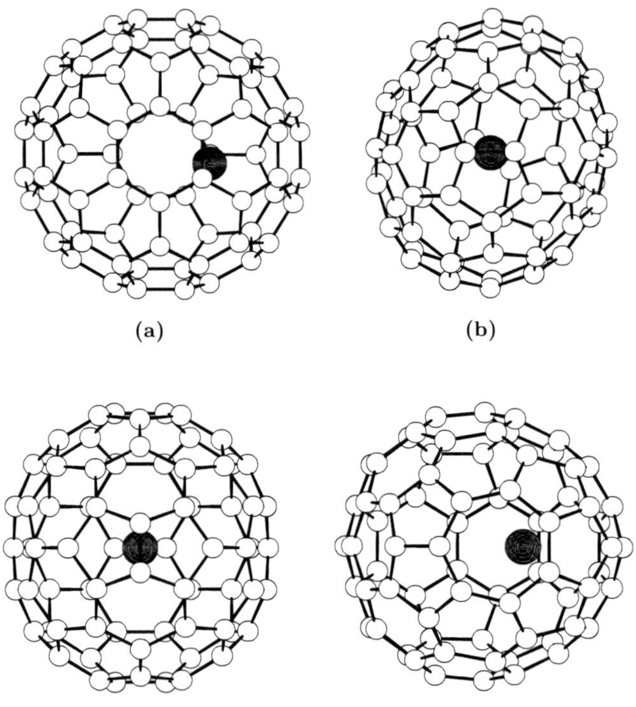

Figure 3. B3LYP/3-21G∼dz optimized structures [150] of Ca@C_{72} isomers: (a) IPR, (b) 5/5 pair, (c) 5/5 pair, (d) 7-membered ring.

also entropy term and thus, all the remaining isomers can act as minor species at best - see Fig. 5.

The third illustrative system, Ca@C_{82}, exhibits the richest isomerism among the Ca endohedrals [147,155,157,366-368]. Shinohara *et al.* [155] isolated four isomers of Ca@C_{82} and labeled the isomers by (I), (II), (III), and (IV). Dennis and Shinohara concluded [158,370] from the ^{13}C NMR spectra of Ca@C_{82}(III) its symmetry as C_2. The ultraviolet photoelectron spectra measured by Hino *et al.* [159] support the finding; a similarity with Tm@C_{82}(II) was also noted [167]. Recently, Achiba *et al.* [160] measured the ^{13}C NMR spectra of the all four isomers and assigned the symmetry of isomers (I), (II), (III), and (IV) as C_s, C_{3v}, C_2, and C_{2v}, respectively. The Ca@C_{82} structures and energetics were also computed [156] and the C_{2v} structure found as the lowest-energy isomer. There were still three other low energy species - C_s, C_2, and C_{3v}. Fig. 6 presents the structures for nine isomers of Ca@C_{82} optimized [161] at the B3LYP/3-21G∼dz level. There are some second-order symmetry reductions present: C_{3v}(b) $\rightarrow$ C_s, C_{2v} $\rightarrow$ C_s, C_2(a) $\rightarrow$ C_1, C_2(b) $\rightarrow$ C_1, C_s(b) $\rightarrow$ C_1.

The relative-stability treatment according to eq. (1) can be further improved by an alternative approach to description of the encapsulate motions (i.e., in place of the RRHO approach considered so far). One can expect that if the encapsulate is relatively free to move within the cage then, at sufficiently high temperatures, its behavior in different cages will bring about the same contribution to the partition functions. The contributions would

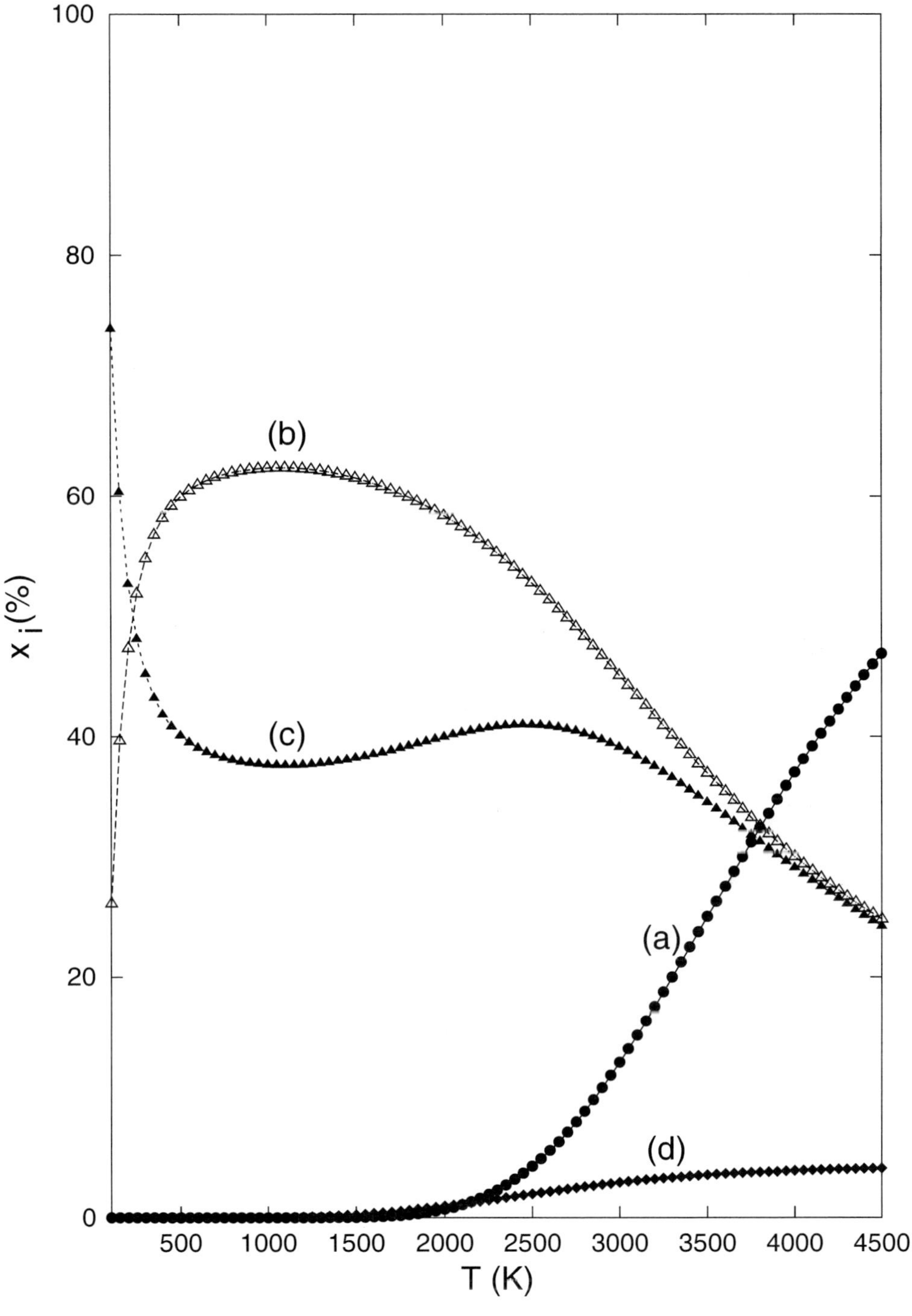

Figure 4. Relative concentrations [150] of the Ca@C_{72} isomers based on the B3LYP/6-31G* energetics and the B3LYP/3-21G∼dz entropy.

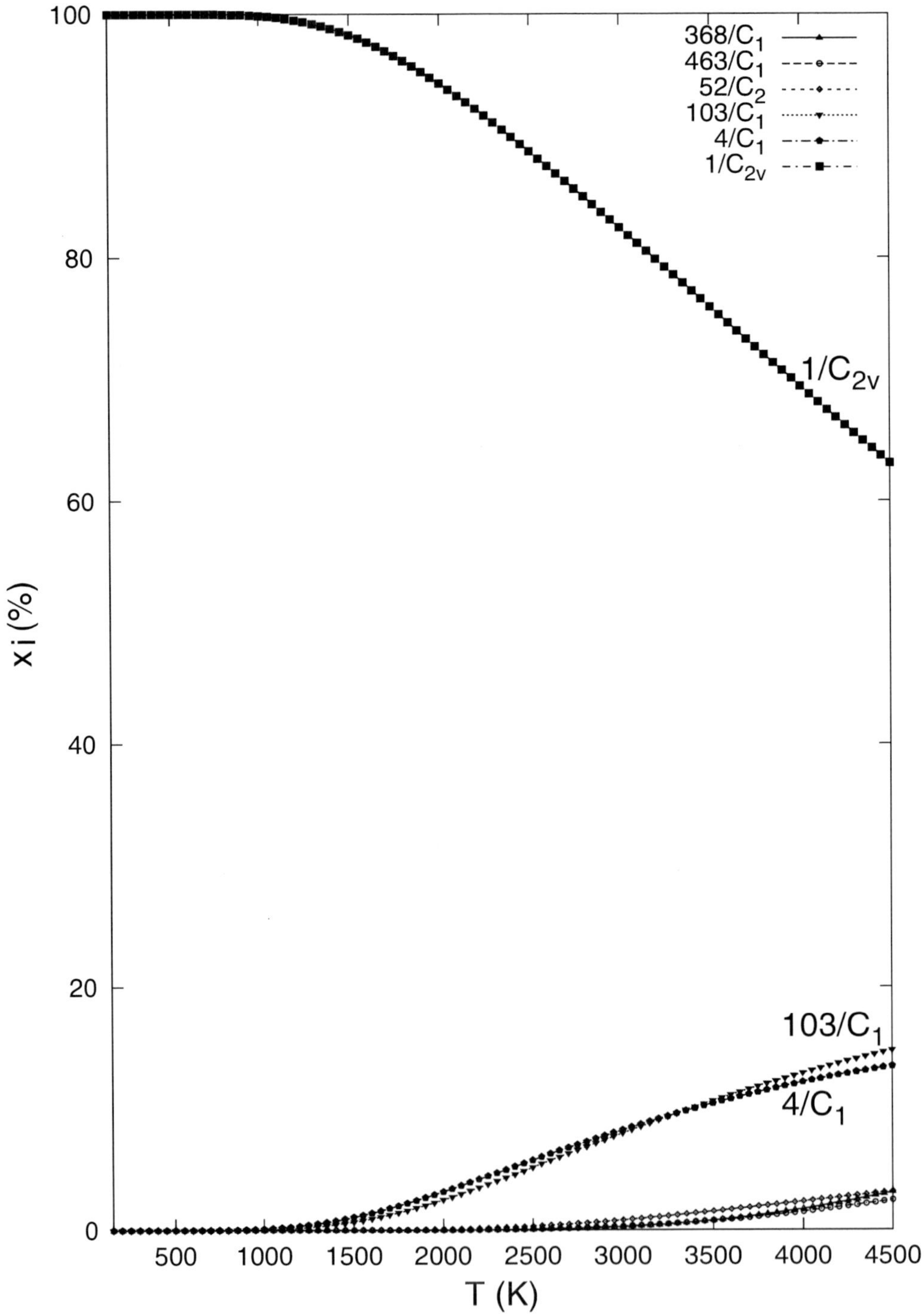

Figure 5. Relative concentrations [154] of the Ca@C_{74} isomers based on the B3LYP/6-31G* energetics and the B3LYP/3-21G∼dz entropy.

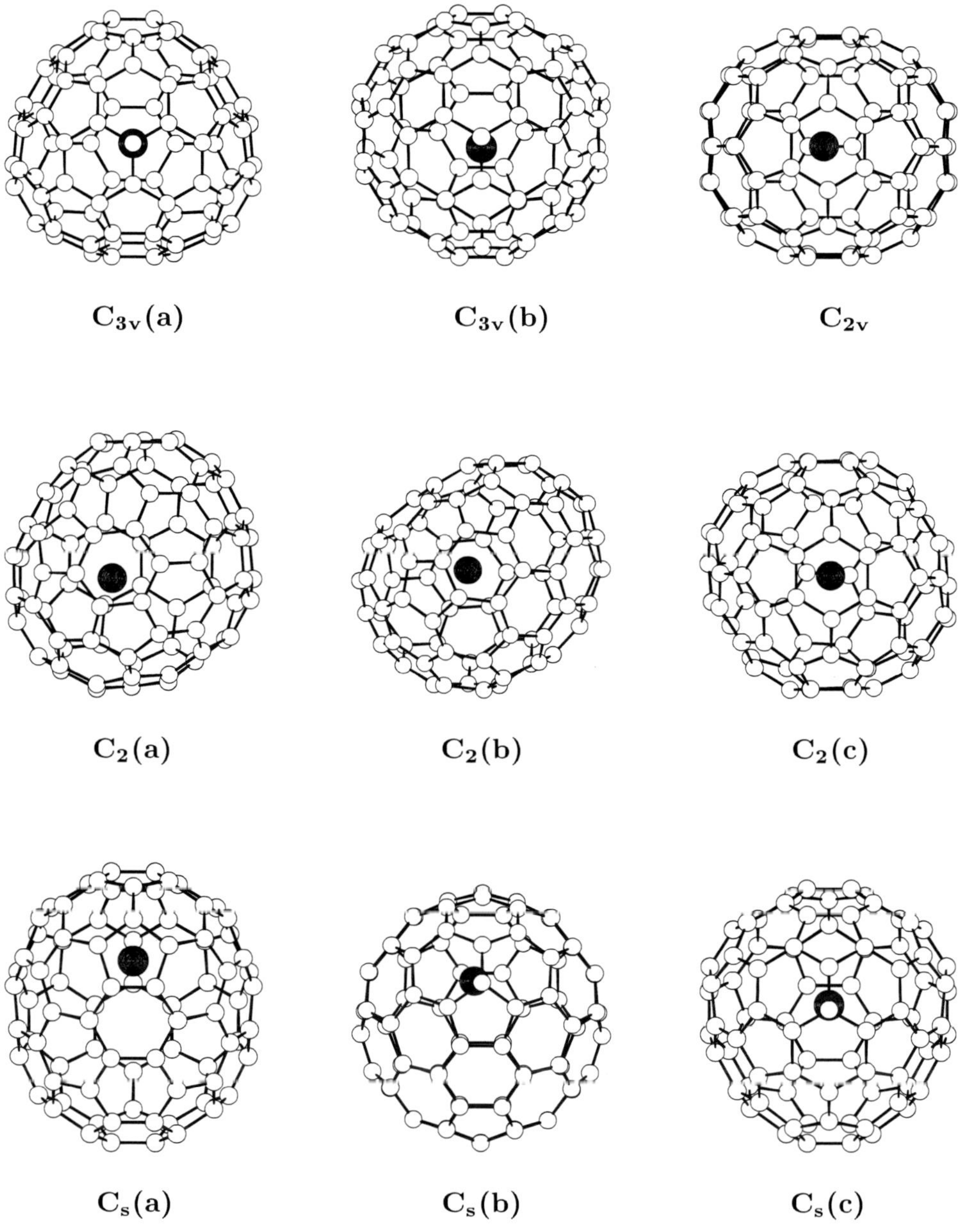

Figure 6. B3LYP/3-21G~dz optimized structures [161] of Ca@C_{82} isomers.

than cancel out in eq. (1). This simplification can be called free, fluctuating, or floating encapsulate model (FEM). In the FEM model, in addition to the removal of the three lowest vibrational frequencies, the symmetries of the cages should be treated as the highest possible, considering the averaging effect of the large amplitude motions of the encapsulate. Fig. 7 presents the temperature development [161] of the relative concentrations of the nine $Ca@C_{82}$ isomers in a wide temperature interval for the RRHO and FEM treatments (the enthalpy part from the B3LYP/6-31G*//B3LYP/3-21G$\sim$dz calculations, the entropy part evaluated at the B3LYP/3-21G$\sim$dz level). The FEM approach is in a better agreement with the available observed concentration information. The observed yields [160] of the isomers were nearly equal except for the considerably less produced C_{3v} species (though the HPLC chromatograms [155] could indicate somewhat larger production differences). The FEM treatment basically reproduces the four isomers, placing the C_{3v}(b) species as the fourth most populated. In the RRHO evaluation, there is also a competing fifth structure, and the C_{3v}(b) cage becomes the third most populated isomer at very high temperatures.

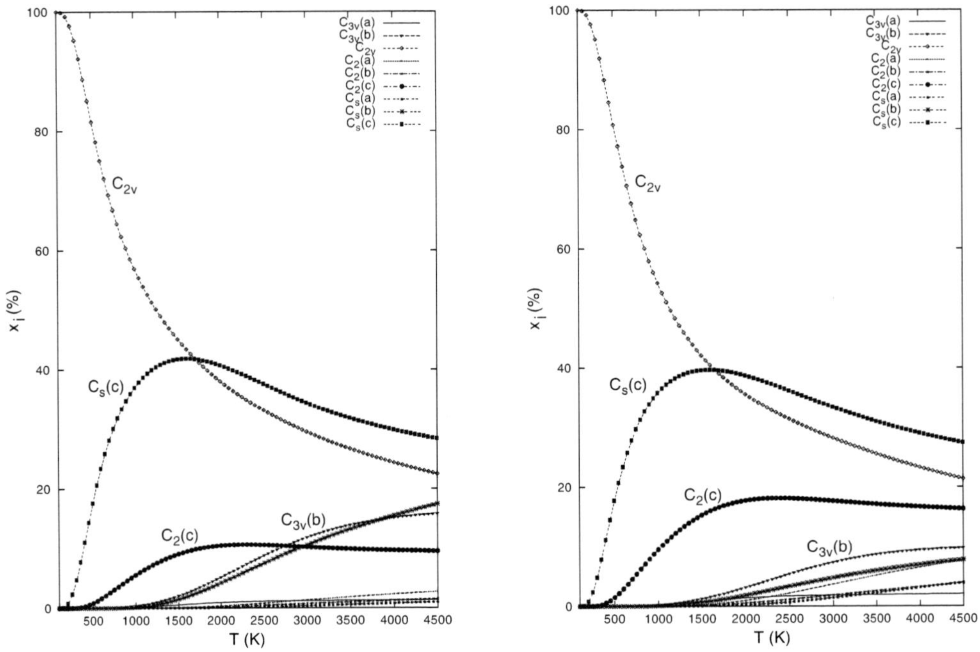

Figure 7. Relative concentrations [161] of the $Ca@C_{82}$ isomers based on the B3LYP/6-31G* energetics and the B3LYP/3-21G$\sim$dz entropy, using the RRHO (left) and FEM (right) treatment.

The last illustrative example deals with $La@C_{82}$, i.e., an electronic open-shell system. The $La@C_{82}$ metallofullerene is one of the very first endohedrals that was macroscopically produced [371]. Recently, structures of two its isomers were clarified [163,164] using ^{13}C NMR spectra of their monoanions generated electrochemically. The major isomer [163] was thus assigned C_{2v} symmetry and the minor species [164] C_s. The C_{2v} structure was moreover confirmed by an X-ray powder diffraction study [372]. The findings stand in a contrast to $Ca@C_{82}$ with four known isomers. Computations at *ab initio* HF and DFT levels pointed out [121,162,373] just three IPR cages with a sufficiently low energy after La atom encapsulation: C_{2v}, C_{3v}(b), and C_s(c). The fourth lowest La endohedral species, C_2(a), is actually already rather high in energy to be significant in experiment. A partial agreement with experiment can be reached [374] for temperatures roughly from 1000 to 1300 K using the RRHO treatment, however, the FEM approximation brings a further improvement. Fig. 8 shows the results for the La@C_{82} isomeric system evaluated just within the FEM approach. A good agreement with the observed facts [163,164] is indeed reached with the latter treatment. There is still one important aspect to be considered. The fullerene and metallofullerene production is not always close to the inter-isomeric equilibrium. This factor may be pertinent to the La@C_{82} case. Lian *et al.* [375] reported a Ni-catalyzed production of La@C_{82} with a considerably variable isomeric ratio, indicating a possible kinetic control.

There is a general stability problem related to fullerenes and metallofullerenes - relative stabilities of clusters with different stoichiometries. Let us consider a metallofullerene formation:

$$\mathrm{X(g) + C_n(g) = X@C_n(g)}. \tag{17}$$

The encapsulation process is thermodynamically characterized by the standard changes of, for example, enthalpy $\Delta H^o_{X@C_n}$ or the Gibbs energy $\Delta G^o_{X@C_n}$. An illustration will be given on a series Ca@C_{74}, Sr@C_{74}, and Ba@C_{74} with the potential-energy changes computed at the B3LYP/6-31G*~dz level and the entropy part at the B3LYP/3-21G~dz level. The equilibrium composition of the reaction mixture is controlled by the encapsulation equilibrium constants $K_{X@C_n,p}$:

$$K_{X@C_n,p} = \frac{p_{\mathrm{X@C_n}}}{p_{\mathrm{X}}p_{\mathrm{C_n}}}, \tag{18}$$

expressed in the terms of partial pressures of the components. The encapsulation equilibrium constant is interrelated with the the standard encapsulation Gibbs energy change:

$$\Delta G^o_{X@C_n} = -RTlnK_{X@C_n,p}. \tag{19}$$

Temperature dependency of the encapsulation equilibrium constant $K_{X@C_n,p}$ is then described by the van't Hoff equation:

$$\frac{dlnK_{X@C_n,p}}{dT} = \frac{\Delta H^o_{X@C_n}}{RT^2} \tag{20}$$

where the $\Delta H^o_{X@C_n}$ term is typically negative so that the encapsulation equilibrium constants decrease with increasing temperature. Let us further suppose that the metal pressure p_{X} is actually close to the respective saturated pressure $p_{X,sat}$ so that we can deal with a special case of clustering under saturation conditions [376]. While the saturated pressures $p_{X,sat}$ for various metals are known from observations [377], the partial pressure of $\mathrm{C_n}$ is less clear (though, $p_{\mathrm{C_n}}$ should exhibit a temperature maximum and then vanish). As already mentioned, the computed equilibrium constants $K_{X@C_n,p}$ have to show a temperature decrease with respect to the van't Hoff equation (20). However, if we consider the combined $p_{X,sat}K_{X@C_n,p}$ term:

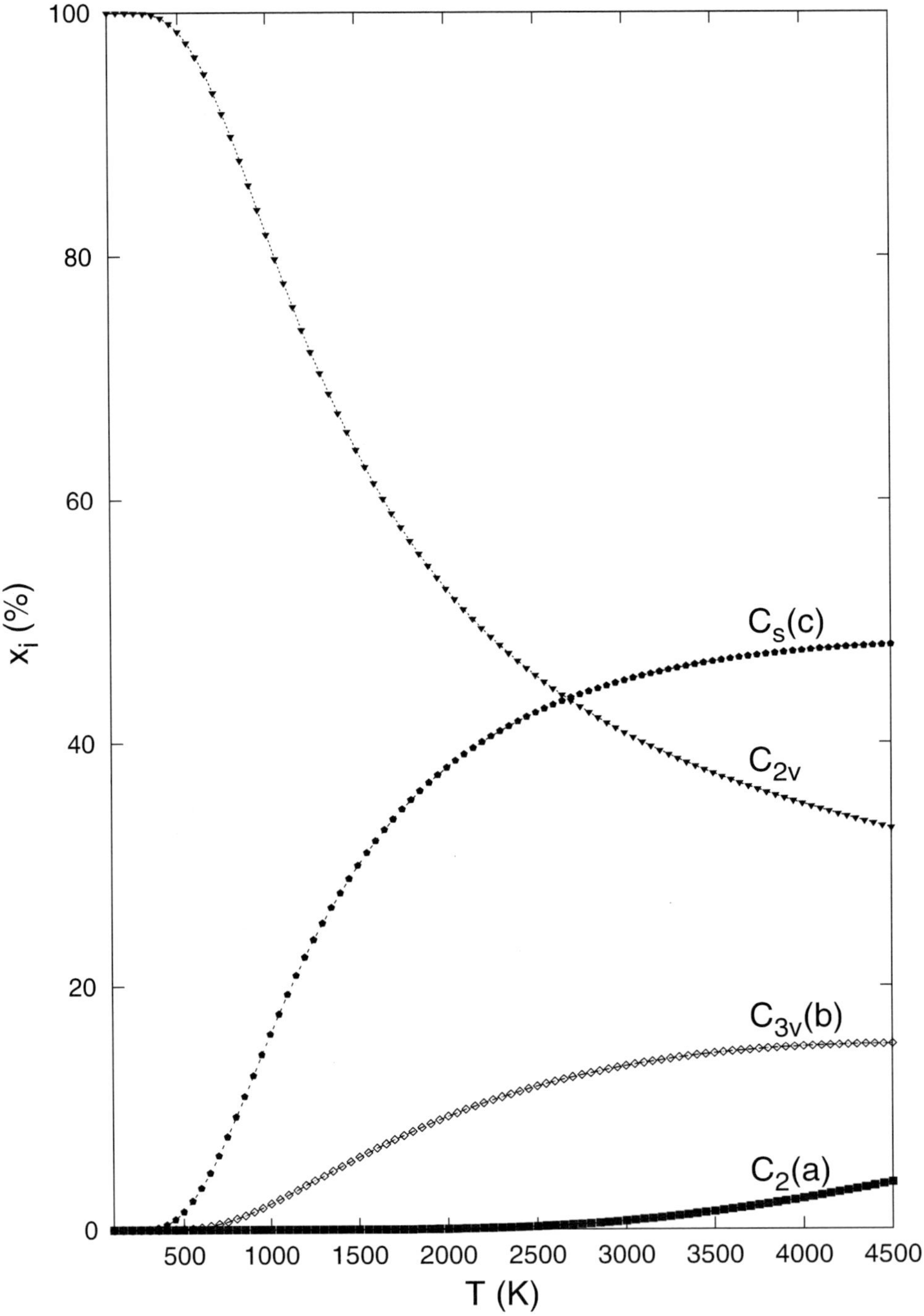

Figure 8. Relative concentrations [374] of the La@C_{82} isomers derived within the FEM approach.

$$p_{X@C_n} \sim p_{X,sat}K_{X@C_n,p}, \tag{21}$$

that directly controls the partial pressures of various X@C$_n$ encapsulates in an endohedral series (based on one common C_n fullerene), we get a different picture. The considered $p_{X,sat}K_{X@C_n,p}$ term can frequently (though not necessarily) be increasing with temperature so that a temperature enhancement of metallofullerene formation in the electric-arc technique is still possible. In fact, while for Ca@C$_{74}$ the $p_{X,sat}K_{X@C_{74},p}$ quotient increases with temperature, it is about constant for Sr@C$_{74}$, and it decreases with temperature for Ba@C$_{74}$. The behavior results from competition between the decreasing encapsulation equilibrium constants and increasing saturated metal pressures. In order to allow for cancellation of various factors introduced by the computational approximations involved, it is better to deal with the relative quotient $\frac{p_{X,sat}K_{X@C_{74},p}}{p_{Ba,sat}K_{Ba@C_{74},p}}$. The computed values of the relative quotient in the series Ca@C$_{74}$, Sr@C$_{74}$, and Ba@C$_{74}$ at a temperature of 1500 K are 4.3×10^{-5}, 5.3×10^{-3}, 1.0, and at 2000 K they read 9.9×10^{-4}, 0.030, 1.0. In principle, an endohedral with lower value of the encapsulation equilibrium constant can still be produced in larger yields if a convenient over-compensation by higher saturated metal pressure can take place. Although the energy terms are likely still not precise enough, their errors could be comparable in the series and thus, they should cancel out in the relative term $\frac{p_{X,sat}K_{X@C_{74},p}}{p_{Ba,sat}K_{Ba@C_{74},p}}$. A similar cancellation should also operate for the higher corrections to the RRHO partition functions, including motions of the encapsulate. Incidentally, the computed stability proportions do correlate with qualitative abundances known from observations. For Ba@C$_{74}$ even microsrystals could be prepared [378] so that a diffraction study was possible, while for Sr@C$_{74}$ at least various spectra could be recorded [379] in solution, and Ca@C$_{74}$ was studied [153] only by NMR spectroscopy. Under some experimental arrangements, under-saturated or perhaps super-saturated metal vapors are also possible. This reservation is applicable not only to the electric-arc treatment but even more likely to newly introduced ion-bombardment production technique [380,381]. Still, eqs. (18) and (21) remain valid, however, the metal pressure has to be described by the values actually relevant. Thus, the saturation regime can give some insight into possible production yields.

Conclusion

This survey depicts interactions between theory and experiment in fullerene research. For fullerenes and metallofullerenes the mutual interaction is indeed essential, given the complexity of the systems. The considerable thermal effects on the relative isomeric populations revealed by the quantum-chemical and statistical-mechanical computations result from a complex interplay between rotational, vibrational, electronic, potential-energy terms, symmetry and chirality factors, etc. Such effects would never be seen if only energetics is considered and entropy neglected. The treatment is however built upon the presumption of the inter-isomeric thermodynamic equilibrium. We do not know yet to which degree this presumption is satisfied in reality though there are a few indications that the equilibrium is not necessarily always established. We can only acknowledge that the thermodynamic-equilibrium treatment has already produced a reasonable computation-observation agreement for the isomeric systems from C_{76} till C_{96}, with no serious failure, and also for a few first metallofullerenes. This relatively large tested set supports the belief in a still wider applicability of the equilibrium treatment. There are also other aspects of the combined treatment that should further be studied - especially anharmonicity of vibrations, motions of encapsulates in cages and reliability of inter-isomeric energetics. The non-IPR structures have not been sufficiently explored yet. Heterofullerenes, like cages based on boron nitride, should also attract more attention. Once the thermodynamic treatment is completely exposed, the interest will gradually shift towards complex kinetic schemes and to fullerene sets with variable stoichiometry. The steadily growing family of fullerene and metallofullerene cages will further attract interest of more applied branches of nanoscience and nanotechnology [382], including quantum computing [383], superconductivity [384], and molecular medicine [385].

Acknowledgements

The reported research has been supported by a Grant-in-aid for NAREGI Nanoscience Project, for Scientific Research on Priority Area (A), and for the Next Generation Super Computing Project, Nanoscience Program, MEXT, Japan, by the National Science Council, Taiwan-ROC, and by the Czech National Research Program 'Information Society' (Czech Acad. Sci. 1ET401110505). Initial phase of the research line was supported by the Alexander von Humboldt-Stiftung and the Max-Planck-Institut für Chemie (Otto-Hahn-Institut). The authors also wish to thank the following organizations for kindly permitting the reprinting of copyrighted material: American Institute of Physics; Elsevier Scientific Publishing Company.

References

[1] H. W. Kroto, J. R. Heath, S. C. O'Brien, R. F. Curl and R. E. Smalley, *Nature* 318 (1985) 162.

[2] J. R. Heath, S. C. O'Brien, Q. Zhang, Y. Liu, R. F. Curl, H. W. Kroto, F. K. Tittel and R. E. Smalley, *Am. Chem. Soc.* 107 (1985) 7779.

[3] W. Krätschmer, L. D. Lamb, K. Fostiropoulos and D. R. Huffman, *Nature* 347 (1990) 354.

[4] *Supercarbon: Synthesis, Properties and Applications* , S. Yoshimura and R. P. H. Chang, Eds., Springer-Verlag, Berlin, 1998.

[5] *Fullerenes and Related Structures*, A. Hirsch, Ed., Springer-Verlag, Berlin, 1999.

[6] *Carbon Nanotubes: Synthesis, Structure, Properties and Applications,* M. S. Dresselhaus, G. Dresselhaus and Ph. Avouris, Eds., Springer-Verlag, Berlin, 2001.

[7] *Nanostructured Carbon for Advanced Applications,* G. Benedek, P. Milani and V. G. Ralchenko, Eds., Kluwer Academic Publishers, Dordrecht, 2001.

[8] M. Buhl and A. Hirsch, *Chem. Rev.* 101 (2001) 1153.

[9] S. Nagase, K. Kobayashi and T. Akasaka, *Bull. Chem. Soc. Jpn.* 69 (1996) 2131.

[10] S. Nagase, K. Kobayashi, T. Akasaka and T. Wakahara, In Fullerenes: *Chemistry, Physics, and Technology*, K. M. Kadish and R. S. Ruoff, Eds., J. Wiley, New York, 2000, p. 395.

[11] S. Iijima, *Nature* 354 (1991) 56

[12] M. H. Ge and K. Sattler, *Chem. Phys. Lett.* 220 (1994) 192.

[13] B. W. Smith, M. Monthioux and D. E. Luzzi, *Nature* 396 (1998) 323.

[14] S. J. Tans, A. R. M. Verschueren and C. Dekker, *Nature* 393 (1998) 49.

[15] M. S. Dresselhaus, G. Dresselhaus, and P. C. Eklund, *Science of Fullerenes and Carbon Nanotubes*, Academic Press, San Diego, CA, 1996.

[16] J. Cioslowski, *Electronic Structure Calculations on Fullerenes and Their Derivatives,* Oxford University Press, Oxford, UK, 1995.

[17] G. E. Scuseria, in *Modern Electronic Structure Theory,* Part I, D. R. Yarkony, Ed. World Scientific, Singapore, Dingapore, 1995, p. 279.

[18] K. Jinno and Y. Saito, *Advan. Chromat.* 36 (1996) 65.

[19] Y. Achiba, K. Kikuchi, Y. Aihara, T. Wakabayashi, Y. Miyake and M. Kainosho, in *The Chemical Physics of Fullerenes* 10 (and 5) Years Later, W. Andreoni, Ed., Kluwer Academic Publishers, Dordrecht, The Netherlands, 1996, p. 139.

[20] M. Saunders, R. J. Cross, H. A. Jiménez-Vázquez, R. Shimshi and A. Khong, *Science* 271 (1996) 1693.

[21] R. Taylor, in *The Chemistry of Fullerenes,* R. Taylor, Ed., World Scientific, Singapore, 1995, p. 1.

[22] K. Heinzinger and Z. Slanina, *MPG-Spiegel* No. 2 (1993) 11.

[23] J. Baggott, Perfect Symmetry. *The Accidental Discovery of Buckminsterfullerene*, Oxford University Press, Oxford, UK, 1994.

[24] H. Aldersey-Williams, *The Most Beautiful Molecule. An Adventure in Chemistry*, Aurum Press, London, UK, 1995.

[25] Z. Slanina, *Chem. Intell.* 4, No. 2. (1998) 52.

[26] Z. Slanina, *Int. J. Hist. Eth. Natur. Sci. Technol. Med. NTM* 9 (2001) 41

[27] J. Mattauch, H. Ewald, O. Hahn and F. Strassmann, *Z. Phys.* 20 (1943) 598.

[28] R. E. Honig, *J. Chem. Phys.* 22 (1954) 126.

[29] J. Drowart, R. P. Burns, G. DeMaria and M. G. Inghram, *J. Chem. Phys.* 31 (1959) 1131.

[30] H. Hintenberger, J. Franzen and K. D. Schuy, *Z. Naturforsch.* 18a (1963) 1236.

[31] K. S. Pitzer and E. Clementi, *J. Am. Chem. Soc.* 81 (1959) 4477.

[32] R. Hoffmann, *Tetrahedron* 22 (1966) 521.

[33] Z. Slanina, *Radiochem. Radioanal. Lett.* 22 (1975) 291.

[34] Z. Slanina and R. Zahradník, *J. Phys. Chem.* 81 (1977) 2252

[35] H. P. Schultz, *J. Org. Chem.* 30 (1965) 1361.

[36] D. E. H. Jones, *New Scientist* 32 (1966) 245.

[37] E. Ōsawa, *Kagaku* 25 (1970) 854; *Chem. Abstr.* 74 (1971) 75698v.

[38] D. A. Bochvar and E. G. Gal'pern, *Dokl. Akad. Nauk SSSR* 209 (1973) 610.

[39] R. A. Davidson, *Theor. Chim. Acta* 58 (1981) 193.

[40] Z. Slanina, X. Zhao, F. Uhlík, S.-L.Lee and L. Adamowicz, *Int. J. Quantum Chem.* 99 (2004) 640.

[41] K. Kobayashi and S. Nagase, in *Endofullerenes - A New Family of Carbon Clusters,* T. Akasaka and S. Nagase, Eds., Kluwer Academic Publishers, Dordrecht, The Netherlands, 2002, p. 99.

[42] Z. C. Wu, D. A. Jelski and T. F. George, *Chem. Phys. Lett.* 137 (1987) 291.

[43] V. Elser and R. C. Haddon, *Nature* 325 (1987) 792.

[44] R. E. Stanton and M. D. Newton, *J. Phys. Chem.* 92 (1988) 2141.

[45] D. E. Weeks and W. G. Harter, *Chem. Phys. Lett.* 144 (1988) 366.

[46] D. E. Weeks and W. G. Harter, *J. Chem. Phys.* 90 (1989) 4744.

[47] Z. Slanina, J. M. Rudziński, M. Togasi and E. Ōsawa, *J. Mol. Struct. (THEOCHEM),* 61 (1989) 169.

[48] D. R. Huffman and W. Krätschmer, *Materials Res. Soc. Proc.* 206 (1991) 601.

[49] T. Braun, *Angew. Chem., Int. Ed. Eng.* 31 (1992) 588.

[50] T. Braun, A. Schubert, H. Maczelka and L. Vasvári, *Fullerene Research* 1985-1993, World Scientific, Singapore, 1995.

[51] D. Bakowies and W. Thiel, *J. Am. Chem. Soc.* 113 (1991) 3704.

[52] M. Häser, J. Almlöf and G. E. Scuseria, *Chem. Phys. Lett.* 181 (1991) 497.

[53] W. Weltner, Jr. and R. J. van Zee *Chem. Rev.* 89 (1989) 1713.

[54] R. F. Curl and R. E. Smalley, *Science* 242 (1988) 1017.

[55] H. W. Kroto, *Science* 242 (1988) 1139.

[56] R. C. Haddon, *Accounts Chem. Res.* 21 (1988) 243.

[57] D. A. Jelski and T. F. George, *J. Chem. Educ.* 65 (1988) 879.

[58] D. J. Klein and T. G. Schmalz, in *Quasicrystals, Networks, and Molecules of Fivefold Symmetry*, I. Hargittai, Ed., VCH Publishers, New York, NY, 1990.

[59] E. Ōsawa, *Kagaku* 45 (1990) 552.

[60] D. R. Huffman, *Phys. Today* 44 (1991) 22.

[61] H. W. Kroto, A. W. Allaf and S. P. Balm, *Chem. Rev.* 91 (1991) 1213.

[62] H. W. Kroto, *Angew. Chem., Int. Ed. Engl.* 31 (1992) 111.

[63] J. R. Heath, in *ACS Symposium Series No 481 - Fullerenes. Synthesis, Properties, and Chemistry of Large Carbon Clusters*, G. S. Hammond and V. J. Kuck, Eds., ACS, Washington, DC, 1992.

[64] J. E. Fischer, P. A. Heiney, D. E. Luzzi and D. E. Cox, in *ACS Symposium Series No 481 - Fullerenes. Synthesis, Properties, and Chemistry of Large Carbon Clusters,* G. S. Hammond and V. J. Kuck, Eds. ACS, Washington, DC, 1992.

[65] Z. Slanina, *Chem. Listy* 86 (1992) 327.

[66] R. E. Smalley, *Accounts Chem. Res.* 25 (1992) 98.

[67] J. P. Hare and H. W. Kroto, *Accounts Chem. Res.* 25 (1992) 106.

[68] J. E. Fischer, P. A. Heiney and A. B. Smith III, *Accounts Chem. Res.* 25 (1992) 112.

[69] F. Diederich and R. L. Whetten, *Accounts Chem. Res.* 25 (1992) 119.

[70] R. C. Haddon, *Accounts Chem. Res.* 25 (1992) 127.

[71] P. J. Fagan, J. C. Calabrese and B. Malone, *Accounts Chem. Res.* 25 (1992) 134.

[72] J. H. Weaver, *Accounts Chem. Res.* 25 (1992) 143.

[73] J. M. Hawkins, *Accounts Chem. Res.* 25 (1992) 150.

[74] F. Wudl, *Accounts Chem. Res.* 25 (1992) 157.

[75] S. W. McElvany, M. M. Ross and J. H. Callahan, *Accounts Chem. Res.* 25 (1992) 162.

[76] R. D. Johnson, D. S. Bethune and C. S. Yannoni, *Accounts Chem. Res.* 25 (1992) 169.

[77] D. Koruga, S. Hameroff, J. Withers, R. Loutfy and M. Sundareshan, *Fullerene C_{60} - History, Physics, Nanobiology, Nanotechnology,* Elsevier, Amsterdam, The Netherlands, 1993.

[78] M. L. Cohen and V. H. Crespi, in *Buckminsterfullerenes*, W. E. Billups and M. A. Ciufolini, Eds., VCH Publishers, New York, NY, 1993.

[79] S. C. Erwin, in *Buckminsterfullerenes*, W. E. Billups and M. A. Ciufolini, Eds., VCH Publishers, New York, NY, 1993.

[80] M. S. Dresselhaus, G. Dresselhaus and P. C. Eklund, *J. Mater. Res.* 8 (1993) 2054.

[81] T. G. Schmalz and D. J. Klein, in *Buckminsterfullerenes*, W. E. Billups and M. A. Ciufolini, Eds., VCH Publishers, New York, NY, 1993.

[82] G. E. Scuseria, in *Buckminsterfullerenes*, W. E. Billups and M. A. Ciufolini, Eds., VCH Publishers, New York, NY, 1993.

[83] C. T. White, J. W. Mintmire, R. C. Mowrey, D. W. Brenner, D. H. Robertson, J. A. Harrison and B. I. Dunlap, in *Buckminsterfullerenes*, W. E. Billups and M. A. Ciufolini, Eds., VCH Publishers, New York, NY, 1993.

[84] R. C. Haddon and K. Raghavachari, in *Buckminsterfullerenes*, W. E. Billups and M. A. Ciufolini, Eds., VCH Publishers, New York, NY, 1993.

[85] J. Cioslowski, *Rev. Comput. Chem.* 4 (1993) 1.

[86] C. Z. Wang, B. L. Zhang, K. M. Ho and X. Q. Wang, *Int. J. Mod. Phys. B* 7 (1993) 4305.

[87] J. González, F. F. Guinea and M. A. H. Vozmediano, *Int. J. Mod. Phys. B* 7 (1993) 4331.

[88] W. Andreoni, in *Electronic Properties of Fullerenes*, H. Kuzmany, J. Fink, M. Mehring and S. Roth, Eds., Springer-Verlag, Berlin, FR Germany, 1993.

[89] Z. Slanina, S.-L. Lee and C.-H. Yu, *Rev. Comput. Chem.* 8 (1996) 1.

[90] J. Cioslowski, N. Rao and D. Moncrieff, *J. Am. Chem. Soc.* 122 (2000) 8265

[91] Y. Achiba, K. Kikuchi, Y. Aihara, T. Wakabayashi, Y. Miyake and M. Kainosho, in *Science* and Technology of Fullerene Materials, P. Bernier, D. S. Bethune, L. Y. Chiang, T. W. Ebbesen, R. M. Metzger and J. W. Mintmire, Eds., Materials Research Society, Pittsburgh, PA, 1995.

[92] Y. Achiba, *Kagaku* 52, No. 5 (1997) 15.

[93] R. Mitsumoto, H. Oji, Y. Yamamoto, K. Asato, Y. Ouchi, H. Shinohara, K. Seki, K. Umishita, S. Hino, S. Nagase, K. Kikuchi and Y. Achiba, *J. Phys. IV* 7 (1997) C2-525.

[94] H. W. Kroto, *Nature* 329 (1987) 529.

[95] T. G. Schmalz, W. A. Seitz, D. J. Klein and G. E. Hite, *J. Am. Chem. Soc.* 110 (1988) 1113

[96] D. E. Manolopoulos and P. W. Fowler, *J. Chem. Phys.* 96 (1992) 7603.

[97] P. W. Fowler and D. E. Manolopoulos, *An Atlas of Fullerenes,* Clarendon Press, Oxford, UK, 1995.

[98] M. Yoshida and E. Ōsawa, *Bull. Chem. Soc. Jpn.* 68 (1995) 2073.

[99] M. Yoshida and E. Ōsawa, *Bull. Chem. Soc. Jpn.* 68 (1995) 2083.

[100] Z. Slanina, M.-L. Sun, S.-L. Lee and L. Adamowicz, in *Recent Advances in the Chemistry and Physics of Fullerenes and Related Materials* , Vol. 2, K. M. Kadish and R. S. Ruoff, Eds., The Electrochemical Society, Pennington, NJ, 1995, p. 1138.

[101] R. Ettl, I. Chao, F. N. Diederich and R. L. Whetten, *Nature* 353 (1991) 149.

[102] D. E. Manolopoulos, *J. Chem. Soc., Faraday Trans.* 87 (1991) 2861.

[103] H. P. Cheng and R. L. Whetten, *Chem. Phys. Lett.* 197 (1992) 44.

[104] Q. Li, F. Wudl, C. Thilgen, R. L. Whetten and F. Diederich, *J. Am. Chem. Soc.* 114 (1992) 3994.

[105] S. Hino, K. Matsumoto, S. Hasegawa, H. Inokuchi, T. Morikawa, T. Takahashi, K. Seki, K. Kikuchi, S. Suzuki, I. Ikemoto and Y. Achiba, *Chem. Phys. Lett.* 197 (1992) 38.

[106] G. Orlandi, F. Zerbetto, P. W. Fowler and D. E. Manolopoulos, *Chem. Phys. Lett.* 208 (1993) 441.

[107] J. M. Hawkins and A. Meyer, *Science* 260 (1993) 1918.

[108] S. J. Austin, P. W. Fowler, G. Orlandi, D. E. Manolopoulos and F. Zerbetto, *Chem. Phys. Lett.* 226 (1994) 219.

[109] R. H. Michel, H. Schreiber, R. Gierden, F. Hennrich, J. Rockenberger, R. D. Beck, M. M. Kappes, C. Lehner, P. Adelmann and J. F. Armbruster, *Ber. Bunsenges. Phys. Chem.* 98 (1994) 975.

[110] R. H. Michel, M. M. Kappes, P. Adelmann and G. Roth, *Angew. Chem., Int. Ed. Engl.* 33 (1994) 1651.

[111] F. Diederich, R. L. Whetten, C. Thilgen, R. Ettl, I. Chao and M. M. Alvarez, *Science* 254 (1991) 1768.

[112] K. Kikuchi, N. Nakahara, T. Wakabayashi, S. Suzuki, H. Shiromaru, Y. Miyake, K. Saito, I. Ikemoto, M. Kainosho and Y. Achiba, *Nature* 357 (1992) 142.

[113] D. Bakowies, A. Geleßus and W. Thiel, *Chem. Phys. Lett.* 197 (1992) 324.

[114] Z. Slanina, J.-P. François, D. Bakowies and W. Thiel, *J. Mol. Struct. (Theochem)* 279 (1993) 213.

[115] S. J. Woo, E. Kim and Y. H. Lee, *Phys. Rev. B* 47 (1993) 6721.

[116] K. Nakao, N. Kurita and M. Fujita, *Phys. Rev. B* 49 (1994) 11415.

[117] M.-L. Sun, Z. Slanina, S.-L. Lee, F. Uhlík and L. Adamowicz, *Chem. Phys. Lett.* 246 (1995) 66.

[118] F. H. Hennrich, R. H. Michel, A. Fischer, S. Richard-Schneider, S. Gilb, M. M. Kappes, D. Fuchs, M. Bürk, K. Kobayashi and S. Nagase, Angew. Chem., *Int. Ed. Engl.* 35 (1996) 1732.

[119] C.-R. Wang, T. Sugai, T. Kai, T. Tomiyama and H. Shinohara, J. Chem. Soc., Chem. Commun. (2000) 557.

[120] K. Kikuchi, N. Nakahara, T. Wakabayashi, M. Honda, H. Matsumiya, T. Moriwaki, S. Suzuki, H. Shiromaru, K. Saito, K. Yamauchi, I. Ikemoto and Y. Achiba, *Chem. Phys. Lett.* 188 (1992) 177.

[121] S. Nagase, K. Kobayashi, T. Kato and Y. Achiba, *Chem. Phys. Lett.* 201 (1993) 475.

[122] G. Orlandi, F. Zerbetto and P. W. Fowler, *J. Phys. Chem.* 97 (1993) 13575.

[123] S. Nagase and K. Kobayashi, *Chem. Phys. Lett.* 214 (1993) 57.

[124] Z. Slanina, S.-L. Lee, K. Kobayashi and S. Nagase, *J. Mol. Struct. (Theochem)* 339 (1995) 89.

[125] M. Saunders, H. A. Jiménez-Vázquez, R. J. Cross, W. E. Billups, C. Gesenberg, A. Gonzalez, W. Luo, R. C. Haddon, F. Diederich and A. Herrmann, *J. Am. Chem. Soc.* 117 (1995) 9305.

[126] F. Diederich, R. Ettl, Y. Rubin, R. L. Whetten, R. Beck, M. Alvarez, S. Anz, D. Sensharma, F. Wudl, K. C. Khemani and A. Koch, *Science* 252 (1991) 548.

[127] R. D. Beck, P. S. John, M. M. Alvarez, F. Diederich and R. L. Whetten, *J. Phys. Chem.* 95 (1991) 8402.

[128] K. Kikuchi, N. Nakahara, M. Honda, S. Suzuki, K. Saito, H. Shiromaru, K. Yamauchi, I. Ikemoto, T. Kuramochi, S. Hino and Y. Achiba, *Chem. Lett.* (1991) 1607.

[129] K. Raghavachari and C. M. Rohlfing, *J. Phys. Chem.* 95 (1991) 5768.

[130] Z. Slanina, J.-P. François, M. Kolb, D. Bakowies and W. Thiel, *Fullerene Sci. Technol.* 1 (1993) 221.

[131] T. J. S. Dennis, T. Kai, T. Tomiyama and H. Shinohara, *Chem. Commun.* (1998) 619.

[132] E. Ōsawa, H. Ueno, M. Yoshida, Z. Slanina, X. Zhao, M. Nishiyama and H. Saito, *J. Chem. Soc., Perkin Trans.* 2 / (1998) 943.

[133] Z. Slanina, S.-L. Lee, M. Yoshida and E. Ōsawa, *Chem. Phys.* 209 (1996) 13.

[134] Z. Slanina, S.-L. Lee, M. Yoshida and E. Ōsawa, in *Physics and Chemistry of Fullerenes and Their Derivatives*, H. Kuzmany, J. Fink, M. Mehring and S. Roth, Eds., World Sci. Publ., Singapore, 1996, p. 389.

[135] Z. Slanina, S.-L. Lee, M. Yoshida and E. Ōsawa, in *Recent Advances in the Chemistry and Physics of Fullerenes and Related Materials*, Vol. 3, K. M. Kadish and R. S. Ruoff, Eds., The Electrochemical Society, Pennington, NJ, 1996, p. 967.

[136] Z. Slanina, S.-L. Lee and L. Adamowicz, *Int. J. Quantum Chem.* 63 (1997) 529.

[137] Z. Slanina, X. Zhao, S.-L. Lee and E. Ōsawa, in *Recent Advances in the Chemistry and Physics of Fullerenes and Related Materials*, Vol. 4, K. M. Kadish and R. S. Ruoff, Eds., The Electrochemical Society, Pennington, NJ, 1997, p. 680.

[138] Z. Slanina, X. Zhao, S.-L. Lee and E. Ōsawa, *Chem. Phys.* 219 (1997) 193.

[139] Z. Slanina, X. Zhao, P. Deota and E. Ōsawa, *J. Mol. Model.* 6 (2000) 312.

[140] X. Zhao, Z. Slanina, H. Goto and E. Ōsawa, *J. Chem. Phys.* 118 (2003) 10534.

[141] X. Zhao and Z. Slanina, in Fullerenes, Vol. 12: The Exciting World of Nanocages and Nanotubes, P. Kamat, D. Guldi and K. Kadish, Eds., The Electrochemical Society, Pennington, NJ, 2002, p. 679.

[142] X. Zhao and Z. Slanina, *J. Mol. Struct. (Theochem)* 636 (2003) 195.

[143] X. Zhao, Z. Slanina, M. Ozawa and E. Ōsawa, P. Deota, K. Tanabe, *Fullerene Sci. Technol.* 8 (2000) 595.

[144] Z. Slanina, X. Zhao and E. Ōsawa, *Chem. Phys. Let.* 290 (1998) 311.

[145] Z. Slanina, F. Uhlík, X. Zhao and E. Ōsawa, *J. Chem. Phys.* 113 (2000) 4933.

[146] S. A. Varganov, P. V. Avramov, S. G. Ovchinnikov and M. S. Gordon, *Chem. Phys. Lett.* 362 (2002) 380.

[147] T. S. M. Wan, H. W. Zhang, T. Nakane, Z. D. Xu, M. Inakuma, H. Shinohara, K. Kobayashi and S. Nagase, *J. Am. Chem. Soc.* 120 (1998) 6806.

[148] K. Kobayashi, S. Nagase, M. Yoshida and E. Ōsawa, *J. Am. Chem. Soc.* 119 (1997) 12693.

[149] S. Nagase, K. Kobayashi and T. Akasaka, *J. Mol. Struct. (Theochem)* 462 (1999) 97.

[150] Z. Slanina, K. Kobayashi and S. Nagase, *Chem. Phys. Lett.* 372 (2003) 810.

[151] Z. Slanina, X. Zhao, X. Grabuleda, M. Ozawa, F. Uhlík, P. M. Ivanov, K. Kobayashi and S. Nagase, *J. Mol. Graphics Mod.* 19 (2001) 252.

[152] Z. Slanina, F. Uhlík, L. Adamowicz, K. Kobayashi and S. Nagase, *Int. J. Quantum Chem.* 100 (2004) 610.

[153] T. Kodama, R. Fujii, Y. Miyake, S. Suzuki, H. Nishikawa, I. Ikemoto, K. Kikuchi and Y. Achiba, in Fullerenes, Vol. 13: *Fullerenes and Nanotubes: The Building Blocks of Next Generation Nanodevices*, D. M. Guldi, P. V. Kamat, and F. D'Souza, Eds., The Electrochemical Society, Pennington, NJ, 2003, p. 548.

[154] Z. Slanina, K. Kobayashi and S. Nagase, *Chem. Phys.* 301 (2004) 153.

[155] Z. D. Xu, T. Nakane and H. Shinohara, *J. Am. Chem. Soc.* 118 (1996) 11309.

[156] K. Kobayashi and S. Nagase, *Chem. Phys. Lett.* 274 (1997) 226.

[157] T. Kimura, T. Sugai and H. Shinohara, *Int. J. Mass Spectrom.* 188 (1999) 225.

[158] T. J. S. Dennis and H. Shinohara, *Appl. Phys. A* 66 (1998) 243.

[159] S. Hino, K. Umishita, K. Iwasaki, M. Aoki, K. Kobayashi, S. Nagase, T. J. S. Dennis, T. Nakane and H. Shinohara, *Chem. Phys. Lett.* 337 (2001) 65.

[160] T. Kodama, R. Fujii, Y. Miyake, K. Sakaguchi, H. Nishikawa, I. Ikemoto, K. Kikuchi and Y. Achiba, *Chem. Phys. Lett.* 377 (2003) 197.

[161] Z. Slanina, K. Kobayashi and S. Nagase, *J. Chem. Phys.* 120 (2004) 3397.

[162] K. Kobayashi and S. Nagase, *Chem. Phys. Lett.* 282 (1998) 325.

[163] T. Akasaka, T. Wakahara, S. Nagase, K. Kobayashi, M. Waelchli, K. Yamamoto, M. Kondo, S. Shirakura, S. Okubo, Y. Maeda, T. Kato, M. Kako, Y. Nakadaira, R. Nagahata, X. Gao, E. van Caemelbecke and K. M. Kadish, *J. Am. Chem. Soc.* 122 (2000) 9316.

[164] T. Akasaka, T. Wakahara, S. Nagase, K. Kobayashi, M. Waelchli, K. Yamamoto, M. Kondo, S. Shirakura, Y. Maeda, T. Kato, M. Kako, Y. Nakadaira, X. Gao, E. van Caemelbecke and K. M. Kadish, *J. Phys. Chem.* B 105 (2001) 2971.

[165] T. Wakahara, S. Okubo, M. Kondo, Y. Maeda, T. Akasaka, M. Waelchli, M. Kako, K. Kobayashi, S. Nagase, T. Kato, K. Yamamoto, X. Gao, E. van Caemelbecke and K. M. Kadish, *Chem. Phys. Lett.* 360 (2002) 235.

[166] K. Kobayashi and S. Nagase, *Mol. Phys.* 101 (2003) 249.

[167] T. Kodama, N. Ozawa, Y. Miyake, K. Sakaguchi, H. Nishikawa, I. Ikemoto, K. Kikuchi and Y. Achiba, *J. Am. Chem. Soc.* 124 (2002) 1452.

[168] B. Cao, M. Hasegawa, K. Okada, T. Tomiyama, T. Okazaki, K. Suenaga and H. Shinohara, *J. Am. Chem. Soc.* 123 (2001) 9679.

[169] S. Nagase and K. Kobayashi, *Chem. Phys. Lett.* 276 (1997) 55.

[170] B. Cao, K. Suenaga, T. Okazaki and H. Shinohara, *J. Phys. Chem.* B 106 (2002) 9295.

[171] C. R. Wang, P. Georgi, L. Dunsch, T. Kai, T. Tomiyama and H. Shinohara, *Curr. Appl. Phys.* 2 (2002) 141.

[172] Y. K. Kwon, D. Tomanek and S. Iijima, *Phys. Rev. Lett.* 82 (1999) 1470.

[173] Z. Slanina, K. Kobayashi and S. Nagase, *Chem. Phys. Lett.* 382 (2003) 211.

[174] Z. Slanina and F. Uhlík, *Chem. Phys. Lett.* 374 (2003) 100.

[175] Z. Slanina, F. Uhlík and L. Adamowicz, *Fullerene Nanotub. Carb. Nanostruct.* 11 (2003) 219.

[176] E. Dietel, A. Hirsch, B. Pietzak, M. Waiblinger, K. Lips, A. Weidinger, A. Gruss and K. P. Dinse, *J. Am. Chem. Soc.* 121 (1999) 2432.

[177] T. Clark, A Handbook of Computational Chemistry. *A Practical Guide to Chemical Structure and Energy Calculations,* J. Wiley, New York, NY, 1985.

[178] D. B. Boyd, *Rev. Comput. Chem.* 1 (1990) 321.

[179] J. J. P. Stewart, *Rev. Comput. Chem.* 1 (1990) 45.

[180] M. C. Zerner, *Rev. Comput. Chem.* 2 (1991) 313.

[181] W. J. Hehre, L. Radom, P. R. von Schleyer and J. A. Pople, Ab Initio Molecular Orbital Theory, J. Wiley, New York, NY, 1986.

[182] M. J. S. Dewar and W. Thiel, *J. Am. Chem. Soc.* 99 (1997) 4899.

[183] M. J. S. Dewar, E. G. Zoebisch, E. F. Healy and J. J. P. Stewart, *J. Am. Chem. Soc.* 107 (1985) 3902.

[184] J. J. P. Stewart, *J. Comput. Chem.* 10 (1989) 209.

[185] M. J. S. Dewar, C. Jie and J. Yu, *Tetrahedron* 49 (1993) 5003.

[186] J. J. P. Stewart, MOPAC 5.0, *QCPE* 455, Indiana University, IN, 1990.

[187] J. J. P. Stewart, MOPAC 2002, Fujitsu Ltd., Tokyo, Japan, 1999.

[188] AMPAC 6.0, Semichem, Shavnee, KS, 1997.

[189] M. J. Frisch, G. W. Trucks, H. B. Schlegel, G. E. Scuseria, M. A. Robb, J. R. Cheeseman, V. G. Zakrzewski, J. A. Montgomery, Jr., R. E. Stratmann, J. C. Burant, S. Dapprich, J. M. Millam, A. D. Daniels, K. N. Kudin, M. C. Strain, O. Farkas, J. Tomasi, V. Barone, M. Cossi, R. Cammi, B. Mennucci, C. Pomelli, C. Adamo, S. Clifford, J. Ochterski, G. A. Petersson, P. Y. Ayala, Q. Cui, K. Morokuma, D. K. Malick, A. D. Rabuck, K. Raghavachari, J. B. Foresman, J. Cioslowski, J. V. Ortiz, B. B. Stefanov, G. Liu, A. Liashenko, P. Piskorz, I. Komaromi, R. Gomperts, R. L. Martin, D. J. Fox, T. Keith, M. A. Al-Laham, C. Y. Peng, A. Nanayakkara, C. Gonzalez, M. Challacombe, P. M. W. Gill, B. Johnson, W. Chen, M. W. Wong, J. L. Andres, C. Gonzalez, M. Head-Gordon, E. S. Replogle and J. A. Pople, *Gaussian* 98, Revision A.11.1, Gaussian, Inc., Pittsburgh, PA, 1998.

[190] M. J. Frisch, G. W. Trucks, H. B. Schlegel, G. E. Scuseria, M. A. Robb, J. R. Cheeseman, J. A. Montgomery, Jr., T. Vreven, K. N. Kudin, J. C. Burant, J. M. Millam, S. S. Iyengar, J. Tomasi, V. Barone, B. Mennucci, M. Cossi, G. Scalmani, N. Rega, G. A. Petersson, H. Nakatsuji, M. Hada, M. Ehara, K. Toyota, R. Fukuda, J. Hasegawa, M. Ishida, T. Nakajima, Y. Honda, O. Kitao, H. Nakai, M. Klene, X. Li, J. E. Knox, H. P. Hratchian, J. B. Cross, C. Adamo, J. Jaramillo, R. Gomperts, R. E. Stratmann, O. Yazyev, A. J. Austin, R. Cammi, C. Pomelli, J. W. Ochterski, P. Y. Ayala, K. Morokuma, G. A. Voth, P. Salvador, J. J. Dannenberg, V. G. Zakrzewski, S. Dapprich, A. D. Daniels, M. C. Strain, O. Farkas, D. K. Malick, A. D. Rabuck, K. Raghavachari, J. B. Foresman, J. V. Ortiz, Q. Cui, A. G. Baboul, S. Clifford, J. Cioslowski, B. B. Stefanov, G. Liu, A. Liashenko, P. Piskorz, I. Komaromi, R. L. Martin, D. J. Fox, T. Keith, M. A. Al-Laham, C. Y. Peng, A. Nanayakkara, M. Challacombe, P. M. W. Gill, B. Johnson, W. Chen, M. W. Wong, C. Gonzalez and J. A. Pople, *Gaussian* 03, Revision C.01, Gaussian, Inc., Pittsburgh, PA, 2003.

[191] W. J. Hehre, L. D. Burke and A. J. Schusterman, Spartan, *Wavefunction*, Inc., Irvine, CA, 1993.

[192] H. B. Schlegel and J. J. W. McDouall, in *Computational Advances in Organic Chemistry*, C. Ögretir and I. G. Csizmadia, Eds., Kluwer Academic Publishers, Dordrecht, The Netherlands, 1991.

[193] F. Jensen, *Introduction to Computational Chemistry*, J. Wiley, Chichester, UK, 1999.

[194] Z. Slanina, F. Uhlík and M. C. Zerner, *Rev. Roum. Chim.* 36 (1991) 965.

[195] Z. Slanina, *Int. Rev. Phys. Chem.* 6 (1987) 251.

[196] Z. Slanina, *Comput. Chem.* 13 (1989) 305.

[197] Z. Slanina, *Theor. Chim. Acta* 83 (1992) 257.

[198] Z. Slanina, *Contemporary Theory of Chemical Isomerism*, Kluwer, Dordrecht, The Netherlands, 1986.

[199] Z. Slanina, F. Uhlík, J.-P. François and E. Ōsawa, *Croat. Chem. Acta* 73 (2000) 1047.

[200] H. S. M. Coxeter, *Regular Polytopes,* Macmillan, New York, NY, 1963.

[201] H. S. M. Coxeter, *Introduction to Geometry,* J. Wiley, New York, NY, 1969.

[202] Z. Slanina, *Chem. Intell.* 4, No. 2 (1998) 52.

[203] R. F. Curl, *Phil. Trans. Roy. Soc. London A* 343 (1993) 19.

[204] G. E. Scuseria, Unpublished results (1993).

[205] M.-L. Sun, Z. Slanina and S.-L. Lee, *Chem. Phys. Lett.* 233 (1995) 279.

[206] C. H. Xu and G. E. Scuseria, *Chem. Phys. Lett.* 262 (1996) 219.

[207] Z. Slanina and E. Ōsawa, *Fullerene Sci. Technol.* 5 (1997) 167.

[208] J. M. Rudziński, Z. Slanina, M. Togasi, E. Ōsawa and T. Iizuka, T. *Thermochim. Acta* 125 (1988) 155.

[209] Z. Slanina, J. M. Rudziński and E. Ōsawa, Carbon 25 (1987) 747.

[210] Z. Slanina, J. M. Rudziński and E. Ōsawa, *Collect. Czech. Chem. Commun.* 52 (1987) 2381.

[211] Z. Slanina, J. M. Rudziński, M. Togasi and E. Ōsawa, *Thermochim. Acta* 140 (1989) 87.

[212] Z. Slanina, X. Zhao, N. Kurita, H. Gotoh, F. Uhlík, J. M. Rudziński, K. H. Lee and L. Adamowicz, *J. Mol. Graphics Mod.* 19 (2001) 216.

[213] T. Kiyobayashi and M. Sakiyama, *Fullerene Sci. Technol.* 1 (1993) 269.

[214] Z. Slanina, *Chem. Phys. Lett.* 145 (1987) 512.

[215] Z. Slanina, L. Adamowicz, D. Bakowies and W. Thiel, *Thermochim. Acta* 202 (1992) 249.

[216] Z. Slanina and L. Adamowicz, *Thermochim. Acta* 205 (1992) 299.

[217] Z. Slanina, *Z. Phys. Chem.* 217 (2003) 1119.

[218] Z. Slanina, *Chem. Phys.* 150 (1991) 321.

[219] P. S. Dardi and J. S. Dahler, *J. Chem. Phys.* 93 (1990) 3562.

[220] Z. Slanina, *Thermochim. Acta.* 207 (1992) 9.

[221] Z. Slanina, *Thermochim. Acta.* 209 (1992) 1.

[222] K. Raghavachari, R. A. Whiteside and J. A. Pople, *J. Chem. Phys.* 85 (1986) 6623.

[223] K. Raghavachari and J. S. Binkley, *J. Chem. Phys.* 87 (1987) 2191.

[224] J. M. L. Martin, J. P. François and R. Gijbels, *J. Chem. Phys.* 90 (1989) 3403.

[225] V. Parasuk and J. Almlöf, *J. Chem. Phys.* 91 (1989) 1137.

[226] C. Liang and H. F. Schaefer III, *J. Chem. Phys.* 93 (1990) 8844.

[227] J. M. L. Martin, J. P. François and R. Gijbels, *J. Chem. Phys.* 93 (1990) 8850.

[228] J. Kurtz and L. Adamowicz, *Astrophys. J.* 370 (1991) 784.

[229] J. M. L. Martin, J. P. François and R. Gijbels, *J. Chem. Phys.* 94 (1991) 3753.

[230] J. M. L. Martin, J. P. François and R. Gijbels, *J. Comput. Chem.* 12 (1991) 52.

[231] D. W. Ewing, *Z. Phys. D* 19 (1991) 419.

[232] Z. Slanina, J. M. Rudziński and E. Ōsawa, *Z. Phys. D* 19 (1991) 431.

[233] J. M. L. Martin, J. P. François, R. Gijbels and J. Almlöf, *Chem. Phys. Lett.* 187 (1991) 367.

[234] J. D. Watts and R. J. Bartlett, *Chem. Phys. Lett.* 190 (1992) 19.

[235] V. Parasuk and J. Almlöf, *Theor. Chem. Acta* 83 (1992) 227.

[236] Z. Slanina, J. Kurtz and L. Adamowicz, *Chem. Phys. Lett.* 196 (1992) 208.

[237] Z. Slanina, J. Kurtz and L. Adamowicz, *Mol. Phys.* 76 (1992) 387.

[238] L. A. Curtiss and K. Raghavachari, in *Quantum Mechanical Electronic Structure Calculations with Chemical Accuracy*, S. R. Langhoff, Ed., Kluwer, Dordrecht, The Netherlands, 1995.

[239] K. Raghavachari and L. A. Curtiss, in *Quantum Mechanical Electronic Structure Calculations with Chemical Accuracy*, S. R. Langhoff, Ed., Kluwer, Dordrecht, The Netherlands, 1995.

[240] J. M. L. Martin and P. R. Taylor, *J. Phys. Chem.* 100 (1996) 6047.

[241] W. Krätschmer, N. Sorg and D. R. Huffman, *Surf. Sci.* 156 (1985) 814.

[242] S. W. McElvany, W. R. Creasy and A. O'Keefe, *J. Chem. Phys.* 85 (1986) 632.

[243] S. W. McElvany, B. I. Dunlap and A. O'Keefe, *J. Chem. Phys.* 86 (1987) 715.

[244] W. Krätschmer and K. Nachtigall, in Polycyclic Aromatic Hydrocarbons and Astrophysics, A. Léger, L. d'Hendecourt, and N. Boccara, Eds., D. Reidel, Dordrecht, The Netherlands, 1987.

[245] R. J. van Zee, R. F. Ferrante, K. J. Zerinque, W. Weltner, Jr. and D. W. Ewing, *J. Chem. Phys.* 88 (1988) 3465.

[246] S. W. McElvany, *J. Chem. Phys.* 89 (1988) 2063.

[247] J. R. Heath, A. L. Cooksy, H. W. Gruebele, C. A. Schmuttenmaer and R. J. Saykally, *Science* 244 (1989) 564.

[248] M. Vala, T. M. Chandrasekhar, J. Szczepanski, R. van Zee and W. Weltner, Jr., *J. Chem. Phys.* 90 (1989) 595.

[249] M. Vala, T. M. Chandrasekhar, J. Szczepanski and R. Pellow, *High Temp. Sci.* 27 (1989) 19.

[250] L. N. Shen and W. R. M. Graham, *J. Chem. Phys.* 91 (1989) 5115.

[251] H. M. Cheung and W. R. M. Graham, *J. Chem. Phys.* 91 (1989) 6664.

[252] J. Kurtz and D. R. Huffman, *J. Chem. Phys.* 92 (1990) 30.

[253] J. R. Heath, R. A. Sheeks, A. L. Cooksy and R. J. Saykally, *Science* 249 (1990) 895.

[254] J. Szczepanski and M. Vala, *J. Phys. Chem.* 95 (1991) 2792.

[255] L. N. Shen, P. A. Withey and W. R. M. Graham, *J. Chem. Phys.* 94 (1991) 2395.

[256] J. R., Heath, A. van Orden, E. Kuo and R. J. Saykally, *Chem. Phys. Lett.* 182 (1991) 17.

[257] J. R. Heath and R. J. Saykally, *J. Chem. Phys.* 94 (1991) 1724.

[258] J. R. Heath and R. J. Saykally, *J. Chem. Phys.* 94 (1991) 3271.

[259] P. A. Withey, L. N. Shen and W. R. M. Graham, *J. Chem. Phys.* 95 (1991) 820.

[260] Q. Jiang and W. R. M. Graham, *J. Chem. Phys.* 95 (1991) 3129.

[261] G. von Helden, M.-T. Hsu, P. R. Kemper and B. T. Bowers, *J. Chem. Phys.* 95 (1991) 3835.

[262] D. Zajfman, H. Feldman, O. Heber, D. Kella, D. Majer, Z. Vager and R. Naaman, *Science* 258 (1992) 1129.

[263] G. von Helden, P. R. Kemper, N. G. Gotts and M. T. Bowers, *Science* 259 (1993) 1300.

[264] N. G. Gotts, G. von Helden and M. T. Bowers, *Int. J. Mass Spectr. Ion Process.* 149/150 (1995) 217.

[265] Z. Slanina, F. Uhlík and L. Adamowicz, *J. Radioanal. Nucl. Chem.* 219 (1997) 69.

[266] Z. Slanina, S.-L. Lee, J.-P. François, J. Kurtz, L. Adamowicz and M. Smigel, *Mol. Phys.* 81 (1994) 1489.

[267] Z. Slanina, S.-L. Lee, J.-P. François, J. Kurtz and L. Adamowicz, *Chem. Phys. Lett.* 223 (1994) 397.

[268] Z. Slanina, S.-L. Lee, M. Smigel, J. Kurtz and L. Adamowicz, in *Science and Technology of Fullerene Materials*, P. Bernier, D. S. Bethune, L. Y. Chiang, T. W. Ebbesen, R. M. Metzger and J. W. Mintmire, Eds., Materials Research Society, Pittsburgh, PA, 1995.

[269] Z. Slanina, X. Zhao, E. Ōsawa and L. Adamowicz, *Fullerene Sci. Technol.* 8 (2000) 369.

[270] T. F. Giesen, A. van Orden, H. J. Hwang, R. S. Fellers, R. A. Provençal and R. J. Saykally, *Science* 265 (1994) 756.

[271] D. A. Plattner and K. N. Houk, *J. Am. Chem. Soc.* 117 (1995) 4405.

[272] G. von Helden, N. G. Gotts and M. T. Bowers, *Nature* 363 (1993) 60.

[273] Z. Slanina, *Chem. Phys. Lett.* 173 (1990) 164.

[274] M. W. Chase, Jr., C. A. Davies, J. R. Downery, Jr., D. J. Frurip, R. A. McDonald and A. N. Syverud, *J. Phys. Chem.* Ref. Data 14, Supplement 1, JANAF Thermochemical Tables, Third Edition, Vols. 1,2 (1985).

[275] M. A. Nygren and L. G. M. Pettersson, *Chem. Phys. Lett.* 191 (1992) 473.

[276] B. G. Johnson, in Modern Density Functional Theory: A Tool for Chemistry, J. M. Seminario and P. Politzer, Eds., Elsevier, Amsterdam, The Netherlands, 1995.

[277] J. D. Watts, J. Gauss, J. F. Stanton and R. J. Bartlett, *J. Chem. Phys.* 97 (1992) 8372.

[278] Z. Slanina, *Chem. Phys. Lett.* 161 (1989) 265.

[279] Z. Slanina, *Chem. Phys. Lett.* 166 (1990) 317.

[280] D. Michalska, H. Chojnacki, B. A. Hess, Jr. and L. J. Schaad, *Chem. Phys. Lett.* 141 (1987) 376.

[281] Z. Slanina, *Thermochim. Acta* 127 (1988) 237.

[282] A. van Orden and R. J. Saykally, *Chem. Rev.* 98 (1998) 2313.

[283] Z. Slanina, *Skept. Inq.* 24, No. 4 (2000) 18.

[284] P. W. Fowler and J. Woolrich, *Chem. Phys. Lett.* 127 (1986) 78.

[285] P. W. Fowler, *Chem. Phys. Lett.* 131 (1986) 444.

[286] P. W. Fowler and J. I. Steer, *J. Chem. Soc., Chem. Commun.* (1987) 1403.

[287] P. W. Fowler, J. E. Cremona and J. I. Steer, *Theor. Chim. Acta* 73 (1988) 1.

[288] A. Ceulemans and P. W. Fowler, *Phys. Rev. A* 39 (1989) 481.

[289] A. Ceulemans and P. W. Fowler, *J. Chem. Phys.* 93 (1990) 1221.

[290] P. W. Fowler, J. Chem. Soc., *Faraday Trans.* 86 (1990) 2073.

[291] P. Fowler, *Nature* 350 (1991) 20.

[292] P. W. Fowler, J. Chem. Soc., *Faraday Trans.* 87 (1991) 1945.

[293] P. W. Fowler, R. C. Batten and D. E. Manolopoulos, *J. Chem. Soc., Faraday Trans.* 87 (1991) 3103.

[294] A. Ceulemans and P. W. Fowler, *Nature* 353 (1991) 52.

[295] P. W. Fowler and D. E. Manolopoulos, *Nature* 355 (1992) 428.

[296] P. W. Fowler, D. E. Manolopoulos and R. P. Ryan, *J. Chem. Soc., Chem. Commun.* (1992) 408.

[297] P. W. Fowler, *J. Chem. Soc., Perkin Trans. II* (1992) 145.

[298] P. W. Fowler, D. E. Manolopoulos and R. P. Ryan, *Carbon* 30 (1992) 1235.

[299] D. E. Manolopoulos, D. R. Woodall and P. W. Fowler, *J. Chem. Soc., Faraday Trans.* 88 (1992) 2427.

[300] P. W. Fowler and V. Morvan, *J. Chem. Soc., Faraday Trans.* 88 (1992) 2631.

[301] P. W. Fowler, D. E. Manolopoulos, D. B. Redmond and R. P. Ryan, *Chem. Phys. Lett.* 202 (1993) 371.

[302] D. E. Manolopoulos and P. W. Fowler, *Chem. Phys. Lett.* 204 (1993) 1.

[303] S.-L. Lee, *Theor. Chim. Acta* 81 (1992) 185.

[304] A. J. Stone and D. J. Wales, *Chem. Phys. Lett.* 128 (1986) 501.

[305] C. Coulombeau and A. Rassat, *J. Chim. Phys.* 88 (1991) 173.

[306] W. N. Lipscomb and L. Massa, *Inorg. Chem.* 31 (1992) 2297.

[307] A. C. Tang, Q. S. Li, C. W. Liu and J. Li, *Chem. Phys. Lett.* 201 (1993) 465.

[308] D. E. Manolopoulos, J. C. May and S. E. Down, *Chem. Phys. Lett.* 181 (1991) 105.

[309] X. Liu, D. J. Klein, T. G. Schmalz and W. A. Seitz, *J. Comput. Chem.* 12 (1991) 1252.

[310] X. Liu, D. J. Klein, W. A. Seitz and T. G. Schmalz, *J. Comput. Chem.* 12 (1991) 1265.

[311] X. Liu, T. G. Schmalz and D. J. Klein, *Chem. Phys. Lett.* 188 (1992) 550.

[312] D. E. Manolopoulos, *Chem. Phys. Lett.* 192 (1992) 330.

[313] X. Liu, T. G. Schmalz and D. J. Klein, *Chem. Phys. Lett.* 192 (1992) 331.

[314] M.-L. Sun, Z. Slanina and S.-L. Lee, *Fullerene Sci. Technol.* 3 (1995) 627.

[315] P.W. Fowler, C. M. Quinn and D. B. Redmond, *J. Chem. Phys.* 95 (1991) 7678.

[316] S. Wei, Z. Shi and A. W. Castleman, Jr., *J. Chem. Phys.* 94 (1991) 3268.

[317] D. J. Klein, W. A. Seitz and T. G. Schmalz, *J. Phys. Chem.* 97 (1993) 1231.

[318] D. J. Klein, *Chem. Phys. Lett.* 217 (1994) 261.

[319] A. T. Balaban, D. J. Klein and C. A. Folden, *Chem. Phys. Lett.* 217 (1994) 266.

[320] J. R. Dias, *Chem. Phys. Lett.* 209 (1993) 439.

[321] K. Balasubramanian, *Chem. Phys. Lett.* 182 (1991) 257.

[322] K. Balasubramanian, *Chem. Phys. Lett.* 183 (1991) 292.

[323] K. Balasubramanian, *J. Chem. Inf. Comput. Sci.* 32 (1992) 47.

[324] K. Balasubramanian, *Chem. Phys. Lett.* 198 (1992) 577.

[325] K. Balasubramanian, *Chem. Phys. Lett.* 197 (1992) 55.

[326] K. Balasubramanian, *J. Mol. Spectr.* 157 (1992) 254.

[327] K. Balasubramanian, *Chem. Phys. Lett.* 201 (1993) 306.

[328] K. Balasubramanian, *Chem. Phys. Lett.* 206 (1993) 210.

[329] K. Balasubramanian, *Chem. Phys. Lett.* 202 (1993) 399.

[330] K. Balasubramanian, *J. Phys. Chem.* 97 (1993) 6990.

[331] K. Balasubramanian, *J. Phys. Chem.* 97 (1993) 4647.

[332] D. Babić, S. Bassoli, M. Casartelli, F. Cataldo, A. Graovac, O. Ori and B. York, *J. Mol. Simulat.* 14 (1995) 395.

[333] B. R. Eggen, M. I. Heggie, G. Jungnickel, C. D. Latham, R. Jones and P. R. Briddon, *Science* 272 (1996) 87.

[334] B. R. Eggen, M. I. Heggie, G. Jungnickel, C. D. Latham, R. Jones and P. R. Briddon, *Fullerene Sci. Technol.* 5 (1997) 727.

[335] Z. Slanina, X. Zhao, F. Uhlík, M. Ozawa and E. Ōsawa, *J. Organomet. Chem.* 599 (2000) 57.

[336] S. W. McElvany, J. H. Callahan, M. M. Ross, L. D. Lamb and D. R. Huffman, *Science* 260 (1993) 1632.

[337] J.-P. Deng, D.-D. Ju, G.-R. Her, C.-Y. Mou, C.-J. Chen, Y.-Y. Lin and C.-C. Han, *J. Phys. Chem.* 97 (1993) 11575.

[338] S. Fujita, *Symmetry and Combinatorial Enumeration in Chemistry* , Springer-Verlag, Berlin, FR Germany, 1991.

[339] F. Wahl, J. Wörth and H. Prinzbach, *Angew. Chem., Int. Ed. Engl.* 32 (1993) 1722.

[340] H. Prinzbach, A. Weller, P. Landenberger, F. Wahl, J. Worth, L. T. Scott, M. Gelmont, D. Olevano and B. von Issendorff, *Nature* 407 (2000) 60.

[341] Z. Wang, X. Ke, Z. Zhu, F. Zhu, M. Ruan, H. Chen, R. Huang and L. Zheng, *Phys. Let. A* 280 (2001) 351.

[342] Z. Iqbal, Y. Zhang, H. Grebel, S. Vijayalakshmi, A. Lahamer, G. Benedek, M. Bernasconi, J. Cariboni, I. Spagnolatti, R. Sharma, F. J. Owens, M. E. Kozlov, K. V. Rao and M. Muhammed, *Eur. Phys. J. B* 31 (2003) 509.

[343] Z. Slanina and L. Adamowicz, *Fullerene Sci. Technol.* 1 (1993) 1.

[344] Z. Slanina and L. Adamowicz, *J. Mol. Struct. (Theochem)* 281 (1993) 33.

[345] J. Lu, S. Re, Y. Choe, S. Nagase, Y. S. Zhou, R. S. Han, L. M. Peng, X. W. Zhang and X. G. Zhao, *Phys. Rev. B* 67 (2003) 125415.

[346] Z. X. Zhang, Z. Y. Pan, Y. X. Wang, Z. J. Li and Q, Wei, *Mod. Phys. Lett.* 17 (2003) 877.

[347] C. Piskoti, J. Yarger and A. Zettl, *Nature* 393 (1998) 771.

[348] R. L. Murry, J. R. Colt and G. E. Scuseria, *J. Phys. Chem.* 97 (1993) 4954.

[349] S. G. Louie, *Nature* 384 (1996) 612.

[350] L. F. Sun, S. S. Xie, W. Liu, W. Y. Zhou, Z. Q. Liu, D. S. Tang, G. Wang and L. X. Qian, *Nature* 403 (2000) 384.

[351] L. C. Qin, X. L. Zhao, K. Hirahara, Y. Miyamoto, Y. Ando and S. Iijima, *Nature* 408 (2000) 50.

[352] N. Wang, Z. K. Tang, G. D. Li and J. S. Chen, *Nature* 408 (2000) 50.

[353] Z. Slanina, F. Uhlík and L. Adamowicz, *J. Mol. Graphics Mod.* 21 (2003) 517.

[354] M. D. Diener and J. M. Alford, *Nature* 393 (1998) 668.

[355] H. Moribe, T. Inoue, H. Kato, A. Taninaka, Y. Ito, T. Okazaki, T. Sugai, R. Bolskar, J. M. Alford and H. Shinohara, *The 25th Fullerene-Nanotubes Symposium,* Awaji, Japan, 2003, paper 1P-1.

[356] H. Kato, A. Taninaka, T. Sugai and H. Shinohara, *J. Am. Chem. Soc.* 125 (2003) 7782.

[357] O. V. Boltalina, E. V. Dashkova and L. N. Sidorov, *Chem. Phys. Lett.* 256 (1996) 253.

[358] O. V. Boltalina, I. N. Ioffe, L. N. Sidorov, G. Seifert and K. Vietze, *J. Am. Chem. Soc.* 122 (2000) 9745.

[359] Z. Slanina, S.-L. Lee, K. Kobayashi and S. Nagase, *J. Mol. Struct. (Theochem)* 312 (1994) 175.

[360] Z. Slanina, X. Zhao, F. Uhlík and E. Ōsawa, in *Electronic Properties of Novel Materials - Science and Technology of Molecular Nanostructures,* H. Kuzmany, J. Fink, M. Mehring and S. Roth, Eds., AIP, Melville, NY, 1999, p. 179.

[361] Z. Slanina, K. Ishimura, K. Kobayashi and S. Nagase, *Chem. Phys. Lett.* 384 (2004) 114.

[362] D. R. Kanis, M. A. Ratner, T. J. Marks and M. C. Zerner, *Chem. Mater.* 3 (1991) 19.

[363] Z. Slanina, F. Uhlík, X. Zhao, L. Adamowicz and S. Nagase, *Fulleren. Nanotub. Carb. Nanostruct.* 15 (2007) 195.

[364] T. Ichikawa, T. Kodama, S. Suzuki, R. Fujii, H. Nishikawa, I. Ikemoto, K. Kikuchi and Y. Achiba, *Chem. Lett.* 33 (2004) 1008.

[365] P. J. Hay and W. R. Wadt, *J. Chem. Phys.* 82 (1985) 299.

[366] M. E. Casida, C. Jamorski, K. C. Casida and D. R. Salahub, *J. Chem. Phys.* 108 (1998) 4439.

[367] L. S. Wang, J. M. Alford, Y. Chai, M. Diener, J. Zhang, S. M. McClure, T. Guo, G. E. Scuseria and R. E. Smalley, *Chem. Phys. Let.* 207 (1993) 354.

[368] Y. Kubozono, T. Ohta, T. Hayashibara, H. Maeda, H. Ishida, S. Kashino, K. Oshima, H. Yamazaki, S. Ukita and T. Sogabe, *Chem. Lett.* (1995) 457.

[369] F. G. Hopwood, K. J. Fisher, P. Greenhill, G. D. Willett and R. Zhang, *J. Phys. Chem.* B 101 (1997) 10704.

[370] H. Shinohara, in *Fullerenes: Recent Advances in the Chemistry and Physics of Fullerenes and Related Materials,* Vol. 4, K. M. Kadish and R. S. Ruoff, Eds., The Electrochemical Society, Pennington, NJ, 1997, p. 467.

[371] Y. Chai, T. Guo, C. Jin, R. E. Haufler, L. P. F. Chibante, J. Fure, L. Wang, J. M. Alford and R. E. Smalley, *J. Phys. Chem.* 95 (1991) 7564.

[372] E. Nishibori, M. Takata, M. Sakata, H. Tanaka, M. Hasegawa and H. Shinohara, *Chem. Phys. Lett.* 330 (2000) 497.

[373] S. Nagase and K. Kobayashi, *Chem. Phys. Lett.* 228 (1994) 106.

[374] Z. Slanina, K. Kobayashi and S. Nagase, *Chem. Phys. Lett.* 388 (2004) 74.

[375] Y. F. Lian, S. F. Yang and S. H. Yang, *J. Phys. Chem.* B 106 (2002) 3112.

[376] Z. Slanina, *J. Cluster Sci.* 15 (2004) 3.

[377] C. B. Alcock, V. P. Itkin and M. K. Horrigan, *Can. Metallurg. Quart.* 23 (1984) 309.

[378] A. Reich, M. Panthofer, H. Modrow, U. Wedig and M. Jansen, *J. Am. Chem. Soc.* 126 (2004) 14428.

[379] O. Haufe, M. Hecht, A. Grupp, M. Mehring and M. Jansen, *Z. Anorg. Allgem. Chem.* 631 (2005) 126.

[380] A. Gromov, N. Krawez, A. Lassesson, D. I. Ostrovskii and E. E. B. Campbell, *Curr. App. Phys.* 2 (2002) 51.

[381] E. E. B. Campbell, *Fullerene Collision Reactions,* Kluwer Academic Publishers, Dordrecht, 2003.

[382] *Fullerenes: From Synthesis to Optoelectronic Properties* , D. M. Guldi and N. Martin, Eds., Kluwer Academic Publishers, Dordrecht, 2002.

[383] A. Ardavan, M. Austwick, S. C. Benjamin, G. A. D. Briggs, T. J. S. Dennis, A. Ferguson, D. G. Hasko, M. Kanai, A. N. Khlobystov, B. W. Lovett, G. W. Morley, R. A. Oliver, D. G. Pettifor, K. Porfyrakis, J. H. Reina, J. H. Rice, J. D. Smith, R. A. Taylor, D. A. Williams, C. Adelmann, H. Mariette and R. J. Hamers, *Phil. Trans. Roy. Soc. London A* 361 (2003) 1473.

[384] C. C. Chancey and M. C. M. O'Brien, *The Jahn-Teller Effect in* C_{60} *and Other Icosahedral Complexes,* Princeton University Press, Princeton, 1997.

[385] R. D. Bolskar, A. F. Benedetto, L. O. Husebo, R. E. Price, E. F. Jackson, S. Wallace, L. J. Wilson and J. M. Alford, *J. Am. Chem. Soc.* 125 (2003) 5471.

In: Progress in Fullerene Research
Editor: Milton Lang, pp. 123-143
ISBN: 978-1-60021-841-5

Chapter 4

EPITAXIAL GROWTH OF UNDOPED AND MULTIVALENT IMPURITY-DOPED C_{60} FILMS BY MOLECULAR BEAM EPITAXY

Jiro Nishinaga*[*] *and Yoshiji Horikoshi
School of Science and Engineering, Waseda University, 3-4-1 Okubo, Shinjuku, Tokyo 169-8555, Japan

Abstract

A variety of investigations has been done for physical and chemical properties of C_{60}, and revealed unique potentialities of C_{60} crystals such as superconductivity and magnetism. C_{60} molecules are reported to crystallize into a face-centered cubic structure on crystalline Si and GaAs substrates, and the crystalline quality is confirmed to be surprisingly high. However, C_{60} crystals are very fragile and chemically unstable due to the fact that C_{60} crystals are formed by van der Waals force, which is very weak compared with other bonding structures such as covalent and ionic bindings. Because of these characteristics, conventional techniques such as etching and ion milling are not suitable for fabricating nanostructures with C_{60}. Area-selective epitaxy techniques make it possible to achieve damage-free fabrications. Therefore, we have carried out area-selective epitaxial growth of C_{60} crystals on SiO_2-masked GaAs substrates by molecular beam epitaxy, and the fabrication of C_{60} nanoscale structures is successfully achieved. During C_{60} deposition, C_{60} molecules landing on SiO_2 masks either evaporate or migrate to the open area enhancing growth rate of C_{60} layers. Mean diffusion length of C_{60} molecules on SiO_2 mask at 200°C is evaluated to be 200-300 nm.

Mechanical strength and chemical stability are very important for practical device applications. To study feasibility of C_{60} layers, it is inevitable to obtain harder and more stable C_{60} layers. It has been reported that metal-C_{60} interactions are stronger than van der Waals interactions between C_{60} molecules. Therefore, metal doping in C_{60} layers may produce much harder and chemically stable C_{60} films. Multivalent impurity-doped C_{60} films (Al, Ga, Ge) are grown on GaAs and quartz glass substrates. All the impurity-doped C_{60} films show no X-ray diffraction peak, indicating that the films have an amorphous structure. Vickers hardness values of all the impurity-doped C_{60} films are considerably enhanced, and the films are confirmed to be undissolved in organic solvents. New photoluminescence peaks of electron transitions between HOMO and LUMO states of C_{60} molecules are found in Al-doped and

[*] E-mail address: jiro247@moegi.waseda.jp. Tel +81-3-5286-3176; Fax +81-3-3209-3450.

Ga-doped C_{60} films. Optimized bonding structures of these impurity atoms to C_{60} molecules are determined by using ab initio calculations. Stable bonds between impurities and C_{60} molecules are verified to be formed.

1. Epitaxial Growth of Undoped C_{60} Crystals

1.1. Introduction

Discovery of C_{60} [1] and synthesis on a microscopic scale of solid C_{60} [2] have stimulated intensive research efforts, and extensive investigations of C_{60} physical and chemical properties have revealed its application of superconductivity [3] and magnetism [4]. C_{60} molecules have a highly symmetric structure and crystallize into a face-centered cubic structure on crystalline substrates such as Si and GaAs [5-8]. Although the heterointerfaces between C_{60} layers and these crystalline substrates have large lattice mismatch, the crystalline quality of high purity C_{60} epitaxial layers is found to be surprisingly high. In this section, we discuss C_{60} crystal growth on GaAs (111)A, (111)B and (001) substrates by molecular beam epitaxy. To establish epitaxial relationships between C_{60} layers and GaAs substrates, the grown samples are investigated by reflection high energy electron diffraction (RHEED) and X-ray diffraction (XRD) $2\theta/\omega$ scan and in-plane scanning measurements [9].

Next, we discuss the fabrication of nanoscale structures such as quantum wires and dots using C_{60} crystals. Since C_{60} films are very fragile and chemically unstable due to the van der Waals force, conventional techniques such as etching and ion milling are not acceptable for nanoscale fabrication processes. Area-selective epitaxy makes it possible to achieve damage-free fabrication. Therefore, we have carried out area-selective epitaxial growth of C_{60} crystals on GaAs (111)A, (111)B and (001) substrates with nanoscale open areas by using SiO_2 mask. These structures are formed by electron beam (EB) lithography. The grown structures of C_{60} crystals and the surface morphology are investigated by scanning electron microscopy (SEM) and atomic force microscopy (AFM) [10].

1.2. Structural Characterizations of C_{60} Layers on GaAs Substrates

C_{60} films are grown on GaAs (111)A, (111)B and (001) substrates by solid source MBE with background pressure of 10^{-10} Torr. GaAs substrates are first degreased in an organic solvent, and etched in an alkaline etchant, and loaded in the growth chamber. Native oxide layers of GaAs are removed by a thermal flash at 580°C in As_4 atmosphere. After growing a 20-nm-thick GaAs buffer layer at 500°C, C_{60} layer growth is performed at substrate temperatures of 100, 150 and 200°C. 99.5% C_{60} powder is used for the C_{60} source. C_{60} beam equivalent pressure (BEP) is fixed at $1.0\text{x}10^{-7}$ Torr with the growth rate at 0.23 Å/s. The layer thickness is fixed at 250nm. Surface structures of C_{60} layers and the substrates are characterized by RHEED with incident electron energy of 13 keV. The grown samples are investigated by XRD with Cu-Kα radiation (λ= 1.54Å) at room temperature. XRD $2\theta/\omega$ scan and in-plane scanning measurements are performed and compared with each other.

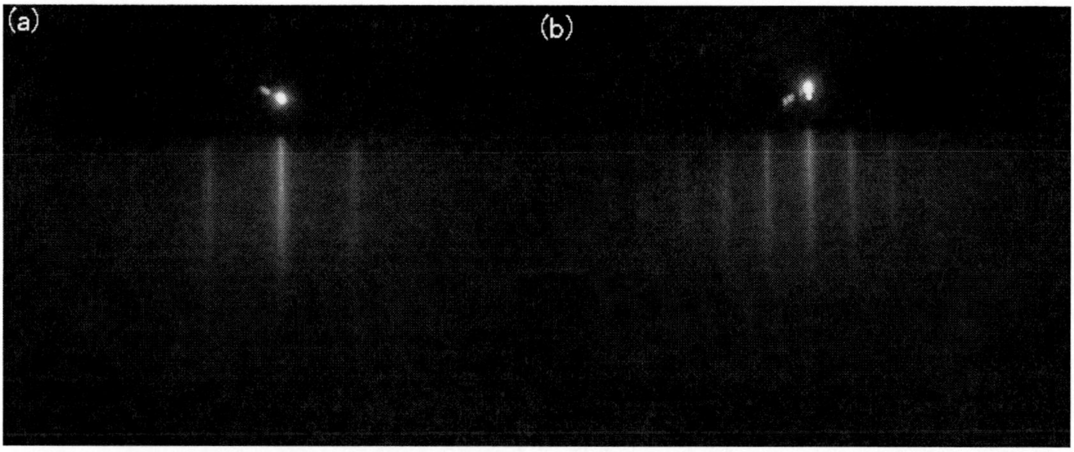

Figure 1. RHEED patterns of a C_{60} layer surface on GaAs (111)A. Electron beam along the [11-2] azimuth (a) and the [01-1] azimuth (b) of the GaAs substrates.

C_{60} film growth has been carried out at substrate temperatures of 100, 150 and 200°C, because the growth rates steeply drop above 200°C, and furthermore, no growth occurs above 250°C. For all these substrate temperatures, no essential change in RHEED patterns is confirmed. Fig. 1 shows the RHEED patterns of a C_{60} layer after 1-hour growth at 200°C on GaAs (111)A. The RHEED pattern of Fig. 1(a) is observed when the incident electron beam azimuth is parallel to the [11-2] direction of GaAs substrate, while the pattern shown in Fig. 1 (b) is obtained when the sample is rotated by 30° from the [11-2] direction. This azimuth is parallel to the [01-1] direction. The streak interval in Fig. 1(a) is approximately $\sqrt{3}$ times longer than that in Fig. 1(b), and the diffraction patterns show a 6-fold symmetry, indicating that the C_{60} epitaxial orientation is [111]. The observed epitaxial relationship is $[11\text{-}2]_{GaAs}$ // $[11\text{-}2]_{C60}$ and $[01\text{-}1]_{GaAs}$ // $[01\text{-}1]_{C60}$. Fig. 2 shows the RHEED patterns of a C_{60} layer grown on a GaAs (111)B substrate. The patterns of Fig. 2(a) is observed when the electron beam azimuth is parallel to the [-1-12] direction of the GaAs substrate. Fig. 2(b) is with the beam azimuth parallel to the [01-1] direction of the substrate. These diffraction patterns also have a 6-fold symmetry, indicating that the epitaxial relationship is $[\text{-}1\text{-}12]_{GaAs}$ // $[11\text{-}2]_{C60}$ and $[01\text{-}1]_{GaAs}$ // $[01\text{-}1]_{C60}$. Fig. 3 shows the RHEED pattern of a C_{60} layer on a GaAs (001) substrate with an incident beam azimuth parallel to the [1-10] direction of the GaAs substrate. This pattern shows a 6-fold symmetry. Therefore, the C_{60} epitaxial layer is oriented the [111] direction as with the films grown on GaAs (111)A and (111)B. This observed streak interval coincides well with that in Fig. 1 (b), indicating that the epitaxial relationship is $[1\text{-}10]_{GaAs}$ // $[01\text{-}1]_{C60}$. From the observed streak intervals in C_{60} layers, the lattice constant of cubic C_{60} is estimated to be approximately 14 Å that coincides well with that of bulk cubic C_{60} crystals [11, 12]. Since these RHEED patterns show very sharp and clear streaks, the crystal quality of grown C_{60} layers is fairly good.

To investigate the crystalline quality more precisely, XRD measurements are performed. Fig. 4 shows the $2\theta/\omega$ diffraction patterns observed on the C_{60} layers grown at 200°C on (111)A, (111)B and (001) GaAs substrates. In these samples, only the (*hhh*) diffractions are observed from the C_{60} films, suggesting that the films are [111] oriented crystals with a fcc structure. From 2θ value of 10.83° at the (111) peak, the cubic lattice constant of the C_{60} films is estimated to be approximately 14.1Å.

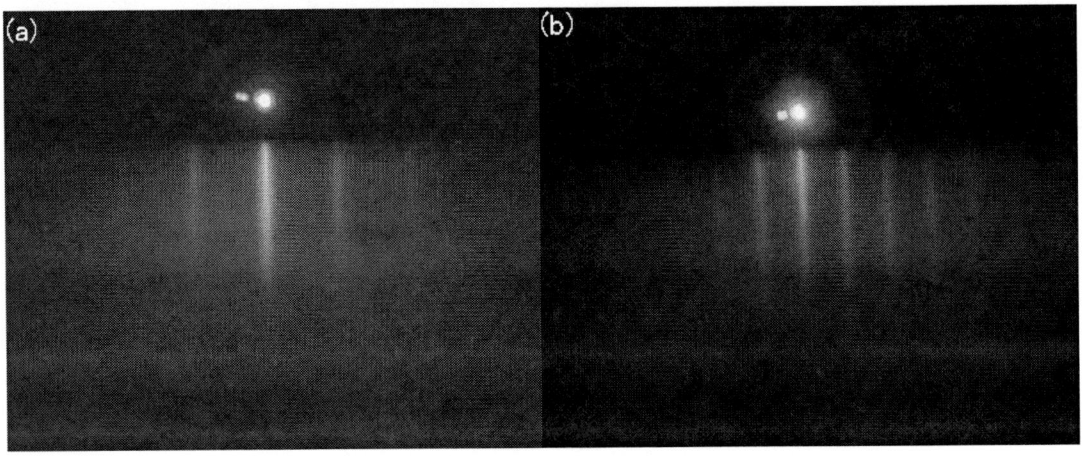

Figure 2. RHEED patterns of a C_{60} layer surface on GaAs (111)B. Electron beam along the [-1-12] azimuth (a) and the [01-1] azimuth (b) of the GaAs substrate.

Figure 3. RHEED pattern of C_{60} layer surface on GaAs (001) with the electron beam along the [1-10] azimuth of the GaAs substrate.

In order to confirm the epitaxial relationships between C_{60} films and GaAs substrates, The XRD-in-plane scan measurements are performed. XRD-in-plane scan patterns along the [01-1] direction of GaAs (111)A and (111)B substrates, and the pattern along the [1-10] direction of a GaAs (001) substrate are demonstrated in Fig. 5. The patterns (a) and (b) exhibit the (220) and (440) diffraction peaks, suggesting that the [01-1] directions of the C_{60} and GaAs are parallel when the layers are grown on GaAs(111)A and (111)B substrates. In the results denoted by (c), the (220) and (440) diffraction peaks are observed as with the diffraction peaks of the layers on GaAs (111)A and (111)B, indicating that the [01-1] direction of the C_{60} is parallel to the [1-10] direction of GaAs (001). These results agree well with the RHEED measurements.

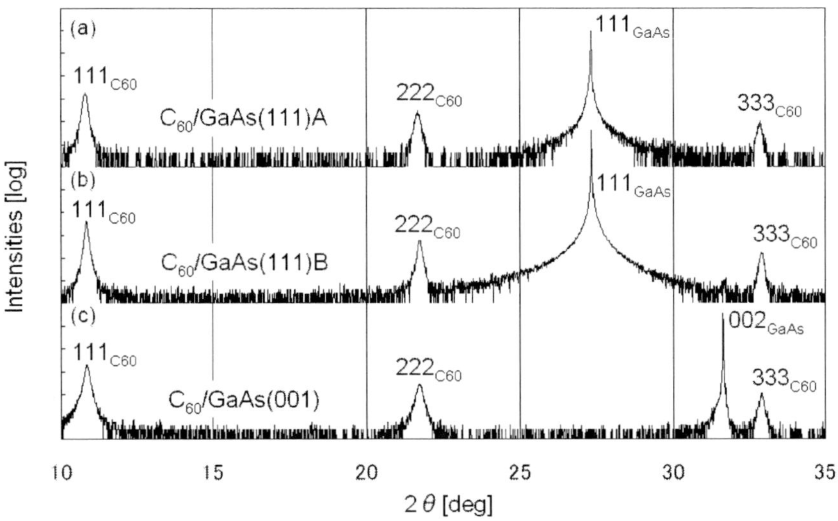

Figure 4. X-ray $2\theta/\omega$ diffraction patterns of C_{60} films on GaAs (111)A, (111)B and (001) substrates.

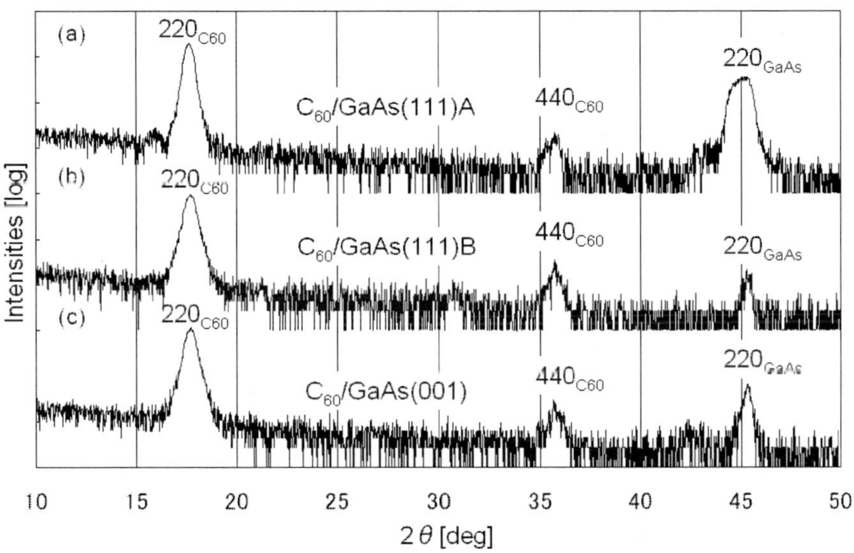

Figure 5. XRD-in-plane scanning patterns along the [01-1] direction of a GaAs (111)A and (111)B, and the pattern along the [1-10] direction of a GaAs (001) substrate.

1.3. Optical Properties of Undoped C_{60} Layers

The optical characteristics of the C_{60} films are investigated by photoluminescence (PL) measurements performed at 4.2K and 300K using the 488nm line of argon ion laser as an excitation source, and the PL spectra of the C_{60} layers grown on GaAs (001) substrates are shown in Fig. 6. The 1.52eV and 1.50eV peaks measured at 4.2K correspond to the edge emission and acceptor related emission from the GaAs substrates, while the other peaks around 1.69eV are caused by the C_{60} layers. The PL spectra from the C_{60} layers are in good agreement with literature [13-15]. The spectral positions of the PL peaks from the C_{60} layers shift only slightly when measurement temperature varies. This is due to the weak intermolecular bonds of C_{60} molecules [16]. The 1.69eV emission line from the C_{60} layers is attributed to the radiative recombination of a self-trapped polaron exciton [13]. Such an exciton complex occurs due to the strong electron-vibration coupling on C_{60} clusters. This excitation is localized, and can be a Frenkel exciton trapped by a lattice dislocation.

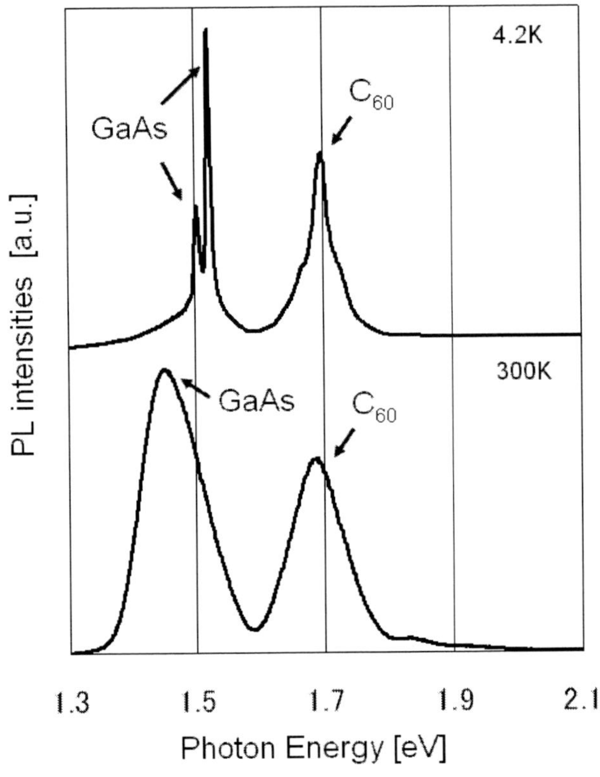

Figure 6. PL spectra of pure C_{60} films grown on GaAs (001) substrates measured at 4.2K and 300K. The sample is excited by the 488nm line of an argon ion laser.

1.4. Area-Selective Epitaxy of C_{60} Crystals

Area-selective epitaxy of C_{60} crystals are carried out on SiO_2 masked substrates by MBE. A 40-nm-thick SiO_2 film is deposited on GaAs (111)A, (111)B and (001) substrates by a conventional sputtering technique. Open area of the SiO_2 mask in various shapes (e.g. triangles, lines & spaces, and dots) is formed by EB lithography and wet chemical etching. These masked GaAs substrates are mounted on a Mo holder alongside an unmasked (planer) GaAs (001) substrate. The unmasked substrates are used to observe RHEED patterns and estimate the C_{60} growth thickness. Native oxide layers of the GaAs surface are removed by a thermal flash at 580°C in As_4 atmosphere. After 2x4 RHEED patterns of the GaAs (001) substrates are confirmed, C_{60} layer growth is carried out at 200°C. C_{60} BEP is fixed at 1.0 x 10^{-7} Torr. The film thickness is fixed at 180nm. The grown structures of the C_{60} crystals are investigated by SEM and AFM.

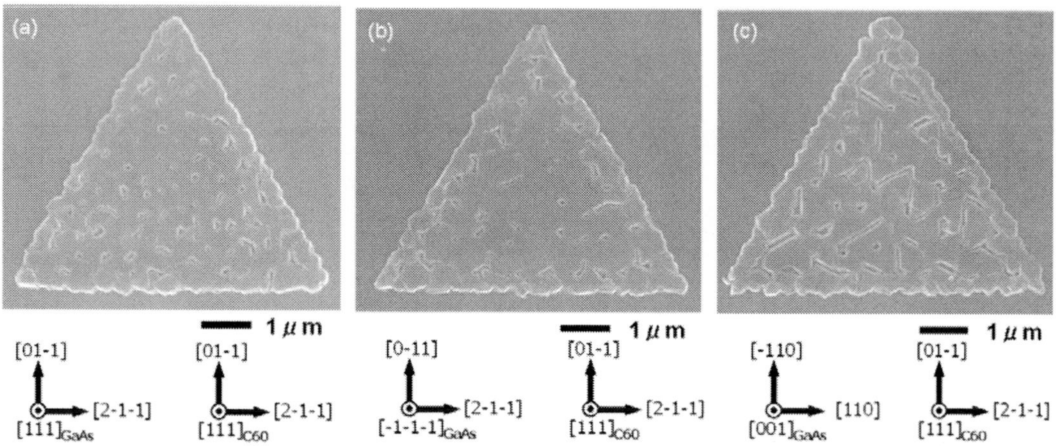

Figure 7. SEM images of triangular C_{60} crystals grown at 200°C. The side length of SiO_2 open area is 3μm. Figure 7(a), (b) and (c) show the C_{60} crystals on GaAs (111)A, GaAs (111)B and GaAs (001) substrates, respectively.

Fig. 7 shows SEM images of the triangular C_{60} crystals grown on GaAs substrates at 200°C. Fig. 7(a), (b) and (c) show the C_{60} crystals grown on GaAs (111)A, (111)B and (001) substrates, respectively. At the substrate temperatures below 150°C, C_{60} layers are grown even on the SiO_2 mask. On the other hand, the C_{60} growth occurs only on the GaAs surface (i.e. SiO_2 open area) grown at 200°C. The facets appearing on the C_{60} crystal surface suggest that [111] oriented C_{60} layers are grown on GaAs (111)A, (111)B and (001) substrates. The indexes of facets formed on the C_{60} surface are found to be {100}, {110}. These images show that surface morphology of C_{60} crystals grown on GaAs (111)A and (111)B is smoother than that grown on GaAs (001). Thus, the crystalline quality of C_{60} films on GaAs (111)A and (111)B may be better than that grown on GaAs (001). These C_{60} crystals are found to be thicker than the C_{60} layer grown on the unmasked substrate by 30nm, and the side length of triangular C_{60} crystals is longer than that of SiO_2 open area. These results suggest that C_{60} molecules landing on SiO_2 mask migrate to the open area, and enhance the growth rate faster. C_{60} lateral growth is confirmed to be less sensitive to crystalline directions, i.e., in both <211> and <011> directions, and the lateral growth is approximately 80nm.

Fig. 8 exhibits a topographic map of a triangular C_{60} crystal on GaAs (111)B measured by AFM. The cross section image along the black line in the figure suggests that a part of C_{60} molecules landing on SiO_2 mask goes to the open area, and enhances the C_{60} growth rate. Since the volume of the C_{60} crystals grown by C_{60} vertical flux is estimated by the growth thickness of unmasked (planer) substrates, the contribution of C_{60} molecules diffused from SiO_2 mask can be estimated by comparing the growth rate. The mean diffusion length of C_{60} molecules on SiO_2 mask at 200°C is estimated as 200-300nm. It is also confirmed that the estimated diffusion length does not depend on size and shape of the opening area.

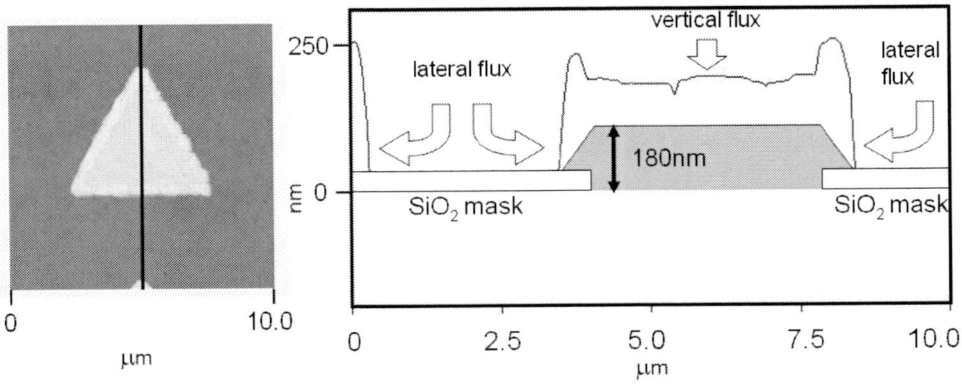

Figure 8. Topographic map of a triangular C_{60} crystals on GaAs (111)B measured by AFM, and the cross section image along the black line.

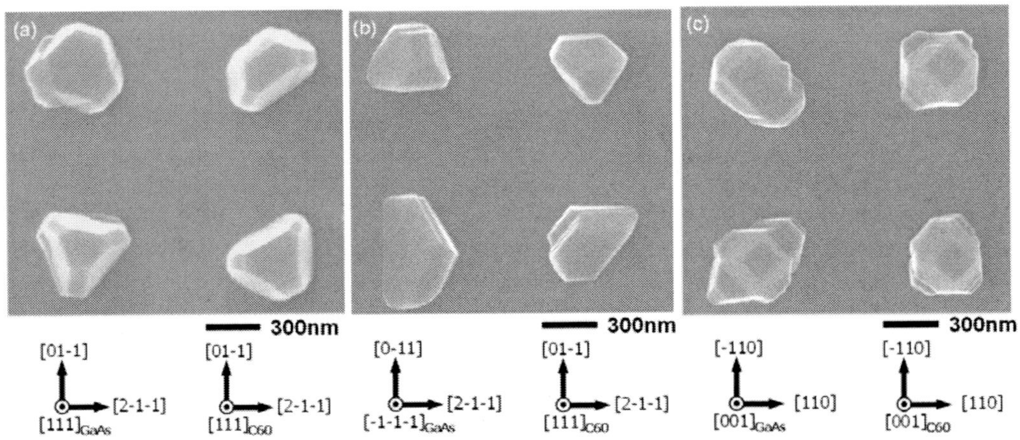

Figure 9. SEM images of the nano-scale C_{60} crystals grown at 200°C on GaAs (111)A (a), (111)B (b) and (001) (c) substrates.

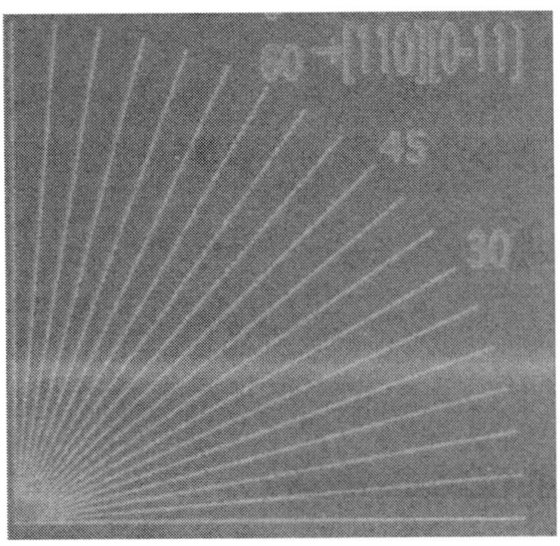

Figure 10. SEM image of C_{60} crystals grown on the radial mask patterns with 150nm-wide windows.

Fig. 9 shows SEM images of C_{60} nano-crystals grown at 200°C on GaAs (111)A (a), (111)B (b), and GaAs (001) (c) substrates. The diameter of dot open area is about 100nm. The size of the grown crystals is larger than that of SiO_2 open area, indicating that C_{60} lateral growth occurs. These images show that C_{60} crystal facets grown on a GaAs (001) substrate are different from those on GaAs (111)A and (111)B. In Fig. 9 (a) and (b), the C_{60} crystals have a 3-fold symmetry, indicating that the epitaxial orientation of C_{60} crystals is [111]. While, the C_{60} crystals in Fig. 9(c) have a 4-fold symmetry, indicating that this epitaxial orientation is [001]. Fig. 10 shows the SEM image of C_{60} crystals grown on the radial mask patterns with 150nm-wide windows. Fig. 11 demonstrates the expanded images of some parts of the radial patterns. Fig. (a) and (b) show the C_{60} crystals on GaAs (111)A, and Fig. (c) and (d) show those grown on GaAs (111)B. Fig. (e) and (f) show those grown on GaAs (001). The {100}, {110} facets of C_{60} crystals easily appear. These images suggest that C_{60} crystals grown on GaAs (111)A and (111)B have a 3-fold symmetry, and those grown on GaAs (001) have a 4-fold symmetry. It is interesting to note that the different epitaxial relationships are observed between the nanoscale growth and planer or wide area growth on GaAs (001) substrates. When the open area size is larger than approximately 250nm, the grown crystals show [111] epitaxial orientation. While, when it is below this value, the grown crystals show [001] orientation. Although it is reported that the different epitaxial orientation of C_{60} crystals occurs on the different surface reconstructions of GaAs (001) substrates [17, 18], the present finding suggests that the crystal size is also one of the determining factors for the C_{60} epitaxial orientation.

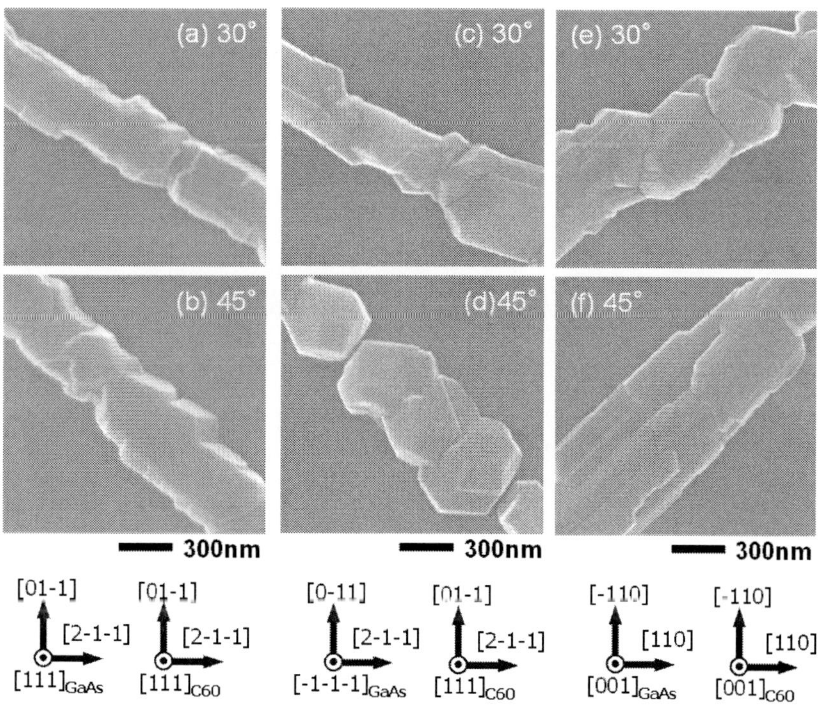

Figure 11. Expanded images of the C_{60} crystals grown on the radial mask patterns. Figure (a) and (b) show the C_{60} crystals on GaAs (111)A. Figure (c) and (d) show those grown on GaAs (111)B. Figure (e) and (f) show those grown on GaAs (001).

2. Growth of Multivalent Impurity-Doped C_{60} Films

2.1. Introduction

Mechanical strength and chemical stability are very important for practical device applications. However, C_{60} crystals are very fragile and chemically unstable due to the fact that C_{60} crystals are formed by van der Waals force [16]. Thus, the films are not suitable for device applications. To investigate the feasibility of C_{60} layers, it is inevitable to obtain harder and more chemically stable C_{60} films, keeping the original properties of C_{60} crystals. It has been reported that the metal-C_{60} interaction is stronger than the van der Waals interaction between C_{60} molecules [19-21]. Therefore, it is expected that metal doping in C_{60} films makes much harder and more chemically stable C_{60} films.

In this section, we discuss C_{60} layers doped with multivalent impurities such as aluminum, gallium and germanium grown on GaAs substrates and quartz glass substrates by MBE. Mechanical and optical properties of the layers are studied by Vickers hardness test, absorption coefficient spectra and PL measurements. All the multivalent impurity-doped C_{60} films are found to be much harder and more chemically stable than pure C_{60} films. A new PL peak with the energy of 1.95eV appears in Al-doped and Ga-doped C_{60} films. This energy coincides with the energy difference between the highest occupied molecular orbital (HOMO) and the lowest unoccupied molecular orbital (LUMO) of C_{60} molecules [22].

In order to obtain optimized structures and the corresponding binding energies between multivalent impurities and C_{60} molecules, ab initio calculations for the systems are performed. Stable covalent bonds between the impurities and C_{60} molecules are verified to be formed. The binding energies are confirmed to be much higher than those of the van der Waals force between C_{60} molecules [23].

2.2. Mechanical and Optical Characteristics of Multivalent Impurity-Doped C_{60} Films

Multivalent impurity-doped C_{60} films are grown on GaAs (001) and quartz glass substrates by solid source MBE. The process of the GaAs substrate preparation is similar to that used in the pure C_{60} layer growth. After growing a 50-nm-thick GaAs buffer layer at 580°C, multivalent impurity-doped C_{60} film growth is carried out at a substrate temperature of 100°C. C_{60} BEP is fixed at 1.0×10^{-7} Torr. The impurity cell temperatures (Al, Ga, Ge) are varied to control the molecular ratios of impurity atoms to C_{60} molecules. Sticking coefficients of the impurities are assumed to be unity, because the substrate temperature is as low as 100°C. The sticking coefficient of C_{60} at 100°C should be unity because the growth rate of C_{60} layers remains constant at substrate temperatures below 150°C. Therefore, the resulting compositions of the grown layer are equal to the flux ratios. Table 1 shows the cell temperatures of the impurities with different molecular ratios of impurity atoms to C_{60} molecules.

Table 1. Cell temperatures with different ratios of impurities to C_{60}.

Flux ratios of impurities to C_{60}	3	10	25
Ga cell temperature (°C)	710	755	805
Al cell temperature(°C)	985	1075	1130
Ge cell temperature(°C)	1070	1150	1190

Quartz glass substrates are degreased by an organic solvent, and the surface contaminations are evaporated by thermal annealing at 600°C for 30 minutes in the growth chamber before deposition. Multivalent impurity-doped C_{60} films are grown in the same manner as those grown on GaAs substrates. The crystalline characteristics are investigated by RHEED and XRD 2θ/ω scan measurements. The aluminum distributions in the Al-doped C_{60} films are investigated by a secondary ion mass spectroscopy (SIMS) measurement using Cs ions. Vickers hardness test is applied to study the mechanical properties. Absorption and reflectance spectra of the films grown on quartz glass substrates are measured at room temperature at the wavelength range between 220 and 800 nm. PL measurements are performed at room temperature by using the 488 nm line of argon ion laser.

RHEED patterns of pure C_{60} films grown on GaAs (001) substrates exhibit well characterized (1x1) and indicate that the epitaxial orientation is [111] direction. However, all the impurity-doped C_{60} films on both GaAs and quartz glass substrates show halo patterns from the beginning of the growth, and the films show no peak in XRD 2θ/ω scan. These results suggest that the C_{60} films doped with the impurity atoms have an amorphous structure.

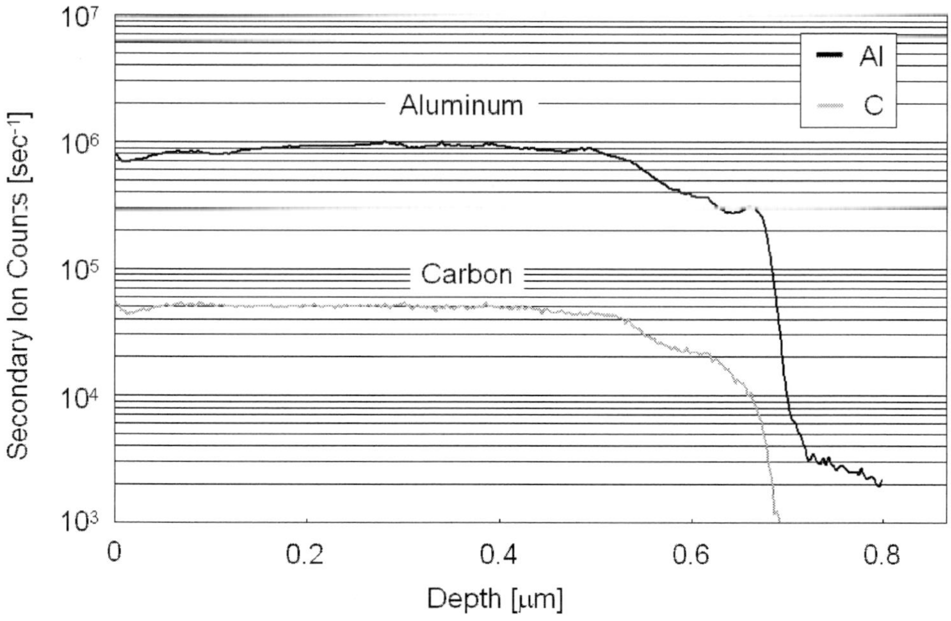

Figure 12. Depth profiles of aluminum and carbon atoms of the Al-doped C_{60} film measured by SIMS measurement. The molecular ratio of Al to C_{60} is 25.

Fig. 12 shows depth profiles of aluminum and carbon atoms evaluated by a SIMS measurement. The thickness of this layer is 700nm, and the molecular ratio of Al to C_{60} is 25. It is confirmed that there is no segregation of aluminum atom in this film.

Table 2. Vickers hardness values of pure and impurity-doped C_{60} films.

Pure C_{60} crystals	Ga-doped C_{60} films (Ga / C_{60}=25)	Al-doped C_{60} films (Al / C_{60}=25)	Ge-doped C_{60} films (Ge / C_{60}=25)
20HV*	70HV0.01	245HV0.01	450HV0.01

*Reference data [11, 24]

Vickers hardness value of pure C_{60} crystals is reported to be around 20HV [11, 24], which is equal to that of pure gold surfaces. To investigate the effect of multivalent impurity incorporation on the hardness, Vickers hardness test for the thick impurity-doped C_{60} films with the molecular ratio of 25 are carried out. Table 2 shows the Vickers hardness of pure C_{60} and impurity-doped C_{60} films. The Vickers hardness of the impurity-doped C_{60} films is found to be dramatically increased. Especially, the value of Ge-doped C_{60} films is as high as 450HV, which is approximately equal to that of plated nickel surfaces. An order of the Vickers hardness is Ge-doped C_{60} > Al-doped C_{60} > Ga-doped C_{60}, and this order should be closely related to the binding energies between impurity atoms and C_{60} molecules.

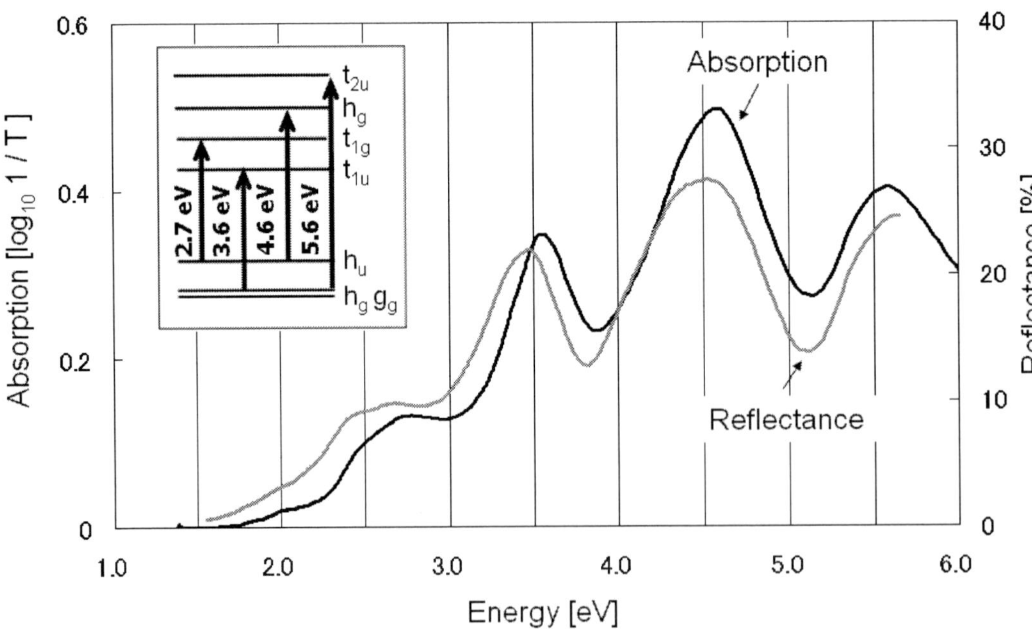

Figure 13. Absorption and reflectance spectra of a 25nm-thick pure C_{60} film deposited on a quartz glass substrate measured at room temperature. The inset shows the simple electron energy levels of C_{60} molecular orbitals.

Pure C_{60} crystals are quickly dissolved in organic solvents due to the nonpolarity and the weak interaction among C_{60} molecules. On the other hand, all the impurity-doped C_{60} films are found to be undissolved in organic solvents. The dramatic structural changes and the observed improvements in hardness and chemical stability are probably induced by the bonding between multivalent impurity atoms and C_{60} molecules. The impurity atoms may act as bridges between C_{60} molecules, and they make large-scale complexes. These complexes probably connect firmly with each other.

Fig. 13 shows absorption and reflectance spectra of a 25-nm-thick pure C_{60} film grown on a quartz glass substrate. Since interactions among C_{60} molecules are very weak, the absorption profiles are governed by the electron structures of a C_{60} molecule. The four absorption peaks are observed at 2.7eV, 3.6eV, 4.6eV and 5.6eV. These absorption profiles are in agreement with reported experimental results [2, 13] and the calculated spectra [16, 25]. These absorption peaks correspond to the electron transitions among the C_{60} molecular orbitals. The simplified electron energy levels of the C_{60} molecular orbitals are shown in an inset of Fig. 13. The h_u is the energy level of HOMO state and the t_{1u} is that of LUMO state. The electron transition between HOMO and LUMO is not observed due to the dipole forbidden transitions [16].

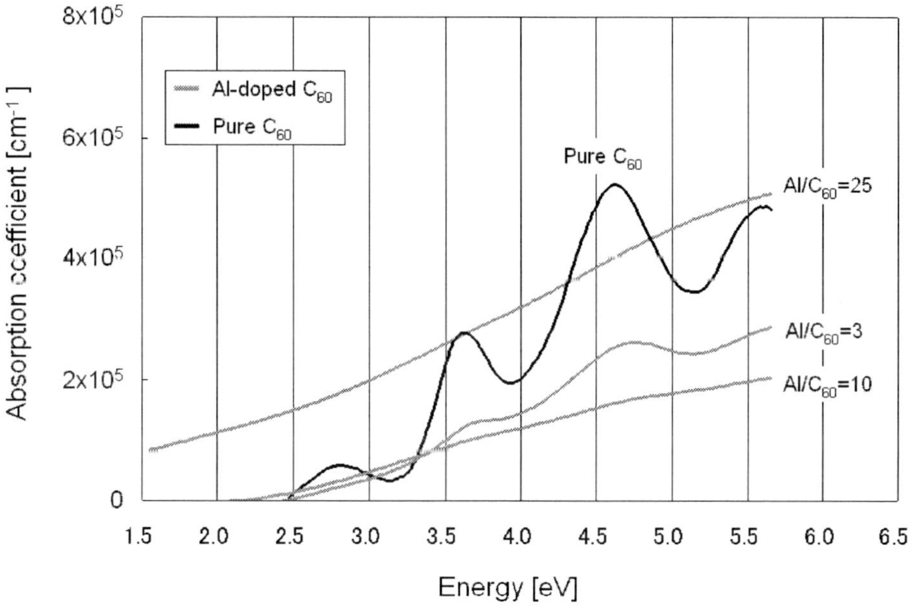

Figure 14. Absorption coefficient spectra at room temperature of a pure C_{60} film and Al-doped C_{60} films with different concentrations

Fig. 14 shows absorption coefficient spectra of a pure C_{60} film and Al-doped C_{60} films grown at several Al cell temperatures on quartz glass substrates. It is confirmed that the absorption peaks in the Al-doped C_{60} films become less pronounced with increasing the Al concentrations. Fig. 15 and Fig. 16 show absorption coefficient spectra of the Ga-doped, Ge-doped C_{60} films respectively, with different concentrations. As with the Al-doped C_{60} films, these absorption peaks become less pronounced with increasing the doping concentrations.

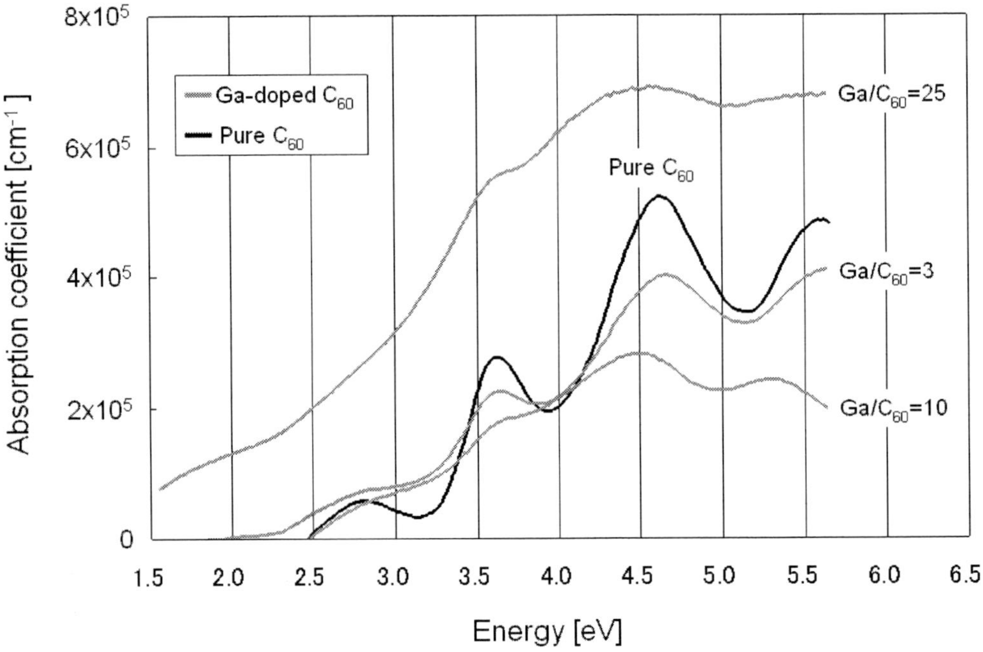

Figure 15. Absorption coefficient spectra at room temperature of a pure C_{60} film and Ga-doped C_{60} films.

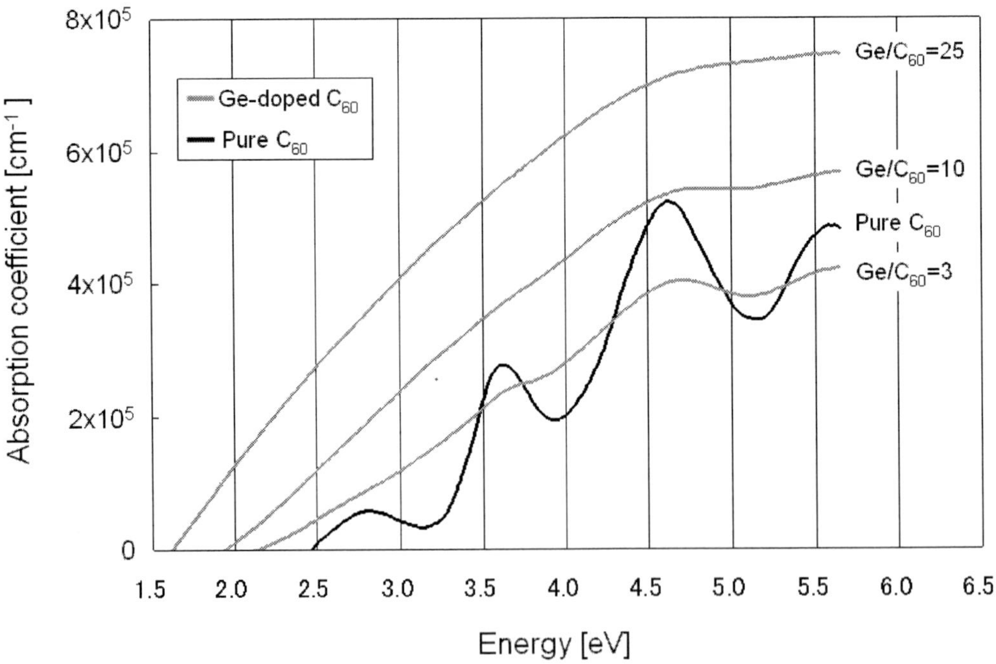

Figure 16. Absorption coefficient spectra at room temperature of a pure C_{60} film and Ge-doped C_{60} films.

Fig. 17 shows PL spectra of pure and Al-doped C_{60} films measured at 300K. In pure C_{60} films, the most dominant emission line (peak 2) is attributed to a self trapped polaron exciton.

The peak around 1.50eV (peak 1) is considered to be a phonon replica of peak 2. The energetic difference between peak 1 and peak 2 is about 180meV. This phonon mode corresponds to the strongest phonon peak in Raman spectrum [26] at 1469cm^{-1} (182meV). This is a total-symmetric A_g mode. The electron transition between HOMO and LUMO is parity forbidden in pure C_{60} films. In the Al-doped C_{60} films, however, new PL emission peaks appear around 1.75eV (peak 3), 1.85eV (peak 4) and 1.95eV (peak 5). The energy of the peak 5 coincides well with the energy difference between HOMO and LUMO states [27], suggesting that the parity forbidden transition is relieved due to Al-C_{60} bonding. The peaks around 1.75eV (peak 3) and 1.85eV (peak 4) are considered to the phonon replicas of peak 5. The energetic difference between peak 4 and peak 5 is about 90meV. This phonon mode corresponds to the phonon peak in Raman spectrum at 709cm^{-1} (88meV). The energetic difference between peak 3 and peak 5 is about 200meV. This phonon mode corresponds to the phonon peak in Raman spectrum at 1575cm^{-1}(196meV).

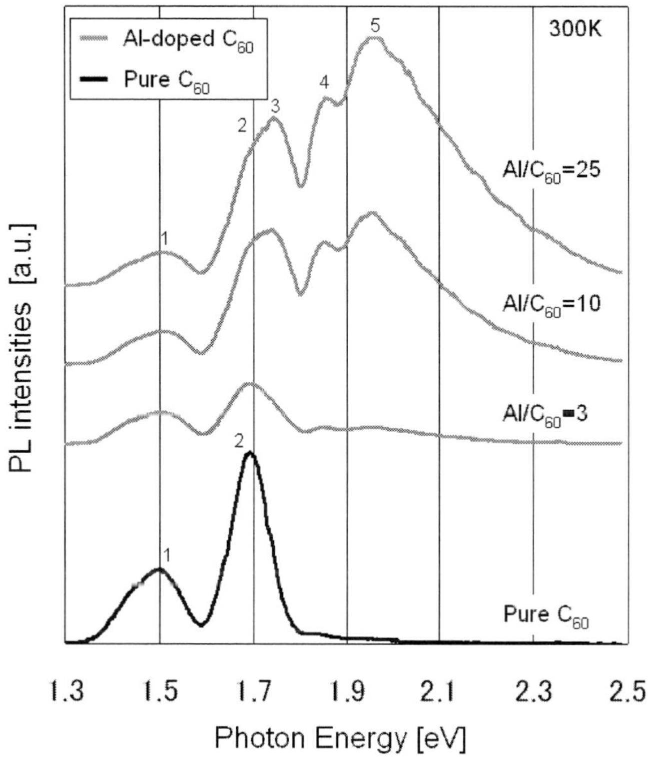

Figure 17. PL spectra of a pure C_{60} film and Al-doped C_{60} films with different concentrations grown on quartz glass substrates measured at 300K.

Fig. 18 shows PL spectra of pure and Ga-doped C_{60} films with different concentrations measured at 300K. As with the Al-doped C_{60} films, new PL emission peaks appear in the Ga-doped C_{60} films. However, Ga-doped C_{60} films show less intense PL than Al-doped films. Fig. 19 shows PL spectra of pure and Ge-doped C_{60} films measured at 300K. The PL is quenched very quickly by increasing the Ge concentrations.

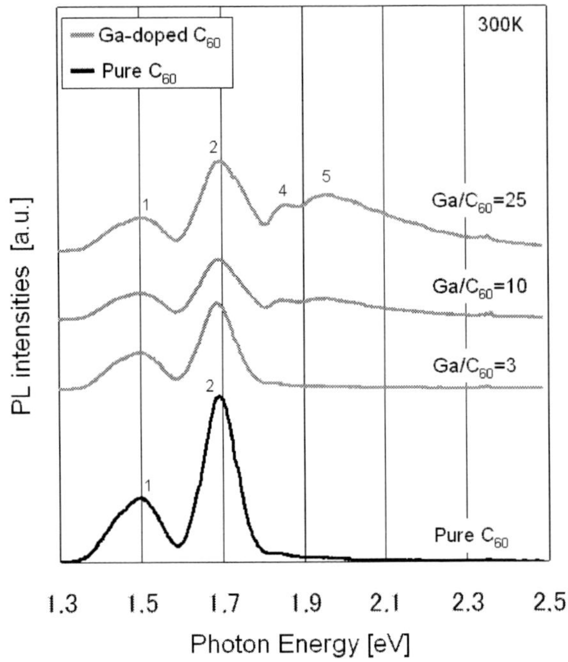

Figure 18. PL spectra of a pure C_{60} film and Ga-doped C_{60} films grown on quartz glass substrates measured at 300K.

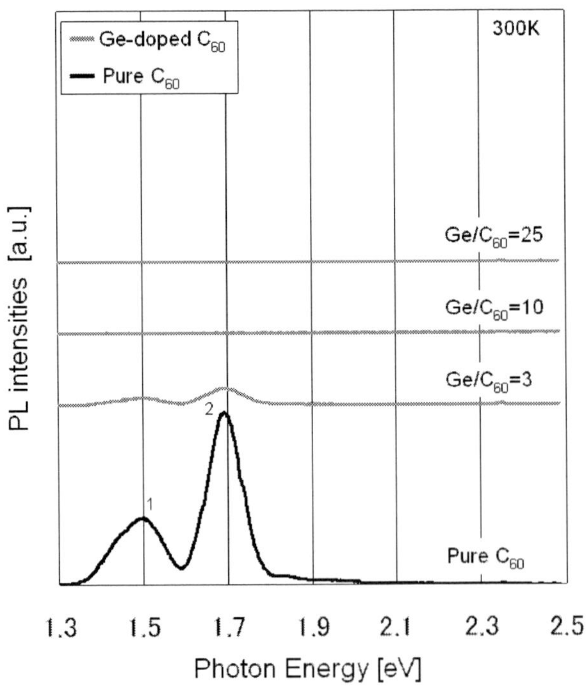

Figure 19. PL spectra of a pure C_{60} film and Ge-doped C_{60} films grown on quartz glass substrates measured at 300K.

2.3. Optimized Structures of Multivalent Impurities and C_{60} by Ab Initio Calculations

To investigate the bonding characteristics between these impurities and C_{60} molecules, the configurations of impurity atoms and C_{60} molecules are optimized by using the Gaussian 03 computer software [28] for the B3LYP molecular calculations. Fig. 20 shows the optimized topography of C_{60}-Ge-C_{60} system with equi-electron-density lines. This result suggests that the electrons are shared between a Ge atom and C_{60} molecules, indicating that stable covalent bonds are formed between them. Stable covalent bonds between the other impurity atoms and C_{60} molecules are confirmed in the same manner. The binding energies between impurities and C_{60} molecules are obtained by calculating the energies of the complexes, and then subtracting energies of the individual impurity atoms and C_{60} molecules. Table 3 shows the estimated binding energies in an impurity atom and a C_{60} molecule system. The binding energies are confirmed to be much higher than that of the van der Waals force between C_{60} molecules [29, 30]. Order of the binding energies in Table 3 (Ge-doped C_{60} > Al-doped C_{60} > Ga-doped C_{60}) coincides with that of the Vickers hardness of the impurity-doped C_{60} films. Thus, the Vickers hardness improvement by multivalent impurity doping seems to be related to these binding energies.

Table 3. Binding energies between a impurity atom and a C_{60} molecule.

system	B3LYP / 6-311+G(d) // B3LYP / 6-31G
C_{60} - C_{60}	0.24 eV*
C_{60} - Ga	0.83 eV
C_{60} - Al	0.96 eV
C_{60} - Ge	1.22 eV

*Reference data [29, 30]

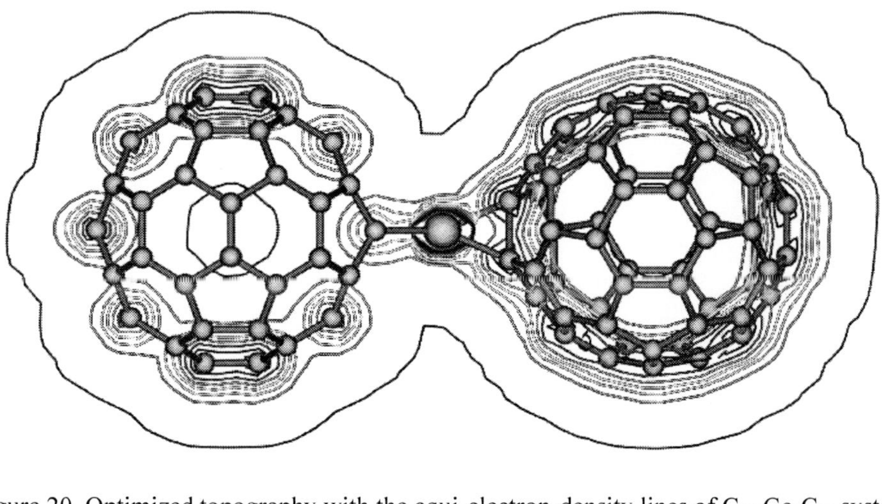

Figure 20. Optimized topography with the equi-electron-density lines of C_{60}-Ge-C_{60} system.

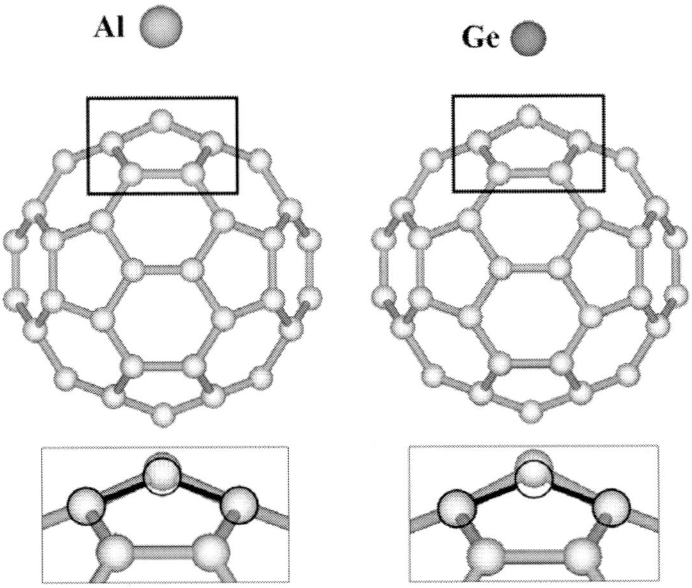

Figure 21. Optimized structures of the Al-C_{60} and Ge-C_{60} systems. In the insets, the magnified images of the distortion of the cages induced by the bonding are shown.

Fig. 21 compares the optimized structure of the Al-C_{60} and Ge-C_{60} systems. Magnified images of the distortion of the cages induced by the bonding are shown in the insets. Discernible differences can be observed between Al-C_{60} and Ge-C_{60} systems. As discussed earlier, the PL intensity of the Al-C_{60} system is much higher than that of the Ge-C_{60} system. Nevertheless, the calculated results indicate that the difference between these two systems is only the small change in the distortion. Thus the experimental PL results and calculated C_{60} distortion can be understood in the following way: In Al-C_{60} system, the distortion of the C_{60} cage is very little. This weak distortion relieves the dipole forbidden transition, resulting in an efficient PL. On the other hand, Ge-C_{60} system creates much higher distortion. This higher distortion probably suppresses the radiative recombination of C_{60} molecules.

Conclusion

High quality [111] oriented C_{60} crystal layers are grown on GaAs (111)A, (111)B and (001) substrates by solid source MBE. The lattice constant of the grown C_{60} layers calculated from XRD results coincides well with that of bulk C_{60} crystals. Efficient PL is observed in the C_{60} layers on GaAs substrates. Fabrications of nanoscale structures of C_{60} crystals are successfully achieved with area-selective epitaxy. During C_{60} deposition, C_{60} molecules landing on SiO_2 masks either evaporate or migrate to the open area enhancing growth rate. The mean diffusion length of C_{60} molecules on SiO_2 mask at 200°C is estimated to be 200-300nm from the volume of the C_{60} crystals grown by the lateral flux. The different epitaxial relationships are observed between the nanoscale growth and planer or wide area growth on GaAs (001). When the open area size is larger than approximately 250nm, the grown crystals show [111] orientation. While, when it is below this value, the epitaxial orientation is [001].

This result indicates that the crystal size may be one of the determining factors for the C_{60} epitaxial orientation.

Multivalent impurity-doped C_{60} films are grown on GaAs and quartz glass substrates. RHEED of the films exhibits halo patterns and all the films show no peak in XRD measurement. SIMS measurements show that there is no segregation of aluminum atoms in the films. Vickers hardness of all the impurity-doped C_{60} films is confirmed to be enhanced, and the films are found to be stable against organic solvents. For pure C_{60} layers, the absorption peaks caused by the electron transition among the C_{60} molecular orbitals are observed. However, absorption peaks of all the impurity-doped C_{60} films become less pronounced due to the bonding between impurity atoms and C_{60} molecules. New PL peaks between HOMO and LUMO states of C_{60} are found in the Al-doped and Ga-doped C_{60} films. Optimized structures of impurity atoms and C_{60} molecules are obtained by ab initio calculations. Stable covalent bonds between impurities and C_{60} molecules are confirmed to be formed. The impurity atoms probably act as bridges between C_{60} molecules.

References

[1] H.W. Kroto, J.R. Heath, S.C. O'Brien, R.F. Curl, R.E. Smalley, "C_{60}: Buckminsterfullerene" *Nature.* 1985, 318, pp 162-163.

[2] W. Krätschmer, L.D. Lamb, K. Fostiropoulos, D.R Huffman, "Solid C_{60}: a new form of carbon" *Nature*. 1990, 347, pp 354-358.

[3] A.F. Hebard, M.J. Rosseinsky, R.C. Haddon, D.W. Murphy, S.H. Glarum, T.T.M. Palstra, A.P. Ramirez, A.R. Kortan, "Superconductivity at 18K in potassium-doped C_{60}" *Nature*, 1991, 350, pp 600-601.

[4] P.M. Allemand, K.C. Khemani, A. Koch, F. Wudl, K. Holczer, S. Donovan, G. Grüner, J. D. Thompson, "Organic molecular soft ferromagnetism in a fullerene C_{60}" *Science*, 1991, 253, pp 301-303.

[5] Koma, "Van der Waals epitaxy – a new epitaxial growth method for a highly lattice-mismatched system" *Thin Solid Films*, 1992, 216, pp 72-76.

[6] T. Sakurai, X.D. Wang, Q.K. Xue, Y. Hasegawa, T. Hashizume, H. Shinohara, "Scanning tunneling microscopy study of fullerenes" *Progress in Surface Science*, 1996, 51, pp 263-408.

[7] Y. Yoneda, K. Sakaue, H. Terauchi, "Epitaxial growth of single-crystal C_{60} films on GaAs(111) substrates" *Journal of the Physical Society of Japan*, 1994, 63, pp 3560-3563.

[8] J.H. Yao, Y.J. Zou, X.W. Zhang, G.H. Chen, "The (111) oriented growth of C_{60} films on GaAs (100) substrates" *Thin Solid Films*, 1997, 305, pp 22-25.

[9] J. Nishinaga, M. Ogawa, Y. Horikoshi, "Selective growth of C_{60} layers on GaAs and their crystalline characteristics" *Thin Solid Films*, 2004, 464-465, pp 323-326.

[10] J. Nishinaga, T. Aihara, T. Toda, F. Matsutani Y. Horikoshi, "Nanoscale selective area epitaxy of C_{60} crystals on GaAs by molecular beam epitaxy" *Journal of Vacuum Science and Technology B*, 2006, 24, pp 1587-1590.

[11] J. Li, S. Komiya, T. Tamura, C. Nagasaki, J. Kihara, K. Kishio, K. Kitazawa, "Growth and properties of pure C_{60} single crystals from vapor" *Physica C*, 1992, 195, pp 205-208.

[12] W.I.F. David, R.M. Ibberson, T. Matsuo, "High-resolution neutron powder diffraction – a case-study of the structure of C_{60}" *Proceedings of the Royal Society of London Series A-Mathematical Physical and Engineering Science*, 1993, 442, pp 129-146.

[13] V. Capozzi, G. Casamassima, G.F. Lorusso, A. Minafra, R. Piccolo, T. Trovato, A. Valentini, "Optical spectra and photoluminescence of C_{60} thin films" *Solid State Communications*, 1996, 98, pp 853-858.

[14] M. Matus, H. Kuzmany, E. Sohmen, "Self-trapped polaron exciton in neutral fullerene C_{60}" *Physical Review Letters*, 1992, 68, pp 2822-2825.

[15] Akimoto, K. Kan'no, "Photoluminescence and near-edge optical absorption in the low-temperature phase of pristine C_{60} single crystals" *Journal of the Physical Society of Japan*, 2002, 71, pp 630-643.

[16] S. Saito, A. Oshiyama, "Cohesive mechanism and energy bands of solid C_{60}" *Physical Review Letters*, 1991, 66, pp 2637-2640.

[17] Q.K. Xue, Y. Ling, T. Ogino, T. Sakata, Y. Hasegawa, T. Hashizume, H. Shinohara, T. Sakurai, "C_{60} single crystal films on GaAs (001) surfaces" *Thin Solid Films,* 1996, 281-282, pp 618-623.

[18] T. Sakurai, Q.K. Xue, T. Hashizume, Y. Hasegawa, "Extraordinary growth of C_{60} on a GaAs (001) As-rich 2x4 surface" *Journal of Vacuum Science and Technology B,* 1997, 15, pp 1628-1632.

[19] A.V. Hamza, J. Dykes, W.D. Mosley, L. Dinh, M. Balooch, "Reaction and passivation of aluminum with C_{60}" *Surface Science*, 1994, 318, pp 368-378.

[20] J.G. Hou, Y. Li, Y. Wang, W. Xu, J. Zuo, Y.H. Zhang, "Study of microstructure and interfacial interaction in Al-C_{60} co-evaporated films" *Physica Status Solidi A - Applied Research*. 1997, 163, pp 403-409.

[21] A.J. Maxwell, P.A. Brühwiler, D. Arvanitis, J. Hasselström, M.K.J. Johansson, N. Martensson, "Electronic and geometric structure of C_{60} on Al (111) and Al (110)" *Physical Review B*, 1998, 57, pp 7312-7326.

[22] J. Nishinaga, T. Aihara, H. Yamagata, Y. Horikoshi, "Mechanical and optical characteristics of Al-doped C_{60} films" *Journal of Crystal Growth*, 2005, 278, pp 633-637.

[23] J. Nishinaga, T. Aihara, A. Kawaharazuka, Y. Horikoshi, "Characteristics of multivalent impurity doped C_{60} films grown by MBE" *Journal of Crystal Growth*, 2007, 301-302, pp 687-691.

[24] M. Tachibana, M. Michiyama, K. Kikuchi, Y. Achiba, K. Kojima, "Temperature dependence of the microhardness of C_{60} crystals" *Physical Review B*, 1994, 49, pp 14945-14948.

[25] F. Negri, G. Orlandi, F. Zerbetto, "Interpretation of the vibrational structure of the emission and absorption spectra of C_{60}" *Journal of Chemical Physics*, 1992, 97, pp 6496-6503.

[26] P.C. Eklund, P. Zhou, K. Wang, G. Dresselhaus, M.S. Dresselhaus, "Optical phonon modes in solid and doped C_{60}" *Journal of Physics and Chemistry of Solids*, 1992, 53, pp 1391-1413.

[27] M.K. Kelly, P. Etchegoin, D. Fuchs, W. Krätschmer, K. Fostiropoulos, "Optical transitions of C_{60} films in the visible and ultraviolet form spectroscopic ellipsometry" *Physical Review B*, 1992, 46, pp 4963-4968.

[28] M.J. Frisch et al., Gaussian 03, Revision C.02, Gaussian Inc., Wallingford, CT, 2004.

[29] L.A. Girifalco, "Molecular properties of C_{60} in the gas and solid phases" *Journal of Physical Chemistry*, 1992, 96, pp 858-861.

[30] P.A. Gravil, M. Devel, Ph. Lambin, X. Bouju, Ch. Girard, A.A. Lucas, "Adsorption of C_{60} molecules" *Physical Review B*, 1996, 53, pp 1622-1629.

In: Progress in Fullerene Research
Editor: Milton Lang, pp. 145-198
ISBN 978-1-60021-841-5

Chapter 5

RECENT ADVANCES IN THE ELECTRONIC AND VIBRATIONAL STUDIES OF CARBON NANOTUBES USING FULLERENES AS LOCAL PROBES

Ferenc Simon*

Budapest University of Technology and Economics, Institute of Physics and Condensed Matter Research Group of the Hungarian Academy of Sciences, H-1521, Budapest P.O.Box 91, Hungary
and
Institut für Materialphysik, Universität Wien
Strudlhofgasse 4, A-1090 Wien, Austria

Abstract

The discovery of fullerenes encapsulated inside single-wall carbon nanotubes (SWCNTs) by Smith *et al.* [Nature 396, 323 (1998)], provides a compelling opportunity to study the nanotubes using fullerenes as local probes. It is reviewed herein, how fullerenes and magnetic fullerenes such as N@C_{60} and $C_{59}N$, and ^{13}C enriched fullerenes can be used to yield unprecedented insight into the nanotube electronic, vibrational, and optical properties and into the growth of SWCNTs. Encapsulated C_{60} fullerenes are transformed to inner tubes by high temperature annealing. The unique, low defect concentration and well shielded nature of inner tubes makes them ideal to study the effect of diameter dependent treatments such as opening and closing of the tubes. Electron-spin resonance studies on the encapsulated magnetic fullerenes indicates a metallic-like finite density of states for an ensemble SWCNT sample, which contradicts the current view of SWCNTs as containing separate metallic and semiconducting nanotubes. The linear spin chains of the encapsulated magnetic fullerenes is suggested as a building element for quantum information processing. The possibility to grow inner tubes from encapsulated fullerenes allows the isotope selective enrichment of the inner tubes if the source fullerene is ^{13}C enriched. Nuclear magnetic resonance on such samples evidences the presence of a low energy spin-gap which is thought to open due to a Peierls transition.

*E-mail address: ferenc.simon@univie.ac.at

Key words: Carbon nanotubes Fullerene encapsulation Double-wall carbon nanotubes Isotope engineered nanotubes Raman spectroscopy Magnetic resonance spectroscopy

1. Introduction

1.1. The Discovery of Peapods

The discovery of fullerenes in [1] in 1985 revolutionized several fields in chemistry, physics and also in biology. Fullerenes gave material scientist a fresh look at carbonaceous systems: it suggested that there may exist a number of other forms of carbon which await discovery. The discovery of multi-walled carbon nanotubes (MWCNT) by Sumio Iijima in 1991 fulfilled this expectation [2]. In fact, he was using the same apparatus which was used for the production of fullerenes. This event was followed by the discovery of single-wall carbon nanotubes (SWCNT) in 1993 [3, 4], which can be grown in similar conditions as the fullerenes and MWCNTs but exclusively in the presence of metal catalysts.

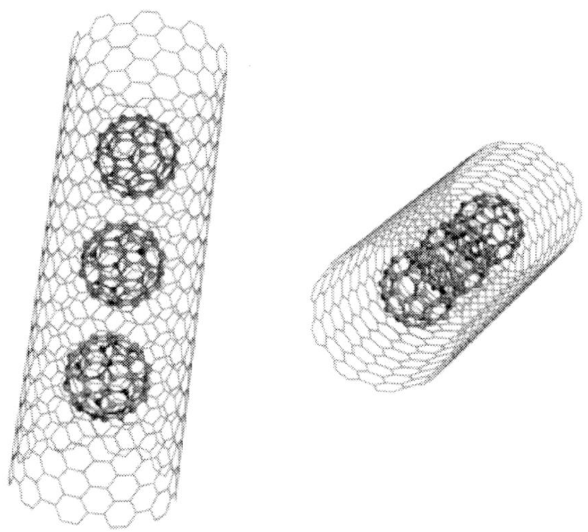

Figure 1. Schematics of the peapod structure from two viewpoints.

Fullerenes not only represent the forerunners of carbon nanotubes but are structurally connected: one of the structurally simplest carbon nanotube, the (5,5) SWCNT, can be represented as an elongated fullerene capped with two half C_{60}s. Related to the growing interest of the scientific community there has been a shift in terminology: originally, tubes were termed "bucky-tubes" for "large, elongated fullerenes", more recently the fullerenes are often mentioned as "small nanotubes". One significant event which further connects nanotubes and fullerenes was the discovery of fullerene peapods. In 1998, Smith, Monthioux and Luzzi reported the observation of fullerenes encapsulated inside single-wall carbon nanotubes using high-resolution transmission electron microscopy (HR-TEM) [5]. This peapod structure is particularly interesting as it combines two fundamental forms of carbon: fullerenes and carbon nanotubes. The schematics of the structure is shown in Fig. 1. It was

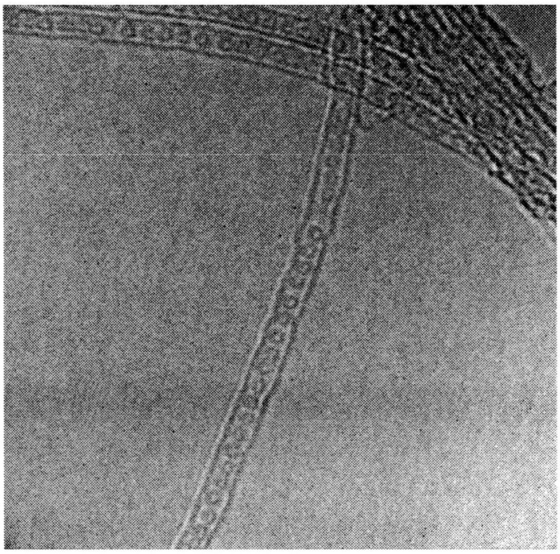

Figure 2. HR-TEM image of highly fullerene filled C_{60}@SWCNT peapods.

suggested that fullerenes which are co-produced with the SWCNTs enter the tubes through openings and remain inside as it is energetically preferred [6, 7, 8, 9].

It was also shown that macroscopic filling with the fullerenes can be achieved [10, 11, 12]. In Fig. 2 we show a HR-TEM image of the resulting structure: the high filling with fullerenes is clearly recognized. For such peapods, the presence of C_{60}s can be observed by scanning tunneling spectroscopy through the influence on the electronic properties of the tubes [13]. The high filling opens the way for magnetic resonance studies through encapsulation of magnetic fullerenes and the results are discussed herein. Such methods usually require that a sizeable amount of electron or nuclear spins are present at well defined locations.

The faith of fullerenes and nanotubes were further linked by the observation that upon intensive electron irradiation [10] or upon heating to 1200 °C [11] the encapsulated fullerenes are transformed to inner tubes. These inner tubes are single-walled such as the host outer tubes and the whole structure forms a double-wall carbon nanotube (DWCNT). It is the aim of this work to review how the encapsulated fullerenes and the peapod based inner tubes can be used to study the electronic and vibrational properties of the single-wall carbon nanotubes.

1.2. Single-Wall Carbon Nanotubes

The principal interest in carbon nanotubes (CNTs) comes from the fact that they contain carbon only and all carbons are locally sp^2 bound, like in graphite, which provides unique mechanical and transport properties. This, combined with their huge, >1000, aspect ratio (the diameters being 1-20 nm and their lengths over 1 micron), endows them with their enormous application potential. The not exhaustive list of applications includes field-emission displays (exploiting their sharp tips) [14], cathode emitters for small sized x-ray tubes for

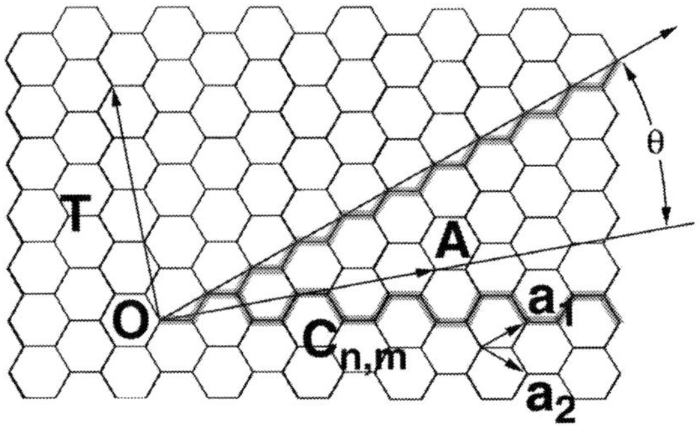

Figure 3. Geometry of a graphene sheet. $\mathbf{a}_1$ and $\mathbf{a}_2$ are the primitive lattice vectors. θ is measured between $\mathbf{a}_1$ and $\mathbf{A}$ and is called chiral angle. The Hamada vectors for an armchair (lower solid curve) and a zig-zag (upper solid curve) are indicated.

medical applications [15], reinforcing elements for CNT-metal composites, tips for scanning probe microscopy [16], high current transmitting wires, cables for a future space elevator, elements of nano-transistors [17], and elements for quantum information processing [18].

Carbon nanotubes can be represented as rolled up graphene sheets, i.e. single layers of graphite. Depending on the number of coaxial carbon nanotubes, they are usually classified into multi-wall carbon nanotubes (MWCNTs) and single-wall carbon nanotubes (SWCNTs). Some general considerations have been clarified in the past 16 years of nanomaterial research related to these structures. MWCNTs are more homogeneous in their physical properties as the large number of coaxial tubes smears out individual tube properties. This makes them suitable candidates for applications where their nanometer size and the conducting properties can be exploited such as e.g. nanometer sized wires. In contrast, SWCNT materials are grown as an ensemble of weakly interacting tubes with different diameters. The physical properties of similar diameter SWCNTs can change dramatically as the electronic structure is very sensitive to the rolling-up direction, the so-called chiral- or Hamada-vector [19, 20]. The chiral vector is characterized by the (n, m) vector components which denote the direction along which a graphene sheet is rolled up to form a nanotube. The chirality uniquely sets the diameter of the tube. The geometry of a graphene sheet is shown in Fig. 3.

Depending on the chiral vector, SWCNTs can be metallic or semiconducting [20]. This provides a richer range of physical phenomena as compared to the MWCNTs, however significantly limits the range of applications. To date, neither the directed growth nor the controlled selection of SWCNTs with a well defined chiral vector has been performed successfully. A real sample contains SWCNT aggregates, called bundles, which contain tubes with random chirality and the samples also contain other carbonaceous contaminations, such as amorphous carbon and graphite. Correspondingly, current research is focused on the post-synthesis separation of SWCNTs with a narrow range of chiralities [21, 22, 23, 24] or on methods which yield information that are specific to SWCNTs with different chiralities. Examples for the latter are the observation of chirality selective band-gap fluorescence in semiconducting SWCNTs [25] and chirality assigned resonant Raman scattering [26, 27].

Clearly, several fundamental questions need to be answered before all the benefits of these novel nanostructures could be fully exploited. Recent theoretical and experimental efforts focused on the understanding of the electronic and optical properties of single-wall carbon nanotubes. It has been long thought that the one-dimensional structure of SWCNTs renders their electronic properties inherently one-dimensional [19, 20]. This was suggested to result in a range of exotic correlated phenomena such as the Tomonaga-Luttinger (TLL) state [28], the Peierls transition [29, 30], ballistic transport [31], and bound excitons [32, 33, 34, 35]. The presence of the TLL state is now firmly established [36, 37, 38], there is evidence for the ballistic transport properties [31] and there is growing experimental evidence for the presence of excitonic effects [39, 40]. The Peierls transition is argued to be observed in NMR investigations reviewed herein.

1.3. Double-Wall Carbon Nanotubes

An interesting follow-up of the peapod structure discovery is that the encapsulated fullerenes can be fused into a smaller diameter inner tube [11, 41] thus producing a double-wall carbon nanotube (DWCNT). DWCNTs were first observed to form under intensive electron radiation [10] in a high resolution transmission electron microscope from C_{60} peapods. Following the synthesis of C_{60} peapods in macroscopic amounts [12], bulk quantities of the DWCNT material are available using a high temperature annealing method [41]. Alternatively, DWCNTs can be produced with usual synthesis methods such as arc-discharge [42] or Chemical Vapor Deposition, CVD, [43] under special conditions. According to the number of shells, DWCNTs are between SWCNTs and MWCNTs. Thus, one expects that DWCNTs may provide a material where improved mechanical stability as compared to SWCNTs coexists with the rich variety of electronic properties of SWCNTs. There are, of course, a number of yet unanswered questions e.g. if the outer tube properties are unaffected by the presence of the inner tube or if the commensurability of the tube structures plays a role. These questions should be answered before the successful application of these materials.

The inner tubes grown inside SWCNTs from peapods turned out to be a particularly interesting system as they are remarkably defect free and are isolated from the environment which results in very long phonon life-times, i.e. very narrow vibrational modes [44]. In addition, their smaller diameters results in a larger energy spread, i.e. larger spectral splitting, for diameter dependent phonon modes such as e.g. the radial breathing mode (RBM). These two effects make the inner tubes very suitable to study diameter dependent physics of the small diameter tubes with precision. Here, we review how to employ the inner tubes as *probes* of the outer tube properties. The additional benefit of the inner tube growth from fullerenes is the possibility to use ^{13}C isotope enriched C_{60} to obtain specifically ^{13}C enriched inner tubes. This way, the ^{13}C concentration on the inner tubes is much larger than on the outer ones or in the other carbonaceous phases, which are inevitably present in the sample. In Fig. 4, we show the schematics of such a DWCNT.

Here, we review the efforts to study the SWCNTs properties through encapsulation of fullerenes using Raman and magnetic resonance spectroscopy. The reviewed phenomena include the precise characterization of diameter distribution of SWCNTs, the study of reversible hole engineering on the SWCNTs, the study of the inner tube growth mechanism

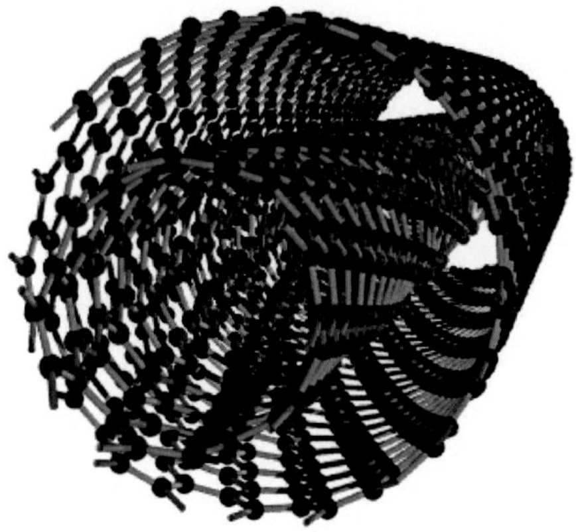

Figure 4. Schematic structure of an isotope engineered DWCNT with (14,6) outer and (6,4) inner tubes. ^{12}C and ^{13}C are shown in black and blue, respectively. The inner tube is 89 % ^{13}C enriched and the outer contains natural carbon (1.1 % ^{13}C abundance), which are randomly distributed for both shells.

with the help of ^{13}C isotope labeling, the study of local density of states on the tubes using nuclear magnetic resonance (NMR) on the ^{13}C isotope enriched inner tubes, and the electron spin resonance (ESR) studies of the SWCNTs using encapsulated magnetic fullerenes. This review is organized as follows. First, we present the general properties of DWCNTs using Raman spectroscopy and we discuss the electronic and vibrational properties of the inner tubes, which are the probes in the subsequent studies. Second, we present the use of the inner tubes to probe the host outer tube diameter distribution and to study the opening and closing of holes on the outer tubes. Third, we present a study on the inner tube growth mechanism using isotope enriched carbon. Fourth, we discuss the efforts related to studying the SWCNT properties by encapsulating magnetic fullerenes using ESR. Fifth, we discuss the NMR results on the isotope enriched inner tubes and in particular we present the observation of a low energy spin-gap in the density of states of SWCNTs.

2. Experimental Methods and Sample Preparation

2.1. Sample Preparation

The starting SWCNT samples

SWCNTs from different sources and prepared by different methods were used. Commercial arc-discharge grown SWCNTs with 50 % weight purity (Nanocarblab, Moscow, Russia) and laser ablation prepared SWCNTs with 10 % weight purity (Tubes@Rice, Houston, USA) were used. The latter material was purified through repeated steps of air oxidation and washing in HCl. Some laser ablation prepared and purified samples were obtained from H. Kataura. SWCNTs in the purified samples are usually well opened to enable fullerene encapsulation. If not, annealing in air at 450 °C for 0.5 hour makes them sufficiently open. SWCNTs grown by the HiPco process (Carbon Nanotechnologies Inc., Houston, USA) were used as reference. Most samples were used in the form of a buckypaper, which is prepared by filtering a suspension of SWCNTs. We found that commercially available SWCNTs already meet a required standard in respect of purity and quality. In addition, for the amount of experimental work described here, reproducible samples i.e. a large amount of SWCNTs from similar quality, were required. Commercial samples meet this requirement, which compensates for their slightly inferior quality compared to laboratory prepared ones.

Synthesis of peapods

Encapsulation of fullerenes at low temperatures inside SWCNTs (solvent method) was performed by sonicating the fullerene and opened SWCNT suspensions together in organic solvents following Refs. [45, 46, 47, 48]. For fullerene encapsulation at high temperatures (the vapor method), the SWCNTs and the fullerenes were sealed under vacuum in a quartz ampoule and annealed at 650 °C for 2 hours [12]. Fullerenes enter the inside of the SWCNTs at this temperature due to their high vapor pressure that is maintained in the sealed environment. Non-encapsulated fullerenes were removed by dynamic vacuum annealing at the same temperature for 1 hour. High purity fullerenes were obtained from a commercial source (Hoechst AG, Frankfurt, Germany). The filling of SWCNTs with the fullerenes was characterized by observing the peapod structure in high-resolution transmission electron microscopy (HR-TEM), by x-ray diffraction studies of the one-dimensional array of fullerenes inside the SWCNTs and by the detection of the fullerene modes from the cages encapsulated inside the SWCNTs using Raman spectroscopy [12, 49].

Synthesis of DWCNTs

DWCNTs were prepared by two routes: from fullerene peapods and using chemical vapor deposition (CVD) growth technique [50]. The peapods were transformed to DWCNTs by a dynamic vacuum treatment at 1250 °C for 2 hours following Ref. [41]. Again, the DWCNT transformation was followed by HR-TEM and by the observation of the DWCNT structure factors using x-ray studies. In addition, new Raman modes emerge after the 1250 °C heat treatment particularly in a frequency range that is clearly upshifted from the outer tube RBMs. For the CVD DWCNT growth [50], the catalyst was a modified version of the Fe/Mo/MgO system developed by Liu *et al.* [51] for SWCNT synthesis.

Both kinds of DWCNTs have advantageous and disadvantageous properties. For peapod template grown DWCNTs, the inner tube is known to fill only up to ~ 70 % of the outer tube length [52]. This is the consequence of insufficient carbon in the fullerenes: the C_{60} peapods have 60 carbon atoms per 1 nm (the lattice constant of the peapod) whereas the (9,0) inner tube with $d = 0.7084$, which is representative of the most abundant 7 nm diameter inner tube contains 36 carbon atoms per the $c_0 = 0.4238$ nm lattice constant [53]. In contrast, CVD grown inner tubes fill up to the total length of the outer tubes, however such samples have usually a less well defined tube diameter distribution: i.e. small diameter SWCNTs and large diameter DWCNTs are also grown [54]. Peapod template grown

DWCNTs can be grown with relatively narrow diameter distribution due to the available narrow diameter distribution of the SWCNT host tubes. This also allows for a good control over the DWCNT diameter as described in Ref. [55] and is discussed below.

Synthesis of isotope engineered DWCNTs

Commercial ^{13}C isotope enriched fullerenes (MER Corp., Tucson, USA) were used to prepare fullerene peapods C_{60},C_{70}@SWCNT with enriched fullerenes. Two supplier specified grades of ^{13}C enriched fullerene mixtures were used: 25 and 89 %. These values were slightly refined using Raman spectroscopy. The 25 % grade was nominally C_{60}, and the 89 % grade was nominally C_{70} with C_{60}/C_{70}/higher fullerene compositions of 75:20:5 and 12:88:$<$ 1, respectively. The above detailed standard routes were performed for the peapod and the DWCNT productions.

2.2. Experimental Methods

Raman spectroscopy

Raman spectra were measured with a Dilor xy triple spectrometer using various lines of an Ar/Kr laser, a He/Ne laser and a tunable Ti:sapphire and Rhodamin dye-laser in the 1.54-2.54 eV (805-488 nm) energy range. Tunable lasers allow to record the so-called Raman map [26, 27] i.e. to detect the SWCNT resonance energies through the Raman resonance enhancement [56], which ultimately allows the chiral index assignment. The spectra can be recorded in normal (NR) and high resolution (HR) mode, respectively ($\Delta\bar{\nu}_{\mathrm{NR}} = 1.3\ \mathrm{cm}^{-1}$ for blue and $\Delta\bar{\nu}_{\mathrm{HR}} = 0.4\ \mathrm{cm}^{-1}$ in the red). The samples in the form of bucky-paper are kept in dynamic vacuum and on a copper tip attached to a cryostat, which allows temperature variation in the 20-600 K temperature range. Raman spectroscopy was used to characterize the diameter distribution of the SWCNTs, to determine the peapod concentrations, and to monitor the DWCNT transformation of the peapod samples.

Electron spin resonance

The peapod and the reference SWCNT materials were mixed with the ESR silent high purity SnO_2 powder in a mortar to separate the pieces of the conducting bucky-papers. The samples were sealed under dynamic vacuum. A typical microwave power of 10 μW and 0.01 mT magnetic field modulation at ambient temperature were used for the measurements in a Bruker Elexsys X-band spectrometer.

Nuclear magnetic resonance

Nuclear magnetic resonance (NMR) is usually an excellent technique for probing the electronic properties at the Fermi level of metallic systems. The examples include conducting polymers, fullerenes, and high temperature superconductors. However the 1.1% natural abundance of ^{13}C with nuclear spin I=1/2 limits the sensitivity of such experiments. As a result, meaningful NMR experiments has to be performed on ^{13}C isotope enriched samples. NMR data were taken with the samples sealed in quartz tubes filled with a low pressure of

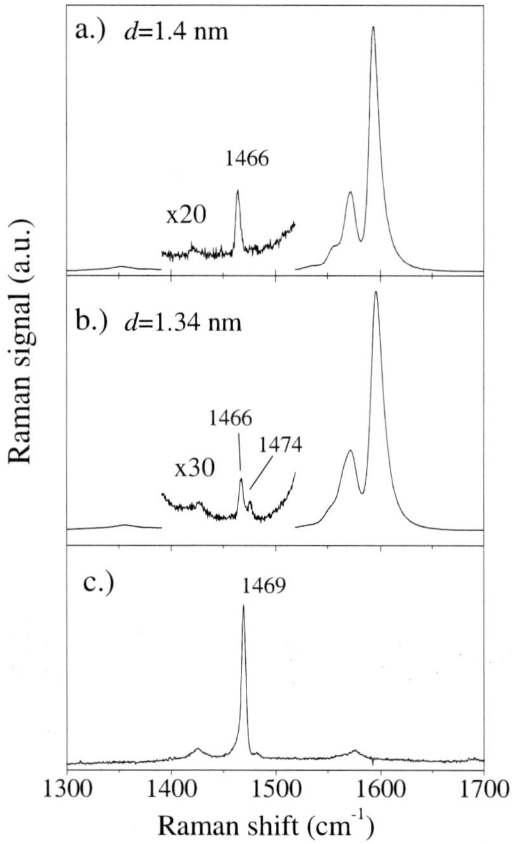

Figure 5. The $A_g(2)$ Raman spectral region and the SWCNT D and G modes of fullerenes encapsulated in different diameter tubes (a-b). We show the corresponding Raman spectrum region of crystalline C_{60} (c). Raman shifts for the peaks and mean diameters, d, of the SWCNTs are also given.

high purity Helium gas [57]. We probed the low frequency spin dynamics (or low energy spin excitations, equivalently) of the inner-tubes using the spin lattice relaxation time, T_1, defined as the characteristic time it takes the ^{13}C nuclear magnetization to recover after saturation. The signal intensity after saturation, $M(t)$, was deduced by integrating the fast Fourier transform of half the spin-echo for different delay times, t.

3. Results and Discussion

3.1. Peapods and Peapod Based Inner Tubes as Local Probes

3.1.1. Structure and Vibrational Properties of Fullerene Peapods

As discussed in the Introduction, peapods were discovered by Smith et al. using HR-TEM [5]. Later the same authors suggested a method which yields abundant nanotube filling [11] and results in structures similar to that in Fig. 2. Kataura et al. suggested an alternative synthesis method of peapods and the use of Raman spectroscopy to characterize the filling efficiency [12]. Raman spectroscopy is a bulk method compared to HR-TEM and is capable to determine whether filling on a macroscopic level has been achieved. In Fig. 5, we compare the Raman spectra of crystalline C_{60} and C_{60} peapod samples. For the latter we show peapods in SWCNTs with two different mean diameters, d.

The Raman spectra of peapod samples in the plotted frequency range consist of the SWCNT G and D modes at 1590 and 1355 cm^{-1}, respectively and narrow lines related to the Raman active modes of the C_{60} inside the SWCNT [58, 49]. Enlarged is the most significant Raman active line of C_{60} peapod, the $A_g(2)$ mode [59]. For the $d = 1.4$ nm SWCNT peapod sample, we observe a single $A_g(2)$ line at 1466 cm^{-1} and for the $d = 1.34$ nm SWCNT the well known doublet peapod signal at 1466 and 1474 cm^{-1}[49]. The 3 cm^{-1} downshift of the main $A_g(2)$ peak and the splitting into two components in small diameter SWCNTs was assigned to the encapsulation and thus to a tube-fullerene interaction [49]. It was found that it is the intensity ratio of the split components and not the Raman shifts which changes as a function of the SWCNT diameter in peapod samples [60]. Two alternatives are suggested for the origin of the $A_g(2)$ mode components at 1474 cm^{-1}: one associates it with immobile C_{60} molecules in the smaller diameter nanotubes [60]. The other scenario associates the upshifted peak with the presence of C_{60} linear chains with pentagonal, i.e. facing pentagons, coordination and the main peak at 1466 cm^{-1} with hexagonally coordinated fullerene chains [61].

Although Raman spectroscopy provides a macroscopic method and a spectroscopic tool to characterize the encapsulation, it cannot give a quantitative estimate of the filling efficiency. This originates in the fact, that Raman intensities are affected by a number of factors such as the so-called Raman matrix elements and the strength of Raman resonance enhancement [56]. The latter effect is discussed below in Section 3.1.3. The solution to this problem is the use of methods which enable a direct separation of fullerene and SWCNT related carbons. We present in Section 3.4. a method employing ^{13}C enriched fullerenes and NMR spectroscopy, which enable such a measurement. Here, we discuss the method developed by Pichler et al. to determine the fullerene filling factors [62] using electron-energy-loss spectroscopy (EELS). The Authors exploited the fact that the C1 s core-level excitation spectrum of C_{60} shows uniquely sharp features, thus the amount of C_{60} related carbon atoms can be measured in a peapod sample. This also allowed a precise calibration of the relative intensity of SWCNT G and C_{60} $A_g(2)$ modes C_{60} [63].

A third alternative method to determine the fullerene filling efficiency exploits the ordered nature of the fullerenes inside the tubes: C_{60}s form a 1D chain inside the tubes. This results in the modulation of the hexagonal Bragg peaks of the SWCNTs in the bundles. In Fig. 6, we show the X-ray diffraction pattern of the unfilled SWCNTs and peapods. The encapsulation with C_{60} strongly modifies the intensity of the peaks, in particular the 10 peak around $q = 4.5$ nm^{-1} is strongly depressed in comparison to the other peaks [52]. The intensity modulation of the hexagonal lattice peaks by the different form factor of filled and empty SWCNT can be related to the filling content of the encapsulated peapods [52, 64]. The first peak (A) decreases for filled nanotubes in comparison to the pristine SWCNT material and a second peak (B) appears, which is related to the one-dimensional lattice of the encapsulated C_{60} molecules. A numerical simulation [52, 64] leads for the inter-C_{60} distance and the coherence length of 0.95 nm and 25 nm, respectively, in agreement with

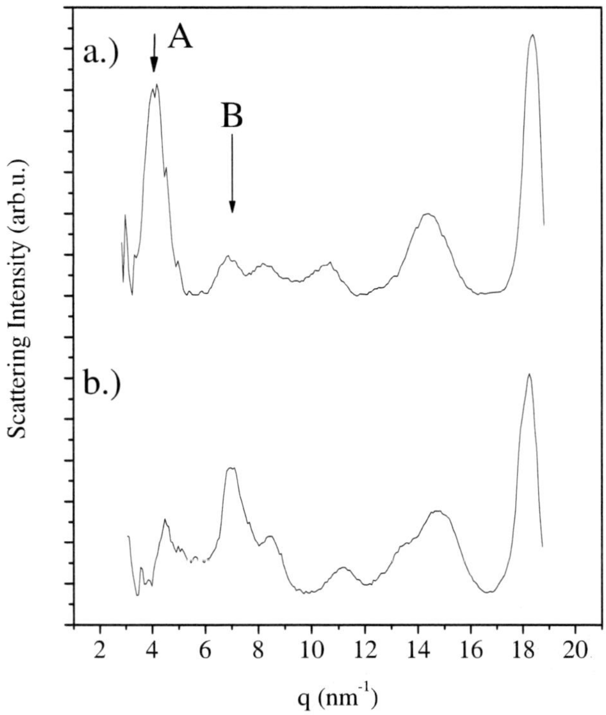

Figure 6. X-ray diffraction profile of a.) pristine SWCNT, b.) C_{60}@SWCNT peapod samples. The relative intensity of the A and B peaks is a measure of the peapod concentration. The scattering peak observed at 18 nm^{-1} comes from residual graphitic carbon. Reprinted figure with permission from Ref. [46], F. Simon *et al.* Chem. Phys. Lett. **383**, 362 (2004). Copyright (2004) by Elsevier.

the electron diffraction studies [62, 65].

3.1.2. Growth and Properties of Peapod Based Inner Tubes

Encapsulating fullerenes and their transformation into inner tubes by the high temperature annealing process [41] provides a unique opportunity to study the properties of the host outer tubes. In Fig. 7 we show the evolution of the SWCNT Raman spectrum upon C_{60} fullerene encapsulation and the DWCNT transformation after Ref. [44]. The series of sharp modes in the peapod spectrum, which are related to the encapsulated fullerenes [49], disappear upon the heat treatment and a series of sharp modes appear in the 250-450 cm^{-1} spectral range. The presence of inner tubes after this protocol have been independently confirmed by HR-TEM [11] and structures similar to that shown in Fig. 8 were observed. The small diameter tubes with $d \sim 0.7$ nm would have RBM modes in the $\sim$ 250-450 cm^{-1} spectral range, which clarifies the identification of these modes. The identification of the inner tube RBMs is possible due to the strong d dependence of this Raman mode [66]. It is shown in Section 3.2. that unambiguous assignment can be given with the help of selective isotope enrichment of the inner walls.

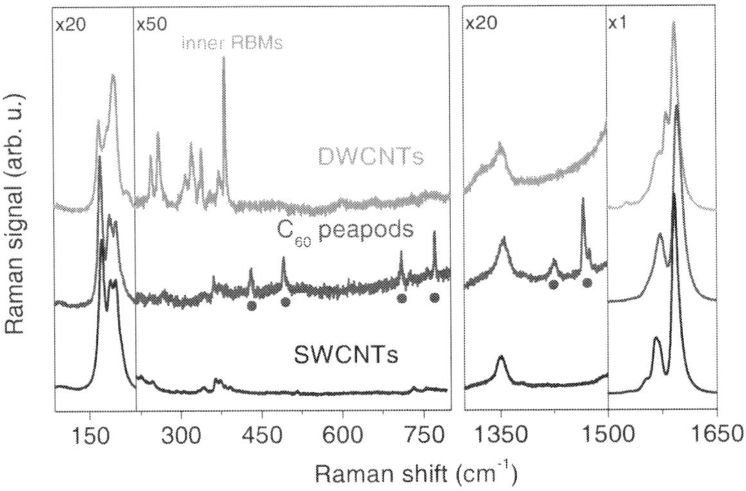

Figure 7. Transformation of fullerene peapods to DWCNTs as followed with Raman spectroscopy at 496.5 nm laser excitation and 90 K. The SWCNT Raman spectra (lower curve) is shown as reference. The fullerene related peapod modes (dots) in the middle curve disappear upon the heat treatment. Note the sharp RBMs appearing in the 250-450 cm^{-1} for the DWCNT sample. Courtesy of R. Pfeiffer and H. Kuzmany.

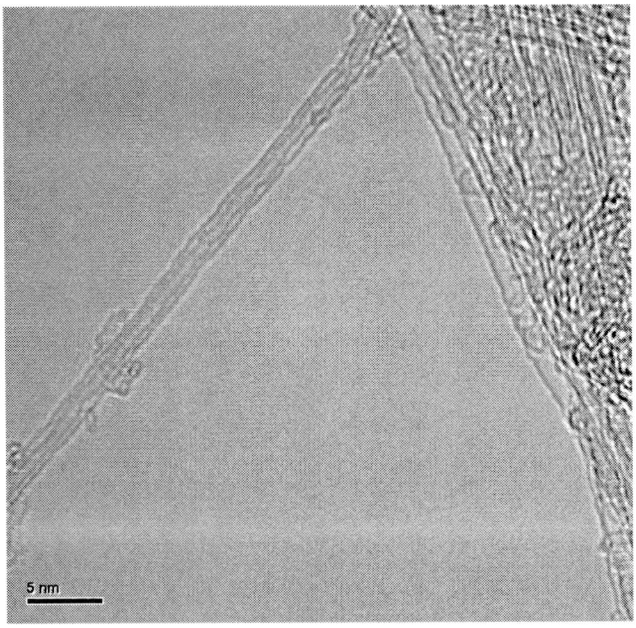

Figure 8. HR-TEM micrograph of a peapod template based DWCNT.

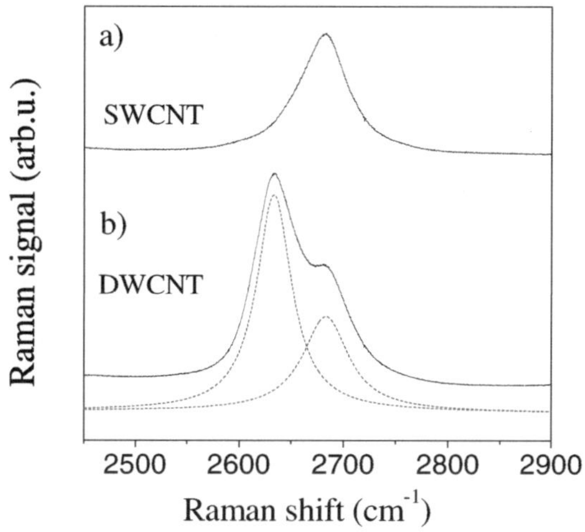

Figure 9. Raman spectra in the G' mode range of a) SWCNT and b) DWCNT samples at DWCNT 515 nm laser excitation. Dashed curves shows deconvolution to inner (lower frequency) and outer (higher frequency) tubes. Reprinted figure with permission from Ref. [67], R. Pfeiffer *et al.* Phys. Rev. B **71**, 155409 (2005). Copyright (2005) by the American Physical Society.

The assignment of the dispersive SWCNT G' mode [20] mode to inner and outer tube modes was also possible due to the (weaker) diameter dependence and the high energy of this mode [60]. In Fig. 9, we show the energy dependence of this mode for DWCNTs. Clearly, the DWCNT G' mode consists of two components. Other modes, such as the SWCNT D and G modes are less diameter sensitive and assignment of inner and outer tube D and G modes requires the use of isotope enriched DWCNTs as shown below.

A variety of additional information can be gained about the inner tube properties when their RBMs are studied using the additive mode, i.e. high-resolution of the Raman spectrometer. In Fig. 10, we show the inner tube RBMs at 90 K with high-resolution in comparison with an SWCNT sample with similar tube diameter prepared by the HiPco process. Three striking observations are apparent in the comparison of the two spectra: i) there are a larger number of inner tube RBMs than geometrically allowed and they appear to cluster around the corresponding modes in the SWCNT sample, ii) the inner tube RBMs are on average an order of magnitude narrower than the SWCNT RBMs in the HiPco sample [44] and iii) the Raman intensity of the inner tubes is large in view of the ~ 3 times less number of carbon atoms on them [60]. Points ii) and iii) are explained by the long phonon and quasi-particle life-times of inner tubes which are discussed further below.

Observation i), i.e. the clustering behavior of the observed inner tube RBMs around SWCNT RBMs, is further evidenced in energy dispersive Raman measurements. In Fig. 11, we show the Raman map for the DWCNTs from Ref. [68]. The advantage of studying Raman maps is that the optical transition energies are also contained in addition to the Raman shifts. These two quantities uniquely identify the chirality of a nanotube [20, 69, 58]. The analogous Raman map for HiPco SWCNTs were measured by Fantini *et al.* [26]

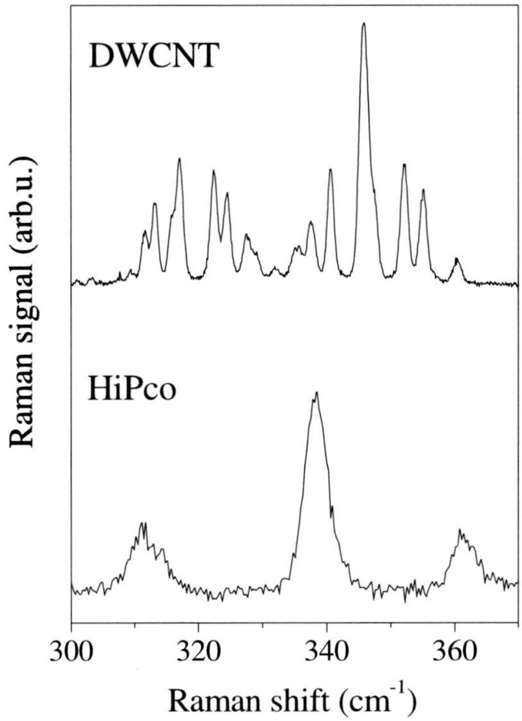

Figure 10. Raman spectra of the RBMs in DWCNT and HiPco (SWCNT) samples at 594 nm laser excitation and 90 K in the high resolution spectrometer mode.

and Telg *et al.* [27]. Their results are also shown in Fig. 11) with squares and circles for metallic and semiconducting tubes, respectively. It turns out that SWCNTs with chiralities for which $2n + m$=const can be grouped into families for which the tube resonance energies and the Raman shifts follow similar trends [25]. The comparison of the HiPco results and the DWCNT Raman map confirms the above statement, i.e. that a number of inner tube modes are observed for the DWCNT where only a few (or one) SWCNT chirality is present. This is best seen for the (6,5) and (6,4) chiralities which are well resolved from other modes.

In Fig. 12, we show the Raman maps for the two samples near the energy and Raman shift regions for the (6,5) and (6,4) tube modes [25, 26, 27]. The comparison of the Raman maps of the two kinds of samples shows that the corresponding tube modes are split into up to 15 components for the inner tube RBMs. This is explained by the inner-outer tube interaction in the DWCNT samples: an inner tube with a particular chirality can be grown in outer tubes with different diameters (chiralities). The varying inner-outer tube spacing can give rise to a different Raman shift for the split components. The large number of split components is a surprising result as it is expected that an inner tube with a given diameter is grown in maximum 1-2 outer tubes where its growth is energetically preferred.

To further prove the origin of the splitting and to quantify this effect, model calculations on the inner-outer tube interactions were performed [68, 70] following the continuum model of Popov and Henrard [71]. These calculations showed that the interaction of inner and

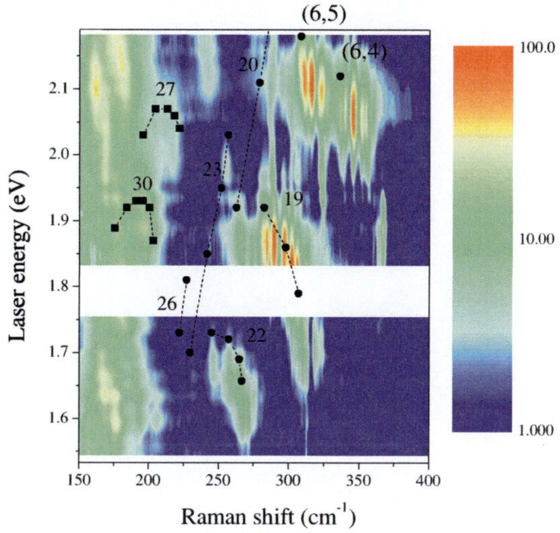

Figure 11. Raman map of DWCNTs. Circles and squares are the E_{22}^{s} and E_{11}^{m} optical transition peaks as measured in a HiPco sample [26], respectively. The family numbers and the chiral indexes for the (6,5) and (6,4) tubes are indicated. Dashed lines join chiralities in the same family. Laser excitation was not available in the missing area. Reprinted figure with permission from Ref. [68], R. Pfeiffer *et al.* Phys. Rev. B **72**, 161404 (2005). Copyright (2005) by the American Physical Society.

outer tubes can gives rise to a shift in the inner tube RBM frequency up to 30 cm $^{-1}$.

3.1.3. Phonon and Quasi-Particle Life-Times of Inner Tubes

We turn to the discussion of the observed very narrow line-widths of the RBMs. This is the most important property of the inner tube RBMs, which will be exploited throughout in this work. Intrinsic line-widths can be determined by deconvoluting the experimental spectra with the spectrometer response fit a Voigtian fit. The Gaussian component describes the spectrometer resolution and the Lorentzian one gives the intrinsic line-width. The Lorentzian component for some inner tube RBMs is as small as 0.4 cm $^{-1}$ [44], which is an order or magnitude smaller than the values obtained for isolated individual tubes in a normal SWCNT sample [72]. The narrow line-widths, i.e. long phonon life-times of the inner tube RBMs was originally associated to the perfectness of the inner tubes grown from the peapod templates [44]. It was found, however, that inner tubes in chemical vapor deposition (CVD) grown DWCNTs have similarly small line-widths [50]. In Fig. 13, the high resolution spectra for the inner tube RBMs in CVD and peapod template grown DWCNTs is shown. This suggests, that the tube environment plays an important role in the magnitude of the observable RBM line-width.

The tube-tube interactions have been shown to give rise to up to $\approx$ 30 cm^{-1} extra shift to the RBMs [68]. The principal difference between SWCNTs and inner tubes in DWCNTs (irrespective whether these are CVD or peapod template grown) is the different surrounding of a small diameter SWCNT with a given chirality: for the SWCNT sample,

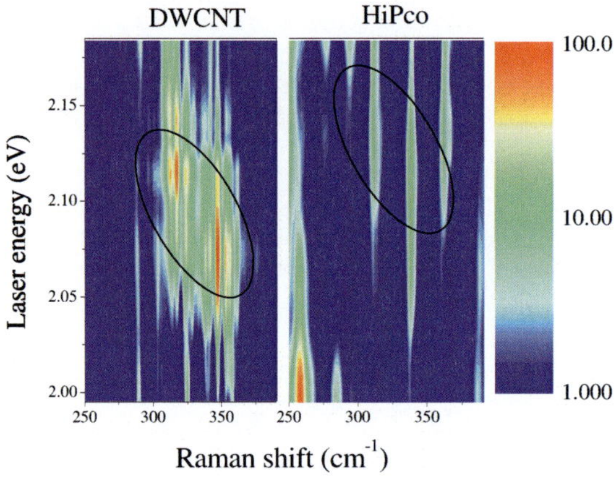

Figure 12. Raman map comparison of the (6,5) and (6,4) tube RBM ranges for DWCNT and SWCNT (HiPco) samples. Ellipsoids indicate the corresponding tube modes. Note the progressive transition energy downshift for the split components of the inner tubes and the 30 meV transition energy difference between the two kinds of samples, which are discussed in the text.

each tube is surrounded by the ensemble of other SWCNTs. For a close packed hexagonal bundle structure [73], this involves 6 nearest neighbors with random chiralities. This causes an inhomogeneous broadening of the RBMs. However, the nearest-neighbor of an inner tube with a given chirality is an outer tube also with a well defined chirality. A given inner tube can be grown in several outer tubes with different diameters, however the chiralities of an inner-outer tube pair is always well defined, therefore the nearest neighbor interaction acting on an inner tube is also well-defined.

In addition to the long phonon life-times of inner tubes, the life-time of optical excitations, i.e. the life-time of the quasi-particle associated with the Raman scattering is unexpectedly long. To demonstrate this, we compare the resonant Raman scattering data for an inner tube and a SWCNT with the same chirality following Ref. [74]. In Fig. 14 we show the energy profile of the resonant Raman scattering at some selected temperatures for two 6,4 tube modes: one is an inner tube in a DWCNT sample, the other is a SWCNT in a sample prepared by the CoMoCat method. Such energy profiles are obtained by taking an energy (vertical) cross section of a Raman map such as shown in Fig. 12. For the DWCNT sample, the strongest (6,4) inner tube component at 347 cm^{-1} and for the SWCNT CoMoCat sample the (6,4) tube mode at 337 cm^{-1} is shown. The temperature dependent resonant Raman data can be fitted with the conventional resonance Raman theory for Stokes Raman modes [56, 75]:

$$I(E_{\mathrm{l}}) = M_{\mathrm{eff}}^4 \left| \frac{\left(E_{\mathrm{l}} - E_{\mathrm{ph}}\right)^4 \left(n_{\mathrm{BE}}(E_{\mathrm{ph}}) + 1\right)}{\left(E_{\mathrm{l}} - E_{\mathrm{ii}} - i\Gamma\right)\left(E_{\mathrm{l}} - E_{\mathrm{ph}} - E_{\mathrm{ii}} - i\Gamma\right)} \right|^2 \tag{1}$$

Here, the electronic density of states of SWCNTs is assumed to be a Dirac function and the effective matrix element, M_{eff}, describing the electron-phonon interactions is taken to

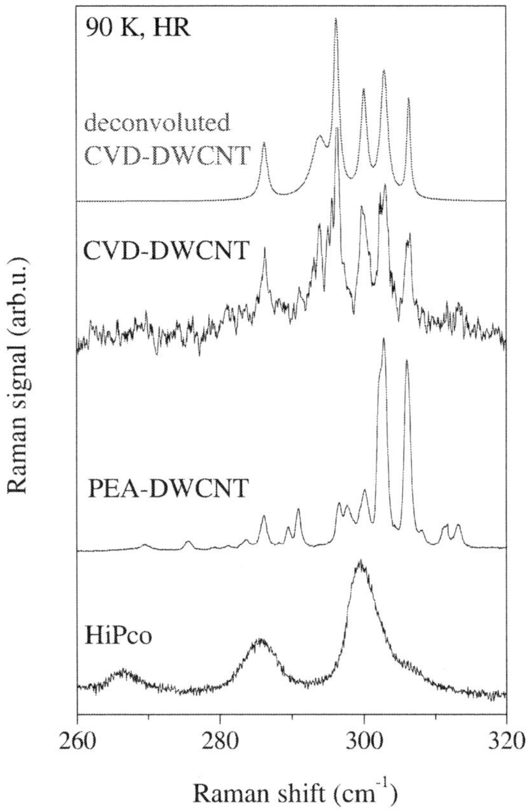

Figure 13. High resolution Raman spectra taken at 676 nm laser excitation and 90 K on the CVD- and PEA-DWCNT and an SWCNT reference (HiPco) sample. The deconvoluted spectrum is also shown for the CVD-DWCNT sample. The narrow line-widths indicate the long RBM phonon life-times of the inner tubes in both DWCNT materials. Reprinted figure with permission from Ref. [50], F. Simon *et al.* Chem. Phys. Lett. **413**, 506 (2005). Copyright (2005) by Elsevier.

be independent of temperature and energy. E_l, E_{22} and E_{ph}, Γ are the exciting laser, the second optical transition, the phonon energies, and the Raman broadening parameter, respectively. $n_{BE}(E_{ph}) = (\exp(E_{ph}/k_BT) - 1)^{-1}$ is the Bose-Einstein function and accounts for the thermal population of the vibrational state [56] and $n_{BE}(E_{ph}) + 1$ changes a factor $\sim$ 2 between 80 and 600 K. The temperature dependence of E_{ph} is ~ 1 % for the studied temperature range [76] thus it can be neglected. The first and second terms in the denominator of Eq. 1 describe the incoming and outgoing resonances, respectively and are indicated on a simulated curve by arrows in Fig. 14. These are separated by E_{ph}. This means the apparent width of the resonance Raman data does not represent Γ.

Clearly, the resonance width is always smaller for the DWCNT than for the SWCNT sample. In other words, the life-time of the optically excited quasi-particle is longer lived for the DWCNT. The quasi-particle life-time is an important parameter for the application of carbon nanotubes in optoelectronic devices [34, 35]. As a result, DWCNTs appear to be

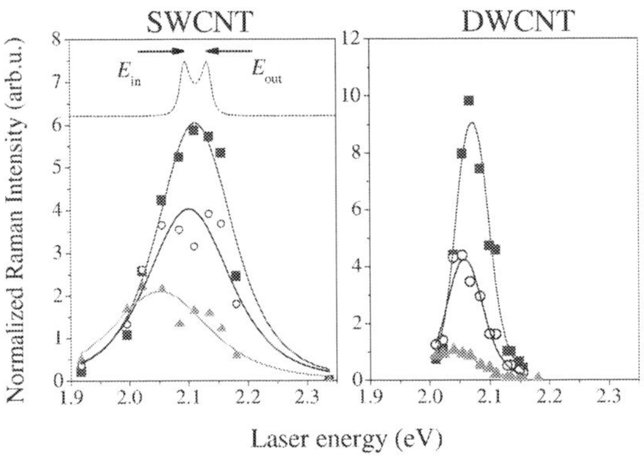

Figure 14. Raman resonance profile for the (6,4) tubes in the SWCNT (CoMoCat) and DWCNT samples, ■: 80 K, ○: 300 K, ▲: 600 K. Solid curves show fits with the RRS theory. Dashed curve is a simulation for the 80 K SWCNT data with $\Gamma = 10$ meV. Arrows indicate the incoming and outgoing resonance energies. Note the much narrower widths for the DWCNT sample. Reprinted figure with permission from Ref. [74], F. Simon *et al.* Phys. Rev. B **74**, 121411(R) (2006). Copyright (2006) by the American Physical Society.

superior in this respect than their one-walled counterparts.

3.1.4. Probing the SWCNT Diameter Distribution through Inner Tube Growth

As discussed above, the Raman spectra of inner tubes have several advantages compared to that of the outer tubes: i) their RBMs have a $\sim$ factor 2 times larger splitting due to the smaller diameters, ii) the line-widths are about 10 times narrower. The larger spectral splitting and narrower line-widths of the inner tube RBMs enable to characterize the inner tube diameter distribution with a spectral resolution that is about 20 times larger as compared to the analysis on the outer tubes. To prove that studying the inner tubes can be exploited for the study of outer ones, here we show that there is a one-to-one correspondence between the inner and outer tube diameter distributions following Ref. [55].

In Fig. 15, we compare the inner tube RBM Raman spectra for four different DWCNT materials based on SWCNTs with different diameters and produced with different methods. The SWCNTs were two arc-discharge grown SWCNTs (SWCNT-N1 and N2) and two laser ablation grown tubes (SWCNT-R and SWCNT-L). The diameter distributions of the SWCNT materials were determined from Raman spectroscopy [75] giving $d_{\mathrm{N1}} = 1.50$ nm, $\sigma_{\mathrm{N1}} = 0.10$ nm, $d_{\mathrm{N2}} = 1.45$ nm, $\sigma_{\mathrm{N1}} = 0.10$ nm, $d_{\mathrm{R}} = 1.35$ nm, $\sigma_{\mathrm{R}} = 0.09$ nm, and $d_{\mathrm{L}} = 1.39$ nm, $\sigma_{\mathrm{L}} = 0.09$ nm for the mean diameters and the variances of the distributions, respectively.

The spectra shown are excited with a 647 nm laser that is representative for excitations with other laser energies. The RBMs of all the observable inner tubes, including the split components [44], can be found at the same position in all DWCNT samples within the ± 0.5 cm^{-1} experimental precision of the measurement for the whole laser energy range studied. This proves that vibrational modes of DWCNT samples are robust against the starting material.

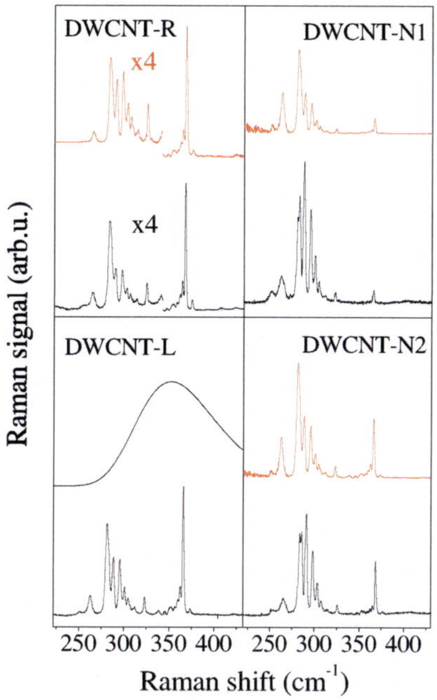

Figure 15. As measured Raman spectra of the inner nanotube RBMs for four DWCNT samples (lower curves in each quarter) at 647 nm laser excitation. The upper spectra (shown in red) are smart scaled from the lower left spectrum. The Gaussian diameter distribution is shown for the DWCNT-L sample. Reprinted figure with permission from Ref. [55], F. Simon *et al.* Phys. Rev. B **71**, 165439 (2005). Copyright (2005) by the American Physical Society.

As the four samples have different diameter distributions, the overall Raman patterns look different. However, scaling the patterns with the ratio of the distribution functions (smart scaling) allows to generate the overall pattern for all systems, starting from e.g. DWCNT-L in the bottom-left corner of Fig. 15. It was assumed that the inner tube diameter distributions follow a Gaussian function with a mean diameter 0.72 nm smaller than those of the outer tubes following Ref. [52] and with the same variance as the outer tubes. The empirical constants from Ref. [77] were used for the RBM mode Raman shift versus inner tube diameter expression. The corresponding Gaussian diameter distribution of inner tubes is shown for the DWCNT-L sample in Fig. 15. A good agreement between the experimental and simulated patterns for the DWCNT-R sample is observed. A somewhat less accurate agreement is observed for the DWCNT-N1, N2 samples, which may be related to the different growth method: arc discharge for the latter, as compared to laser ablation for the R and L samples. The observed agreement has important consequences for the understanding of the inner tube properties. As a result of the photoselectivity of the Raman experiment, it proves that the electronic structure of the inner tubes is identical in the different starting SWCNT materials.

The scaling of the inner tube Raman spectra with the outer tube distribution shows that the inner tube abundance follows that of the outer ones. This agrees with the findings of

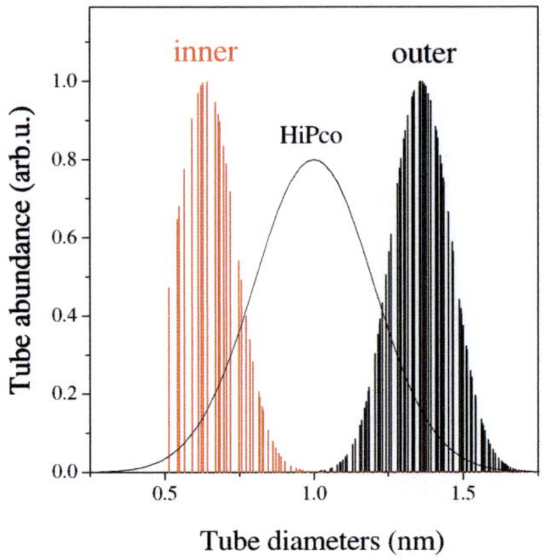

Figure 16. Schematic diameter distribution of inner and outer tubes. Vertical lines denote a tube with a given chirality. Inner-outer tube relative abundances are not to scale. Note the inner tube abundance cut-off at $\sim$ 0.5 nm. The diameter distribution for a HiPco SWCNT sample is shown for comparison.

x-ray diffractomery on DWCNTs [52] and is natural consequence of the growth of inner tubes inside the outer tube hosts. As a result, the inner-outer tube diameter distributions are similar to that shown schematically in Fig. 16: the width of the inner outer tube distributions are the same, and the mean values differ by 0.72 nm. A cut-off in the inner tube diameter distribution was imposed at $d \approx 0.48$ nm due to the inability of C_{60} to enter into outer tubes smaller than 1.2 nm [6, 7, 8, 9].

In summary, the RBM frequencies of inner tubes with neighboring diameters are on the average four times more separated from each other than those of the outer ones [77]. In addition, the outer tube RBMs have a typical FWHM of 10 cm^{-1} [72] as compared to the FWHM of the inner tube RBMs of 0.4-0.5 cm^{-1} [44]. As a result, the available spectral resolution is at least 20 times larger when the RBMs of the inner tubes are investigated. This allows to study chirality or at least diameter selective reactions of the outer tubes by analyzing the inner tube response.

3.1.5. Studying the Reversible Hole Engineering of Nanotubes Using Peapods

Soon after the discovery of the peapods [5], it was recognized [12, 78] that opening the SWCNTs by oxidation in air or by treating in acids is a prerequisite for good filling. Good filling means a macroscopic filling where the peapods are observable not only by local microscopic means such as HR-TEM but also by spectroscopy such as Raman scattering. On the other hand, a heat treatment around 1000 °C was known to close the openings which results in a low or no fullerene filling. It was also shown that the geometrically possible maximum of filling can be achieved when purified SWCNTs were subject to a 450 °C heat treatment in flowing oxygen [62]. However, these studies have concerned the overall fullerene filling, with no knowledge on the precise dependence on the thermal treatment or tube diameter specificity.

The high diameter and chirality sensitivity of Raman spectroscopy for the inner tubes allows to study the behavior of tube openings when subject to different treatments. More precisely, openings which allow fullerenes to enter the tubes can be studied. This is achieved by studying the resulting inner tube RBM pattern when the outer tube host was subject to some closing or opening treatments prior to the fullerene encapsulation [79]. Annealing of as-purchased or opened tubes was performed at various temperatures between 800 °C and 1200 °C in a sealed and evacuated quartz tube at a rest gas pressure of 10^{-6} mbar. Opening of the tubes was performed by exposure to air at various temperatures between 350 °C and 500 °C.

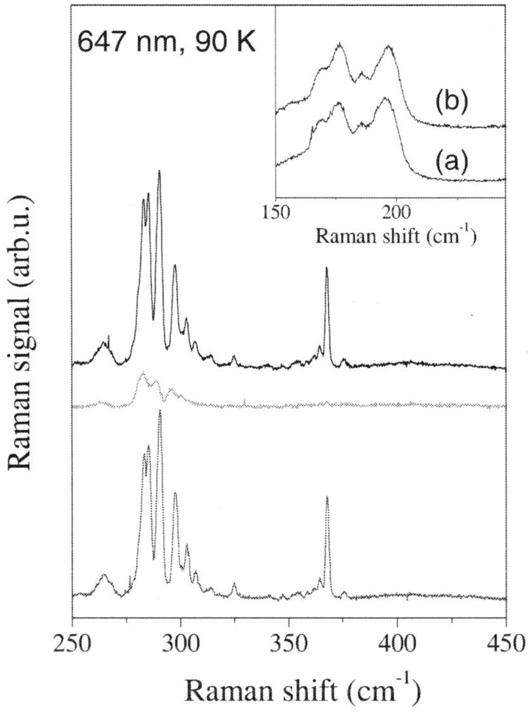

Figure 17. Raman spectra in the spectral range of the inner shell tube RBM for nanotubes after special pre-treatment. Bottom: after filling as-purchased tubes with C_{60} and standard transformation; Center: after annealing the as-purchased tubes at 1000 °C for two hours, filling with C_{60} and standard transformation; Top: after re-opening the annealed samples, filling with C_{60}, and standard transformation. All spectra recorded at 90 K for 647 nm excitation and normalized to the response of the outer tube RBM of the pristine tubes. Insert: the RBM of the outer tubes before (a) and after (b) annealing at 1000 °C. Reprinted figure with permission from Ref. [79], F. Hasi *et al.* J. Nanosci. Nanotechn. **5**, 1785 (2005). Copyright (2005) by the American Scientific Publishers.

Figure 17 shows the Raman response of tubes after the standardized DWCNT transformation conditions but different pre-treatment. Only the spectral range of the inner tube is shown in the main part of the figure. The spectrum at the center was recorded under identical conditions but the SWCNT was pre-annealed before the standardized filling and standardized transformation. Almost no response from inner shell tubes is observed for this material, which means no fullerenes had entered the tubes: the tubes were very efficiently closed by the annealing process. The small response just below 300 cm $^{-1}$ could originate from some small leaks in the cage were a few C_{60} molecules or even more likely some carbon contamination had entered the primary tubes on a level which leads to less than 20 % inner tubes for the larger diameter tubes and less than 3 % for the smaller diameter tubes. The insert in Fig. 17 depicts the RBM response from the outer tubes before and after annealing. The two spectra are almost identical, which proves that no outer tube coalescence had occurred at the temperature applied. The spectrum at the top in Fig. 17 was recorded after reopening the annealed tubes at 500 °C on air and standard filling and transformation. The spectra derived from the pristine and from the reopened tubes are identical in all details. This means no dramatic damages by cutting a large number of holes into the sidewalls have happened. Consequently, the sidewalls of the tubes remain highly untouched by the opening process. Thus, it is suggested that fullerenes enter the tubes through holes at the tube ends.

This technique can also be used to study the diameter selective tube opening process. Fig. 18 depicts Raman spectra of the RBM recorded for 647 nm excitation for tubes which were closed at 1000 °C for two hours and then reopened by exposure to air at the temperatures indicated. As an overall result, the opening process becomes less efficient with decreasing temperature. However, the reduction in efficiency hardly depends on the tube diameter. The latter behavior is more evidently seen from an analysis of the response for the individual peaks as depicted in part (b) of Fig. 18. For opening at 500 °C 95 % of the closed tubes could be reopened, nominally independent of tube diameter. For opening at 400 °C only about 60 % of the tubes could be opened. Surprisingly, as plotted this fraction is even slightly decreasing with decreasing tube diameter. Similarly, for opening at 350 °C only 45 % of the tubes could be opened again, with a slight decrease of this fraction for small diameter tubes.

In summary, the reaction activity of holes in SWCNTs can be investigated by studying the process of filling the tubes with C_{60} fullerenes and transforming the latter to a set of small diameter inner tubes. Reversible closing and opening of the tubes can be achieved. The tube opening process is sensitive to temperature but insensitive to tube diameter. The almost identical Raman spectra for the pristine and for the opened tubes with respect to line positions, line-widths and line intensities suggest that the tube walls remain untouched and the holes for the tube filling are located at the tube ends. After tube closing, a small amount of inner tubes can be grown mostly in high defect areas of the tubes. In contrast to tube opening, the tube closing process is diameter sensitive.

3.2. Growth Mechanism of Inner Tubes Studied with Isotope Labeled Fullerenes

The growth of inner tubes from fullerenes raises the question, whether the fullerene geometry plays an important role in the inner tube growth or it acts as a carbon source only. Theoretical results suggest the first possibility [80, 81]. In addition, it needs clarification whether carbon exchange occurs between the two tube walls. Here, we review the studies using ^{13}C isotope labeling, which were aimed at answering these two open questions. ^{13}C

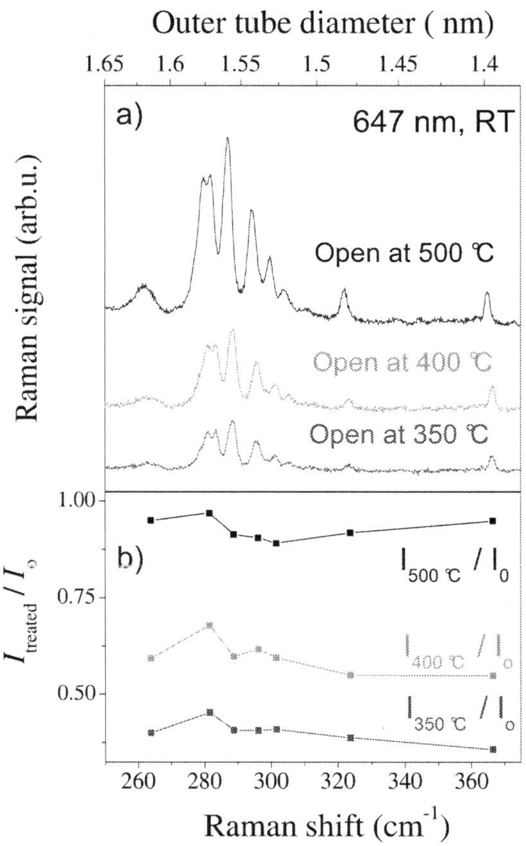

Figure 18. Raman response of the RBM of inner shell CNTs for standard filling and transformation procedures after the pristine tubes had been closed and reopened at temperatures indicated (a). Part (b) depicts evaluated relative intensities versus Raman shift for the three different opening conditions. The upper scale provides outer tube diameters. Reprinted figure with permission from Ref. [79], F. Hasi *et al.* J. Nanosci. Nanotechn. **5**, 1785 (2005). Copyright (2005) by the American Scientific Publishers.

is a naturally occurring isotope of carbon with 1.1 % abundance. In general, isotope substitution provides an important degree of freedom to study the effect of change in phonon energies while leaving the electronic properties unaffected. This have helped to unravel phenomena such as e.g. the phonon-mediated superconductivity [82].

First, we discuss the inner tube growth from isotope labeled fullerenes [57], and secondly we present the growth of inner tubes from isotope labeled organic solvents [83].

Commercial ^{13}C isotope enriched fullerenes with two different enrichment grades were used to grow isotope enriched inner tubes. Fullerene encapsulation [12] and inner tube growth was performed with the conventional methods [41]. This results in a compelling isotope engineered system: double-wall carbon nanotubes with ^{13}C isotope enriched inner walls and outer walls containing natural carbon [57].

In Fig. 19a, we show the inner tube RBM range Raman spectra for a natural DWCNT

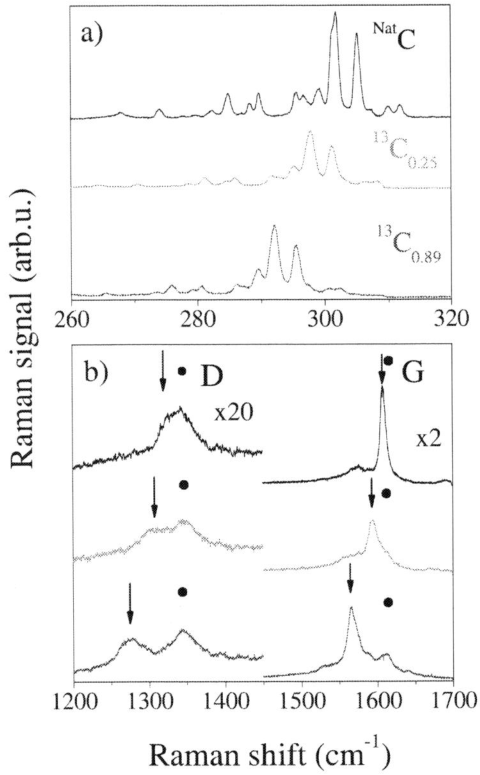

Figure 19. Raman spectra of DWCNTs with natural carbon and ^{13}C enriched inner tubes at 676 nm laser excitation and 90 K. The inner tube RBM (a) and D and G mode spectral ranges (b) are shown. Arrows and filled circles indicate the D (left) and G (right) modes corresponding to the inner and outer tubes, respectively. Reprinted figure with permission from Ref. [57], F. Simon *et al.* Phys. Rev. Lett. **95**, 017401 (2005). Copyright (2005) by the American Physical Society.

and two DWCNTs with differently enriched inner walls, 25 % and 89 %. These two latter samples are denoted as $^{13}C_{25}$- and $^{13}C_{89}$-DWCNT, respectively. The inner wall enrichment is taken from the nominal enrichment of the fullerenes used for the peapod production, whose value is slightly refined based on the Raman data. An overall downshift of the inner tube RBMs is observed for the ^{13}C enriched materials accompanied by a broadening of the lines. The downshift is clear evidence for the effective ^{13}C enrichment of inner tubes. The magnitude of the enrichment and the origin of the broadening are discussed below.

The RBM lines are well separated for inner and outer tubes due to the $\nu_{\mathrm{RBM}} \propto 1/d$ relation and a mean inner tube diameter of $d \sim 0.7$ nm [52, 55]. However, other vibrational modes such as the defect induced D and the tangential G modes strongly overlap for inner and outer tubes. Arrows in Fig. 19b indicate a gradually downshifting component of the observed D and G modes. These are the components of the D and G modes of the inner tubes. The sharper appearance of the inner tube G mode, as compared to the response from the outer tubes, is related to the excitation of semiconducting inner tubes and metallic outer tubes [44, 55].

The shifts for the RBM, D and G modes can be analyzed for the two grades of enrichment. The average value of the relative shift for these modes was found to be $(\nu_0 - \nu)/\nu_0 = 0.0109(3)$ and $0.0322(3)$ for the $^{13}C_{0.25}$- and $^{13}C_{0.89}$-DWCNT samples, respectively. Here, ν_0 and ν are the Raman shifts of the same inner tube mode in the natural carbon and enriched materials, respectively. In the simplest continuum model, the shift originates from the increased mass of the inner tube walls. This gives $(\nu_0 - \nu)/\nu_0 = 1 - \sqrt{\frac{12+c_0}{12+c}}$, where c is the concentration of the ^{13}C enrichment on the inner tube, and $c_0 = 0.011$ is the natural abundance of ^{13}C in carbon. The resulting values of c are $0.277(7)$ and $0.824(8)$ for the 25 and 89 % samples, respectively.

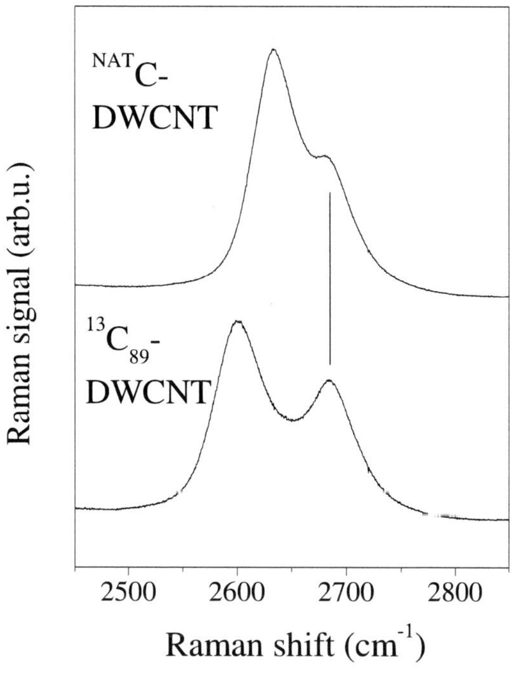

Figure 20. G' spectral range of DWCNTs with natural carbon and ^{13}C enriched inner walls with 515 nm laser excitation. Note the unchanged position of the outer tube G' mode indicated by a vertical line.

The growth of isotope labeled inner tubes allows to address whether carbon exchange between the two walls occurs during the inner tube growth. In Fig. 20, we show the G' spectral range for DWCNTs with natural carbon and ^{13}C enriched inner walls with 515 nm laser excitation. The G' mode of DWCNTs is discussed in detail in Ref. [70] and it was introduced in Fig. 9: the upper G' mode component corresponds to the outer tubes and the lower to the inner tubes. The outer tube G' components are unaffected by the ^{13}C enrichment within the 1 cm^{-1} experimental accuracy. This gives an upper limit to the extra ^{13}C in the outer wall of 1.4 %. This proves that there is no sizeable carbon exchange between the two walls as this would result in a measurable ^{13}C content on the outer wall, too. This is important for the contrast of the NMR signal between the two walls as it is discussed further below: were the outer shell also enriched, one could not obtain a reliable information about the density of states on the inner shell alone.

The narrow RBMs of inner tubes and the freedom to control their isotope enrichment allows to precisely compare the isotope related phonon energy changes in the experiment and in *ab-initio* calculations. This was performed by J. Kürti and V. Zólyomi in Ref. [57]. The validity of the above simple continuum model for the RBM frequencies was verified by performing first principles calculations on the (n, m)=(5,5) tube as an example. In the calculation, the Hessian matrix was determined by DFT using the Vienna Ab Initio Simulation Package [84]. Then, a large number of random ^{13}C distributions were generated and the RBM vibrational frequencies were determined from the diagonalization of the corresponding dynamical matrix for each individual distribution. The distribution of the resulting RBM frequencies can be approximated by a Gaussian where center and variance determine the isotope shifted RBM frequency and the spread in these frequencies. The difference between the shift determined from the continuum model and from the *ab-initio* calculations is below 1 %.

The broadening for the ^{13}C enriched inner tubes is best observed for the narrow RBMs. In Fig. 21a-b we show the RBMs of some inner tubes for the ^{Nat}C and $^{13}\mathrm{C}_{0.25}$ samples. Curves (1) are the line-shapes after deconvolution with the Gaussian response of our spectrometer. In Fig. 21a, the resulting line-shapes are Lorentzians, but in Fig. 21b, the line-shape still contains a Gaussian component, as discussed below. The FWHMs of the resulting line-shapes are $0.76(4)$, $0.76(4)$, $0.44(4)$, $0.54(4)$ and $1.28(6)$, $1.30(6)$, $1.12(6)$, $1.16(6)$ for the inner tube RBMs shown in Fig. 21 of the ^{Nat}C and $^{13}\mathrm{C}_{0.25}$ materials, respectively. The origin of the extra broadening is due to the random distribution of ^{12}C and ^{13}C nuclei. Calculations indicate that the ratio between the half width of extra broadening and the shift, $\Delta\nu/\left(\nu_0-\nu\right)$, is approximately 0.19 for a 30 % ^{13}C enriched sample. The corresponding broadened line-shapes are shown in Fig. 21b as curve (3). When the magnitude of the Gaussian randomness related broadening was fit (curve (2) in Fig.21b), we found that $\Delta\nu/\left(\nu_0-\nu\right) = 0.15$. Similar broadening was observed for the 89 % sample which can also be reproduced by the calculation.

The known characteristics of isotope labeled inner tubes allow to study the possibility of inner tube growth from non-fullerene carbon sources [83]. For this purpose, we choose organic solvents containing aromatic rings, such as toluene and benzene. These are known to wet the carbon nanotubes and are appropriate solvents for fullerenes. As described in the following, the organic solvents indeed contribute to the inner tube growth, however only in the presence of C_{60} "stopper" molecules [83]. In the absence of co-encapsulated fullerenes the solvents alone give no inner tube.

The fullerene+organic solvents encapsulation was performed by dissolving typically 150 μg fullerenes in 100 μl solvents and then sonicating with 1 mg SWCNT in an Eppendorf tube for 1 h. The weight uptake of the SWCNT is ~15 % [57] that is shared between the solvent and the fullerenes. The peapod material was separated from the solvent by centrifuging and it was then greased on a sapphire substrate. The solvent prepared peapods were treated in dynamic vacuum at 1250 °C for 2 hours for the inner tube growth. The inner tube growth efficiency was found independent of the speed of warming.

The growth of inner tubes from the solvents can be best proven by the use of C_{60} containing natural carbon and a solvent mixture consisting of ^{13}C enriched and natural

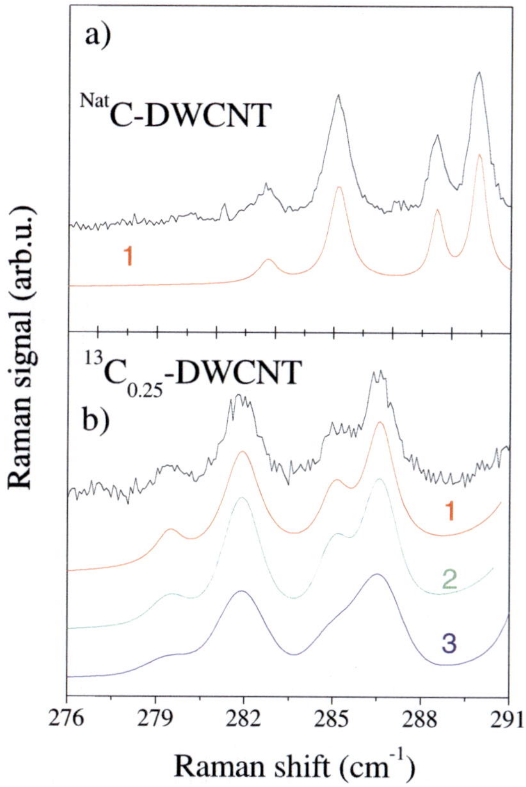

Figure 21. RBMs of some inner tubes at 676 nm laser excitation with 0.5 cm^{-1} spectral resolution. a) natural carbon-, b) 25 % ^{13}C enriched inner shell DWCNT. Curves are the line-shapes after deconvolution by the spectrometer response (1), the simulated line-shape with an extra Gaussian broadening to the intrinsic lines of the ^{Nat}C material (2), and the calculated line-shape (see text) (3). Reprinted figure with permission from Ref. [57], F. Simon *et al.* Phys. Rev. Lett. **95**, 017401 (2005). Copyright (2005) by the American Physical Society.

carbon containing solvents with varying concentrations. Toluene was a mixture of ring ^{13}C labeled ($^{13}C_6H_6$-$^{NAT}CH_3$) and natural toluene ($^{NAT}C_7H_8$). Benzene was a mixture of ^{13}C enriched and natural benzene. The labeled site was > 99 % ^{13}C labeled for both types of molecules. The ^{13}C content, x, of the solvent mixtures was calculated from the concentration of the two types of solvents and by taking into account the presence of the naturally enriched methyl-group for the toluene. In Fig. 22a, we show the G' modes of DWCNTs with varying ^{13}C labeled content in toluene+C_{60} based samples and in Fig. 22b, we show the same spectra after subtracting the outer SWCNT component. A shoulder appears for larger values of x on the low frequency side of the inner tube mode, whereas the outer tube mode is unchanged. Similar behavior was observed for the benzene+C_{60} based peapod samples (spectra not shown) although with a somewhat smaller spectral intensity of the shoulder. The appearance of this low frequency shoulder is evidence for the presence of a sizeable ^{13}C content in the inner tubes. This proves that the solvent indeed contributes to

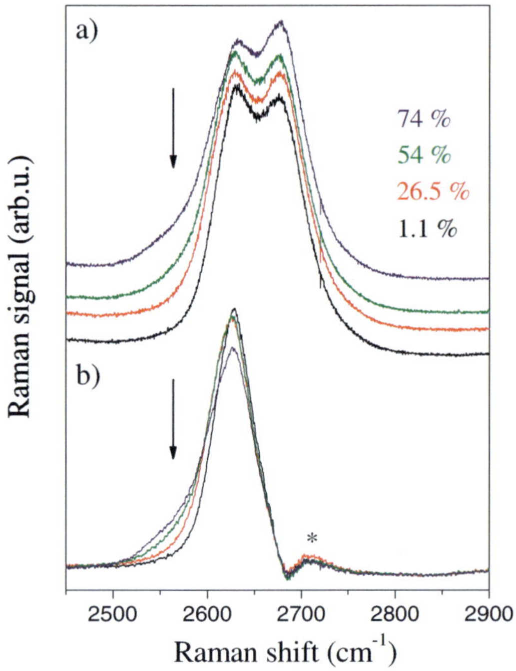

Figure 22. a) The G' mode of toluene+C_{60} peapod based DWCNTs with varying ^{13}C enrichment at 515 nm laser excitation. From top to bottom: 74 %, 54 %, 26.5 % and natural ^{13}C content. b) The G' mode of the inner tubes after subtracting the experimental SWCNT spectrum. A small residual peak is observed around 2710 cm^{-1} (denoted by an asterisk) due to the imperfect subtraction. Arrows indicate the spectral weight shifted toward lower frequencies. Reprinted figure with permission from Ref. [83], F. Simon and H. Kuzmany, Chem. Phys. Lett. **425**, 85 (2006). Copyright (2006) by Elsevier.

the inner tube formation as it is the only sizeable source of ^{13}C in the current samples. The appearance of the low frequency shoulder rather than the shift of the full mode indicates an inhomogeneous ^{13}C enrichment. A possible explanation is that smaller diameter nanotubes might be higher ^{13}C enriched as they retain the solvent better than larger tubes.

To quantify the ^{13}C enrichment of the inner tubes, the downshifted spectral weight of the inner tube G' mode was determined from the subtracted spectra in Fig. 23b. The subtraction does not give a flat background above 2685 cm^{-1}, however it is the same for all samples and has a small spectral weight, thus it does not affect the current analysis. The line-shapes strongly deviate from an ideal Lorentzian profile. Therefore the line positions cannot be determined by fitting, whereas the first moments are well defined quantities. The effective ^{13}C enrichment of the inner tubes, c, is calculated from $(\nu_0 - \nu)/\nu_0 = 1 - \sqrt{\frac{12+c_0}{12+c}}$, where ν_0 and ν are the first moments of the inner tube G' mode in the natural carbon and enriched materials, respectively, and $c_0 = 0.011$ is the natural abundance of ^{13}C in carbon. The validity of this "text-book formula" is discussed above and it was verified by *ab-initio* calculations for enriched inner tubes in Ref. [57]. In Fig. 23, we show the effective ^{13}C

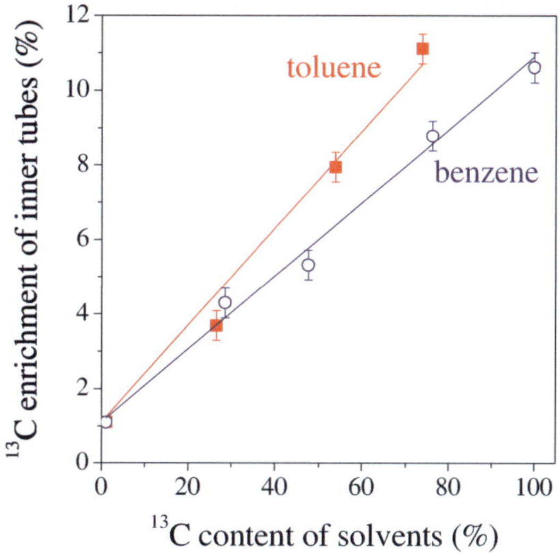

Figure 23. ^{13}C content of inner tubes based on the first moment analysis as explained in the text as a function of ^{13}C enrichment of benzene and toluene. Lines are linear fits to the data and are explained in the text. Reprinted figure with permission from Ref. [83], F. Simon and H. Kuzmany, Chem. Phys. Lett. **425**, 85 (2006). Copyright (2006) by Elsevier.

content in the inner tubes as a function of the ^{13}C content in the starting solvents. The scaling of the ^{13}C content of the inner tubes with that in the starting solvents proves that the source of the ^{13}C is indeed the solvents. The highest value of the relative shift for the toluene based material, $(\nu_0 - \nu)/\nu_0 = 0.0041(2)$, corresponds to about 11 cm^{-1} shift in the first moment of the inner tube mode. The shift in the radial breathing mode range (around 300 cm^{-1}) [69] would be only 1 cm^{-1}. This underlines why the high energy G' mode is convenient for the observation of the moderate ^{13}C enrichment of the inner tubes. When fit with a linear curve with $c_0 + A * x$, the slope, A directly measures the carbon fraction in the inner tubes that originates from the solvents.

The synthesis of inner tubes from organic solvent proves that any form of carbon that is encapsulated inside SWCNTs contributes to the growth of inner tubes. As mentioned above, inner tubes are not formed in the absence of fullerenes but whether the fullerene is C_{60} or C_{70} does not play a role. It suggests that fullerenes act only as a stopper to prevent the solvent from evaporating before the synthesis of the inner tube takes place. It also clarifies that the geometry of fullerenes do not play a distinguished role in the inner tube synthesis as it was originally suggested [80, 81]. It also proves that inner tube growth can be achieved irrespective of the carbon source, which opens a new prospective to explore the in-the-tube chemistry with other organic materials.

3.3. ESR Studies on Encapsulated Magnetic Fullerenes

Observation of the intrinsic ESR signal of pristine SWCNTs remains elusive [85, 86]. Now, it is generally believed that intrinsic ESR of the tubes can not be observed as conduction

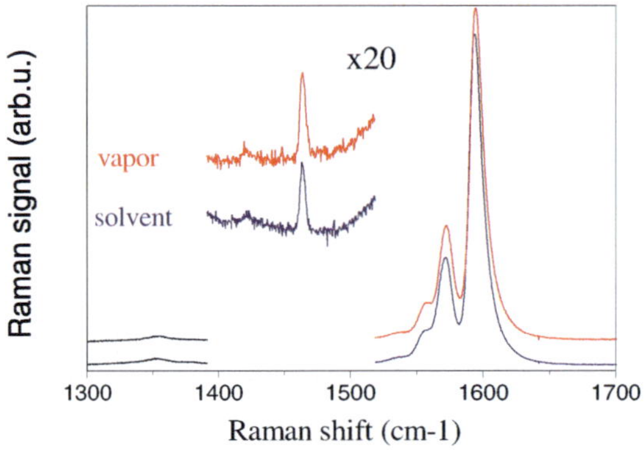

Figure 24. Raman spectra of vapor and solvent filled C_{60}@SWCNT at 488 nm excitation and 90 K. The spectra are normalized to the amplitude of the SWCNT G mode. Reprinted figure with permission from Ref. [46], F. Simon *et al.* Chem. Phys. Lett. **383**, 362 (2004). Copyright (2004) by Elsevier.

electrons on metallic tubes are relaxed by defects too fast to be observable. In addition, one always observes a number of ESR active species in a sample, such as graphitic carbon or magnetic catalyst particles, which prevent a meaningful analysis of the signal. In contrast, local probe studies could still allow an ESR study of tubes, provided the local spin probe can be selectively attached to the tubes. This goal can be achieved by using magnetic fullerenes, such as e.g. N@C_{60} or C_{59}N, since fullerenes are known to be selectively encapsulated inside SWCNTs [5] and can be washed from the outside using organic solvents [12]. As the properties and handling of the two magnetic fullerenes are quite different, the synthesis of the corresponding peapods and the results are discussed separately.

3.3.1. ESR on N@C_{60}@SWCNT

N@C_{60} is an air stable fullerene [87] but decays rapidly above $\sim$ 200 °C [88] which prevents the use of the conventional vapor method of peapod preparation which requires temperatures above 400 °C. To overcome this limitation and to allow in general the synthesis of temperature sensitive peapod materials, low temperature peapod synthesis (solvent method) was developed independently by four groups [45, 46, 47, 48]. These methods share the common idea of mixing the opened SWCNTs with C_{60} in a solvent with low fullerene solubility such as methanol [45] or n-pentane [46]. The encapsulation is efficient as it is energetically preferred for C_{60} to enter the tubes rather than staying in the solution. After the solvent filling, excess fullerenes can be removed by sonication in toluene, which is a good fullerene solvent as it was found that fullerenes enter the tube irreversibly. HR-TEM has shown an abundant filling with the fullerenes [45, 47] but here we discuss macroscopic characterizations using Raman spectroscopy, which has proven that peapods prepared by the solvent method are equivalent to the vapor prepared peapods.

In Fig. 24, we show the comparison of the Raman spectra of vapor- and solvent-filled C_{60} peapod samples. The Raman spectra of peapod samples in the plotted frequency range consist of the SWCNT G and D modes at 1590 and 1355 cm^{-1}, respectively and narrow

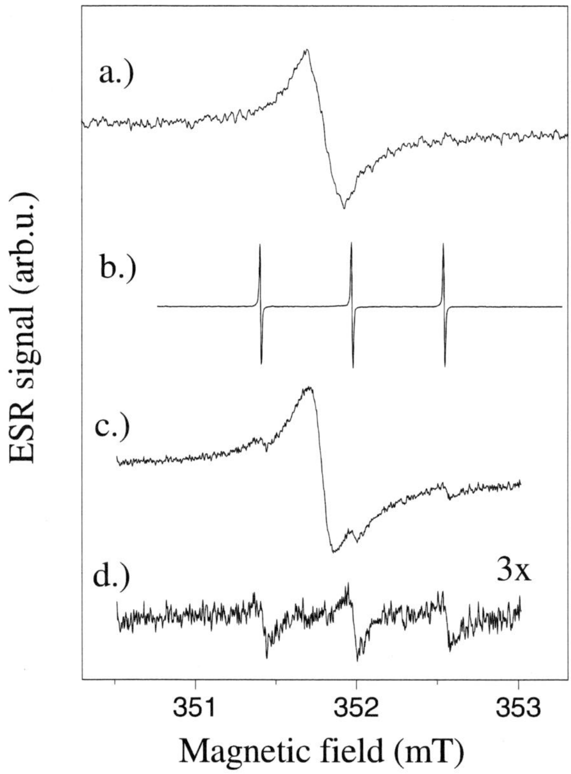

Figure 25. X-band electron spin resonance spectrum of the a.) pristine SWCNT, b.) crystalline N@C_{60}:C_{60}, c.) (N@C_{60}:C_{60})@SWCNT and d.) the triplet component of the (N@C_{60}:C_{60})@SWCNT ESR spectrum at ambient temperature. Reprinted figure with permission from Ref. [46], F. Simon *et al.* Chem. Phys. Lett. **383**, 362 (2004). Copyright (2004) by Elsevier.

lines related to the Raman active modes of the C_{60} inside the SWCNT [12, 49]. We show enlarged the most significant Raman active line of C_{60} peapod, the $A_g(2)$ mode. We observe a single $A_g(2)$ line at 1466 cm^{-1} for both peapod materials which is consistent with the relatively large, $d = 1.4$ nm, tube diameter [60]. The observation of the 1466 cm^{-1} C_{60} $A_g(2)$ mode is characteristic [49] for the encapsulated C_{60}. Its observation for the solvent prepared sample with the same intensity as for the vapor prepared sample proves the effective macroscopic filling. The absence of an extra line at 1469 cm^{-1}, that is the Raman shift of the $A_g(2)$ of crystalline C_{60} is evidence that no C_{60} is present apart from those encapsulated in the nanotubes.

The low temperature synthesis allows to encapsulate the N@C_{60} fullerene. The N@C_{60}:C_{60} endohedral fullerene: fullerene solid solution can be produced in a N_2 arc-discharge tube following Ref. [89] with a typical yield of a few 10 ppm [90]. In Fig. 25., the ESR spectra of the starting SWCNT, (N@C_{60}:C_{60})@SWCNT, and N@C_{60}:C_{60} are shown. The ESR spectrum of the pristine SWCNT for the magnetic field range shown is dominated by a signal that is assigned to some residual carbonaceous material, probably graphite. Fig.

25c. shows, that after the solvent encapsulation of $N@C_{60}:C_{60}$ in the NCL-SWCNT, a hyperfine N triplet ESR is observed, similar to that in pristine $N@C_{60}:C_{60}$, superimposed on the broad signal present in the pristine nanotube material. Fig. 25d. shows the triplet component of this signal after subtracting the signal observed in pristine SWCNT. The hyperfine triplet in $N@C_{60}:C_{60}$ is the result of the overlap of the $^4S_{3/2}$ state of the three 2p electrons of the N atom and the ^{14}N nucleus, with nuclear spin, $I = 1$. The isotropic hyperfine coupling of $N@C_{60}:C_{60}$ is unusually high as a result of the strongly compressed N atomic $2p^3$ orbitals in the C_{60} cage thus it unambiguously identifies this material [87].

The hyperfine coupling constant observed for the triplet structure in the encapsulated material, $A = 0.57 \pm 0.01$ mT, agrees within experimental precision with that observed in $N@C_{60}:C_{60}$ [87], which proves that the encapsulated material is $(N@C_{60}:C_{60})@SWCNT$. The ESR line-width for the encapsulated material, $\Delta H_{pp} = 0.07$ mT, is significantly larger than the resolution limited ΔH_{pp} =0.01 mT in the pristine $N@C_{60}:C_{60}$ material, the lines being Lorentzian. The most probable cause for the broadening is static magnetic fields from residual magnetic impurities in the SWCNT [91]. The ESR signal intensity is proportional to the number of N spins, and this allows the quantitative comparison of N concentrations in $(N@C_{60}:C_{60})@SWCNT$ and $N@C_{60}:C_{60}$. Taking the filling value from the EELS measurement, we obtain that the N spin concentration in $(N@C_{60}:C_{60})@SWCNT$ is $\sim$2.5 times smaller than in the starting $N@C_{60}:C_{60}$ material. This cannot be due to a loss of N spins during the synthesis as $N@C_{60}:C_{60}$ is stable below 100 °C and N loss becomes rapid above 220 °C only [88]. The observed difference may be caused by the limited accuracy of the ESR intensity measurement and could also be caused by microwave losses as the conducting bucky-paper pieces are separated from each-other in a poorly controlled way.

3.3.2. ESR on $C_{59}N@SWCNT$

As seen from the ESR results on encapsulated $N@C_{60}$, relatively limited information can be deduced about the tubes themselves. This stems from the fact that the N spins are well shielded in $N@C_{60}$ [92] and are thus are relatively insensitive to the SWCNT properties. In contrast, $N@C_{60}@SWCNT$ peapods might find another application as building elements of a quantum computer as proposed by Harneit *et al.* [18]. A better candidate for magnetic fullerene encapsulation is the $C_{59}N$ monomer radical as here the unpaired electron is on the cage and is a sensitive probe of the environment. This material can be chemically prepared [93], however it forms a non-magnetic dimer crystal $(C_{59}N)_2$. It appears as a spinless monomer in an adduct form [94] or attached to surface dangling bonds [95]. The magnetic $C_{59}N$ monomer radical is air sensitive but it can be stabilized as a radical when it is dilutely mixed in C_{60}[96]. As a result, a different strategy has to be followed to encapsulate $C_{59}N$ inside SWCNT, which is discussed in the following along with preliminary ESR results [97].

To obtain $C_{59}N$ peapods, air stable $C_{59}N$ derivatives, ($C_{59}N$-der in the following) were prepared chemically by A. Hirsch and F. Hauke following standard synthesis routes [94, 98]. The $C_{59}N$-der was 4-Hydroxy-3,5-dimethyl-phenyl-hydroazafullerene whose structure is shown in Fig. 26. The $C_{59}N$ derivatives were encapsulated either pure or mixed with C_{60} as $C_{59}N$-der:C_{60} with 1:9 concentrations using a modified version of the low temperature encapsulation method. In brief, the mixture of the dissolved fullerenes and SWCNTs were

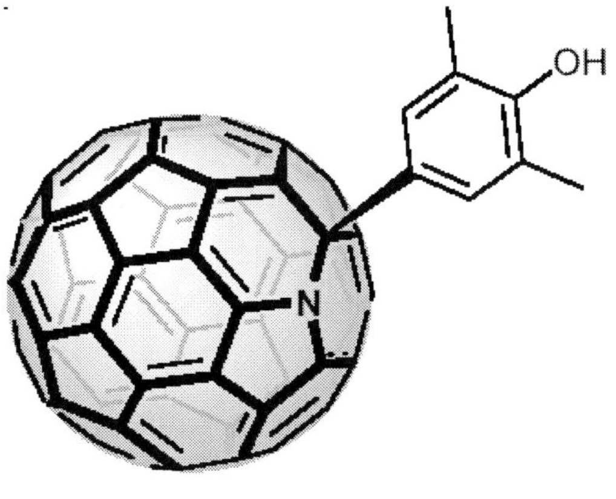

Figure 26. Schematic structure of the 4-Hydroxy-3,5-dimethyl-phenyl-hydroazafullerene. Reprinted figure with permission from Ref. [97], F. Simon *et al.* Carbon, **44**, 1958 (2006). Copyright (2006) by Elsevier.

sonicated in toluene and filtered. It is expected that the $C_{59}N$ monomer radical can be obtained after a heat treatment in dynamic vacuum, which is discussed below.

Raman spectroscopy was performed to characterize the SWCNT filling with the $C_{59}N$-der. In Fig. 27, the Raman spectra of the pristine and encapsulated $C_{59}N$-der is shown. The Raman spectra of the peapod sample (lower curve in Fig. 27) in the plotted frequency range consist of the SWCNT G modes around 1550 cm^{-1} and additional lines related to the Raman active modes of the encapsulated azafullerene derivative [49]. The major Raman modes of the pristine $C_{59}N$-der are similar to those of the $(C_{59}N)_2$ dimer [99]. The strongest mode is observed at 1459.2 cm^{-1} which is derived from the C_{60} $A_g(2)$ mode and is downshifted to 1457 cm^{-1} after the encapsulation procedure. The 2.2 cm^{-1} downshift proves the encapsulation of the molecule inside the SWCNT. When encapsulated inside SWCNTs, the corresponding $A_g(2)$ mode of C_{60} downshifts with 3 cm^{-1} , which is assigned to the softening of the C_{60} $A_g(2)$ vibrational mode due to the interaction between the ball and the SWCNT wall [49]. The slight difference between the downshift for the azafullerene and for the C_{60} peapods might be attributed to the different structure of the two molecules. The encapsulation also manifests in a line broadening: the main component of the 1457 cm^{-1} mode is broadened from 4.5 cm^{-1} FWHM in the pristine material to 10 cm^{-1} FWHM in the encapsulated one. This is similar to the values found for encapsulated C_{60} [49].

The integrated intensity of the observed $A_g(2)$ derived mode of the $C_{59}N$ is approximately 5 times larger than that of a C_{60} peapod prepared identically when normalized by the SWCNT G mode intensity. This, however, cannot be used to measure the encapsulation efficiency as Raman intensities depend on the strength of the Raman resonance enhancement and the Raman scattering matrix elements [56]. For C_{60} peapods the Raman signal was calibrated with independent and carbon number sensitive measurements: EELS studies gave the total number of C_{60} related and non-C_{60} related carbons [62] and the mass of encapsulated C_{60}s was determined from NMR studies using ^{13}C enriched fullerenes [57, 100]. In the current case, neither methods can be employed and we determined the filling efficiency

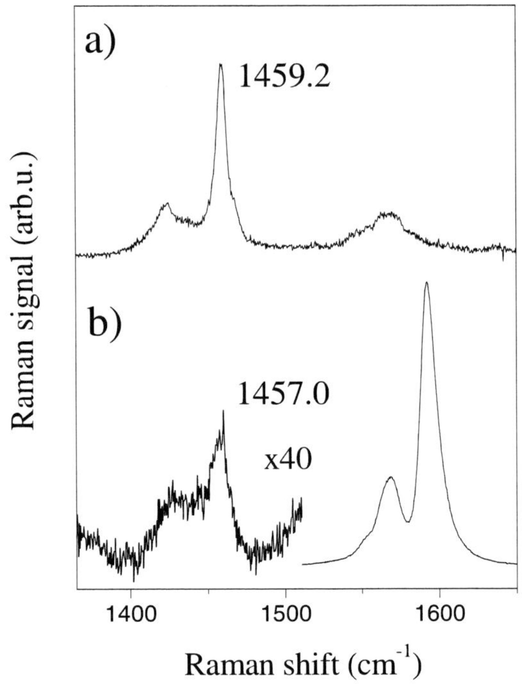

Figure 27. Tangential mode of the Raman spectra of the $C_{59}N$-der before (a) and after encapsulation (b) excited with a 488 nm laser. Labels mark the position of the strongest $C_{59}N$-der mode in the two samples. The Raman G-mode of SWCNT dominates the peapod spectrum in the 1550-1650 cm^{-1} range. Reprinted figure with permission from Ref. [97], F. Simon *et al.* Carbon, **44**, 1958 (2006). Copyright (2006) by Elsevier.

for the azafullerene by encapsulating a mixture of the azafullerene and C_{60}. In Fig. 28., the Raman spectra of the encapsulated $C_{59}N$-der:C_{60} mixture with weight ratios of 1:9 in the starting solvent is shown. The Raman spectrum of the encapsulated mixture was simulated with a weighted sum of the separately recorded spectra for encapsulated $C_{59}N$-der and C_{60}. The best agreement between the simulated and the experimental spectra is for a $C_{59}N$-der content of 0.12(2). This value is close to the expected value of 0.1 and it proves that the azafullerene enters the tubes with the same efficiency as C_{60}.

Fig. 29 shows the room temperature ESR spectra of $C_{59}N$:C_{60}@SWCNT after 600 °C vacuum annealing from Ref. [101]. The spectra of $C_{59}N$:C_{60}, a $C_{59}N$ monomer embedded in C_{60} [96], is also shown for comparison. This latter spectrum was previously assigned to the superposition of rotating $C_{59}N$ monomers and bound $C_{59}N$-C_{60} heterodimers [102]. The large spin density at the ^{14}N nucleus of the rotating $C_{59}N$ molecule results in an ESR triplet signal and the $C_{59}N$-C_{60} heterodimer has a singlet signal (arrow in Fig. 29) as the spin density resides on the C_{60} molecule. ^{14}N triplet structures are observed in the peapod samples with identical hyperfine coupling as in the crystalline sample and are thus identified as the ESR signals of rotating $C_{59}N$ monomer radicals encapsulated inside SWCNTs. The additional component (arrow in Fig. 29) observed for sample B, which contains co-

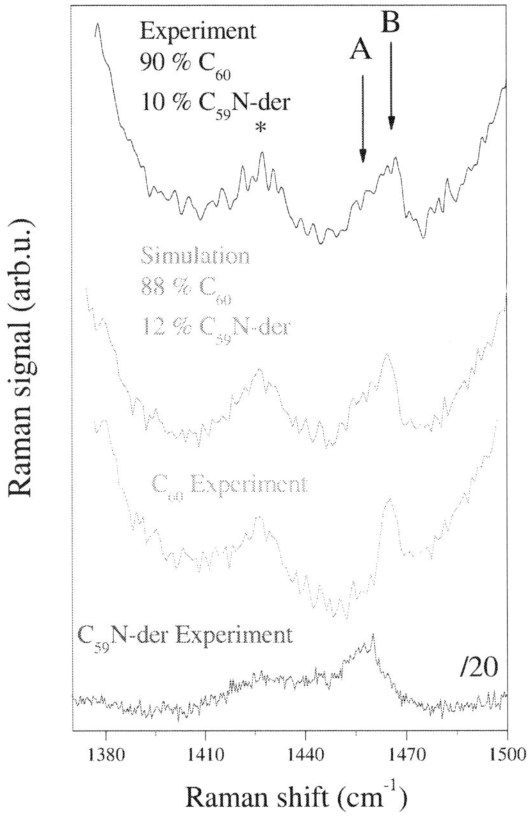

Figure 28. Raman spectra of the encapsulated $C_{59}N$-der:C_{60} mixture at the 488 nm laser excitation. The spectra for the $C_{59}N$-der and C_{60} peapods is shown together with their weighted sum as explained in the text. A and B mark the components coming nominally from the superposing two phases. The asterisk marks a mode that is present in the pristine SWCNT material. Note the different scale for the $C_{59}N$-der peapod material. Reprinted figure with permission from Ref. [97], F. Simon *et al.* Carbon, **44**, 1958 (2006). Copyright (2006) by Elsevier.

encapsulated C_{60}, is identified as $C_{59}N$-C_{60} heterodimers encapsulated inside SWCNTs since this signal has the same g-factor as in the crystalline material. This singlet line is absent in sample A which does not contain C_{60}. For both peapod samples a broader line with HWHM of $\Delta H \sim 0.6$ mT is also observed. The broader component appears also on heat treatment of reference samples without encapsulated $C_{59}N$-der and is identified as a side-product. Annealing at 600 °C is optimal: lower temperatures result in smaller $C_{59}N$ signals and higher temperatures increase the broad impurity signal without increasing the $C_{59}N$ intensity.

The observation of the ESR signal of $C_{59}N$ related spins proves that after the 600 °C heat treatment, a sizeable amount of rotating $C_{59}N$ monomer radicals are present in the sample. This is not surprising in the view of the ability to form $C_{59}N$ monomers from

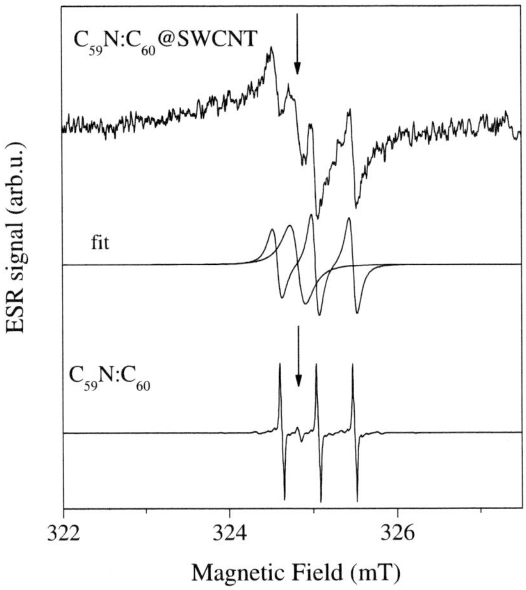

Figure 29. ESR spectra of crystalline $C_{59}N$:C_{60} (a) and ($C_{59}N$:C_{60})@SWCNT obtained by annealing the ($C_{59}N$-der:C_{60})@SWCNT. Solid curves show the deconvolution of the different ESR components for the encapsulated material. Reprinted figure with permission from Ref. [101], F. Simon *et al.* Phys. Rev. Lett. **97**, 136801 (2006). Copyright (2006) by the American Physical Society.

$C_{59}N$ at similar temperatures [103], however the current process is not reversible and the remnants of the side-groups are probably removed by the dynamic pumping.

The coexistence of bound $C_{59}N$-C_{60} heterodimers and rotating $C_{59}N$ molecules was understood for $C_{59}N$:C_{60} as a thermal equilibrium between the ground state heterodimer and the rotating monomers [102]. The inset in Fig. 30 shows a similar behavior for $C_{59}N$:C_{60}@SWCNT: the heterodimer dominates the low temperature spectrum and vanishes at higher temperatures, however the relative intensity of the heterodimer is much larger in this material. The heterodimer signal intensity normalized by the total (heterodimer+triplet) intensity gives the heterodimer concentration and is shown in Fig. 30. Similarly to crystalline $C_{59}N$:C_{60}, the heterodimer concentration can be fitted with:

$$\frac{I_{C_{59}N-C_{60}}}{I_{\text{total}}} = \frac{1}{\left(1 + e^{(-E_a/T + \Delta S)}\right)} \tag{2}$$

where E_a is the binding energy of the heterodimer and ΔS is the entropy difference between the rotating monomer and the static heterodimer states. A fit with Eq. 2 for the peapod material is shown in Fig. 30 as a solid curve and gives $E_a(\text{peapod}) = 2800(200)$ K and $\Delta S(\text{peapod}) = 9(1)$. This compares to the results for the crystalline material with $E_a(\text{cryst}) = 2400(600)$ K and $\Delta S(\text{cryst}) = 11(2)$ [102]. The higher heterodimer concentrations for the peapod material is caused by the larger E_a and smaller ΔS values. The latter is explained by the limited rotational freedom of encapsulated $C_{59}N$.

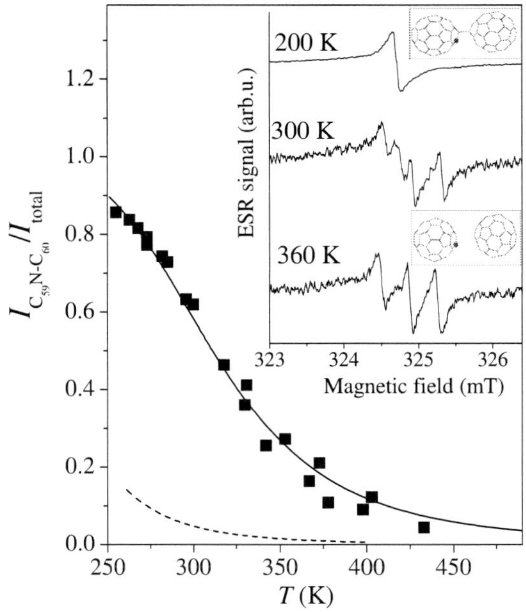

Figure 30. Concentration of $C_{59}N$-C_{60} bound heterodimers in $C_{59}N$:C_{60}@SWCNT. Solid curve is a fit with parameters explained in the text. Dashed curve shows the same quantity for crystalline $C_{59}N$:C_{60} above the 261 K phase transition. Note the much higher heterodimer concentration for the peapod material. Inset shows the temperature evolution of the spectra and the schematics of the heterodimer and monomers which correspond to the low and high temperature states, respectively. Reprinted figure with permission from Ref. [101], F. Simon *et al.* Phys. Rev. Lett. **97**, 136801 (2006). Copyright (2006) by the American Physical Society.

Similar to NMR spectroscopy, the ESR spin-lattice relaxation time, T_1, of the localized heterodimer spins yields information on the electronic structure of the host SWCNT material [104]. T_1 can be measured by time-resolved ESR measurements or in continuous wave ESR spectroscopy from the line-width, ΔH, by separating the homogeneous, relaxation related line-width from the inhomogeneous one. In Fig. 31a, we show ΔH for the heterodimer signal as determined from fits with derivative Lorentzian lines. Clearly, ΔH has a temperature dependent component in addition to a $\Delta H_0 = 0.089(2)$ mT residual line-width, which is obtained by averaging the line-widths below 50 K. The line-shape of the heterodimer signal does not change with temperature, which indicates a uniform, homogeneous broadening in addition to the inhomogeneous residual width.

To obtain the homogeneous line-width, ΔH_{Hom}, we subtracted ΔH_0 from the line-width data: $\Delta H_{\mathrm{Hom}} = \sqrt{\Delta H^2 - \Delta H_0^2}$. Fig. 31b shows ΔH_{Hom} and $1/T_1 = \gamma \Delta H_{\mathrm{Hom}}$, where $\gamma = 1.761 \cdot 10^{11} (\mathrm{sT})^{-1}$ is the electron gyromagnetic ratio. $1/T_1$ as a function of T is linear, with $(T_1 T)^{-1} = 4.2(2) \cdot 10^4 (\mathrm{sK})^{-1}$ (fit shown in Fig 31b), which suggests that Korringa relaxation, i.e. the interaction with conduction electrons [104] gives the relaxation of the heterodimer. The coupling constant, A, of localized spins and conduction electrons is 11 meV as determined from the Korringa relation [104]:

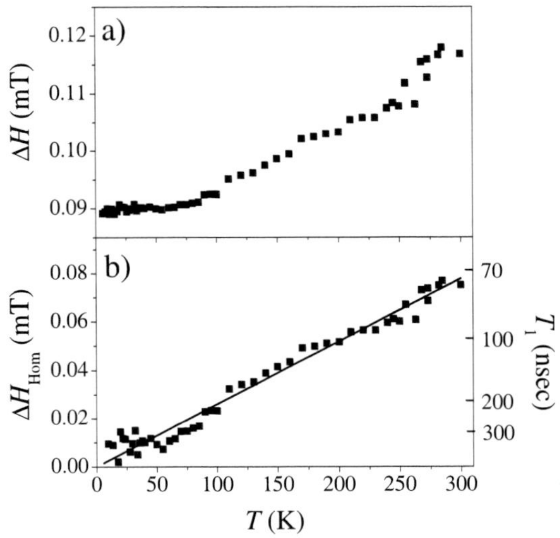

Figure 31. a) Temperature dependence of the ESR line-width for the heterodimer signal. b) The homogeneous contribution to the line-width with a linear fit (solid line). The corresponding T_1 values are shown on the right axis. Reprinted figure with permission from Ref. [101], F. Simon *et al.* Phys. Rev. Lett. **97**, 136801 (2006). Copyright (2006) by the American Physical Society.

$$\frac{1}{T_1 T} = \left(\frac{4\pi k_B}{\hbar}\right) A^2 \overline{n}(E_F)^2 \tag{3}$$

where $\overline{n}(E_F) = 0.014$ states/eV/atom is the DOS at the Fermi level for a $d \approx 1.4$ nm metallic tube in the tight-binding approximation [69]. The above discussed uniformity of the homogeneous broadening suggests that the heterodimer spins do not sense separate metallic and semiconducting tubes as it would be expected based on the geometry of tubes alone [69]. This can be explained by charge transfer in the SWCNTs bundles, which shifts the Fermi level and renders all tubes metallic. Indeed, several transport studies have indicated that tubes in the bundles [105, 106] and tubes exposed to small molecules such as O_2 [107] show metallic behavior.

Summarizing the ESR studies of encapsulated $C_{59}N$, we observed $C_{59}N$ monomer radicals encapsulated in SWCNTs. At low temperatures, bound $C_{59}N$-C_{60} heterodimers are observed when mixtures of the two fullerenes are encapsulated. Electron spin-relaxation of the heterodimer shows an overall metallic behavior of the tubes. The material is a step toward the realization of confined linear spin-chains, which might find application in e.g. quantum information processing.

3.4. NMR Studies on Isotope Engineered Heteronuclear Nanotubes

The growth of the "isotope engineered" nanotubes, i.e. DWCNTs with highly enriched inner wall allows to study these samples with NMR with an unprecedented specificity for

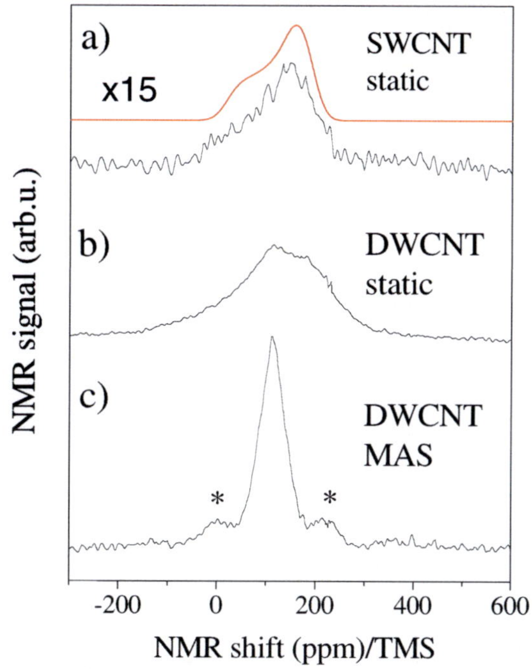

Figure 32. NMR spectra normalized by the total sample mass, taken with respect to the tetramethylsilane (TMS) shift. (a) Static spectrum for non-enriched SWCNT enlarged by 15. Smooth solid line is a chemical shift anisotropy powder pattern simulation with parameters published in the literature [91]. (b) Static and (c) MAS spectra of $^{13}C_{0.89}$-DWCNT, respectively. Asterisks show the sidebands at the 8 kHz spinning frequency. Reprinted figure with permission from Ref. [57], F. Simon *et al.* Phys. Rev. Lett. **95**, 017401 (2005). Copyright (2005) by the American Physical Society.

the small diameter carbon nanotubes. This allows a study of the electronic properties of such nanotubes. For normal SWCNTs, either grown from natural or ^{13}C enriched carbon, the NMR signal originates from all kinds of carbon like amorphous or graphitic carbon.

NMR allows to determine the macroscopic amount of enriched tubes in our sample as it is sensitive to the number of ^{13}C nuclei in the sample. In Fig. 32, we show the static and magic angle spinning, MAS, spectra of ^{13}C enriched DWCNTs, and the static spectrum for the SWCNT material. The mass fraction which belongs to the highly enriched phase can be calculated from the integrated signal intensity by comparing it to the signal intensity of the 89 % ^{13}C enriched fullerene material. We found that the mass fraction of the highly enriched phase relative to the total sample mass is 13(4) %. The expected mass ratio of inner tubes as compared to the total sample mass is 15 %, which is obtained from the SWCNT purity (50 %), the $\sim$70 % volume filling for peapod samples [62], and the mass ratio of encapsulated fullerenes to the mass of the SWCNTs. Thus, the measured mass fraction of the highly enriched phase is very similar to that of the calculated mass fraction of inner tubes. This suggests that the NMR signal comes nominally from the inner tubes, and other carbon phases such as amorphous or graphitic carbon are non ^{13}C enriched.

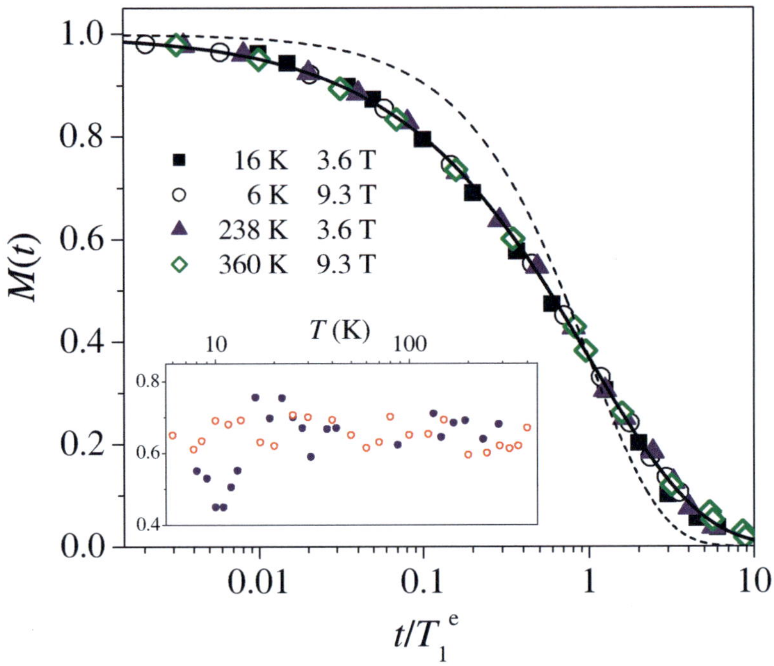

Figure 33. Reduced nuclear magnetization recovery, $M(t)$, as a function of the scaled delay time t/T_1^e (see Eq. (4)), for various temperature and magnetic field values. Both axes are dimensionless. Solid curve shows stretched exponential fit with $\beta = 0.65$ and dashed curve shows single exponential with $\beta = 1$. Inset shows temperature dependence of the best fit values of β at 3.6 Tesla (•) and 9.3 Tesla (○), and average value of the data set $\beta = 0.65$ (solid line). Reprinted figure with permission from Ref. [100], P. M. Singer *et al.* Phys. Rev. Lett. **95**, 236403 (2005). Copyright (2005) by the American Physical Society.

The typical chemical shift anisotropy (CSA) powder pattern is observed for the SWCNT sample in agreement with previous reports [91, 108]. However, the static DWCNT spectrum cannot be explained with a simple CSA powder pattern even if the spectrum is dominated by the inner tube signal. The complicated structure of the spectrum suggests that the chemical shift tensor parameters are highly distributed for the inner tubes. It is the result of the higher curvature of inner tubes as compared to the outer ones: the variance of the diameter distribution is the same for the inner and outer tubes [109] but the corresponding bonding angles show a larger variation [110]. In addition, the residual line-width in the MAS experiment, which is a measure of the sample inhomogeneity, is 60(3) ppm, i.e. about twice as large as the ~35 ppm found previously for SWCNT samples [91, 108]. The isotropic line position, determined from the MAS measurement, is 111 (2) ppm. This value is significantly smaller than the isotropic shift of the SWCNT samples of 125 ppm [91, 108]. However, recent theoretical *ab-initio* calculations have successfully explained this anomalous isotropic chemical shift [111]. It was found that diamagnetic demagnetizing currents on the outer walls cause the diamagnetic shift of the inner tube NMR signal.

In addition to the measurement of the line position, the dynamics of the nuclear relaxation is a sensitive probe of the local electronic properties [104]. The electronic properties of the nanotubes was probed using the spin lattice relaxation time, T_1, defined as the characteristic time it takes the ^{13}C nuclear magnetization to recover after saturation [100]. The signal intensity after saturation, $S(t)$, was deduced by integrating the fast Fourier transform of half the spin-echo for different delay times t. The data were taken with excitation pulse lengths $\pi/2 = 3.0\ \mu$s and short pulse separation times of $\tau = 15\ \mu$s [104]. The value of T_1 was obtained by fitting the t dependence of $S(t)$ to the form $S(t) = S_a - S_b \cdot M(t)$, where $S_a \simeq S_b$ (> 0) are arbitrary signal amplitudes, and

$$M(t) = \exp\left[-\left(t/T_1^e\right)^{\beta}\right], \tag{4}$$

is the reduced magnetization recovery of the ^{13}C nuclear spins. Fig. 33 shows the results of $M(t)$ for the inner-tubes as a function of the scaled delay time t/T_1^e, under various experimental conditions listed in the Figure. We find that $M(t)$ does not follow the single exponential form with $\beta = 1$ (dashed line), but instead fits well to a stretched exponential form with $\beta \simeq 0.65(5)$, implying a distribution in the relaxation times T_1. The data in Fig. 33 is displayed on a semi-log scale for the time axis in order to emphasize the data for earlier decay times and to illustrate the collapse of the data set for the upper 90 % of the NMR signal. For a broad range of experimental conditions, the upper 90 % of the $M(t)$ data is consistent with constant $\beta \simeq 0.65(5)$ (see inset), implying a field and temperature independent underlying distribution in T_1. The lower 10 % of the $M(t)$ data, corresponding to longer delay times, comes from the non-enriched outer-walls which have much longer relaxation times under similar experimental conditions [91, 108].

The collapse of the data set in Fig. 33 to Eq. (4) with constant $\beta = 0.65(5)$ is a remarkable experimental observation. From an experimental point of view, it implies that all one needs in order to characterize the T and H dependence of the underlying T_1 distribution is the bulk (or average) value, T_1^e (Eq. (4)). From an interpretational point of view, it implies that each inner-tube in the powder sample has a different value of T_1, yet *all* the T_1 components and therefore all the inner-tubes follow the same T and H dependence within experimental uncertainty. This finding is in contrast to earlier reports in SWCNTs where $M(t)$ fits well to a bi-exponential distribution, 1/3 of which had a short T_1 value characteristic of fast relaxation from metallic tubes, and the remaining 2/3 had long T_1 corresponding to the semiconducting tubes [91, 108, 112, 113], as expected from a macroscopic sample of SWCNTs with random chiralities. The data for the inner-tubes in DWCNTs differ in that a similar bi-exponential fit to $M(t)$ is inconsistent with the shape of the recovery in Fig. 33. Furthermore, if there were 1/3 metallic and 2/3 truly semi-conducting inner-tubes in the DWCNT sample, one would expect the ratio of T_1 between semiconducting and metallic tubes to increase exponentially with decreasing T below the semiconducting gap (~ 5000 K). As a consequence, one expects an increasingly large change in the underlying distribution in T_1 with decreasing T. This change would manifest itself as a large change in the shape of the $M(t)$, however this is not the case as shown in Fig. 33. The possibility of two components in T_1 with different T dependence can therefore be ruled out, and instead it could be concluded that all T_1 components (corresponding to distinct inner-tubes) exhibit the same T and H dependence within experimental scattering.

The experimentally observed uniform metallicity of inner tubes is a surprising observation. This is suggested to be caused by the shifting of the inner tube Fermi levels due

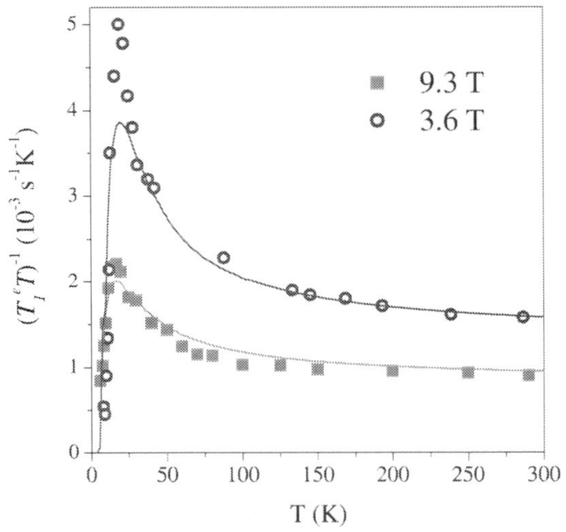

Figure 34. Temperature dependence of spin-lattice relaxation rate divided by temperature, $1/T_1^eT$, in units of ($10^3\times$ s^{-1} K $^{-1}$). Grey curves are best fits to Eq. (6) with $2\Delta = 46.8(40.2)$ K for $H = 3.6(9.3)$ Tesla, respectively. Reprinted figure with permission from Ref. [100], P. M. Singer *et al.* Phys. Rev. Lett. **95**, 236403 (2005). Copyright (2005) by the American Physical Society.

to charge transfer between the two tube walls. Indeed, using *ab-initio* calculations Okada and Oshiyama have found that DWCNTs made of non-metallic zig-zag inner-outer tubes, such as the (7,0)@(16,0) DWCNT, are metallic [114]. The direction of the charge transfer goes against the Faraday effect as inner tubes are electron and outer tubes are hole doped. Although, calculations are difficult if not impossible for an arbitrary inner-outer tube pair, this result confirms that two non-metallic tubes when forming an inner tube can render the electronic structure metallic.

With these arguments, the bulk average T_1^e defined in Eq. (4) can be considered and its uniform T and H dependence can be followed. The $M(t)$ data can be fitted with the constant exponent $\beta = 0.65(5)$, which reduces unnecessary experimental scattering in T_1^e. In Fig. 34 we show the temperature dependence of $1/T_1^eT$ in for two different values of the external magnetic field H. The data can be separated into two temperature regimes; the high temperature regime $\gtrsim$ 150 K, and the low T regime $\lesssim$ 150 K. At high temperatures, $1/T_1^eT$ is independent of T which indicates a metallic state [104] for all of the inner tubes. A strong magnetic field dependence for T_1 was also observed, which was explained by a 1D spin diffusion mechanism for T_1 [100].

The origin of the unusual T dependence of $1/T_1^eT$ in the low temperature regime ($\lesssim 150$ K) is very peculiar. Some trivial explanations can be ruled out as its origin. Firstly, one can rule out the possibility of an activation type mechanism where T_1 is dominated by fluctuating hyperfine fields with a characteristic time scale τ which increases with decreasing T (i.e. glassy slowing). This would result in a peaked relaxation with a strongly field dependent peak value [104], which is clearly not the case. Furthermore at low T, $1/T_1^eT$ drops below its high temperature value, which rules out the possibility of a T independent com-

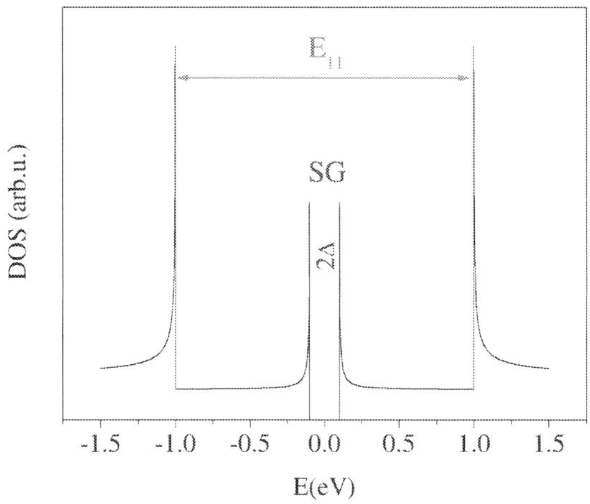

Figure 35. Schematics of the density of states proposed to explain the relaxation results. A small energy, 2Δ a few meV secondary gap opens at the Fermi level of metallic inner tubes, which is displayed not to scale. Note the van Hove singularities (vHs) at $\pm\Delta$. The unaffected vHs of the original band structure are shown for the first optical transition (E_{11}).

ponent in $1/T_1^eT$ plus an activated component on top. Secondly, the possibility of a simple $1/T$ Curie-like T dependence in $1/T_1^eT$ as a result of paramagnetic centers in the sample can be ruled out. This can be inferred from the pronounced gap in $1/T_1^eT$, together with the fact that no loss of ^{13}C NMR signal intensity in the entire temperature range of the experiment was observed. This implies that all the ^{13}C nuclei measure a gap in the spin excitation spectrum which cannot be explained by a Curie-like mechanism.

Following these considerations, one is lead to consider the simplest possible explanation for the experimental data in a non-interacting electron model of a 1D semiconductor with a small secondary gap (SG). The SG may be a result of the finite inner-wall curvature [19, 115, 116, 53]. The $1/T_1^eT$ data can be fitted using this non-interacting model with only one free parameter, the homogeneous SG, 2Δ. The normalized form of the gapped 1D density-of-states $n(E)$

$$n(E) = \begin{cases} \frac{E}{\sqrt{E^2-\Delta^2}} & \text{for } |E| > \Delta \\ 0 & \text{otherwise} \end{cases} \tag{5}$$

here, E is taken with respect to the Fermi energy). The total DOS of an inner tube is shown schematically in Fig. 35. Eq. (5) is used to calculate $1/T_1^eT$ [117] as such

$$\frac{1}{T_1^eT} = \alpha(\omega) \int_{-\infty}^{\infty} n(E)n(E+\omega) \left(-\frac{\delta f}{\delta E}\right) dE, \tag{6}$$

where E and ω are in temperature units for clarity, f is the Fermi function $f = [\exp(E/T)+1]^{-1}$, and the amplitude factor $\alpha(\omega)$ is the high temperature value for $1/T_1^eT$. The results of the best fit of the data to Eq. (6) are presented in Fig. 34, where $2\Delta = 43(3)$ K ($\equiv$ 3.7 meV) is H independent within experimental scattering between 9.3 and 3.6 Tesla.

We note that at the largest external field of 9.3 Tesla, $\omega = 4.5$ mK $\ll \Delta, T$, however, ω must be retained inside the integral. This is a direct consequence of the one dimensionality which gives a logarithmic divergence inside the integral (solved numerically) of the form $\ln(T/\omega)$ for $T \lesssim \Delta$. At the lowest temperatures T ($\lesssim 5$ K), T_1^e ($\gtrsim 300$ s) becomes so long that the low energy excitations specific to the homogeneous properties of the inner-tubes become inefficient, and other excitations take over, possibly defect related. In such cases we find that the recovery shape is no longer universal and the underlying distribution in T_1 is smoothed out, possibly due to spin-diffusion effects.

The origin of the experimentally observed gap still remains to be clarified. Tight binding calculations predict that applied magnetic fields can induce SG's of similar magnitude for metallic SWCNT [118]. However, such a scenario can be excluded due to the absence of any field dependence of the gap. The NMR data would be more consistent with a curvature induced SG for metallic tubes [19, 115, 116, 53], however for the typical inner-tubes the predicted values, ~ 100 meV, are over an order of magnitude larger than our experimental data. Other scenarios, such as quantization of levels due to finite short lengths of the nanotubes could be considered as well, however, in all these cases a behavior independent of tube size and chirality is certainly not expected.

This suggests that electron-electron interactions play an important role for the metallic inner tubes. It was predicted that electron-electron correlations and a Tomonaga-Luttinger (TLL) state leads to an increase in $1/T_1T$ with decreasing T [119], which is a direct consequence of the 1D electronic state. The correlated 1D nature may also lead to a Peierls instability [20] with the opening of a small collective gap 2Δ and a sharp drop in $1/T_1T$ below $\Delta \sim 20$ K. Therefore, the presence of both a TLL state and a Peierls instability could possibly account for the data.

Summarizing the NMR studies on DWCNTs, it was shown that T_1 has a similar T and H dependence for all the inner-tubes have, with no indication of a metallic/semiconducting separation due to chirality distributions. At high temperatures ($T \gtrsim 150$ K) $1/T_1^eT$ of the inner tubes exhibit a metallic 1D spin diffusion state. Below $\sim$150 K, $1/T_1^eT$ increases dramatically with decreasing T, and a gap in the spin excitation spectrum is found below $\Delta \simeq 20$ K. A 1D correlated state was suggested to be responsible for the opening of a small collective gap in the excitations, possibly a Peierls instability [20, 69]. Certainly, theoretical work on the electronic properties of inner wall nanotubes in DWCNT, and on the incidence of 1D correlation effects are needed to clarify the exact nature of the ground states.

4. Summary

In summary, we reviewed how molecules encapsulated inside SWCNTs can be used to study various properties of the tubes themselves. Inner tubes grown from encapsulated fullerenes were shown to be an excellent probe of diameter dependent reactions on the outer tubes. Inner tubes grown from isotope labeled fullerenes and organic solvents allowed to understand the role of the different carbon phases in the growth of the inner tubes. In addition, isotope labeled inner tubes were shown to yield an unparalleled precision to study the density of states near the Fermi level using NMR. It was reviewed how magnetic fullerenes can be encapsulated inside SWCNTs yielding linear spin chains with sizeable spin concentrations and also to allow ESR studies of the tube properties.

Acknowledgements

This work was supported by the Hungarian State Grants No. TS049881, F61733 and NK60984, by the EU project MERG-CT-2005-022103, and by the Magyary and Bolyai postdoctoral fellowships. H. Kuzmany, R. Pfeiffer, Á. Kukovecz, Z. Kónya, T. Pichler, H. Rauf, C. Kramberger, F. Hasi, E. Obraztsova, H. Alloul, P. M. Singer, J. Bernardi, P. Wzietek, F. Fülöp, A. Rockenbauer, A. Jánossy, T. Fehér, L. Forró, F. Hauke, A. Hirsch, V. N. Popov, V. Zólyomi and J. Kürti are acknowledged for their participation in the reviewed works. Dario Quintavalle is acknowledged for preparing some of the figures and for the careful reading of the manuscript.

References

[1] H. W. Kroto, J. R. Heath, S. C. O'Brien, R. F. Curl, and R. E. Smalley. C_{60}: Buckminsterfullerene. *Nature*, 318:162–163, 1985.

[2] Sumio Iijima. Helical microtubules of graphitic carbon. *Nature*, 354:56–58, 1991.

[3] Sumio Iijima and Toshinari Ichihashi. Single-shell carbon nanotubes of 1-nm diameter. *Nature*, 363:603–605, 1993.

[4] D. S. Bethune, C. H. Kiang, M. S. DeVries, G. Gorman, Savoy R., and R. Beyers. Cobalt-catalysed growth of carbon nanotubes with single-atomic-layer walls. *Nature*, 363:605, 1993.

[5] Brian W. Smith, Marc Monthioux, and David E. Luzzi. Encapsulated C_{60} in carbon nanotubes. *Nature*, 396:323–324, 1998.

[6] M. Melle-Franco, H. Kuzmany, and F. Zerbetto. Mechanical interaction in all-carbon peapods. *J. Phys. Chem. B*, 109:6986–6990, 2003.

[7] S. Berber, Y.-K. Kwon, and D. Tománek. Microscopic Formation Mechanism of Nanotube Peapods. *Phys. Rev. Lett.*, 88:185502, 2002.

[8] M. Otani, S. Okada, and A. Oshiyama. Energetics and electronic structures of one-dimensional fullerene chains encapsulated in zigzag nanotubes. *Phys. Rev. B*, 68:125424–1–8, 2003.

[9] A. Rochefort. Electronic and transport properties of carbon nanotube peapods. *Phys. Rev. B*, 67:115401–1–7, 2003.

[10] B. W. Smith, M. Monthioux, and D.E. Luzzi. Carbon nanotube encapsulated fullerenes: a unique class of hybrid materials. *Chem. Phys. Lett.*, 315:31–36, 1999.

[11] B. W. Smith and D.E. Luzzi. Formation mechanism of fullerene peapods and coaxial tubes: a path to large scale synthesis. *Chem. Phys. Lett.*, 321:169–174, 2000.

[12] H. Kataura, Y. Maniwa, T. Kodama, K. Kikuchi, K. Hirahara, K. Suenaga, S. Iijima, S. Suzuki, Y. Achiba, and W. Krätschmer. High-yield fullerene encapsulation in single-wall carbon nanotubes . *Synthetic Met.*, 121:1195–1196, 2001.

[13] D. J. Hornbaker, S. J. Kahng, S. Misra, B. W. Smith, A. T. Johnson, E.J. Mele, D. E. Luzzi, and A. Yazdani. Mapping the one-dimensional electronic states of nanotube peapod structures. *Science*, 295:828–831, 2002.

[14] A. N. Obraztsov, I. Pavlovsky, A. P. Volkov, E. D. Obraztsova, A. L. Chuvilin, and V. L. Kuznetsov. *J. Vac. Sci. Techn. B*, 18:1059, 2000.

[15] G. Z. Yue, Q. Qiu, B. Gao, Y. Cheng, J. Zhang, H. Shimoda, S. Chang, J. P. Lu, and O. Zhou. Generation of continuous and pulsed diagnostic imaging x-ray radiation using a carbon-nanotube-based field-emission cathode. *Appl. Phys. Lett.*, 81:355–368, 2002.

[16] J. H. Hafner, C. L. Cheung, and C. M. Lieber. Growth of nanotubes for probe microscopy tips. *Nature*, 398:761, 1999.

[17] Adrian Bachtold, Peter Hadley, Takeshi Nakanishi, and Cees Dekker. Logic circuits with carbon nanotube transistors. *Science*, 294:1317–1320, 2001.

[18] W. Harneit, C. Meyer, A. Weidinger, D. Suter, and J. Twamley. *Phys. St. Solidi B*, 233:453, 2002.

[19] N. Hamada, S. Sawada, and A. Oshiyama. New one-dimensional conductors: Graphitic microtubules. *Phys. Rev. Lett.*, 68:1579.1581, 1992.

[20] R. Saito, G. Dresselhaus, and M.S. Dresselhaus. *Physical Properties of Carbon Nanotubes*. Imperial College Press, 1998.

[21] D. Chattopadhyay, L. Galeska, and F. Papadimitrakopoulos. A route for bulk separation of semiconducting from metallic single-wall carbon nanotubes. *J. Am. Chem. Soc.*, 125:3370–3375, 2003.

[22] R. Krupke, F. Hennrich, H. von Lohneysen, and M. M. Kappes. Separation of Metallic from Semiconducting Single-Walled Carbon Nanotubes. *Science*, 301:344–347, 2003.

[23] Z. H. Chen, X. Du, M. H. Du, C. D. Rancken, H. P. Cheng, and A. G. Rinzler. Bulk Separative Enrichment in Metallic or Semiconducting Single-Walled Carbon Nanotubes. *Nano Lett.*, 3:1245–1259, 2003.

[24] M. Zheng, A. Jagota, M. S. Strano, A. P. Santos, P. Barone, S. G. Chou, G Diner, M. S. B. A. Dresselhaus, R. S. McLean, G. B. Onoa, G. G. Samsonidze, E. D. Semke, M. Usrey, and D. J. Walls. Structure-Based Carbon Nanotube Sorting by Sequence-Dependent DNA Assembly. *Science*, 302:1545–1548, 2003.

[25] Sergei M. Bachilo, Michael S. Strano, Carter Kittrell, Robert H. Hauge, Richard E. Smalley, and R. Bruce Weisman. Structure-Assigned Optical Spectra of Single-Walled Carbon Nanotubes. *Science*, 298:2361–2366, 2002.

[26] C. Fantini, A. Jorio, M. Souza, M. S. Strano, M. S. Dresselhaus, and M. A. Pimenta. Optical Transition Energies for Carbon Nanotubes from Resonant Raman Spectroscopy: Environment and Temperature Effects. *Phys. Rev. Lett.*, 93:147406, 2004.

[27] H. Telg, J. Maultzsch, S. Reich, F. Hennrich, and C. Thomsen. Chirality Distribution and Transition Energies of Carbon Nanotubes. *Phys. Rev. Lett.*, 93:177401, 2004.

[28] R. Egger and A. O. Gogolin. Effective low-energy theory for correlated carbon nanotubes. *Nature*, 79:50825085, 1997.

[29] K. P. Bohnen, R. Heid, H. J. Liu, and C. T. Chan. Lattice Dynamics and Electron-Phonon Interaction in (3,3) Carbon Nanotubes. *Phys. Rev. Lett.*, 93:245501–1–4, 2004.

[30] D. Connétable, G.-M. Rignanese, J.-C. Charlier, and X. Blase. Room Temperature Peierls Distortion in Small Diameter Nanotubes. *Phys. Rev. Lett.*, 94:015503–1–4, 2005.

[31] Sander J. Tans, Michel H. Devoret, Hongjie Dai, Andreas Thess, Richard E. Smalley, L. J. Geerligs, and Cees Dekker. Individual single-wall carbon nanotubes as quantum wires. *Nature*, 386:474 – 477, 1997.

[32] C.L. Kane and E.J. Mele. Ratio Problem in Single Carbon Nanotube Fluorescence Spectroscopy. *Phys. Rev. Lett.*, 90:207401–1–4, 2003.

[33] Catalin D. Spataru, Sohrab Ismail-Beigi, Lorin X. Benedict, and Steven G. Louie. Excitonic Effects and Optical Spectra of Single-Walled Carbon Nanotubes. *Phys. Rev. Lett.*, 92:077402–1–4, 2004.

[34] Vasili Perebeinos, J. Tersoff, and Phaedon Avouris. Scaling of Excitons in Carbon Nanotubes. *Phys. Phys. Lett.*, 92:257402–1–4, 2004.

[35] Vasili Perebeinos, J. Tersoff, and Phaedon Avouris. Effect of Exciton-Phonon Coupling in the Calculated Optical Absorption of Carbon Nanotubes. *Phys. Phys. Lett.*, 94:027402–1–4, 2005.

[36] M. Bockrath, D. H. Cobden, Jia Lu, Rinzler A. G., R. E. Smalley, L. Balents, and P. L. McEuen. Luttinger-liquid behaviour in carbon nanotubes. *Nature*, 397:598 – 601, 1999.

[37] H. Ishii, H. Kataura, H. Shiozawa, H. Yoshioka, H. Otsubo, Y. Takayama, T. Miyahara, S. Suzuki, Y. Achiba, M. Nakatake, T. Narimura, M. Higashiguchi, K. Shimada, H. Namatame, and M. Taniguchi. Direct observation of Tomonaga-Luttinger-liquid state in carbon nanotubes at low temperatures. *Nature*, 426:540–544, 2003.

[38] H. Rauf, T. Pichler, M. Knupfer, J. Fink, and H. Kataura. Transition from a Tomonaga-Luttinger Liquid to a Fermi Liquid in Potassium-Intercalated Bundles of Single-Wall Carbon Nanotubes. *Phys. Rev. Lett.*, 93:096805–1–4, 2004.

[39] Feng Wang, Gordana Dukovic, , Louis E. Brus, and Tony F. Heinz. The Optical Resonances in Carbon Nanotubes Arise from Excitons. *Science*, 308:838–841, 2005.

[40] J. Maultzsch, R. Pomraenke, S. Reich, E. Chang, D. Prezzi, A. Ruini, E. Molinari, M. S. Strano, C. Thomsen, and C. Lienau. Exciton binding energies in carbon nanotubes from two-photon photoluminescence. *Phys. Rev. B*, 72:241402, 2005.

[41] S. Bandow, M. Takizawa, K. Hirahara, M. Yudasaka, and S. Iijima. Raman scattering study of double-wall carbon nanotubes derived from the chains of fullerenes in single-wall carbon nanotubes. *Chem. Phys. Lett.*, 337:48–54, 2001.

[42] J. L. Hutchison, N. A. Kiselev, E. P. Krinichnaya, A. V. Krestinin, R. O. Loutfy, A. P. Morawsky, V. E. Muradyan, E. D. Obraztsova, J. Sloan, S. V. Terekhov, and D. N. Zakharov. Double-walled carbon nanotubes fabricated by a hydrogen arc discharge method. *Carbon*, 39:761–770, 2001.

[43] Wencai Ren, Feng Li, Jian Chen, Shuo Bai, and Hui-Ming Cheng. Morphology, diameter distribution and Raman scattering measurements of double-walled carbon nanotubes synthesized by catalytic decomposition of methane. *Chem. Phys. Lett.*, 359:196–202, 2002.

[44] R. Pfeiffer, H. Kuzmany, Ch. Kramberger, Ch. Schaman, T. Pichler, H. Kataura, Y. Achiba, J. Kürti, and V. Zólyomi. Unusual High Degree of Unperturbed Environment in the Interior of Single-Wall Carbon Nanotubes. *Phys. Rev. Lett.*, 90:225501–1–4, 2003.

[45] M. Yudasaka, K. Ajima, K. Suenaga, T. Ichihashi, A. Hashimoto, and S. Iijima. Nano-extraction and nano-condensation for C_{60} incorporation into single-wall carbon nanotubes in liquid phases. *Chem. Phys. Lett.*, 380:42–46, 2003.

[46] F. Simon, H. Kuzmany, H. Rauf, T. Pichler, J. Bernardi, H. Peterlik, L. Korecz, F. Fülöp, and A. Jánossy. Low temperature fullerene encapsulation in single wall carbon nanotubes: synthesis of N@C_{60}@SWCNT. *Chem. Phys. Lett.*, 383:362–367, 2004.

[47] M. Monthioux and L. Noé. Room temperature synthesis of C_{60}@SWNT (peapods). In *XVIIIth International Winterschool on Electronic Properties of Novel Materials.* , 2004. AIP Proceedings.

[48] A. N. Khlobystov, D. A. Britz, J. W. Wang, S. A. O'Neil, M. Poliakoff, and G. A. D. Briggs. Low temperature assembly of fullerene arrays in single-walled carbon nanotubes using supercritical fluids. *J. Mat. Chem.*, 14:2852–2857, 2004.

[49] T. Pichler, H. Kuzmany, H. Kataura, and Y. Achiba. Metallic Polymers of C_{60} Inside Single-Walled Carbon Nanotubes. *Phys. Rev. Lett.*, 87:267401–1–4, 2001.

[50] F. Simon, Á Kukovecz, Z. Kónya, R. Pfeiffer, and H. Kuzmany. Highly defect-free inner tubes in CVD prepared double wall carbon nanotubes. *Chem. Phys. Lett.*, 413:506–511, 2005.

[51] B. C. Liu, S. C. Lyu, S. I. Jung, H. K. Kang, C.-W. Yang, J. W. Park, Park C.Y., and C. J. Lee. Single-walled carbon nanotubes produced by catalytic chemical vapor deposition of acetylene over FeMo/MgO catalyst. *Chem. Phys. Lett.*, 383:104–108, 2004.

[52] Masatoshi Abe, Hiromichi Kataura, Hiroshi Kira, Takeshi Kodama, Shinzo Suzuki, Yohji Achiba, Ken-ichi Kato, Masaki Takata, Akihiko Fujiwara, Kazuyuki Matsuda, and Yutaka Maniwa. Structural transformation from single-wall to double-wall carbon nanotube bundles. *Phys. Rev. B*, 68:041405(R), 2003.

[53] V. Zólyomi and J. Kürti. First-principles calculations for the electronic band structures of small diameter single-wall carbon nanotubes. *Phys. Rev. B*, 70:085403–1–8, 2004.

[54] M. Endo, H. Muramatsu, T. Hayashi, Y. A. Kim, M. Terrones, and M. S. Dresselhaus. Nanotechnology: 'buckypaper' from coaxial nanotubes. *Nature*, 433:476, 2005.

[55] F. Simon, Á. Kukovecz, C. Kramberger, R. Pfeiffer, F. Hasi, H. Kuzmany, and H. Kataura. Diameter selective characterization of single-wall carbon nanotubes. *Phys. Rev. B*, 71:165439–1–5, 2005.

[56] H. Kuzmany. *Solid-State Spectroscopy, An Introduction*. Springer Verlag, Berlin, 1998.

[57] F. Simon, Ch. Kramberger, R. Pfeiffer, H. Kuzmany, J. Zólyomi, V. Kürti, P. M. Singer, and H. Alloul. Isotope Engineering of Carbon Nanotube Systems. *Phys. Rev. Lett.*, 95:017401–1–4, 2005.

[58] H. Kataura, Y. Kumazawa, Y. Maniwa, I. Umezu, S. Suzuki, Y. Ohtsuka, and Y. Achiba. Optical properties of single-wall carbon nanotubes. *Synthetic Met.*, 103:2555–2558, 1999.

[59] M.S. Dresselhaus, G. Dresselhaus, and P. C. Ecklund. *Science of Fullerenes and Carbon Nanotubes*. Academic Press, 1996.

[60] R. Pfeiffer, H. Kuzmany, T. Pichler, H. Kataura, Y. Achiba, M. Melle-Franco, and F. Zerbetto. Electronic and mechanical coupling between guest and host in carbon peapods. *Phys. Rev. B*, 69:035404, 2004.

[61] M. C. Martin, J. Fabian, J. Godard, P. Bernier, J. M. Lambert, and L. Mihály. Nanotube field of C_{60} molecules in carbon nanotubes: Atomistic versus continuous approach. *Phys. Rev. B*, 74:045421–1–4, 2006.

[62] X. Liu, T. Pichler, M. Knupfer, M. S. Golden, J. Fink, H. Kataura, Y. Achiba, K. Hirahara, and S. Iijima. Filling factors, structural, and electronic properties of C_{60} molecules in single-wall carbon nanotubes. *Phys. Rev. B*, 65:045419–1–6, 2002.

[63] H. Kuzmany, R. Pfeiffer, C. Kramberger, T. Pichler, X. Liu, M. Knupfer, J. Fink, H. Kataura, Y. Achiba, B. W. Smith, and D. E. Luzzi. Analysis of the concentration

of C_{60} fullerenes in single wall carbon nanotubes. *Appl. Phys. A*, 76(4):449–455, 2003.

[64] H. Kataura, Y. Maniwa, M. Abe, A. Fujiwara, T. Kodama, K. Kikuchi, J. Imahori, Y. Misaki, S. Suzuki, and Y. Achiba. Optical properties of fullerene and non-fullerene peapods. *Appl. Phys. A*, 74:349–354, 2002.

[65] K. Hirahara, K. Bandow, S. Suenaga, H. Kato, T. Okazaki, T. Shinohara, and S. Iijima. Electron diffraction study of one-dimensional crystals of fullerenes. *Phys. Rev. B*, 64:115420–1–5, 2001.

[66] J. Kürti, G. Kresse, and H. Kuzmany. First-principles calculations of the radial breathing mode of single-wall carbon nanotubes. *Phys. Rev. B*, 58:R8869–R8872, 1998.

[67] R. Pfeiffer, H. Kuzmany, F. Simon, S. N. Bokova, and E. Obraztsova. Resonance Raman scattering from phonon overtones in double-wall carbon nanotubes. *Phys. Rev. B*, 71:155409–1–8, 2005.

[68] R. Pfeiffer, F. Simon, H. Kuzmany, and V. N. Popov. Fine structure of the radial breathing mode of double-wall carbon nanotubes. *Phys. Rev. B*, 72:161404 –1–4, 2005.

[69] M. S. Dresselhaus, G. Dresselhaus, and Ph. Avouris. *Carbon Nanotubes: Synthesis, Structure, Properties, and Applications*. Springer, Berlin, Heidelberg, New York, 2001.

[70] R. Pfeiffer, Ch. Kramberger, F. Simon, H. Kuzmany, V. N. Popov, and H. Kataura. Interaction between concentric Tubes in DWCNTs. *Eur. Phys. J. B*, 42:345–350, 2004.

[71] V. N. Popov and Luc Henrard. Breathinglike phonon modes of multiwalled carbon nanotubes. *Phys. Rev. B*, 65:235415, 2002.

[72] A. Jorio, R. Saito, J. H. Hafner, C. M. Lieber, M. Hunter, T. McClure, G. Dresselhaus, and M. S. Dresselhaus. Structural (n, m) Determination of Isolated Single-Wall Carbon Nanotubes by Resonant Raman Scattering. *Phys. Rev. Lett.*, 86:1118–1121, 2001.

[73] Andreas Thess, Roland Lee, Pavel Nikolaev, Hongjie Dai, Pierre Petit, Jerome Robert, Chunhui Xu, Young Hee Lee, Seong Gon Kim, Andrew G. Rinzler, Daniel T. Colbert, Gustavo E. Scuseria, David Tománek, John E. Fischer, and Richard E. Smalley. Crystalline Ropes of Metallic Carbon Nanotubes. *Science*, 273:483–487, 1996.

[74] F. Simon, R. Pfeiffer, and H. Kuzmany. Temperature dependence of optical excitation life-time and band-gap in chirality assigned semiconducting single-wall carbon nanotubes. *Phys. Rev. B*, 74:121411(R)–1–4, 2006.

[75] H. Kuzmany, W. Plank, M. Hulman, Ch. Kramberger, A. Grüneis, Th. Pichler, H. Peterlik, H. Kataura, and Y. Achiba. Determination of SWCNT diameters from the Raman response of the radial breathing mode. *Eur. Phys. J. B*, 22(3):307–320, 2001.

[76] Nachiket R. Raravikar, Pawel Keblinski, Apparao M. Rao, Mildred S. Dresselhaus, Linda S. Schadler, and Pulickel M. Ajayan. Temperature dependence of radial breathing mode Raman frequency of single-walled carbon nanotubes. *Phys. Rev. B*, 66:235424–1–9, 2002.

[77] Ch. Kramberger, R. Pfeiffer, H. Kuzmany, V. Zólyomi, and J. Kürti. Assignment of chiral vectors in carbon nanotubes. *Phys. Rev. B*, 68:235404, 2003.

[78] Brian W. Smith and David E. Luzzi. Formation mechanism of fullerene peapods and coaxial tubes: a path to large scale synthesis. *Chem. Phys. Lett.*, 321:169–174, 2000.

[79] F. Hasi, F. Simon, and H. Kuzmany. Reversible Hole Engineering for Single-Wall Carbon Nanotubes. *J. Nanosci. Nanotechn.*, 5:1785–1791, 2005.

[80] Y. Zhao, B. I. Yakobson, and R. E. Smalley. Dynamic Topology of Fullerene Coalescence. *Phys. Rev. Lett.*, 88:185501–1–4, 2002.

[81] S. W. Han, M. Yoon, S. Berber, N. Park, E. Osawa, J. Ihm, and D. Tománek. Microscopic mechanism of fullerene fusion. *Phys. Rev. B*, 70:113402–1–4, 2004.

[82] J. Bardeen, L. N. Cooper, and J. R. Schrieffer. Theory of Superconductivity. *Phys. Rev.*, 108:1175–1204, 1957.

[83] F. Simon and H. Kuzmany. Growth of single-wall carbon nanotubes from ^{13}C isotope labelled organic solvents inside single wall carbon nanotube hosts. *Chem. Phys. Lett.*, 425:85–88, 2006.

[84] G. Kresse and D. Joubert. From ultrasoft pseudopotentials to the projector augmented-wave method. *Phys. Rev. B*, 59:17581775, 1999.

[85] A. S. Claye, N. M. Nemes, A. Jánossy, and J. E. Fischer. Structure and electronic properties of potassium-doped single-wall carbon nanotubes. *Phys. Rev. B*, 62:4845–4848 (R), 2000.

[86] J.-P. Salvetat, T. Fehér, C. LHuillier, F. Beuneu, and L. Forró. Anomalous electron spin resonance behavior of single-walled carbon nanotubes. *Phys. Rev. B*, 72:075440–1–6, 2005.

[87] T. Almeida Murphy, Th. Pawlik, A. Weidinger, M. Hhne, R. Alcala, and J.-M. Spaeth. *Phys. Rev. Lett.*, 77:1075, 1996.

[88] M. Waiblinger, K. Lips, W. Harneit, A. Weidinger, E. Dietel, and A. Hirsch. Thermal stability of the endohedral fullerenes N@C_{60}, N@C_{70}, and P@C_{60}. *Phys. Rev. B*, 64:159901–1–4, 2001.

[89] B. Pietzak, M. Waiblinger, T.A. Murphy, A. Weidinger, M. Hohne, E. Dietel, and A. Hirsch. *Chem. Phys. Lett.*, 279:259, 1997.

[90] A. Jánossy, S. Pekker, F. Fülöp, F. Simon, and G. Oszlányi. In *Proceedings of the IWEPNM*, page 199, 2000.

[91] X.-P. Tang, A. Kleinhammes, H. Shimoda, L. Fleming, K. Y. Bennoune, S. Sinha, C. Bower, O. Zhou, and Y. Wu. Electronic Structures of Single-Walled Carbon Nanotubes Determined by NMR. *Science*, 288:492, 2000.

[92] K.-P. Dinse. EPR investigation of atoms in chemical traps. *Phys. Chem. Chem. Phys.*, 4:5442, 2002.

[93] J. C. Hummelen, B. Knight, J. Pavlovich, R. Gonzalez, and F. Wudl. Isolation of the Heterofullerene $C_{59}N$ as Its Dimer $(C_{59}N)_2$. *Science*, 269:1554–1556, 1995.

[94] J. C. Hummelen, C. Bellavia-Lund, and F. Wudl. *Heterofullerenes*, volume 199, page 93. Springer, Berlin, Heidelberg, 1999.

[95] M. J. Butcher, F. H. Jones, P. H. Beton, P. Moriarty, B. N. Cotier, M. D. Upward, K. Prassides, K. Kordatos, N. Tagmatarchis, F. Wudl, V. Dhanak, T. K. Johal, C. Crotti, C. Comicioli, and C. Ottaviani. $C_{59}N$ Monomers: Stabilization through Immobilization. *Phys. Rev. Lett.*, 83:3478–3481, 1999.

[96] F. Fülöp, A. Rockenbauer, F. Simon, S. Pekker, L. Korecz, S. Garaj, and A. Jánossy. Azafullerene $C_{59}N$, a Stable Free Radical Substituent in Crystalline C_{60}. *Chem. Phys. Lett.*, 334:223, 2001.

[97] F. Simon, H. Kuzmany, J. Bernardi, F. Hauke, and A. Hirsch. Encapsulating $C_{59}N$ azafullerene derivatives inside single-wall carbon nanotubes. *Carbon*, 44:1958–1962, 2006.

[98] A. Hirsch and B. Nuber. Nitrogen heterofullerenes. *Acc. Chem. Res.*, 32:795–804, 1999.

[99] H. Kuzmany, W. Plank, J. Winter, O. Dubay, N. Tagmatarchis, and K. Prassides. Raman spectrum and stability of $(C_{59}N)_2$. *Phys. Rev. B*, 60:1005–1012, 1999.

[100] P. M. Singer, P. Wzietek, H. Alloul, F. Simon, and H. Kuzmany. NMR Evidence for Gapped Spin Excitations in Metallic Carbon Nanotubes. *Phys. Rev. Lett.*, 95:236403–1–4, 2005.

[101] F. Simon, H. Kuzmany, B. Nafradi, T. Feher, L. Forró, F. Fülöp, A. Jánossy, A. Rockenbauer, L. Korecz, F. Hauke, and A. Hirsch. Magnetic fullerenes inside single-wall carbon nanotubes. *Phys. Rev. Lett.*, 97:136801–1–4, 2006.

[102] A. Rockenbauer, G. Csányi, F. Fülöp, S. Garaj, L. Korecz, R. Lukács, F. Simon, L. Forró, S. Pekker, and A. Jánossy. Electron delocalization and dimerization in solid $C_{59}N$ doped C_{60} fullerene. *Phys. Rev. Lett.*, 94:066603, 2005.

[103] F. Simon, D. Arčon, N. Tagmatarchis, S. Garaj, L. Forró, and K. Prassides. ESR signal in azafullerene $(C_{59}N)_2$ induced by thermal homolysis. *J. Phys. Chem. A.*, 103:6969, 1999.

[104] C. P. Slichter. *Principles of Magnetic Resonance*. Spinger-Verlag, New York, 3rd ed. 1996 edition.

[105] R. Martel, V. Derycke, C. Lavoie, J. Appenzeller, K. K. Chan, J. Tersoff, and Ph. Avouris. Ambipolar Electrical Transport in Semiconducting Single-Wall Carbon Nanotubes . *Phys. Phys. Lett.*, 87:256805–1–4, 2001.

[106] M. Radosavljevic, J. Lefebvre, and A. T. Johnson. High-field electrical transport and breakdown in bundles of single-wall carbon nanotubes. *Phys. Rev. B*, 64:241307–1–4, 2001.

[107] P. G. Collins, K. Bradley, M. Ishigami, and A. Zettl. Extreme oxygen sensitivity of electronic properties of carbon nanotubes. *Science*, 287:1801–1804, 2000.

[108] C. Goze-Bac, S. Latil, P. Lauginie, V. Jourdain, J. Conard, L. Duclaux, A. Rubio, and P. Bernier. Magnetic interactions in carbon nanostructures. *Carbon*, 40:1825–1842, 2002.

[109] F. Simon, R. Pfeiffer, C. Kramberger, M. Holzweber, and H. Kuzmany. *The Raman response of double wall carbon nanotubes*, pages 203–224. Springer New York, 2005.

[110] J. Kürti, V. Zólyomi, M. Kertész, and S. Guangyu. The geometry and the radial breathing mode of carbon nanotubes: beyond the ideal behaviour. *New. J. Phys.*, 5:125.1125.21, 2003.

[111] M. A. L. Marques, M. d'Avezac, and F. Mauri. Magnetic response of carbon nanotubes from ab initio calculations, 2006.

[112] H. Shimoda, B. Gao, X. P. Tang, A. Kleinhammes, L. Fleming, Y. Wu, and O. Zhou. Lithium Intercalation into Opened Single-Wall Carbon Nanotubes: Storage Capacity and Electronic Properties. *Phys. Rev. Lett.*, 88:15502, 2002.

[113] A. Kleinhammes, S.-H. Mao, X.-J. Yang, X.-P. Tang, H. Shimoda, J. P. Lu, O. Zhou, and Y. Wu. Gas adsorption in single-walled carbon nanotubes studied by NMR. *Phys. Rev. B*, 68:75418–1–6, 2003.

[114] S. Okada and A. Oshiyama. Curvature-Induced Metallization of Double-Walled Semiconducting Zigzag Carbon Nanotubes. *Phys. Rev. Lett.*, 91:216801–1–4, 2003.

[115] C.L. Kane and E.J. Mele. Size, Shape, and Low Energy Electronic Structure of Carbon Nanotubes. *Phys. Rev. Lett.*, 78:1932–1935, 1997.

[116] J.W. Mintmire and C.T. White. Universal Density of States for Carbon Nanotubes. *Phys. Rev. Lett.*, 81:2506, 1998.

[117] T. Moriya. The Effect of Electron-Electron Interaction on the Nuclear Spin Relaxation in Metals. *J. Phys. Soc. Jpn.*, 18:516, 1963.

[118] Jian Ping Lu. Novel Magnetic Properties of Carbon Nanotubes. *Phys. Rev. B*, 74:1123, 1995.

[119] H. Yoshioka. NMR relaxation rate in metallic carbon nanotubes. *J. Phys. Chem. Solids.*, 63:1281, 2002.

In: Progress in Fullerene Research
Editor: Milton Lang, pp. 199-233
ISBN: 978-1-60021-841-5

Chapter 6

ENDOHEDRAL FULLERENES: FROM CHEMICAL PHYSICS TO NANOTECHNOLOGY AND NANOMEDICINE

Vitaly K. Koltover*

Institute of Problems of Chemical Physics, Russian Academy of Sciences, Chernogolovka, Moscow Region, 142432, Russia

Abstract

At the present time the "atomic park" of endohedral fullerenes, $A@C_{2n}$, counts approximately 50 elements of the Periodic Table. The purpose of this review is to highlight examples of progress in chemical physics of endohedral fullerenes and point out fertile grounds for future applications of ENF in engineering and medicine. In large part, we concentrate on paramagnetic metallofullerenes, $M@C_{2n}$, since the efforts of our group were aimed mainly at synthesis and study of these objects. Atoms entrapped inside the fullerene cage manifest unusual, even unique, structure and electron properties. In paramagnetic $La@C_{2n}$, for example, there is the partial runoff of electron spin density outside the fullerene cage ("spin leakage") testified by EPR, ENDOR and NMR data. In $Be@C_{60}$, $N@C_{60}$, $P@C_{60}$, on the other hand, there is the accumulation of electron spin density on the entrapped atoms as if the electronic shells of the trapped atoms are compressed ("spin compression"). There is a strong tendency of ENF to form nanoparticles in liquid and solid states. The MAS (magic angle spinning) NMR study of solid complexes of La-ENF and Y-ENF with hexamethylphosphoramide has revealed the vast range of chemical shifts for the ^{31}P NMR, along with the enormous shifts for the ^{139}La NMR. Being unchanged with temperature, these shifts suggest their Knight nature caused by the density of conduction electrons at the positions of the NMR observed nuclei. Furthermore, the fullerene cages can be modified in plenty of organic-functionalized derivatives. Yet, chemical and physical properties of ENF are still far from being clear since these compounds are not easily accessible. Paramagnetic ENF hold much promise as the basic elements for realizing nanometer scale devices, amongst them – molecular magnets, superconductors with low critical temperatures, molecular memory devices, field effect transistors, quantum dots, fullerene-based quantum computers. Furthermore, there are the fields in medicine where the employment of ENF appears to have considerable promise too. For example, the water proton relaxivity of some $Gd@C_{82}(OH)_n$ is

* E-mail address: koltover@icp.ac.ru

much higher than that of Gd(III)-DTPA, which is in current clinical use as the contrast agent for magnetic resonance imaging. $M@C_{2n}$ with the appropriate particle-emitting radionuclide inside, like β-emitting ^{90}Y, ^{47}Sc, ^{149}Pr, ^{153}Sm, ^{166}Ho or ^{177}Lu, can be synthesized in macroscopic quantities, while advances in the molecular biotechnology provide targeting vectors to deliver therapeutic doses of the ionizing radiation with high specificity for treatment of cancer cells. Yet, neither of these ideas has been realized up to date.

Keywords: endohedral fullerenes, metallofullerenes, spin leakage, spin compression, nanoparticles, magnetic resonance, EPR, NMR, nanotechnology, spin electronics, nanomedicine.

Introduction

Endohedral fullerenes (ENF, $A@C_{2n}$) first discovered in 1985, are defined as compounds that contain an atom (sometimes, two and more atoms) inside the fullerene cage C_{2n} [1]. The @-symbol, that is in common use nowadays, was advocated by Richard Smalley and his co-workers in [2]. In the formula $A@C_{2n}$, all atoms listed to the right of @ symbol are assumed to be part of the cage network, and all to the left are situated inside the cage. Under this notation[1], a Buckminsterfullerene cage with a single lanthanum atom inside of the cage is written $La@C_{60}$, a C_{82} cage with three scandium atom trapped inside is written $Sc_3@C_{82}$, etc. $La@C_{60}$, $La@C_{82}$ and $Sc_3@C_{82}$ are depicted on Figure 1. $La@C_{60}$, that was the first $M@C_{2n}$ detected in the mass-spectra, is thermodynamically instable at higher temperatures and, thereafter, cannot be isolated in macroscopic quantity. $La@C_{82}$ has been recognized as the "archetype" after its first extraction in 1991 [2]. Mono-metallofullerenes with the C_{82} cage are most plentiful. The unique stability of $M@C_{82}$ (82 is a special "magic" number) has been confirmed experimentally and explained theoretically. In short, by dropping down to the C_{82} fullerene and adding two extra electrons from the 6*s* shell of an internal atom, one regains the 84 electron count, and there may be an arrangement of the pentagons and hexagons that

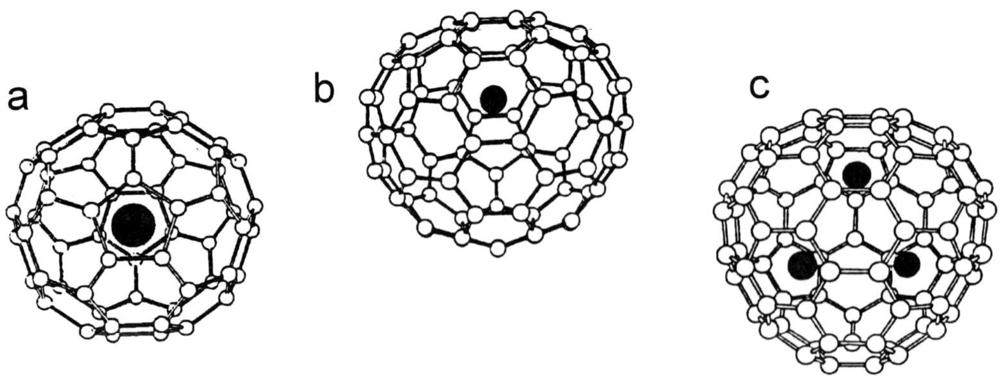

Figure 1. Endohedral metallofullrenes: (a) $La@C_{60}$, (b) $La@C_{82}$, (c) Sc3@C82.

[1] In the acknowledgment of the cited paper, the authors noted that "the delightful suggestion of the use of the @ symbol in formulas for complex fullerenes is due to Ori Chesnovsky" [2].

preserves this as a shell closing. Sc is the only element of the Ln-group small enough to fit three of its ions inside of the C_{82} [2-7]. Interestingly, $U@C_{2n}$ are formed in similar relative abundance to their lanthanide analogues, i.e. – with magic numbers 60, 70, 74, 82 (the last one being "most magic") [4].

Now, in the early 21st century, this man-caused "atomic park" runs to tens of elements, mainly of the second group and of the third group of the Periodic Table. Among them, paramagnetic ENF hold much promise as new potential materials with unique magnetic, electrical and optical properties for applications in nanoelectronics including fullerene-based quantum computers, in nanomedicine and so on.

The goal of this mini review was to digest the basic results of fundamental research in chemical physics of endohedral fullerenes and to overview the basic ideas of their possible applications in engineering and medicine. In large part, we concentrate on paramagnetic metallofullerenes, $M@C_{2n}$, since the efforts of our group were aimed mainly at synthesis and study of these objects.

Preparation of Endohedral Fullerenes

1. Synthesis

ENF with a single lanthanum atom trapped on the inside of the fullerene carbon cage were first produced by laser vaporization of a lanthanum oxide/graphite composite rode in a flow argon gas, extracted with toluene and certified by Fourier transform ion cyclotron resonance and mass-spectroscopic technique [1, 2]. Next principal stride forward was done owing to development of the electric-arc generator in which graphite electrodes were doped with the material to be encapsulated, e.g. lanthanum or scandium [3]. Electric-arc reactors are now widely used to produce ENF in macroscopic amounts, predominantly ENF of metals which are able to form carbides and acetylinides, i.e. – the metals mostly of calcium and scandium subgroups including radioactive U, Th and Pu [4-23].

Figure 2 represents a scheme of an electric-arc reactor designed by Dr. V. P. Bubnov in Institute of Problems of Chemical Physics (IPCP RAS) [14-17]. This reactor is equipped with graphite composite electrodes containing metals or hydrides of metals, MH_x (M = La, Y, etc., $x \leq 3$).

Anodes are spectral pure graphite rods (6 mm in diameter, 160 mm in length) in which a round hole (3mm in diameter) was drilled through the centers from the both sides. The holes were filled with mixture of a metal or its hydride, graphite cement, and graphite powder. During evaporation of the composite electrodes, there is the large-scale dispersion of the composite charge by the convective streams in the arc that steeply decreases formation of fullerenes and almost totally suppresses formation of endohedral fullerenes. To solve this problem, some graphite cement is introduced in the charge. It makes the composite electrodes more homogeneous, the central part of which evaporates more stable in the arc. The implementation of hydrides, instead of pure metals, enables to achieve more homogeneous distribution of the metal in the graphite rods. It further stabilizes the process of evaporation of the electrodes in the arc resulting in the essential increase of the $M@C_{2n}$ content in the soot.

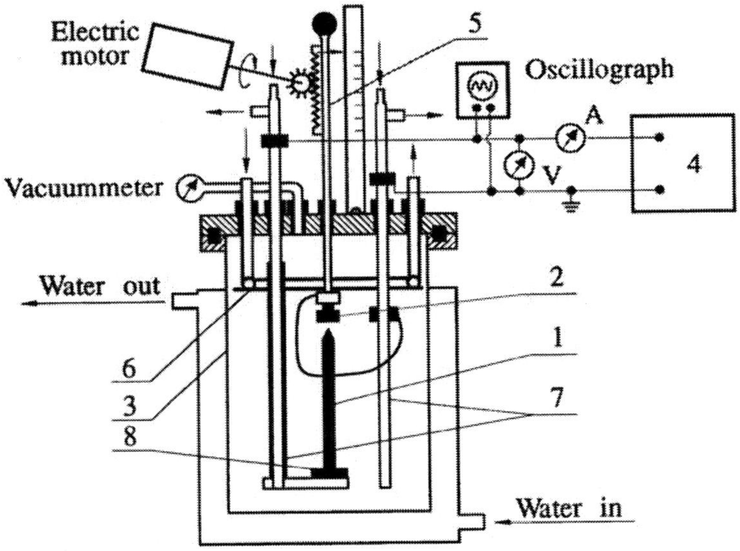

Figure 2. A schematic representation of the high-efficiency electric-arc reactor for synthesis of $M@C_{2n}$, designed in the IPCP RAS: 1, composite graphite electrode (anode) to be volatilized; 2, nonvolatilized graphite electrode (cathode); 3, stainless steel cavity cooled by water from outside; 4, direct current arc source; 5, stainless steel rod; 6, water-cooled copper screen; 7, inlets cooled by water; 8, graphite ring [16, 17].

The composite electrodes were prepared as follows. The relevant metal filing was thoroughly stirred with graphite powder and graphite cement (analytic grade, GCS trademark, Dylan Industries Inc.) as a binder in the proportion 1:0.57:1.43, and the charge was packed in the hole of the graphite electrode up to 30% of the total weight of the rode. The atomic ratio of metal to carbon varied from 0.5 to 1.5 %. It was found that the yield of $M@C_{2n}$ was maximal at M/C = 1.0%. Namely, the yield of the DMF (*N,N*-dimethylformamide) extracts was 5.0% of the soot weight for $Ce@C_{2n}$, 6.0% for $La@C_{2n}$ and 8% for $Y@C_{2n}$, i.e. one can obtain 0.8 g of $Y@C_{2n}$ from 10 g of $Y@C_{2n}$-containing soot. To make the composite electrode rods more homogeneous, they were exposed to the three stage thermal treatment: (1) treatment in the vacuum furnace at 130 °C for 4–5 h; (2) heating at 1100 °C for 4 h; (3) treatment just in the reactor in vacuum (10^{-3} Torr) at 1800–1900 ^{0}C for 1 h by passing the direct current (DC) of 190–200 A through the electrodes.

Cathodes, 12–16 mm in diameter, were made of MPG-8 graphite. It is interesting to note that the $M@C_{2n}$ yield was twice higher when a mobile cathode was placed above the anodes while the reactor itself was vertically oriented. This might be due to the most uniform precipitation of the soot in the case. To maximize the yield of $M@C_{2n}$, the conditions of evaporation of the electrodes in the arc were found to be as follows: the helium pressure 120 Torr, the electric arc current 90 A at the voltage 28–30 V, the arc length 5 mm, the distance between the arc and a cooling surface of the reactor 50 mm, the evaporation rate 1 mm per min [16, 17].

An arc-discharge anode made of a $LaNi_2$ alloy and graphite was used in [18]. Nickel exhibited a catalytic effect on production of $La@C_{2n}$, especially the production of the so-called "minor" isomer, $La@C_{82}$ (II). The yield of the latter was higher than that of the "major isomer", $La@C_{82}$ (I) [18]. Furthermore, Shinohara's group has achieved the significant increase in yield of $Eu@C_{2n}$ using the composite rods doped with nickel as anode [19]. Recently, the electric-arc sputtering of graphite with deposition of lanthanum compounds on surface of the graphite electrode has been proposed but the obtained yield of $La@C_{2n}$ was low [20].

The electric-arc technique has been smartly extrapolated by Sueki et al. to elements of other groups of the Periodic Table [21]. A multielement radioactive tracer containing radioisotopes of 23 elements was prepared after that a large amount of lanthanum nitrate was added to the multitracer solution as a carrier. This carrier solution was then used to make the carbon rods suitable for preparation of $M@C_{2n}$ by the arc discharged method. Relative yields of the $M@C_{2n}$ (M = Ln, Zr, Hf, etc.) were found to decrease exponentially as a function of the group number of the element, being estimated for Fe and Co less than one hundredth of that for $La@C_{82}$, in practice, trace quantities [21-22]. Yet, macroscopic amounts of titanium metallofullerenes $Ti_2@C_{80}$, $Ti_2@C_{84}$ were produced in DC arc reactor with Ti/C composite electrode rods [23].

Introduction of small amounts of nitrogen (1-3 Torr N_2/300 Torr He) into an electric-arc reactor allows for the efficient production of a new family of trimetallic nitride template (TNT) $A_xB_{3-x}N@C_{2n}$ ($x = 0–3$, $n = 34$, 38-44), where A and B are metals, such as Sc, Y, Ce, Gd, Tb, Dy, Ho, Er, Tm, Lu. This process generates milligram quantities of TNT-ENF products and makes it possible to isolate the material, determine its crystal structure, and study its physical and chemical properties. The first, "archetypal", representative of this new class was $Sc_3N@C_{80}$ containing a triscandium nitride cluster encapsulated in an icosahedron C_{80} cage. It seems to be the most abundant in all the fullerene family, only exceeded by C_{60} and C_{70} [8].

Putting the noble atoms into fullerenes has been arranged by heating fullerenes at 650 ^{0}C for a few hours under 3000 atmospheres of the noble gas. Helium, neon, argon, krypton are introduced into the C_{60} fullerene cage in about one in 1000 molecules (0.1%). For Xe the yield was even less (0.03%) but it was still possible to separate $Xe@C_{60}$ from the empty C_{60} by using HPLC and to study $^{129}Xe@C_{60}$ by ^{13}C NMR. Presumably, one or more of the carbon-carbon bonds of the cage is broken to let the atoms go inside of the fullerene cage through the so-called "window mechanism" [6].

Putting the atoms of groups I and V of the Periodic Table into fullerenes has been realized in molecular beam machines. A beam of ions or fast atoms hits a surface where a small fraction of them enters the fullerene molecules [24]. For example, C_{60} is evaporated from an effusion cell onto a copper substrate, being simultaneously irradiated with implantation ions from the ion source. The irradiated material is accumulated, removed from the substrate and dissolved in an appropriate solvent, like toluene. The alternative is a glow discharge reactor, i.e. – a quartz tube with two electrodes and a continuous stream of low-pressure gas to be inserted through the tube. A fullerene powder is deposited in the central part of the tube, which is surrounded by a furnace, and the glow discharge is started after the vapor of the fullerene gets high enough pressure for its sufficient sublimation to the water-cooled cathode.

Endohedral alkali metal fullerenes were produced in a molecular beam machine in which a monolayer of C_{60} was repeatedly exposed to an intensive beam of the alkali ion at an energy chosen so that the ions can penetrate the carbon cage but cannot destroy it, approximately 6 eV for Li^+ and 40 eV for K^+. Approximately 15 microgram of $Li@C_{60}$ can be deposited per hour [25]. The collisional insertion technique was used to implant ions of cesium into C_{60} and C_{70} by scattering Cs^+ ion beams from the fullerene layer deposited on gold [26]. Besides, Mu@C_{60} was synthesized after irradiating the solid C_{60} films with positive muoniums [27].

A similar ion implantation technique, i.e. – the bombarding of C_{60} with ions, is used for production of endohedral fullerenes of nitrogen and phosphorus [28]. Kaufman ion source and glow discharge are employed to produce the species of this kind and both are low productive. Milligrams of C_{60} containing N@C_{60} in a concentration of 10^{-4} to 10^{-5} are obtained, so that 1 of 10^4 C_{60} is, at most, filled with nitrogen. For N@C_{70} and P@C_{60}, the yields are even lower. The concentrations of the ion-bombarded samples in typical saturated solutions in CS_2 or toluene are of about 10^{-7} M [29]. Fortunately, nitrogen and phosphorous atoms trapped inside the cage, possess unpaired electrons with a total electron spin $S = 3/2$ and EPR, due to selectivity for paramagnetic centers, has turned out to be sensitive enough to detect trace amounts of these species, in the range of 1 ppm. Additionally, owing to the high symmetry of the cage, most of the conventional relaxation mechanisms otherwise affecting electron spins in solution are absent. Correspondingly, the fluctuating zero field splitting (ZFS) interaction induced by collision-induced deformations of the carbon shell constitutes the dominant relaxation mechanism. It results in the exceptionally narrow line-widths, of only $\approx$1.5 μT even in solid C_{60}, providing extremely sensitive EPR detection of these species [29].

Besides, the introduction of negatively charged fullerenes into a low temperature metal plasma column by a strong axial magnetic field was used to produce Li@C_{60}, Li@C_{70}, Gd@C_{60}, Gd@C_{70}, Nd@C_{60}, Nd@C_{70}, Y@C_{60}, Y@C_{70} and even ^{99m}Tc@C_{60} and ^{99m}Tc@C_{70} [30]. Furthermore, the radio-frequency (RF) reactor technique has been applied. In such a reactor, RF plasma that is generated between a centered anode and two metal cathodes in a vacuum chamber, serves a source of ions. C_{60} is evaporated from an effusion cell and the ions are put in the monolayer of the growing fullerene film. For example, Cu@C_{60} and Ba@C_{74} were synthesized by the RF-plasma method [31, 32].

To put tritium (^{3}H) into C_{60}, a hot-atom chemistry method, the gamma recoil generating the atom at high kinetic energy in a nuclear reaction, is employed. For example, ^{6}Li salt of C_{60} is irradiated by thermal neutrons in a low-flux nuclear reactor. ^{6}Li reacts with thermal neutrons producing ^{3}H, some of which after moving through the sample and loosing energy stop inside C_{60} molecules (see review [6]). Another way was followed by Indian authors who produced ^{3}H@C_{60} by irradiating ^{3}He@C_{60} with neutrons in a high-flux reactor [33]. Activation of holmium-165 at C_{82} by high-flux thermal neutrons was used to generate ^{166}Ho@C_{82} which undergo β^- decay to produce stable ^{166}Er@C_{2n} [34]. A similar nuclear recoil implantation technique was used to produce ^{7}Be@C_{60} [35]. During irradiation of the mixture of C_{60} and Li_2CO_3 with protons about eight hours, the reaction ^{7}Li(p, n)^{7}Be leads to the production of ^{7}Be. After the kinetic energy of ^{7}Be had decreased to an appropriate value, it penetrated into C_{60} cage producing ^{7}Be@C_{60} [35]. Furthermore, radioactive isotopes of ^{79}Kr, ^{133}Xe, ^{81}Rb, ^{24}Na, ^{65}Zn, ^{83}Sr, ^{153}Sm, ^{161}Dy, ^{177}Lu, ^{69}Ge, ^{86}Zr, ^{168}Hf, ^{71}As, ^{120}Sb, $^{72\text{-}75}$Se, ^{121}Te, ^{210}Po, ^{34m}Cl, ^{77}Br, ^{99m}Tc@C_{60}, ^{99}T@C_{70} were implanted in the fullerene cages by nuclear recoil implantation technique [30, 33-38]. In principal, heavy ion beams of accelerators might be applied to put any ions into fullerene cages. However, yields of ENF in the nuclear implantation experiments usually do not exceed trace quantities.

A chemical method was successfully used to put molecular hydrogen into the fullerene cage [39, 40]. First, an orifice large enough to insert an atom or molecule was constructed on the surface of C_{60}. Next, the orifice was enlarged to a 13-membered ring by insertion of a sulfur atom on the rim and 100% encapsulation of molecular hydrogen was attained. Then, the 13-membered ring orifice was reduced to the 12-membered ring by extrusion of the sulfur atom at the rim of the orifice and the ring was further reduced into an eight-member ring by reductive coupling of two carbonyl groups also at the orifice. Finally, the orifice was closed by a thermal reaction without the loss of incorporated hydrogen [39, 40]. A "molecular surgery" as the method of producing ENF has been reviewed in [41].

2. Isolation and Purification

The electric-arc method is the only one that provides ENF in preparative amounts. However, the resultant soot is complex and contain 5-8% of ENF, at the most, the remainder being empty (pristine) fullerenes and debris like the fullerene black [16, 17]. Moreover, under the arc conditions, metal oxides are readily reduced to the neutral metal, atoms of which may not go only into the carbon cages. For example, the metal atoms may form nanoparticles. Indeed, having extracted $M@C_{2n}$ containing soot (M = La, Y, Yb) with DMF, we detected an unusual singlet EPR signal in the primary DMF extract with the *g*-value close to that of free electrons and extremely narrow line-width (ΔH_{pp} = 0.006 mT) [42]. EPR lines of the similar sharpness at room temperature arise from electrons of conductivity in the case of metal nanoparticles or from solvated electrons in the case of liquid ammonia solutions of alkali metals (see refs. in paper [42]). Thus, isolation and purification of $M@C_{2n}$ from the soot present a real challenge to chemical physics of fullerenes.

The encapsulated atom in $M@C_{2n}$ locates at an off-centered position adjacent to a six-member ring of the carbon cage, oscillates around the minimum energy potential well at room temperature charge and even performs a larger amplitude boat-shaped motion inside the C_{82} cage when the temperature is high enough [5-12]. Furthermore, $M@C_{2n}$ molecule has a rather specific electron structure. For example, electron structure of $Ca@C_{60}$ can essentially be expressed as $Ca^{2+}@C_{60}{}^{2-}$ because the entrapped atom donates both valence 4*s*-electrons to C_{60} cage [7, 12]. A similar structure appears to be true for paramagnetic $Cu@C_{60}$ [43]. For $Sc@C_{2n}$, $La@C_{2n}$ and $Y@C_{2n}$ there are data for a nearby complete transfer of their both valence *s*-electrons and unpaired *d*-electrons from the entrapped atom, Sc($3d^14s^2$), La($5d^16s^2$) or Y($4d^15s^2$), to the carbon cage that results in the formal electron structure as $M^{+3}@C_{2n}{}^{3-}$. This specific electron structure, a "superatom", was unambiguously testified by various structural and spectroscopic measurements, among them – synchrotron x-ray powder diffraction, EXAFS (extended X-ray absorption fine structure), transmission electron microscopy, ultra-high vacuum scanning tunnelling microscopy, and EPR. Mono-metallofullerenes of Ce, Pr, Nd, Gd, Tb, Dy, Ho, Er, Lu are also paramagnetic due to *f*-electrons of the trapped atoms. For example, the encaged Gd, Dy and Er ion is in a formally trivalent state configuration, $4f^7$, $4f^9$, and $4f^{11}$ for Gd, Dy and Er, respectively. For Sm, Eu, and Tm, the formal state of the encaged atom seems to be M^{+2} and, therefore, these $M@C_{2n}$ are diamagnetic. Interestingly, that Tm takes the trivalent state in $HoTm@C_{82}$ and $Tm_2@C_{82}$, contrary to its divalent state in $Tm@C_{82}$ [3, 5, 7, 9-12, 44-48]. The $Hf_2@C_{80}$ and $Ti_2@C_{84}$ isomers possess unique electronic structures, namely an effective charge transfer of two electrons at most per entrapped atom to the cage takes place [22, 23]. In $CeSc_2N@C_{80}$, the oxidation state (+3) for cerium has been well documented [49]. The oxidation state of the actinide atoms, U, Th, and Pu in the C_{82} fullerene cage was estimated to be 3+ with the formal structure of the molecule being $M^{3+}@C_{82}^{3-}$ [13]. As a result, $M@C_{2n}$ molecules possess much greater electric dipole moments than empty fullerenes. For example, $La@C_{82}$ has P = 3.8 *D* while C_{60} has P = 0.35 *D*.

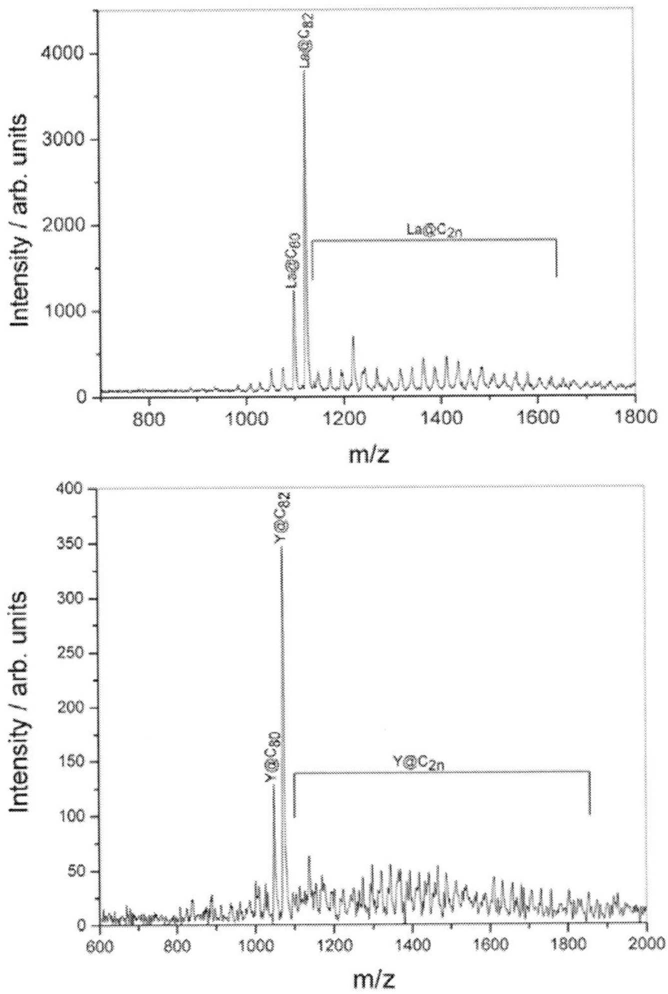

Figure 3. Typical mass spectra of the final DMF-extracts of $La@C_{2n}$ (a) and $Y@C_{2n}$ (b) obtained after the stepwise selective extraction from the fullerene-containing soot. The mass spectroscopic analysis was performed by a laser-desorption time-of-flight mass spectrometry (LD-TOF MS) on "Bruker TOF-1" [16].

Fuchs et al. were first who found that the absolute yield of $La@C_{82}$ was 2-3 times higher when the polar azeotropic CS_2/CH_3OH mixture was used as the extraction solvent instead of less polar toluene [50]. We have developed the stepwise procedure for the selective extraction of $M@C_{2n}$ from the fullerene-containing soot based on the much greater dipole moments of $M@C_{2n}$ by comparison with empty fullerenes. Following this line, soot is first treated with a non-polar solvent to extract empty fullerenes. Then, the rest of the soot is treated with a polar solvent to extract $M@C_{2n}$ [15]. Typically, the weighed sample of the soot (≈10 g) was placed in a filter-paper vessel which, in its turn, was placed for 3 h into a glass flask containing 500 ml of *o*-xylene (P = 0.62 *D*) to extract empty fullerenes. The resultant solution was thoroughly filtered off, the solvent was evaporated on a rotary evaporator and the extract was dried for 1 h at 90 ^{o}C in vacuum. The cycle was repeated five times, the fifth portion of *o*-xylene emerging virtually colorless. This colorless fraction served as a performance criterion in deciding when to end the *o*-xylene extraction stage. After the *o*-xylene treatment, the soot was subjected to the second extraction with DMF (P = 3.86 *D*). Again, the soot sample in a paper-made vessel was placed in a flask containing 500 ml of the solvent and the extraction procedure was performed for 3 h. The obtained solution was filtered off, the solvent was evaporated, and the residue was dried for 1 h at 90 ^{o}C in vacuum and weighed. The cycle with the polar solvent was repeated six times. The results of the X-ray fluorescent analysis showed that the contents of La and Y in the final DMF extracts were 11.8–12.3% and 7.9–8.1%, respectively, i.e. the metal contents almost coincided with the theoretical lanthanum content in $La@C_{82}$ (12.4%) and yttrium content in $Y@C_{82}$ (8.3%) [16].

Figure 3 depicts typical mass spectra of the final DMF extracts. In the case of $La@C_{2n}$ the most intense line belongs to the molecular cation of $La@C_{82}$ (M/z=1123). This spectrum also shows a less intense line of $La@C_{80}^{+}$ and a number of minor lines arising from other $La@C_{2n}^{+}$, from $La@C_{74}^{+}$ (M/z=1027) to $La@C_{106}^{+}$ (M/z=1411). The mass spectrum of the $Y@C_{2n}$ preparation exhibits mainly the line of $Y@C_{82}$. It is worthy to note that there are no peaks of empty fullerenes (M/z=720 of C_{60}^{+}, M/z=840 of C_{70}^{+} and so on) in either mass spectra, even though the HPLC technique was not used in the works.

3. Nanoparticles or Anions?

In general, apart from DMF, any solvent with the value of electric dipole moment close to that of $M@C_{2n}$ must have been favorable for the $M@C_{2n}$ isolation, including DMSO, aniline, carbon disulphide, chlorobenzene, toluene/tetrahydrofuran, triethylamine/acetone, pyridine, (see [15-17, 50-54] and refs therein). The question, however, arises as to the state of $M@C_{2n}$ molecules in such extracts. Having dissolved our DMF extracts of $La@C_{82}$ and $Y@C_{82}$ in DCB (*o*-dichlorobenzene), we obtained EPR spectra [14-17] with no principal difference in hyperfine coupling constants (a_{iso}) and *g*-values from the EPR spectra published by other groups, who obtained their $M@C_{2n}$ without using DMF [3, 5, 10, 11]. However, the integral intensities of the EPR signals of the concentrated $M@C_{2n}$ solutions were not as intense as expected, assuming each molecule has one unpaired electron with S=1/2, but conformed about one spin per 30–50 molecules. We have suggested that paramagnetic $M@C_{2n}$ molecules at high concentrations form clusters (nanoparticles) in which the exchange coupling with partial quenching of the electron spins takes place [55]. One would expect the clustering since the "superatom" electron structure results in the large dipole moments of $M@C_{2n}$ and, thereby, in strong dipole–dipole and charge transfer interactions of $M@C_{2n}$.

An alternative proposal has been made by Solodovnikov et al. [56]. These authors called attention to the similarity of the visible–near infrared (vis–NIR) spectra of $La@C_{82}$ in pyridine, DMF and hexamethylphosphoramide (HMPA) with the vis-NIR spectra of anion

$La@C_{82}^-$ obtained in an electrochemical cell. Based on this similarity, they suggested that the paramagnetic $La@C_{82}$ is reduced by the solvent into diamagnetic anion, $La@C_{82}^-$ [56].

However, the similarity in the vis-NIR spectra, by themselves, cannot serve as the basis for the anion hypothesis as the electron states and the absorption spectra of fullerenes undergo the dramatic changes after formation of donor-acceptor solvates, complexes or clusters. The formation of complexes of $La@C_{2n}$ with DMF and HMPA has been directly confirmed by IR spectroscopy [15, 57]. Moreover, much as the charge-transfer complexes, $M@C_{2n}$ form the spin-transfer complexes in which the electron spin density partially localizes on atoms of the matrix in which $M@C_{2n}$ are embedded [58, 59]. The formation of host-guest supramolecular complex of $Dy@C_{82}$ with *p*-tert-butylcalix[8]arene was found [60]. Besides, it was recently found that $La@C_{82}$ forms charge-transfer complexes with azacrown ethers [61]. Moreover, selective isolation of $La@C_{82}$ from the soot was accomplished by utilizing its complexation with 1,4,7,10,13,16-hexaazacyclooctadecane [61].

Actually, the origin of the hypothetical $M@C_{82}^-$ has remained unclear in the paper [56]. From thermodynamics, it is hardly believed that $M@C_{2n}$ may be reduced by DMF or another solvent of this kind in the absence of the high reductive additives. Besides, the anion hypothesis raises the question of how the anion of $M@C_{2n}$ is supplied with a counter-ion? It is not the question for those who obtain anions in electrochemical cells when a cathode serves an electron donor. But that is the question for those who care for "solvent reduction": if $M@C_{82}^-$ is generated via reduction of $M@C_{82}$ by the solvent molecules, then a counter-ion must be generated too. Since $M@C_{2n}$ molecules are actually free radicals having unpaired electrons, a hypothetic counter-ion should have been a cation-radical of the solvent. Meanwhile, the nature of the counter-ions has left unanswered by the followers of the solvent reduction hypothesis. Presently, there is no evidence in the unambiguous support of the anion hypothesis. On the contrary, the hypothesis that $M@C_{2n}$ form nanoparticles in the solvents of high polarity has been recently proved [62].

Jakes and Dinse, who reported EPR data for the radical anion $[Sc_3N@C_{80}]^-$, had to employ Na or K as the reducing agents to reduce $Sc_3N@C_{80}$ in terahydrofuran [63]. Yet, the I_p values for $M@C_{82}$ are 1.2−1.6 eV smaller than those for C_{60} and C_{70} while their E_a values are 0.4−0.7 eV larger than those for C_{60} and C_{70} [9, 48]. Therefore, the possibility of reduction by the solvents cannot be ignored entirely. As far as the presence of admixture of water in the solvent must not be entirely ruled out, a possible hydrolysis of DMF into the anion of formate and the cation of dimethylamine, that might proceed during extraction of $M@C_{2n}$ by DMF, was considered in our paper [57]:

$$H(CO)N(CH_3)_2 + H_2O \rightarrow$$
$$HCOOH + NH(CH_3)_2 \leftrightarrow HCOO^- + NH_2(CH_3)_2^+.$$

Then, the formate-anion might reduce $M@C_{2n}$ into the anion, after which the recombination of the radicals $HCOO^{\bullet}$ would give oxalic acid that breaks down with formation of H_2O, CO and CO_2 as the final products, along with $M@C_{2n}^-$. In principle, this hypothetical reaction scheme could solve both problems of the anion hypothesis, the genesis of anion $M@C_{82}^-$ and the nature of its counter-ion. However, if this were so, then all of the $M@C_{2n}$ should have been reduced into anions by the formate admixture during the extraction process, since the initial amount of $M@C_{2n}$ in the soot is low. Consequently, we should have observed the total loss of the paramagnetic $M@C_{2n}$ but the $M@C_{2n}$ extracts always exhibited the EPR signals. Moreover, after dissolving our DMF-extracts of $M@C_{2n}$ in *o*-xylene, DCB or toluene we always detected the "standard" octet EPR signals which did not differ from the EPR signals of the $M@C_{2n}$ of other researchers, who obtained their $M@C_{2n}$ without using DMF.

Besides, one should have expected for an increase of the EPR signal of the $M@C_{2n}$ after oxidation of $M@C_{2n}^{-}$ anions into $M@C_{2n}$ paramagnetic molecules but no accretion in the EPR signal was observed after addition of oxidizers, even iodine, to the extracts. Furthermore, the reduction of $M@C_{2n}$ by dimethylamine, if at all, is an endothermic reaction, equilibrium of which is shifted to the initial reagents. Additionally, such things as Coulomb interaction of ions and exothermic solvation of the ions by polar solvents should have been taken into account. While the solvent (DMF) is being evaporated, the ions quit their solvation shells and the reaction of reduction of $M@C_{2n}$ by dimethylamine, which is volatile gas, is reverted. If this is the case, we get again neutral $M@C_{2n}$ molecules instead of the anions.

Furthermore, it is well known that fullerenes molecules behave as electron-deficient spherical polyolefins to which amines can add on the carbon-carbon double bonds very readily and multiply (see refs. in [9, 64]). Therefore, if the anion-reaction scheme had taken place, then the $M@C_{2n}$ should have reacted with dimethylamine during the high-temperature extraction procedure giving the final products like $M@C_{82}H_nR_m$. However, nothing of this kind was detected in the mass spectra of our $M@C_{2n}$ preparations as well as in the preparations of our colleagues.

Applying electrospray ionization mass spectrometry (ESI-MS) technique in the negative ionization mode, Kareev et al. detected the anions of $La@C_{2n}$ and $Y@C_{2n}$ in the DMF extracts [65]. However, a cathode was the donor of electrons in these experiments. On evaporating the negatively charged drops of the solution, no wonder that some of electrons localize on molecules of $M@C_{2n}$. Hence, the finding of the $M@C_{82}^{-}$ anions at the negative ionization mode in the ESI-MS experiments came as no surprise. Therefore, the finding of the anions in the ESI-MS experiments at the negative ionization mode cannot be considered as evidence in support of formation of anionic endohedral metallofullerenes during their extraction with DMF. Likewise, the finding of the $M@C_{82}^{+}$ cations in the LD-TOF mass-spectra can be no means considered as a proof that the $M@C_{2n}$ are present in the extracts as the positive diamagnetic cations. As for the formation of the "black precipitate" after addition of water solution of bis(triphenylphosphoranylidene)ammonium chloride, PPN(Cl), into the DMF solution of $M@C_{2n}$, it is also not of crucial importance since both, $M@C_{2n}$ and PPN(Cl), have limited solubility in DMF and DMF/H_2O [65].

It may be inferred that $M@C_{2n}^{-}$ anions, if they do exist, might constitute a minor portion of the DMF extracts. On the contrary, there is good reason to believe that the dominant bulk of $M@C_{2n}$ exists as nanoparticles rather than anions [62]. An additional point to emphasize is that it is easy to think of the $M@C_{2n}$ nanoparticles as a collection of electrically neutral molecules whereas the nanoparticles collected from anions are difficult if not possible to realize without occurrence of relevant counter-ions.

In the meantime, the "anion hypothesis" initiated a new approach to production of $M@C_{2n}$ in preparative amounts [53]. After extraction of $M@C_{2n}$ from the carbon soot with TCB (1,2,4-trichlorobenzene), they were electrochemically reduced into $M@C_{2n}^{-}$ anions by utilizing the bulk electrolysis of the extract. The lower reduction potentials of $M@C_{2n}$ by comparison with empty fullerenes provide a selectivity of the electrochemical reduction. Besides, $M@C_{2n}^{-}$ anions have essentially higher solubility in polar solvents than empty fullerenes. Correspondingly, the following chemical oxidation of the obtained $M@C_{2n}^{-}$ anions into $M@C_{2n}$ makes it possible to obtain $M@C_{2n}$ in a high yield without using the multistage HPLC [53]. It is worthy to note that Bolskar and Alford showed usefulness of chemical oxidation for solubilizing and separating endometallofullerenes, as demonstrated by the separation of three species of $Gd@C_{2n}$ and by the solubilization of $Tm@C_{60}^{+}$ and $Tm@C_{70}^{+}$ [66].

Furthermore, Tsuchiya et al. have recently proposed an advanced TCB/pyridine/DMF scheme of extraction of $La@C_{2n}$ with using of n-Bu_4NClO_4 (tertbutylammonium perchlorate) [54]. After the extracting solvent was removed, the precipitate was washed by acetone/CS_2 in order to extract, as it was assumed, anions of two isomers of $La@C_{82}^{-}$ with anion of $La_2@C_{80}^{-}$. Then, $CHCl_2COOH$ (dichloracetic acid) was added to the acetone/CS_2 solution to yield $La@C_{82}$ and $La_2@C_{80}$. Finally, CS_2 solvent was replaced with toluene and the one-stage HPLC made it possible to separate pure isomers of $La@C_{82}$ and $La_2@C_{80}$. The authors believe that the preferential extraction of $M@C_{2n}$ was achieved "through the use of reduction in the extraction process" [54]. However, the likelihood of reduction of $M@C_{2n}$ by the solvent molecules, is little or no, as it was pointed out above. $CHCl_2COOH$ cannot serve as reducer of $M@C_{2n}$ for thermodynamics reasons. The presumable role of Bu_4NClO_4 as a source of counter-ions stabilizing the $M@C_{2n}^{-}$ seems to be unbelievable since it was added to the combined extracts after their extraction from the soot. We believe that the additives, such as n-Bu_4NClO_4 and $CHCl_2COOH$, prevent aggregation of the $M@C_{2n}$ nanoparticles, thus, providing their better solubility and, as a result, higher yields. Accordingly, the following separation of the $M@C_{2n}$ by one-stage HPLC becomes more effective.

Be it as it may, $M@C_{2n}$ extracts contain broad assortments of the compounds, for example, from $La@C_{74}$ to $La@C_{106}$ as it is depicted on Figure 3. Furthermore, there are known 9 different isomers for $Ca@C_{82}$ and $La@C_{82}$, 2 isomers for $Y@C_{82}$, $Ce@C_{82}$, $Pr@C_{82}$, $Gd@C_{82}$, $Nd@C_{82}$, and $Dy@C_{82}$, 3 isomers for $Sc_2@C_{84}$, $Sm@C_{82}$, $Sm@C_{84}$, $Eu@C_{82}$, $Er@C_{82}$, $Tm@C_{82}$, and $Yb@C_{82}$, 4 isomers for $Er_2@C_{92}$ and $Yb@C_{84}$, 5 isomers for $Tb@C_{82}$ and $Tm_2@C_{82}$ and so on [3, 5-12, 14-18, 44, 45, 48, 67-69]. Meanwhile, pure substances are required for technological applications, be it nanoelectronics or nanomedicine. It is possible, using an arc reactor with the optimized parameters and a large column HPLC, to produce up to 100 mg of purified metallofullerenes isomers in a couple of weeks [5, 9, 69]. No matter how expensive laborious and tedious the multistage HPLC may be, it seems to be indispensable but feasible.

Preservation of Paramagnetic Endohedral Fullerenes

Paramagnetic ENF are, in nature, free radicals. As such, they are chemically reactive and, among other reactions, undergo oxidative destruction on exposure to atmosphere oxygen with loss of unpaired electrons and paramagnetism. It is primarily pertinent to paramagnetic endometallofullerenes $M@C_{2n}$ (M = La, Y, etc.) the "superatom" structure of which, $M^{3+}@C_{2n}^{3-}$, results in localization of unpaired electrons almost entirely on the fullerene cages. For example, $La@C_{2n}$ samples on exposure to laboratory atmosphere for few days give endohedral fullerene oxides, such as $La@C_nO$, as the precursors to cage opening and metal loss with formation of the metal oxides and carbonates [70]. As a consequence, $M@C_{2n}$ should have been kept under a noble gas or in vacuum.

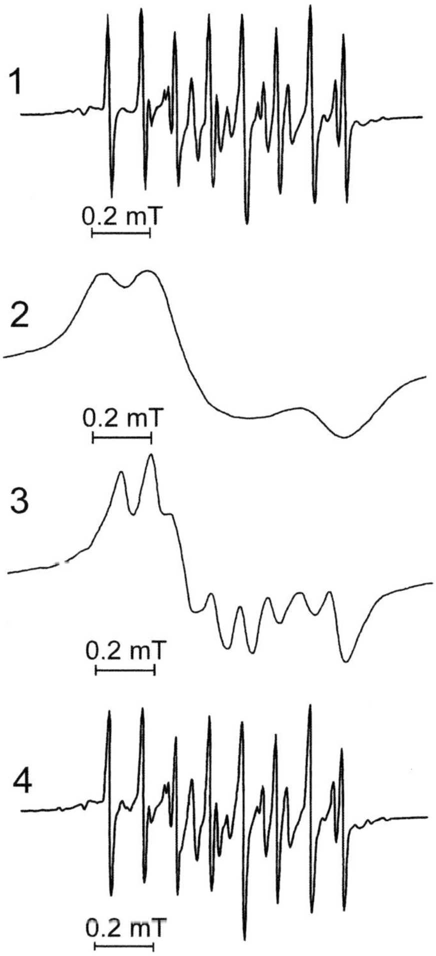

Figure 4. X-band EPR spectra of La@C_{82} in the polycarbonate (PC) samples: 1 – initial solution of PC in DCB (gel), 293 K; 2 – PC film, 293 K; 3 – PC film, 425 K; 4 –solution of PC in DCB (gel) after 1-year's storage of the doped film on air, 293 K. The spectra (1) and (4) were recorded in the absence of oxygen (10^{-3} Hg). Microwave power P = 1 mw, amplitude of modulation of the spectrometer magnetic field H_m = 0.01 mT (modulation frequency 100 kHz) [71].

However, it was discovered that, on embedding in poly(bisphenol A carbonate) film, M@C_{2n} features the exclusive stability against oxygen [71]. Figure 4 displays EPR spectra of La@C_{82} in the polycarbonate samples. A solution of PC in DCB makes a gel, the EPR spectrum of which at room temperature exhibits two octet partially overlapping each other (spectrum 1). The usual octet splitting appears due to the hyperfine coupling (hfc) of the unpaired electron with the nuclear magnetic moment of ^{139}La (I = 7/2). Both octets are essentially identical to the signals of La@C_{82} in pure DCB (with no PC added) and belong to the so-called "major" and "minor" isomers, La@C_{82} (I) and La@C_{82} (II), which differ in the symmetry of their fullerene cages (see, for example, [3, 10, 11, 14-17, 46]). The spectra of La@C_{82} in polymer films (spectra 2 and 3) considerably differ from that in the gel. The octet splitting becomes observable above the temperature of transition of polycarbonate from the glassy state into the high-elasticity state (≈400 K), when mobility of the polymer segments becomes high enough.

On air, EPR signals in the PC gel have vanished during a two-week's period due to the oxidative destruction of the endometallofullerene. By contrast, if La@C_{82} was put inside the polymer film, the EPR signal has undergone no changes during storage on air under usual laboratory conditions. When this film was dissolved in DCB after the storage on air for more than one year, it formed the gel, the EPR signal of which was essentially identical to the signal of the afresh La@C_{82} in DCB (compare the spectra 4 and 1 on Figure 4).

The amazing stability of M@C_{2n} in the polycarbonate film is explained by the hampered diffusion of oxygen in the glassy state of the polymer. This finding enables to employ the polymer films for storage of these and other paramagnetic endohedral fullerenes and for further usage of them [72, 73]. A similar strategy has been tested recently for encapsulation of Dy@C_{82} into conducting polymer polythiophene [74].

Spin Leakage from Paramagnetic Endohedral Fullerenes

Another important question concerns the space of localization of unpaired electrons in paramagnetic M@C_{2n}. It is generally known that the isotropic hfc constant, the so-called Fermi contact coupling constant a_{iso}, is a measure of the spin density of unpaired electron at the nucleus:

$$a_{iso} = (8\pi/3)\mu_e\mu_N\,|\psi_0|^2,$$

where μ_e and μ_N are the electron and nuclear magnetic moments, $|\psi_0|^2$ is the square of the electron wave function at the nucleus. The latter is the spin density of unpaired electron on the spherically symmetric *s*-orbitals since only these orbitals have a nonzero magnitude of $|\psi_0|^2$.

The hfc values observed in EPR spectra of La@C_{82} does not exceed ≈0.1 mT (see, for example, Figure 4). For other Ln@C_{2n}, the hfc values are even less [3, 5, 10, 11]. It is very small compared, for example, with that (5 mT) observed in EPR spectra of La^{2+} in CaF_2. The small hfc values for paramagnetic Ln@C_{2n} (M = Sc, La, Y, etc.) indicate that the unpaired electron is almost totally delocalized over the C atoms of the fullerene cage, in accordance with the "superatom" electron structure, M^{3+}@$C_{2n}{}^{3-}$ [5, 10, 11].

However, the question arises whether the electron spin density is restricted by the fullerene cage or may further penetrate, outside the fullerene shell? Studying La@C_{82} in DMF solution, we have revealed a symmetric quasi-doublet signal with $g = 2.001$ and the distance between its extreme lines about 0.04 mT at room temperature (Figure 5, spectrum 1). Meanwhile, EPR spectra of La@C_{82} in solvents of low polarity, like DCB, toluene, etc., exhibit octet hyperfine structures with the distances between the extreme lines about 0.84 mT for La@C_{82}(I) and 0.57 mT for La@C_{82} (II) (Figure 4, spectrum 1). A similar doublet signal, instead of the ordinary octet, was observed for solution of this M@C_{2n} in DMSO [55]. We have showed that the doublet signal is actually a poorly resolved octet from La@C_{82} solvated by the solvent molecules. The quasi-doublet pattern of the EPR spectra appears as the result of the decrease in a_{La} to the value comparable with the broadening of the individual lines of the octet [55]. Since the isotropic hfc is proportional to electron spin density on the nucleus, these findings testify that, in the case of DMF, the spin density on the La atom is essentially less than in the case of DCB or toluene. It suggests a greater transfer of unpaired electron from the central metal onto the solvate shell of the $M@C_{2n}$ in the case of the polar solvents. Furthermore, it can be seen from Figure 5 that the distance between the extreme lines of the "doublet" increases with decrease in temperature, so that the octet hyperfine structure with a_{La} = 0.006(3) mT becomes observable at the lower temperature. It gives evidence that the transfer of the electron spin density requires energy of activation. As a result, at lower temperatures a "spin leakage" from La becomes less probable and the appropriate increase of a_{La} occurs [55].

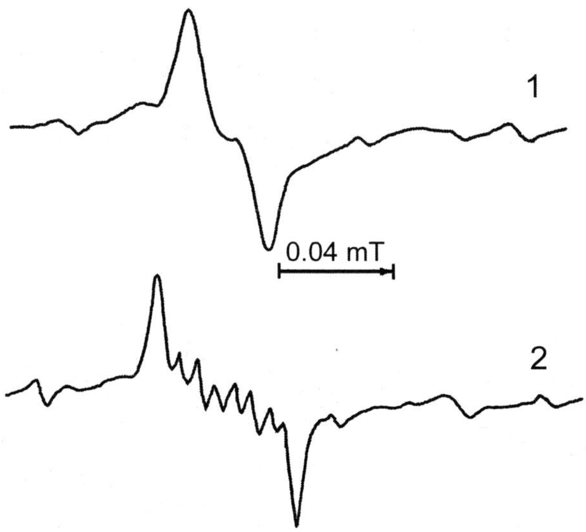

Figure 5. X-band EPR spectra of $La@C_{82}$ in DMF (*N,N*-dimethylformamide). The spectra were recorded in the absence of oxygen at 295 K (1) and 215 K (2) at H_m = 0.002 mT and P = 0.5 mW [55].

Next demonstration of the "spin leakage" has been obtained in the study of $La@C_{82}$ embedded in polycarbonate films by ENDOR (electron-nuclear double resonance). ENDOR is a special magnetic resonance technique that enables direct monitoring of spin density of unpaired electrons (see refs. in [58, 59]). In an ENDOR experiment, the spectrometer magnetic field and microwave frequency are maintained at fairly constant values fitting to the electron spin resonance condition of the paramagnetic center, and the microwave irradiation power is provided sufficiently large to saturate the EPR signal. Simultaneously, the sample is exposed to a variable radio-frequency field and the intensity of the saturated EPR signal is being taken as a function of the radio-frequency. If the radio-frequency corresponds to a magnetic resonance frequency of a nuclear spin, then the reorientation of the nuclear spin takes place. As a result, the saturation of the EPR signal is relieved and the latter builds up [58, 59].

The detected ENDOR signal of $La@C_{82}$ (Figure 6), while is poorly resolved, consists of two lines that is typical for the nuclear spin number I = 1/2. Moreover, this signal has been detected in the radio-frequency range ≈50 MHz, in which NMR of protons takes place at the given value of the spectrometer magnetic field. Since $La@C_{82}$ does not contain hydrogen atoms, the observed ^{1}H-ENDOR spectrum testifies the hyperfine interaction of the unpaired electron of M@C_{2n} with the matrix hydrogen atoms.

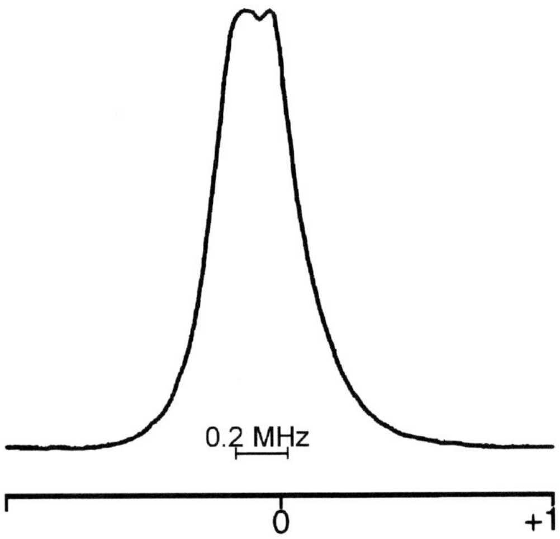

Figure 6. A Q-band ^{1}H-ENDOR spectrum of La@C_{82} in a polycarbonate film recorded at 2 K. The doublet is centered at ν_0 = 53.691 MHz. The microwave frequency is 35.218 GHz, the spectrometer magnetic field strength H_0 = 1261 mT, radio-frequency power is 15 W, sweep rate is 0.02 MHz/s. A decrease of temperature down to 2 K provided a better saturation of EPR signals while the enhancement of the microwave frequency from X to Q-band increased the Larmor frequency of protons from ≈14 to ≈53 MHz, thus eliminating spectral overlap of ^{1}H-ENDOR signal with those of other nuclei [58, 59].

The resonance radio-frequencies for the ^{1}H-ENDOR signal are determined by the formula:

$$\nu_{\pm} = \nu_0 \pm a_H/2,$$

where $\nu_0 = g_N\beta_N H_0/h$ is the proton Larmor frequency, g_N, β_N, H_0 and h are the proton g-value, nuclear Bohr magneton, the magnetic field strength and Planck's constant, a_H is the proton hfc constant. In general, a_H depends on both types of interactions, isotropic contact and dipolar, the former being directly proportional to the electron spin density on the nuclei. Hence, from the splitting of the doublet, it is easy to estimate that the Fermi contact coupling constant $a_H \geq 0.0790 \pm 0.002$ MHz. If a_H normalized to one unit of the s-electron spin density on a hydrogen atom is 1420 MHz, then one can estimate that in our case the fraction of the electron spin density on the matrix protons $|\rho_H| > 0.079/1420 \approx 0.006\%$ [58, 59]. Thus, it is evident from ENDOR that the unpaired electrons may go as far as the polymer protons.

Besides, to elucidate how far the unpaired electron may localize outside the fullerene cage, we investigated NMR spectra of liquid solutions of the same La-ENF in HMPA (hexamethylphosphoramide) molecules of which contain the NMR active nucleus, phosphorus-31, thereby providing great scope for NMR experimentation. We have revealed the paramagnetic shift in the position of the ^{31}P NMR signal of HMPA from 22.900 ppm in neat solvent to 23.048 ppm in the presence of 0.01 M of La@C_{82} that was caused by hyperfine coupling of nuclear spins of ^{31}P with electron spins of paramagnetic molecules of La@C_{82} [59].

The exchange of the solvent molecules between the complexes and the bulk solvent phase results in the paramagnetic shift δ of the bulk solvent's NMR line:

$$\delta = \Delta\omega_p/\omega_0 = -a_N p_L(\gamma_e^2/\gamma_N)/4kT,$$

where $\Delta\omega_p$ and ω_0 are the observable frequency shift of the NMR line and the operating frequency of the spectrometer, a_N is the hfc constant deriving from the electron spin density localized on the detected nucleus, p_L is the molar fraction of the complexes, γ_e and γ_N are the gyromagnetic ratios for the electron and the nucleus, k is the Boltzmann constant, T is the temperature (see refs. in 59]). With $\delta = 0.148$ ppm, providing each molecule of La@C_{82} can attach up to 10 molecules of HMPA capable to trade with the bulk solvent, one can estimate that $|a_P| \geq 0.135$ MHz. If a_P normalized to one unit of the *s*-electron spin density on a phosphorous atom is 10178 MHz, one can estimate that the fraction of the electron spin density on the matrix phosphorus $|\rho_P| \geq 0.135/10178 \sim 0.001\%$. However, it has also been revealed that molecules of M@C_{82} form nanoparticles [62]. Considering that the nanoparticle EPR signal has the integral intensity of no more than one spin per 30–50 molecules of M@C_{82}, a true concentration of the paramagnetic molecules in HMPA should be much below the primary concentration of 0.01 M. In the proper proportion, the electron spin density on the phosphorus obviously ranges up to ≈0.05 % [75].

Thus, EPR, ENDOR and NMR unambiguously testify a partial transfer of the spin density from the entrapped endohedral atom beyond the fullerene cage that is a spin leakage of the fullerene shell. The spin leakage of the fullerene shell is obviously worth consideration as the important factor in sight of possible applications of the lanthanide endometallofullerenes as the basic elements for realizing molecular magnets and devices for spintronics.

Nanoparticles of Paramagnetic Endometallofullerenes

The clustering of empty fullerenes is a well known phenomenon. A large electric dipole moment of M@C_{2n} certainly favors the formation of nanoparticles in liquid and solid phases [76]. Having studied the dynamic light scattering of solutions of Y@C_{82} in DMF, HMPA and other solvents, we discovered that Y@C_{82} indeed form nanoparticles, the mean size of which ranges up to 100 nm [62]. The reversible de-aggregation of the particles at high laser powers (Figure 7) points out a possibility of the laser control over the clustering of M@C_{2n}. Furthermore, Y@C_{82} demonstrated a very strong fluorescence, especially in DMSO, the decay kinetics of which revealed two components with the half-lifetimes about 2.5 ns and 10 ns attributed to individual Y@C_{82} molecules and to nanoparticles, correspondingly [62].

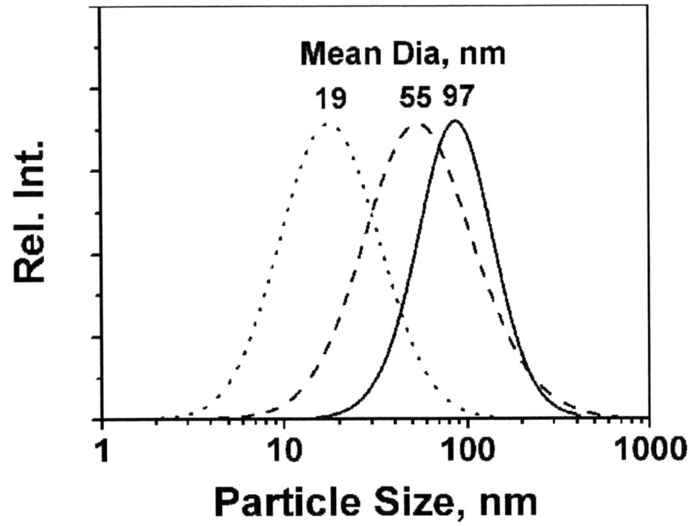

Figure 7. Light scattering intensity versus size of the nanoparticles of Y@C_{82} in DMSO (dimethylsulphoxide) at the different laser power: (——) 0.5 W, (-------) 1.0 W and (·····) 1.5 W [62].

Aggregation processes are common to a wide variety of M@C_{2n}. The X-ray and STM studies of the Y@C_{82} crystals directly confirmed that Y@C_{82} molecules are aligned along the [001] direction in a head-to-tail order with the observed intermolecular distance (11.2 A) shorter than that the simple van der Waals distance (11.4 A), suggesting a relatively stronger interaction than in the case of a simple dispersion type [5]. Gd@C_{82} molecules form dimers in condensed solutions and solid powders. Upon dimerization, the radical spins on the cage are canceled and the resultant spin state of the Gd@C_{82} dimer is described by two independent octet (S = 7/2) spins of the gadolinium [44].

A similar behavior was observed for endohedral alkali metal fullerenes. According to the IR and Raman spectroscopy, the Li atom locates in the carbon cage at a distance of approximately 1.5 A from the center. The chromatographically isolated endohedral species Li@C_{60} and Li@C_{70} demonstrate a strong tendency to form dimers and oligomers with "locking-up" of the unpaired electrons [25]. Correspondingly, there have been observed no additional signals in the EPR spectra of Li@C_{2n} compared to the C_{60} matrix in which the Li@C_{2n} molecules were embedded [77].

We have also studied powder samples of complexes of La@C_{82} and Y@C_{82} with HMPA by the solid state NMR spectroscopy with magic angle spinning (MAS NMR) [57]. The spinning of solid samples at the spin rates up to 20 kHz allows separation of resonances in NMR spectra with overlapping powder patterns so that unique spectral lines can be obtained for chemically distinct, magnetically nonequivalent, sites in the sample. The range of the chemical shifts of ^{31}P NMR lines from HMPA molecules bound to the endometallofullerenes has been found to spread from approximately 1 to −30 ppm. Meanwhile, for ^{31}P in all different compounds of pentavalent phosphorus, including HMPA, the entire range of tabulated values of the diamagnetic chemical shifts usually does not exceed 100 ppm (see [57] and references therein). In this regard, the spread of the chemical shifts of the ^{31}P for the HMPA molecules in the complexes with the endometallofullerenes appears a rather large in comparison with other diamagnetic pentavalent phosphorus compounds. The shifts were not inversely proportional to the absolute temperature, as it should have been in the case of usual paramagnetic shifts. Moreover, EPR signals, the intensities of which corresponded about 1 spin per 40–50 molecules of $M@C_{82}$, had essentially no dependence on temperature. It suggests the strong exchange coupling of unpaired electrons with the loss of the molecular paramagnetism.

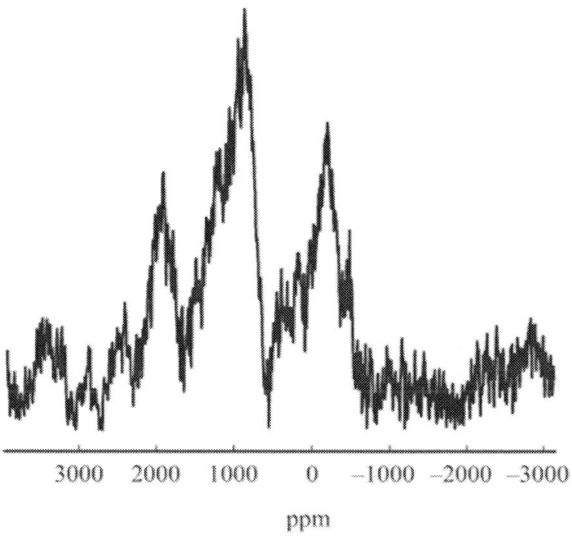

Figure 8. A ^{139}La MAS-NMR spectrum of the powder sample of $La@C_{82}$/HMPA complex recorded at room temperature. The operating frequency is 70.6 MHz, the static magnetic field H_0 = 11.7 T, the spin rate is 15 kHz, the number of accumulations is 80 000, the spectrum width is 500 kHz. The shifts of ^{139}La are externally referenced to an aqueous solution of $LaCl_3$ [57].

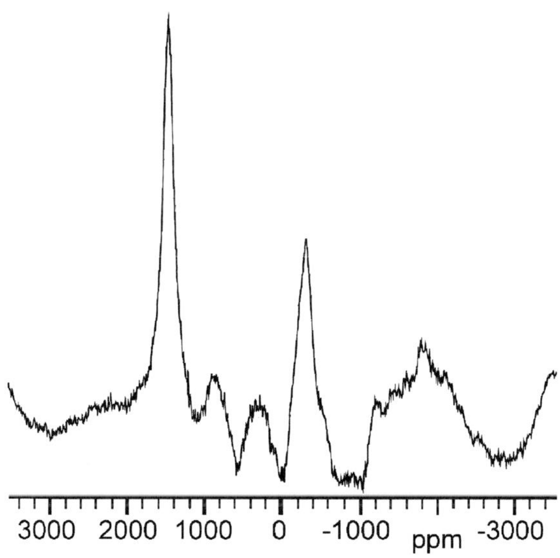

Figure 9. A ^{139}La MAS-NMR spectrum of solid polycarbonate film donated with $La@C_{82}$ recorded at room temperature. The operating frequency is 70.6 MHz, the static magnetic field H_0 = 11.7 T, the spin rate is 6 kHz, the number of accumulations is 4096, the spectrum width is 500 kHz. The shifts of ^{139}La are externally referenced to an aqueous solution of $LaCl_3$ [73].

Additionally, we have acquired well-resolved ^{139}La MAS NMR spectra in the same powder samples of La-ENF/HMPA [57]. The ^{139}La NMR spectra of these samples demonstrated three resonance lines at 1650, 900 and –400 ppm (Figure 8). Furthermore, the solid polycarbonate films donated with La@C_{82} have been studied by ^{139}La MAS NMR as well [73]. In this case, the ^{139}La NMR spectra demonstrated two distinct peaks at approximately –350 and 1500 ppm (Figure 9). The resonance line at 1650 ppm (La-ENF/HMPA powders) and the resonance line at 1500 ppm (La-ENF/polycarbonate films) are far beyond the scope of the chemical shifts of ^{139}La in usual diamagnetic substances which ranges from –130 to 1100 ppm (see refs. in [57, 73]). The enormously large NMR shifts of ^{139}La and the vast spreading of the shifts of ^{31}P may be, presumably, classified as the Knight shifts caused by the density of Landau-Pauli conduction electrons at the position of the NMR observed nuclei in the nanoparticles.

On this point, it is pertinent to note the results of Shinohara and co-workers who performed EPR experiments on the solvent free powder samples of La@C_{82} between 7 and 300 K and found that the paramagnetic component was almost temperature independent above 80 K, indicating Landau-Pauli paramagnetism of the conducting electrons [5].

Electron Compression in Paramagnetic Endohedral Fullerenes

A different situation arises with paramagnetic N@C_{60}, N@C_{70}, and P@C_{60}, where a "spin compression" seems to be the case instead of a spin leakage. The highly reactive atoms of nitrogen and phosphorous, being enclosed in the fullerene cage, keep their free atom ground state configuration and are localized in the center of the fullerene cages. According to EPR and ENDOR data, these compounds show no electron or spin transfer to the carbon shell. The enclosed atoms exhibit properties resembling those of ions in electromagnetic traps, i.e. sharp spectroscopic transitions, long life times and no chemical reactivity. It occurs as if the enclosed atom is freely suspended in a "chemical Faraday cage" [29]. It has been revealed from the ENDOR data that the isotropic hyperfine coupling constant (a_{iso}) for ^{14}N (I = 1) trapped in N@C_{60} is increased by approximately 50% at 5 K compared to the free ^{14}N atom's value. For ^{31}P (I = 1/2) inside C_{60}, the increase by a factor of 2.5 as compared with the free atom was discovered. In addition, the increase of a_{iso} with temperature, rather than a decrease as often observed, has been detected. The authors offered a model in which the small admixtures of exited states with larger hyperfine interactions are invoked to explain the increase of a_{iso} with temperature and the large increase in a_{iso} compared to the free atom values (see refs. [29]).

Yet, there is another explanation that might be put forward. Since a_{iso} is known to be directly proportional to the electron spin density on the nuclei, the abovementioned effects exhibit the redistribution of the spin density in the atoms entrapped by the fullerene cages, as if the electronic shells of the trapped atoms are compressed under high pressure [78]. The quantum-chemical calculations of the spin density distributions and a_{iso} values confirmed that, when N is encapsulated inside C_{60}, the electron spin density at nucleus N increases. In P@C_{60}, the corresponding increase is much larger [79, 80]. The quantum-chemical calculations show that the similar compression of atomic orbitals and the redistribution of electron spin density may be also expected for noble gas endohedral atoms since they are localized in the centre of the fullerene cage and only van der Waals interactions occur between the noble gases and the fullerene cages [80-82].

The "spin compression" of endohedral atoms raises again the old question of whether it is feasible to influence on nuclear reactions via electronic shells [83]. Of course the electronic shells and nuclear "drops" are of too different and incommensurable energy scales. However, a positive answer to the abovementioned question is not excluded for β^- and β^+ decays and for electron capture (K, L-decays). That is the question of interaction between the nucleon and electron-neutrino fields. The case in point is an extent to which the nucleus electric field is shielded by the atom's electrons. For this reason, probability of K-capture (or β^-/β^+ decays) depends upon the density of electron wave function within the nucleus [83].

There are about 200 isotopes in nature, from ^{7}Be to ^{256}Md, which decay through K-capture. The calculated correction coefficients of the shielding for K-capture do not exceed 1% [83]. However, inasmuch as the atomic wave function depends upon physicochemical state of matter, one can expect for much higher effects in the case of extremely high pressures, such as those found in the interior of a hot star or under the dynamic compression in shock waves. The compressed endohedral atoms represent one more realm of searching for the external electron density effects on nuclear processes [78].

This idea has been validated in the experiments with radioactive ^{7}Be entrapped in C_{60} [84, 85]. Beryllium is suitable for this purpose because its atom has the simple electronic structure, $1s^22s^2$. The ^{7}Be nucleus decays through electron capture to ^{7}Li, the exited state of which relaxes to the ground state by the γ quantum emission. Besides, it was found that ^{7}Be implanted in C_{60} goes to the center of the cage. The comparative measurements of the decay rate of ^{7}Be entrapped in C_{60} and that of ^{7}Be in beryllium metal have shown that the endohedral beryllium decays almost 1% faster than exohedral one. The half-lives were 52.68±0.05 and 53.12±0.05 days for ^{7}Be@C_{60} and for the Be metal sample, respectively. Moreover, the speedup of the radioactive decay at 5 K has been detected recently [84]. Actually, the 1-percentage difference is the most dramatic electron effect achieved in nuclear physics so far. Thus, the beryllium atom entrapped in the fullerene cage is endured a compression that results in the increase of electron density at the ^{7}Be nucleus and, as a consequence, in the increased probability of K-capture, as it was first suggested by A. L. Buchachenko [78].

The Potential Applications of The Endofullerene Park in XXI Century

Table 1 is a summary table of the endohedral fullerenes synthesized, by now, at least in trace amounts. From Zero group on, it embraces almost all chemical elements of the periodic system.

Group Zero Endohedral Fullerenes

The noble gas atoms are localized in the centre of the fullerene cage. Only van der Waals interactions occurs between the noble gas atom and the fullerene cage. In the same time, a ring current produced by π-electrons on the C_{60} reduces the magnetic field seen by nuclear

Table 1. Endohedral fullerenes synthesized and identified in the experiments from 1985 to 2006.

Group	Endohedral fullerenes	Refs.
0	$^{4}He@C_{60}$, $^{3}He@C_{60}$, $^{3}He_2@C_{70}$, $Ne@C_{2n}$, ($n = 30, 35$), $Ne_2@C_{70}$, $^{3}He^{22}Ne@C_{70}$, $Ar@C_{60}$, $Kr@C_{60}$, $^{79}Kr@C_{60}$, $Kr@C_{82}$, $^{129}Xe@C_{60}$, $^{133}Xe@C_{2n}$ ($n = 30, 35, 38, 42$)	[6, 35, 37, 86, 87]
1	$Mu@C_{60}$, $^{3}H@C_{60}$, $^{3}H_2@C_{60}$, $H_2@C_{60}$, $Li@C_{60}$, $Na@C_{60}$, $^{24}Na@C_{60}$, $K@C_{60}$, $^{81}Rb@C_{60}$, $Cs@C_{60}$, $Cu@C_{60}$	[24-27, 30, 31, 33, 35, 37, 39, 40, 43, 77, 90]
2	$^{7}Be@C_{60}$, $Mg@C_{2n}$ ($n = 36, 40$), $Ca@C_{2n}$ ($n = 30, 35\text{-}37, 40\text{-}45$), $Sr@C_{2n}$ ($n = 30, 36, 38, 40\text{-}42$), $Ba@C_{2n}$ ($n = 30, 37, 40\text{-}42$), $Zn@C_{60}$	[9, 12, 32, 35, 37, 69, 84, 85]
3	$Sc_m@C_{2n}$ ($m = 1, 2, 3, 4$; $n = 30, 33, 37, 41\text{-}43$), $Y@C_{2n}$ ($n = 30, 35, 40, 41$), $Y_2@C_{2n}$ ($n = 41, 42$), $La@C_{2n}$ ($n = 30, 35, 37\text{-}53$), $La_2@C_{2n}$ ($n = 36, 40, 41$), $Ce@C_{2n}$ ($n = 30, 40, 41, 50$), $Ce_2@C_{80}$, $Pr@C_{60}$, $Pr@C_{82}$, $Pr_2@C_{80}$, $Nd@C_{60}$, $Nd@C_{82}$, $Sm@C_{2n}$, ($n = 30, 37\text{-}42$), $Eu@C_{2n}$ ($n = 30, 35, 36\text{-}49$), $Gd@C_{2n}$ ($n = 30, 35, 37\text{-}41$), $Gd_2@C_{80}$, $Tb@C_{2n}$ ($n = 40\text{-}43$), $Tb_3@C_{80}$, $Dy@C_{2n}$ ($n = 30, 40\text{-}42$), $Dy_2@C_{2n}$ ($n = 41\text{-}47$), $Ho_{1,2,3}@C_{82}$, $Ho_{2,3}@C_{84}$, $HoTm@C_{82}$, $Er@C_{2n}$ ($n = 30, 37, 41$), $Er_2@C_{82}$, $Er_2@C_{92}$, $Tm@C_{2n}$ ($n = 30, 35, 41\text{-}48$), $Tm_2@C_{2n}$, ($n = 40, 41$) $Yb@C_{2n}$ ($n = 30, 36\text{-}42, 44\text{-}48$), $Lu@C_{60}$, $Lu@C_{82}$, $Lu_2@C_{82}$, $Th@C_{84}$, $U@C_{60}$, U_2C_{60}, $U@C_{82}$, $^{237}Pu@C_{82}$	[1-21, 30, 34, 38, 42, 44-48, 50-62, 65-76, 142]
4	$Ti_2@C_{80}$, $Ti_2@C_{82}$, $Zr@C_{82}$, $Hf@C_{84}$, $Hf_2@C_{80}$, $Si@C_{60}$, $Ge@C_{60}$, $^{212}Pb@C_{82}$	[21-23, 35, 37, 141]
5	$N@C_{60}$, $N@C_{70}$, $N@C_{84}$, $N_2@C_{60}$, $N_2@C_{70}$, $P@C_{60}$, $P@C_{70}$, $As@C_{60}$, $Sb@C_{60}$, $^{213}Bi@C_{60}$, $Nb@C_{82}$	[21, 28, 29, 35, 37, 43, 141]
6	$Mo@C_{82}$, $Se@C_{60}$, $Te@C_{60}$, $^{210}Po@C_{60}$	[35-37]
7	$Mn@C_{2n}$, $Mn_2@C_{2n}$, $Tc@C_{60}$, $Tc@C_{70}$	[21, 30, 37]
8	$Fe@C_{2n}$, $Co@C_{2n}$	[21]

spins inside the fullerene cage. As a result, NMR of the guest ^{3}He atom shows a line upfield from helium gas [6]. In principal, NMR signals of ^{3}He inside C_{2n} can be used as tracers for fullerene structures and reactions, for example, to monitor chemical functionalization of the cage [6]. Yet, purification of these compounds from empty fullerenes is a difficult problem since the relevant chromatographic separation factors are very small, from 1.09 to 1.12. In the meantime, hydrophilic fullerenols, $^{133}Xe@C_{60}OH_x$ and $^{133}Xe@C_{70}OH_x$, which were recently synthesized and isolated from hydrophobic $^{133}Xe@C_{2n}$, have turned out to be solved and stable in 0.9% NaCl solution at 20 ^{0}C and, thereafter, may be used in nuclear medicine [86].

The synthesis of a mg amount of pure argon containing fullerene gives promise that superconductors with low critical temperatures may be attained [87]. Besides, these compounds turned out to be of interest for astrophysicists. The existence of fullerenes with the noble gases inside has been conjectured in the interstellar dust from which comets and meteors are formed. Fullerenes containing "planetary" helium and argon were discovered in the carbonaceous chondrite, the Tagish Lake meteorite felt in Canada [88].

Group I Endohedral Fullerenes

In the case of alkaline metal doped C_{60}, the so-called fullerites, three-electrons filling into three-fold degenerate t_{1u} LUMO orbital results in the superconductivity. On this basis, superconductivity is presumably expected for alkaline metals $M@C_{60}$ if the valence of the metal atom would be +3 and, besides, if macroscopic amounts of pure $M@C_{60}$ will be obtained [89]. Another interesting idea, a bipolar electro-fluid shuttle memory device consisting of $K^{+}@C_{60}$ and $F^{-}@C_{60}$ encapsulated in a carbon nanocapsule, has been proposed on the basis of calculations of energetics and operating responses of such a device by classical molecular dynamics simulations [90].

Group II Endohedral Fullerenes

The physical fact that ^{7}Be entrapped in the C_{60} fullerene cage demonstrates the increased probability of K-capture [78, 84, 85] may supply the way toward realization of X-ray laser on this basis.

Group III Endohedral Fullerenes

From the organic chemistry point of view, a fullerene shell of an endohedral fullerene and that of an empty one have no principal difference in chemical reactivity. Therefore, there is a fast growth of papers on various chemical modifications of $M@C_{2n}$. For instance, water-soluble endohedral metallofullerenols of Pr, Gd, etc., having up to 30-40 hydroxyl groups in the fullerene shell, were synthesized [91, 92]. Another Japan group obtained pyrrolidines of $M@C_{82}$ (M = La, Y, Gd) through the reaction of addition with azomethine ylide [93]. The similar reaction was accomplished by Chinese chemists [94]. Synthesis of exohedrally silylated $Y@C_{82}$ and $La@C_{82}$ was described in [95]. The reaction of $La@C_{82}$ with diethyl bromomalonate in the presence of base (the Bingel reaction) has been studied recently. One of the five monoadducts was found to be “EPR-active” [96]. Two different isomeric forms of

$Y@C_{82}(CF_3)_5$ were prepared by Kareev et al. via trifluoromethylation of the relevant isomers of $Y@C_{82}$ [97].

Nowadays there is a strong trend in engineering towards nanotechnologies, novel electronic devices based on single molecules, etc. In this respect, paramagnetic metallofullerenes hold much promise.

Much effort went, first of all, into preparation and research of mono- and multilayer films of lanthanide ENF on various support surfaces. The photoelectrochemical cell (PEC) based on the Langmuir-Blodgett (LB) films of poly(3-hexylthiophene) (P3HT) doped with $Dy@C_{82}$ has been constructed. For this composite PEC, the authors observed a dramatic enhancement of cathodic photocurrent with the power-conversion efficiency and quantum yield being up to 16-fold and 9-fold higher, respectively, than those for pure P3HT cells. It was attributed to the facile photoinduced electron transfer between P3HT and $Dy@C_{82}$ [98]. The ordered films of $M@C_{82}$ (M = La, Y, Ce, Gd, Tb, Dy), metallic, semimetallic and semiconducting, were grown on Si, Cu, Ag, Au surfaces [5, 99-104]. In experiments with $Tb@C_{82}$ on the Au(111) substrate, the authors observed a switching of the $Tb@C_{82}$ molecular orientation caused by the interaction between its electric dipole moment and an external electric field [103]. It worthy to note as well the paper of Nuttall et al. who observed the complicated temperature dependence of the dielectric dipole dynamics in thin films of $La@C_{82}$ and suggested that the dielectric behavior in this molecular solid with even bistability in the molecular rotational states may provide a control of molecular orientations with electric fields leading to molecular memory devices [105].

The single-walled carbon nanotubes (SWNTs) encapsulating $M@C_{2n}$, named as peapods, are relatively new in the fullerene science. The peapods seem to be of interest for many applications ranging from catalysis to quantum computing. For example, the SWCNT material was impregnated with $La_2@C_{80}$. It was achieved by adding a drop of toluene-CS_2 solution of $La_2@C_{80}$ to a piece of the acid purified SWNT paper. Acid is known to open the ends of nanotubes and make defects in their walls, through which $La_2@C_{80}$ molecules diffused into the 1.4 nm diameter SWNTs [106]. Another research group has prepared SWCNTs encapsulating chain of $Gd@C_{82}$ by heating SWCNTs and $Gd@C_{82}$ under a vacuum. The *in situ* high resolution transmission electron microscopy (HRTEM) showed a clear image of the GdC_{82} chain inside the SWCNT, the Gd atoms being separated by 1-2 nm in $Gd@C_{82}$@SWCNT. The dimerization and polymerization of the C_{60} molecules were observed in C_{60}@SWCNT after electron irradiation at 300 keV [107]. In addition, the nanopeapods encapsulating $Sc_2@C_{84}$, $Ce@C_{82}$, $Sm@C_{82}$, $Eu@C_{82}$ and DyC_{82} fullerenes were synthesized and investigated by the same Japan group [19, 108]. Peapods with tunable electronic properties can presumably be obtained by using encapsulated $M@C_{2n}$ of different sorts.

In addition, a smart method has been reported for dividing the semiconductor nanotube into multiple quantum dots with lengths of about 10 nm by inserting $Gd@C_{82}$ in SWNTs. The nanotube electronic bandgap of similar to 0.5 eV turned out to be narrowed down to similar to 0.1 eV at sites where $Gd@C_{82}$ are inserted. This change in the bandgap was explained by local elastic strains and charge transfers at the metallofullerene sites. As a matter of fact, the technique of fabricating an array of quantum dots for nanoelectronic and nanooptoelectronic devices was proposed in the cited work [109].

Transport properties of the fullerene nanopeapods used as channels of field effect transistors (FET) have been studied by the same group [110]. C_{60} and C_{78} fullerene peapods and Gd@C_{82} and Dy@C_{82} metallofullerenes peapods FET showed both p- and n-type ambipolar behavior. In case of C_{90} peapods, all devices have shown metallic properties. The electronic properties of the peapods are presumably explained by the so-called band-gap modulation originated from the encaging fullerenes and metallofullerenes in nanotubes [110].

Furthermore, C_{60} field effect transistors with interfacial modifications of electrodes using C_{60}/La@C_{82}-FETs have been invented by another Japan group. These FETs operated without any annealing processes even once the fabricated devices were exposed to air, in contrast to the conventional C_{60}-FETs. Besides, the improvement in the field effect mobilities was observed with little change in the threshold voltage. Thus, the trapping levels at the interface between C_{60} thin film surface and the gold electrodes may be essentially reduced due to the interfacial surface modifications on the electrodes using M@C_{2n} [111]. Another field-effect transistor device was fabricated with thin films of C_{2v} isomer of Pr@C_{82}. The field-effect mobility for this FET was $1.5x10^{-4}$ cm^2 V^{-1} s^{-1} at 320 K, being comparable to those of other endohedral metallofullerene FET devices [112].

In the context of possible applications of ENF in nanoelectronics, the spin leakage of the fullerene shells can be drawn on to make good use of lanthanide metallofullerenes. For example, the high-spin molecules of Gd@C_{2n} hold promise as building blocks for realizing molecular magnets while the low-spin molecules La@C_{2n} could be of use for realizing "spin shims" and "spin-spacers" in devices for spintronics.

There have been conceived some interesting ideas on applications of M@C_{2n} in biology and medicine. For instance, La@C_{2n} and Y@C_{2n} have been proposed as EPR probes for measurement of oxygen in biomedical systems. This idea was based on the finding that the line-widths in the EPR are linearly broadened with growth of oxygen tension and that the effect of the broadening is reversible [113].

The magnetic resonance imaging (MRI) is the field where endometallofullerenes hold, probably, the greatest promise. To reveal anatomical details for diagnosis, the so-called contrast agents are required to enhance NMR relaxation of water protons. Commercial MRI contrast agents currently used in clinics, are mostly "GADO"-based compounds, such as Gd-diethylenetriaminepentaacetic acid (Magnevist) and so on, which employ Gd^{3+} (electron spin $S = 7/2$) to enhance the relaxation rate of water protons. Suffice it to say that the *in vitro* water proton relaxivity for Gd@$C_{82}(OH)_n$ was found to be 10-times higher than that of Magnevist [92]. However, this compound tends to aggregate in water due to its low solubility. Recently synthesized Gd@$C_{60}[C(COOH)_2]_{10}$ does not have the disadvantage of intermolecular aggregation in water solution while demonstrating the relaxivity comparable to that of Magnevist [114]. These findings are encouraging for the research of GADO-fullerenols as potential MRI contrast agents. Another potential field of application of M@C_{2n} for diagnostics is X-ray imaging. For this purpose, one can utilize the scattering effects of heavy metal atoms trapped inside the fullerene cage.

Furthermore, endohedral fullerenes with the appropriate particle-emitting radionuclide inside offer promise as radiotracers and radiopharmaceuticals. A number of β-emitting nuclides, among them ^{47}Sc, ^{90}Y, ^{149}Pm, ^{153}Sm, ^{166}Ho, ^{177}Lu, ^{64}Cu, ^{67}Cu, ^{111}Ag, ^{89}Sr, ^{32}P, are used for radiation therapy as complexes with chelating ligands, like 1,4,7,10-tetraazacyclododecane-1,4,7,10-tetraacetic acid (DOTA) [115]. M@C_{2n} based pharmaceuticals clearly offer some essential advantages over the usual chelating ligands. High kinetic stability of M@C_{2n} should result in low release of toxic metals and higher resistance to metabolic degradation. Moreover, a fullerene cage can be modified in plenty of organic-functionalized derivatives. This being so, the future advances in molecular biotechnology will provide targeting vectors to deliver therapeutic doses of the ionizing radiation to cancer metastasis cells with high specificity.

However, it is to be mentioned that the question of possible toxicity of $M@C_{2n}$ is also far from being clear. For example, $M@C_{82}$ (M = Dy, Gd, La) and $Dy_2@C_{2n}$ (n = 42-47) efficiently sensitize the reaction of olefins with molecular oxygen. Namely, in the presence of minute amounts of the metallofullerenes, alkenes, 2-methylbut-2-ene and alpha-terpinene undergo ene- and Diels-Alder photooxygenation reactions producing peroxides [116]. Additionally, the inhibitory activity of nanoparticles of gadolinium endohedral fullerenol, $Gd@C_{82}(OH)_{22}$, on growth of hepatoma cells in mice was observed [117]. The tendency to clustering is another critical deterrent to the use of $M@C_{2n}$ [76]. Unfortunately, the aggregation behavior is peculiar even for water soluble fullerenols [118].

Group V Endohedral Fullerenes

Pure $N@C_{60}$ can be prepared in milligram quantities using multi-step HPLC. It is a rather laborious procedure, up to eleven sequential separations, since the retention times and chromatographic separation factors are very small. Yet, it has given impetus to syntheses and studies of various $N@C_{60}$ derivatives, such as monoadducts, $N@C_{60}O$, and pyrrolidine functionalized derivative $N@C_{69}H_{10}N_2O_2$ [119-121].

An intriguing idea has been put forward that $N@C_{60}$ and $P@C_{60}$ could be used as the basis elements for quantum computing. In a quantum computer, the information is encoded in qubits which are formed by different states of a quantum system, e.g. the ground state and an exited state of an atom or molecule. In a spin quantum computer, the qubits are represented by the spin states in an external magnetic field and computation and read-out are done by EPR/NMR pulse sequences. The qubits have to be well resolved from the heat bath in order to maintain quantum mechanical coherence during logic operations. $N@C_{60}$ and $P@C_{60}$ have promising properties, deduced from their spin system and extraordinary long electron spin relaxation times, to become qubits in an electron spin quantum computer [122-123]. For example, $N@C_{60}$ exhibits spin-lattice relaxation time about 0.1 ms at room temperature, while the spin-spin relaxation time for magnetically dilute powder samples is about 20 μs independently on temperature between 5 and 300 K [29]. Some hundreds of basic logic gates might be possible using the electron spins of $N@C_{60}$ or $P@C_{60}$ as the qubits and their nuclear spins for addressing purposes. Moreover, the electron spin of the endohedral atom seems to be well shielded from environmental electrical noise by the fullerene shell that serves as a Faraday cage [29, 122].

This idea is in progress now. $N@C_{60}$, $N@C_{70}$ and $P@C_{60}$ were included in the crystalline host matrix of 2,4,6-tris-(4-bromophenoxy)-1,3,5-triazine. Due to an axial deformation of the fullerene cage in the ordered arrangement, the zero-field splitting of the quartet ground state of the encapsulated atoms is observed, the small size of which leads to well-separated lines for all transitions, opening the possibility of transition selective pulses for future quantum computing [124]. A system composed by two endohedral fullerene molecules, which interact through the spin dipole interaction, was proposed to be used for the realization of quantum gates with pi-gate operational times [125]. Furthermore, the $A@C_{2n}$-based elementary units may be formed into large arrays corresponding to the functional quantum architecture [126].

There is certainly a long way to go for an actual construction of a fullerene based quantum computer. Many serious problems remain to be solved, among them – the problem of chemical stability of the ENF "basic elements". Although the nitrogen containing $A@C_{2n}$ are rather stable and their mass-spectrometric signals are not lost even after several hours heating at 200 ^{0}C, yet the enclosed atom starts to form chemical bonds with carbons of the cage through the aza-bridged intermediates above 50-100 ^{0}C [127]. Thus, the challenges are enormous, as is true for all present concepts of realizing the quantum computers, though.

Non-Classical Endofullerenes

The first representative of this class, $Sc_3N@C_{80}$, is diamagnetic. Later, $Ho_3N@C_{80}$ and $Tb_3N@C_{80}$ were synthesized which have net magnetic moments of 20 and 18 μ_B per cage, correspondingly. The magnetization of these compounds was measured for magnetic fields up to 5 T and temperatures down to 1.8 K [128]. The graphs of magnetization depending on the ratio H/T were successfully described by Brillouin functions with J=16 and g=1.25 for $Ho_3N@C_{80}$ and J=12, g=1.5 for $Tb_3N@C_{80}$. Presumably, a strong ligand field exists within the M_3N cluster in these ENF owing to which the individual magnetic moments of the metal ions (Ho, Tb) are not parallel or antiparallel to each other but parallel to the M-N bond [128].

The single-crystal X-ray diffraction studies showed that the nitride ion is located at the center of the carbon cage in most TNT-ENF of this type but there are exceptions, for example, $CeSc_2N@C_{80}$, in which the nitride ion is offset by 0.36 angstrom in order to accommodate the large Ce^{3+} ion. Moreover, for $CeSc_2N@_{80}$, the Sc-45 NMR spectrum exhibited a signal with a small temperature-dependent Curie shift. This Curie shift and the oxidation state (+3) for cerium are consistent with a weak paramagnetism of the system with a single buried *f* electron spin [129].

By now, the TNT-ENF family has been significantly replenished. $Sc_2C_2@C_{82}$, $Sc_2C_2@C_{84}$, $Sc_3C_2@C_{80}$, and $Y_2C_2@C_{82}$ were synthesizes and systematically studied by using the high resolution synchrotron X-ray radiation in Japan. The Raman spectroscopy of $Sc_2C_2@C_{84}$ has revealed that the C_2 unit in this ENF behaves like a diatomic plane molecular rotor, a "fullerene quantum gyroscope" (see refs. in [130]). Furthermore, four heteroatom $ScYErN@C_{80}$ was made and studied in China. The ScYErN cluster rapidly rotates inside the fullerene cage, the symmetry of which was assigned as I_hC_{80}. Six electrons are presumably transferred from this cluster onto the fullerene that leads to a closed-shell electronic structure of the fullerene shell and, thereby, excellent stability of this molecule [131].

In addition, various exohedral derivatives of $M_3N@C_{2n}$ (M = Sc, Y, Er, Lu) have been synthesized, amongst them – Diels-Alder adducts, mono- and biscycloadducts and bis-silylated $Sc_3N@C_{80}$ [132-135]. It was shown that the chemical reactivity of the fullerene shell essentially depends upon the nature of the enclosed metal cluster [135]. Furthermore, it was found that $Ln_3N@C_{80}$ (Ln = Er, Gd, Ho, Lu, Sc, Tb, Tm, Y), manifest extraordinary kinetic chemical stability with respect to other fullerenes in Diels-Alder reactions with a cyclopentadiene-functionalized resin [136]. This finding has made it possible to obtain these TNT-ENF in pure form directly from the as-prepared soot in a single facile step. It has been also shown that careful control of stoichiometry, time, and temperature allows separation of fullerenes with different cage sizes as well as isomeric species. Moreover, since the Diels-Alder reaction is thermally reversible, the authors demonstrated that the bound empty fullerenes and classical endohedral metallofullerenes can be recovered by displacement with maleic anhydride [136].

An alternative effort to minimize the HPLC-dependence of isolation and purification of ENF has been accomplished in [137]. The authors suggested the approach based on the use of cyclopentadienyl and amino functionalized silica to bind selectively contaminant fullerenes. The purified TNT-ENF samples were obtained at room temperature using amino, diamino, and triamino silica at recoveries of 93% (11 h), 76% (9 h), and 50% (6 h), respectively [137].

Successful exohedral functionalizations of TNT-ENF and peapods filled with TNT-ENF seem to be useful precursor steps on the way to future applications of these compounds in nanotechnology. SWCNTs filled with the trimetal nitride fullerene $Dy_3N@C_{80}$ ("major isomer") were prepared and studied by photoemission spectroscopy [138]. The large chiral cage fullerenes $Tb_3N@C_{88}$ and $Tb_3N@C_{86}$ as well as the isomers of $Tb_3N@C_{80}$. have been synthesized and co-crystallized with Ni(II)-octaethylporphyrin recently [139]. Another important step on this way was taken by Deak et al. who showed that $Er_3N@C_{80}$ can be assembled on a strontium titanate surface template into four distinctive arrangements including close-packed assemblies, one-dimensional chains and two-dimensional grids [140].

Conclusion

As it was said about fullerenes, "fullerene chemistry is still at an early stage and one may anticipate future surprises" [64]. Of even greater valid seem these anticipations to be for endohedral fullerenes. This man-made "atomic park" now counts approximately 50 elements, mainly the elements of calcium and scandium subgroups of the Periodic. Atoms entrapped in the fullerene cages manifest unusual, even unique, structure and electron properties. In paramagnetic $La@C_{2n}$ there is the partial runoff of electron spin density outside the fullerene cage ("spin leakage") testified by EPR, ENDOR and NMR data. In $Be@C_{60}$, $N@C_{60}$, $P@C_{60}$ there is, contrariwise, the accumulation of electron spin density on the entrapped atoms as if their electronic shells are compressed ("spin compression"). Endohedral fullerenes tend to form nanoparticles in liquid and solid state. Furthermore, the fullerene cages can be modified in plenty of organic-functionalized derivatives. Endohedral fullerenes hold much promise as the basic elements for realizing nanometer scale electronic and optoelectronic devices, amongst them – molecular magnets, superconductors with low critical temperature, molecular memory devices, field effect transistors, quantum dots, fullerene-based quantum computers, even X-ray lasers. There are the fields in medicine where the employment of $A@C_{2n}$ appears to have considerable promise too, amongst them – magnetic resonance imaging, X-ray imaging, and radiopharmaceuticals. Yet, neither of these ideas has been realized up to date. Chemical and physical properties of $A@C_{2n}$ are still far from being clear since these compounds are not easily accessible. It concerns, in part, electron and spin-transfer complexes formed by these molecules, chemical reactivity in redox reactions, properties of nanoparticles. If the applications in medicine are concerned, there are some additional problems to go about, such as the clustering in aqueous medium, poorly investigated bio-distribution in tissues, possible toxicity. Until then, this field of research remains largely an academic one.

Acknowledgment

The research of endohedral metallofullerenes in IPCP RAS was initiated by Prof. E.B. Yagubskii. I am grateful to all co-workers of my group in IPCP RAS, L.T. Kasumova, V.P. Bubnov, Y.I. Estrin, E.E. Laukhina, V.P. Lodygina, G.A. Mirontseva, T.A. Parnyuk and K.S. Tarasevich-Laukhina for their interest to this research, excellent assistance and helpful discussions. I acknowledge the Russian Foundation for Basic Research for the support. ENDOR experiments were performed in cooperation with Dr. R.M. Davydov (Northwestern University, Evanston, USA). NMR experiments on the liquid solutions were performed in cooperation with Professor P.T. Manoharan and Dr. M. Subramoni (Regional Scientific Instrumental Center, Chennai, India). The experiments on dynamic light scattering were performed in cooperation with Dr. S. Nath, H. Pal and A.V. Sapre (Radiation Chemistry and Chemical Dynamics Division, Bhabha Atomic Research Centre, Mumbai, India). MAS-NMR experiments were performed in the group of Professor A. Pines, Lawrence Berkeley National Laboratory and University of California in cooperation with Dr. H. Heise and Dr. J.W. Logan. I greatly acknowledge all of them for the fruitful cooperation.

References

[1] Heath, J. R.; O'Brien, S. C.; Zhang, Q.; Liu, Y.; Curl, R. F.; Kroto, H. W.; Smalley, R. E. *J. Amer. Chem. Soc.* 1985, 107, 7779-7780.

[2] Chai, Y.; Guo, T.; Jin, C.; Haufler, R. E.; Chibante, L. P. F.; Fure, J.; Wang, L.; Alford, J. M.; Smalley, R. E. *J. Phys. Chem.* 1991, 95, 7564-7568.

[3] Bethune, D. S.; Johnson, R. D.; Salem, J. R.; de Vries, M. S.; Yannoni, C. S. *Nature.* 1993, 366, 123-128.

[4] Diener, M. D.; Smith, C. A.; Veirs, D. K. *Chem. Mater.* 1997, 9, 1773-1777.

[5] Shinohara, H. *Rep. Prog. Phys.* 2000, 63, 843-892.

[6] Saunders, M.; Cross, R. J. In *Endofullerenes: A New Family of Carbon Clusters*; Akasaka, N.; Nagase, S.; Eds.; Kluwer Academic Publishers: Dordrecht-Boston-London, 2003; pp. 1-11.

[7] Kobayashi, K.; Nagase, S. In *Endofullerenes: A New Family of Carbon Clusters*; Akasaka, N.; Nagase, S.; Eds.; Kluwer Academic Publishers: Dordrecht-Boston-London, 2003; pp. 99-120.

[8] Dorn, H. C.; Balch, A. L.; Dunchamp, J. C. In *Endofullerenes: A New Family of Carbon Clusters*; Akasaka, N.; Nagase, S.; Eds.; Kluwer Academic Publishers: Dordrecht-Boston-London, 2003; pp. 121-131.

[9] Diener, M.; Alford, M.; Bolskar, R. In *Endofullerenes: A New Family of Carbon Clusters;* Akasaka, N.; Nagase, S.; Eds.; Kluwer Academic Publishers: Dordrecht-Boston-London, 2003; pp. 133-151.

[10] Kato, T. In *Endofullerenes: A New Family of Carbon Clusters*; Akasaka, N.; Nagase, S.; Eds.; Kluwer Academic Publishers: Dordrecht-Boston-London, 2003; pp. 153-167.

[11] Kikuchi, K. In *Endofullerenes: A New Family of Carbon Clusters*; Akasaka, N.; Nagase, S.; Eds.; Kluwer Academic Publishers: Dordrecht-Boston-London, 2003; pp. 217-229.

[12] Kubozono, Y. In *Endofullerenes: A New Family of Carbon Clusters*; Akasaka, N.; Nagase, S.; Eds.; Kluwer Academic Publishers: Dordrecht-Boston-London, 2003; pp. 253-272.

[13] Akiyama, K.; Sueki, K.; Haba, H.; Tsukada, K.; Asai, M.; Yaita, T.; Nagame, Y.; Kikuchi, K.; Katada, M.; Nakahara, H. *J. Radioanal. Nucl. Chem.* 2003, 255, 155–158.

[14] Bubnov, V. P.; Koltover, V. K.; Laukhina, E. E.; Estrin, Ya. I.; Yagubskii, E. B. *Russ. Chem. Bull.* 1997, 46, 240-244.

[15] Laukhina, E. E.; Bubnov, V. P.; Estrin, Ya. I.; Golub, Y. I.; Khodorkovskii, M. A.; Koltover, V. K.; Yagubskii, E. B. *J. Mater. Chem.* 1998, 8, 893-895.

[16] Bubnov, V. P.; Laukhina, E. E.; Kareev, I. E.; Koltover, V. K.; Prokhorova, T. G.; Yagubskii, E. B.; Kozmin, Y. P. *Chem. Mater.* 2002, 14, 1004-1008.

[17] Kareev, I. E.; Bubnov, V. P.; Laukhina, E. E.; Koltover, V. K.; Yagubskii, E. B. *Carbon.* 2003, 41, 7, 1375-1380.

[18] Lian, Y.; Yang, S.; Yang, S. *J. Phys. Chem. B.* 2002, 106, 3112-3117.

[19] Sun, B. Y.; Inoue, T.; Shimada, T.; Okazaki, T.; Sugai, T.; Suenaga, K.; Shinohara, H. *J. Phys. Chem. B.* 2004, 108, 9011-9015.

[20] Kuz'min, S. M.; Borovkov, N. Y.; Kolker, A. M.; Brazhyunas, V. L. Russ. *J. Appl. Chem.* 2005, 78, 559-564.

[21] Sueki, K.; Kikuchi, K.; Akiyama, K.; Sawa, T.; Katada, M.; Ambe, S.; Ambe, F.; Nakahara, H. *Chem. Phys. Lett.* 1999, 300, 140-144.

[22] Akiyama. K.; Sueki, K.; Kodama, T.; Kikuchi, K.; Takigawa, Y.; Nakahara, H.; Ikemoto, I.; Katada, M. *Chem. Phys. Lett.* 2000, 317, 490-496.

[23] Hino, S.; Wanita, N.; Kato, M.; Iwasaki, K.; Yoshimura, D.; Inoue, T.; Okazaki, T.; Shinohara, H. *J. Electron Spectrosc. Relat. Phenom.* 2005, 144, Sp. Iss., 239-242.

[24] Krawez, N.; Tellgmann, R.; Hertel, I. V.; Campbell, E. E. B.; Gromov, A.; Kratschmer, W. *Mol. Cryst. Liq. Crys.* 1998, 10, 209-212.

[25] Campbell, E. E. B. In *Endofullerenes: A New Family of Carbon Clusters*; Akasaka, N.; Nagase, S.; Eds.; Kluwer Academic Publishers: Dordrecht-Boston-London, 2003; pp. 67-97.

[26] Kaplan, A.; Bekkerman, A.; Tsipinyuk, B.; Kolodney, E. *J. Chem. Phys.* 2002, 117, 3484-3491.

[27] Percival, P. W.; Addison, J. B.; Brodovitch, J. C.; Sun-Mack, S. *Appl. Magn. Reson.* 1996, 11, 315-323.

[28] Murphy, T. A.; Pawlik, T.; Weidinger, A.; Hohne, M.; Alcala, R.; Spaeth, J. M. *Phys. Rev. Lett.* 1996, 77, 1075-1078.

[29] Pietzak, B.; Weidinger, A.; Dinse, K.-P.; Hirsch, A. In *Endofullerenes: A New Family of Carbon Clusters*; Akasaka, N.; Nagase, S.; Eds.; Kluwer Academic Publishers: Dordrecht-Boston-London, 2003; pp. 13-65.

[30] Cveticanin, J.; Djustebek, J.; Veljkovic, M.; Velickovic, S.; Djordjevic, V.; Neskovic, O. *J. Optoelectron. Adv. Mater.* 2006, 8, 1892-1894.

[31] Huang, H.J.; Ata, M.; Yoshimoto, Y. *Chem. Comm.* 2004, 10, 1206-1207.

[32] Reich, A.; Panthöfer, M.; Modrow, H.; Wedig, U.; Jansen, M. *J. Am. Chem. Soc.* 2004, 126, 14428-14434.

[33] Sahoo, R.R.; Patnaik. A. *Chem. Phys. Lett.* 2001, 349, 201-209.

[34] Cagle, D. W.; Thrash, Th. P.; Alford, M.; Chibante, L. P. F.; Ehrhardt, G. J.; Wilson, L. J. *J. Am. Chem. Soc.* 1996, 118, 8043-8047.

[35] Ohtsuki, T.; Ohno, K. J. *Radioanal. Nuclear Chem. 2004*, 262, 165-173.
[36] Ohtsuki, T.; Ohno, K. *Phys. Rev. B*. 2005, 72, 15, AR 153411.
[37] Saha, S. K.; Chowdhury, D. P.; Das, S. K.; Guin, R. *Nucl. Instrum. Methods Phys. Res. Sect. B-Beam Interact. Mater. Atoms*. 2006, 243, 277-281.
[38] Grushko, Y. S.; Khodorkovski, M. A.; Kozlov, V. S.; Shilin, V. A.; Grachev, S. A.; Artamonova, T. O. *Fuller. Nanotub. Carbon Nanostruct*. 2006, 14, 249-259.
[39] Iwamatsu, S. I.; Murata, S.; Andoh, Y.; Minoura, M.; Kobayashi, K.; Mizorogi, N.; Nagase, S. *J. Org. Chem*. 2005, 70, 4820-4825.
[40] Komatsu, K.; Murata, M.; Murata, Y. *Science*. 2005, 307, 238-240.
[41] Rubin, Y. *Fuller. Rel. Struct*. 1999, 199, 67-91.
[42] Koltover, V. K.; Bubnov, V. P.; Laukhina, E. E.; Estrin, Y. I. *Mol. Mater*. 2000, 13, 239-242.
[43] Knapp, C.; Weiden, N.; Dinse, K.-P. *Magn. Reson. Chem*. 2005, 43, S199-S204.
[44] Furukawa, K.; Okubo, S.; Kato, H.; Kato, T. *J. Phys. Chem. A*. 2003, 107, 10933-10937.
[45] Matsuoka, H.; Ozawa, N.; Kodama, T.; Nishikawa, H.; Ikemoto, I.; Kikuchi, K.; Furukawa, K.; Sato, K.; Shiomi, D.; Takui, T.; Kato, T. *J. Phys. Chem. B*. 2004, 108, 13972-13976.
[46] Weiden, N.; Kato, T.; Dinse, K.-P. *J. Phys. Chem. B*. 2004, 108, 9469-9474.
[47] Morley, G. W.; Herbert, B. J.; Lee, S. M.; Porfyrakis, K.; Dennis, T. J. S.; Nguyen-Manh, D.; Scipioni, R.; van Tol, J.; Horsfield, A. P.; Ardavan, A.; Pettifor, D. G.; Green, J. C.; Briggs, G. A. D. *Nanotechnology*. 2005, 16, 2469-2473.
[48] Guha, S.; Nakamoto, K. *Coord. Chem. Rev*. 2005, 249, 1111-1132.
[49] Wang, X. L.; Zuo, T. M.; Olmstead, M. M.; Duchamp, J. C.; Glass, T. E.; Cromer, F.; Balch, A. L.; Dorn, H. C. *J. Am. Chem. Soc*. 2006, 128, 8884-8889.
[50] Fuchs, D.; Rietschel, H.; Michel, R. H.; Fischer, A., Weis, P.; Kappes, M. M. *J. Phys. Chem*. 1996, 100, 725-729.
[51] Kodama, T.; Higashi, K.; Ichikawa, T.; Suzuki, S.; Nishikawa, H.; Ikemoto, I.; Kikuchi, K.; Achiba, Y. *Chem. Lett*. 2005, 34, 464-465.
[52] Lu, X.; Li, H.; Sun, B.; Shi, Z.; Gu, Z. *Carbon*. 2005, 43, 1546-1549.
[53] Tsuchiya, T.; Wakahara, T.; Shirakura, S.; Maeda, Y.; Akasaka, T.; Kobayashi, K.; Nagase, S.; Kato, T.; Kadish, K. M. *Chem. Mater*. 2004, 16, 4343-4346.
[54] Tsuchiya, T.; Wakahara, T.; Lian, Yongfu; Maeda, Y.; Akasaka, T.; Kato, T.; Mizorogi, N.; Nagase, S. *J. Phys. Chem. B*. 2006, 110, 22517-22520.
[55] Koltover, V. K.; Estrin, Y. I.; Bubnov, V. P.; Laukhina, E. E. *Russ. Chem. Bull*. 2000, 49, 10, 1745-1748.
[56] Solodovnikov, S. P.; Tumanskii, B. L.; Bashilov, V. V.; Lebedkin, S. F.; Sokolov, V. I. *Russ. Chem. Bull*. 2001, 50, 2242-2244.
[57] Koltover, V. K.; Logan, J. W.; Heise, H.; Bubnov, V. P.; Estrin, Y. I.; Kareev, I. E.; Lodygina, V. P.; Pines, A. *J. Phys. Chem. B*. 2004, 108, 33, 12450-12455.
[58] Koltover, V. K.; Parnyuk, T. A.; Bubnov, V. P.; Davydov, R. M. *Dokl. Akad Nauk. Physics*. 2002, 384, 329-331.
[59] Koltover, V. K.; Bubnov, V. P.; Estrin, Y. I.; Lodygina, V. P.; Davydov, R. M.; Subramoni, M.; Manoharan, P. T. *Phys. Chem. Chem. Phys*. 2003, 5, 2774-2777.
[60] Yang, S. F., Yang, S. H. *Langmuir*. 2002, 18, 8488-8495.

[61] Tsuchiya, T.; Sato, K.; Kurihara, H.; Wakahara, T.; Nakahodo, T.; Maeda, Y.; Akasaka, T.; Ohkubo, K.; Fukuzumi, S.; Kato, T.; Mizorogi, N.; Kobayashi, K.; Nagase, S. *J. Am. Chem. Soc.*, 2006, 128, 6699-6703.

[62] Nath, S.; Pal, H.; Sapre, A. V.; Bubnov, V. P.; Estrin, Y. I.; Parnyuk, T. A.; Koltover, V. K. *Fuller. Nanotub. Carbon Nanostruct.* 2004, 12, 1/2, 53-57.

[63] Jakes, P.; Dinse, K.-P. *J. Am. Chem. Soc.*, 2001, 123, 8854-8855.

[64] Hirsch, A. *Angew. Chemie. Int. Ed.* 2001, 40, 1195-1197.

[65] Kareev, I. E.; Bubnov, V. P.; Laukhina, E. E.; Dodonov, A. F.; Kozovski, V. I.; Yagubskii, E. B. *Fuller. Nanotub. Carbon Nanostruct.* 2004, 12, 1/2, 65-69.

[66] Bolskar, R. D.; Alford, J. M. *Chem. Comm.*, 2003, 11, 1292-1293.

[67] Rikiishi, Y.; Kubozono, Y.; Hosokawa, T.; Shibata, K.; Haruyama, Y.; Takabayashi, Y.; Fujiwara, A.; Kobayashi, S.; Mori, S.; Iwasa, Y. *J. Phys. Chem. B.* 2004, 108, 7580-7585.

[68] Leigh, D. F.; Owen, J. H. G.; Lee, S. M.; Porfyrakis, K.; Ardavan, A.; Dennis, T. J. S.; Pettifor, D. G.; Briggs, G. A. D. *Chem. Phys. Lett.* 2005, 414, 307-310.

[69] Bucher, K.; Epple, L.; Mende, J.; Mehring, M.; Jansen, M. *Phys. Status Solidi B-Basic Solid State Phys.* 2006, 243, 3025-3027.

[70] Hettich, R.; Lahamer, A.; Zhou, L., Compton, R. *Inter. J. Mass. Spectrom.* 1999, 183, 335-348.

[71] Koltover, V. K.; Parnyuk, T. A.; Bubnov, V. P.; Laukhina, E. E.; Estrin, Ya. I.; Yagubskii, E.B. *Phys. Solid State.* 2002, 44, 529-530.

[72] Koltover, V. K. *Chinese J. Light Scattering.* 2005, 17, 248-250.

[73] Koltover, V. K. *J. Moll. Liquids.* 2005, 120, 151-154.

[74] Fan, L., Yang, S., Yang, S. *Thin Solid Films.* 2005, 483, 95-101.

[75] Koltover V. K. *Carbon.* 2004, 42, 5/6, 1179-1183.

[76] Koltover, V. K. *J. Moll. Liquids.* 2006, 127, 139-141.

[77] Popok, V. N.; Azarko, I. I.; Gromov, A. V.; Jönsson, M.; Lassesson, A.; Campbell, E. E. B. *Solid State Comm.* 2005, 133, 499-503.

[78] Buchachenko, A.L. *J. Phys. Chem. B.* 2001, 105, 5839-5846.

[79] Kobayashi, K.; Nagase, S.; Dinse, K.-P. *Chem. Phys. Lett.* 2003, 377, 93-98.

[80] Breslavskaya, N. N.; Buchachenko, A. L. *Fuller. Nanotub. Carbon Nanostruct.* 2004, 12, 47-52.

[81] Levin, A. A.; Breslavskaya, N. N. *Russ. Chem. Bull.* 2005, 54, 1999-2002.

[82] Breslavskaya, N. N.; Levin, A. A.; Buchackenko, A. L. *Russ. Chem. Bull.* 2006, 55, 16-21.

[83] Segrè, E.; Wiegand, C. E. *Phys. Rev.* 1949, 75, 1, 39- 43.

[84] Ohtsuki, T.; Ohno, K.; Morisato, T.; Mitsugashira, T.; Hirose, K.; Yuki, H.; Kasagi, J. *Phys. Rev. Lett.* 2007, 98, 25, AR 252501.

[85] Ray, A.; Das, P.; Saha, S. K.; Das, S. K.; Das, J. J.; Madhavan, N.; Nath, S.; Sugathan, P.; Rao, P. V. M.; Jhingan, A. *Phys. Rev. C.* 2006, 73, AR 034232.

[86] Watanabe, S.; Katabuchi, T.; Ishioka, N.S.; Matsuhashi, S.; Muramatsu, H. *J. Radioanal. Nucl. Chem.* 2007, 272, 467-469.

[87] Takeda, A.; Yokoyama, Y.; Ito, S.; Miyazaki, T.; Shimotani, H.; Yakigaya, K.; Kakiuchi, T.; Sawa, H.; Takagi, H.; Kitazawa, K.; Dragoe, N. *Chem. Comm.* 2006, 8, 912-914.

[88] Pizzarello, S., Huang, Y. S., Becker, L., Poreda, R. J., Nieman, R. A., Cooper, G., Williams, M. *Science*. 2001, 293, 2236-2239.

[89] Kratschmer, W. *Nanostruct. Mater*. 1995, 6, 65-72.

[90] Hwang, H. J.; Byun, K. R.; Lee, J. Y.; Kang, J. W. *Curr. Appl. Phys.* 2005, 5, 609-614.

[91] Huang, H. J.; Yang, S. H.; Liu, Z. Y.; Liu, S. Y.; Sun, D. Y. *Chem. Mater*. 1999, 11, 1003-1006.

[92] Mikawa, M.; Kato, H.; Okumura, M.; Narazaki, M.; Kanazawa, Y.; Miwa, N.; Shinohara, H. *Bioconjugate Chem*. 2001, 12, 510-514.

[93] Cao, B. P.; Wakahara, T.; Maeda, Y.; Han, A.H.; Akasaka, T.; Kato, T.; Kobayashi, K.; Nagase, S. *Chem. Europ. J.* 2004, 10, 716-720.

[94] Feng, L.; Lu, X.; He, X. R.; Shi, Z. J.; Gu, Z. N. *Inorg. Chem. Comm*. 2004, 7, 1010-1013.

[95] Yamada, M.; Feng, L.; Wakahara, T.; Tsuchiya, T.; Maeda, Y.; Lian, Y.; Kako, M.; Akasaka, T.; Kato, T.; Kobayashi, K.; Nagase, S. *J. Phys. Chem. B*. 2005, 109, 6049-6051.

[96] Feng, L.; Wakahara, T.; Nakahodo, T.; Tsuchiya, T.; Piao, Q.; Maeda, Y.; Lian, Y.; Akasaka, T.; Horn, E.; Yoza, K.; Kato, T.; Mizorogi, N.; Nagase, S. *Chem. Eur. J.* 2006, 12, 5578-5586.

[97] Kareev, I. E., Lebedkin, S. F., Bubnov, V. P., Yagubskii, E. B., Ioffe, I. N., Khavrel, P. A., Kuvychko, I. V., Strauss, S. H., Boltalina, O. V. *Angew. Chemie. Int. Ed.* 2006, 44, 1846-1849.

[98] Yang, S.; Fan, L.; Yang, S. *J. Phys. Chem. B*. 2004, 108, 4394-4404.

[99] Butcher, M. J.; Nolan, J. W.; Hunt, M. R. C.; Beton, P. H.; Dunsch, L.; Kuran, P.; Georgi, P., Dennis, T. J. S. *Phys. Rev*. 2003, 67, AR 125413.

[100] Fujiki, S.; Kubozono, Y.; Hosokawa, T.; Kanbara, T.; Fujiwara, A.; Nonogaki, Y.; Urisu, T. *Phys. Rev. B*. 2004, 69, AR 045415.

[101] Ton-That, C.; Welland, M. E.; Larsson, J. A.; Greer, J. C.; Shard, A. G.; Dhanak, V. R.; Taninaka, A.; Shinohara, H. *Phys. Rev. B*. 2005, 71, AR 045419.

[102] Schulte, K.; Wang, L.; Moriarty, P. J.; Purton, J.; Patel, S.; Shinohara, H.; Kanai, M.; Dennis, T. J. S. *Phys. Rev. B*. 2005, 71, AR 115437.

[103] Yasutake, Y.; Shi, Z.; Okazaki, T.; Shinohara, H.; Majima, Y. *Nano Lett*. 2005, 5, 1057-1060.

[104] Su, G. J.; Gan, L. H.; Yang, Z. Y.; Pan, G. B.; Wan, L. J.; Wang, C. R. *J. Phys. Chem. B*. 2006, 110, 5559-5562.

[105] Nuttall, Ch. J.; Hayashi, Y.; Yamazaki, H.; Mitani,T. Iwasa, Y. *Adv. Mater*. 2002, 14, 293-296.

[106] Smith, B. W., Luzzi, D. E.; Achiba, Y. *Chem. Phys. Lett*. 2000, 331, 137-142.

[107] Hirahara, K.; Suenaga, K.; Bandow, S.; Kato, H.; Okazaki, T.; Shinohara, H.; Iijima, S. *Phys. Rew. Lett*. 2000, 85, 5384-5387.

[108] Khlobystov, A. N.; Porfyrakis, K.; Kanai, M.; Britz, D. A.; Ardavan, A.; Shinohara, H.; Dennis, T. J. S.; Briggs, G. A. D. *Angew. Chemie. Int. Ed.* 2004, 43, 1386-1389.

[109] Lee, J.; Kim, H.; Kahng, S. J.; Kim, G.; Son, Y. W.; Ihm, J.; Kato, H.; Wang, Z. W.; Okazaki, T.; Shinohara, H.; Kuk, Y. *Nature*. 2002, 415, 1005-1008.

[110] Shimada, T.; Ohno, Y.; Okazaki, T.; Sugai, T.; Suenaga, K.; Kishimoto, S.; Mizutani, T.; Inoue, T.; Taniguchi, R.; Fukui, N.; Okubo, H.; Shinohara, H. *Physica E*. 2004, 21, 1089-1092.

[111] Hiroshiba, N.; Tanigaki, K.; Kumashiro, R.; Ohashi, H.; Wakahara, T.; Akasaka, T. *Chem. Phys. Lett.* 2004, 400, 235-238.

[112] Nagano, T.; Kuwahara, E.; Takyanagi, T.; Kubozono, Y.; Fujiwara, A. *Chem. Phys. Lett.* 2005, 409, 187-191.

[113] Koltover, V. K.; Laukhina, E. E.; Estrin, Y. I.; Bubnov, V. P.; Yagubskii, E. B. *Dokl. Akad Nauk. Physical Chemistry*. 1997, V. 353, No. 1-3, 85-86.

[114] Toth, E.; Bolskar, R. D.; Borel, A.; Gonzalez, G.; Helm, L.; Merbach, A. E.; Sitharaman, B.; Wilson, L. J. *J. Am. Chem. Soc.* 2005, 127, 799-805.

[115] Volkert, W. A.; Hoffman, T. J. *Chem. Rev.* 1999, 99, 2269-2292.

[116] Tagmatarchis, N.; Kato, H.; Shinohara, H. *Phys. Chem. - Chem. Phys.* 2001, 3, 3200-3202.

[117] Wang, J. X., Chen, C. Y.; Li, B.; Yu, H. W.; Zhao, Y. L.; Sun, J.; Li, Y. F.; Xing, G. M.; Yuan, H.; Tang, J.; Chen, Z.; Meng, H.; Gao, Y. X.; Ye, C.; Chai, Z. F.; Zhu, C. F.; Ma, B. C.; Fang, X. H.; Wan, L. J. *Biochem. Pharmacol.* 2006, 71, 872-881.

[118] Shu, C. Y.; Zhang, E. Y.; Xiang, J. F.; Zhu, C. F.; Wang, C. R.; Pei, X. L.; Han, H. B. *J. Phys. Chem. B.* 2006, 110, 15597-15601.

[119] Jones, M. A. G.; Britz, D. A,; Morton, J. J. L.; Khlobystov, A. N.; Porfyrakis, K.; Ardavan, A.; Briggs, G. A. D. *Phys. Chem. Chem. Phys.* 2006, 8, 2083-2088.

[120] Franco, L.; Ceola, S.; Corvaja, C.; Bolzonella, S.; Harneit, W.; Maggini, M. *Chem. Phys. Lett.* 2006, 422, 100-105.

[121] Zhang, J.; Morton, J. J. L.; Sambrook, M. R.; Porfyrakis, K.; Ardavan, A.; Briggs, G. A. D. *Chem. Phys. Lett.* 2006, 432, 523-527.

[122] Harneit, W.; Meyer, C.; Wedinger, A.; Suter, D.; Twamley, J. *Phys. Stat. Sol. B.* 2002, 233, 453-461.

[123] Feng, M.; Twamley, J. *Phys. Rev. A.* 2004, 70, 3, AR 032318.

[124] Naydenov, B.; Spudat, Ch.; Scheloske, M.; Suess, H.I.; Hulliger, J.; Harneit, W. *Phys. Status Solidi B-Basic Solid State Phys.* 2006, 243, 2995-2998.

[125] Garelli, M. S.; Kusmartsev, F. V. *Eur. Phys. J. B.* 2005, 48, 199-206.

[126] Benjamin, S. C.; Ardavan, A.; Andrew, G.; Briggs, D.; Britz, D. A.; Gunlycke, D.; Jefferson, J.; Leigh, D. F.; Lovett, B. W.; Khlobystov, A. N.; Lyon, S. A.; Morton, J. J. L.; Porfyrakis, K.; Sambrook, M. R.; Tyryshkin, A. M. *J. Phys.-Condens. Matter.* 2006, 18, S867-S883.

[127] Cao, B. P.; Peres, T.; Lifshitz, C.; Cross, R. J.; Saunders, M. *Chem. Eur. J.* 2006, 12, 2213-2221.

[128] Wolf, M.; Muller, K. H.; Skourski, Y.; Eckert, D.; Georgi, P.; Krause, M.; Dunsch, L. *Angew. Chemie Int. Ed.* 2005, 44, 3306-3309.

[129] Wang, X. L.; Zuo, T. M.; Olmstead, M. M.; Duchamp, J. C.; Glass, T. E.; Cromer, F.; Balch, A. L.; Dorn, H. C. *J. Am. Chem. Soc.* 2006, 128, 8884-8889.

[130] Nishibori, E.; Terauchi, I.; Sakata, M.; Takata, M.; Ito, Y.; Sugai, T.; Shinohara, H. *J. Phys. Chem. B.* 2006, 110, 19215-19219.

[131] Chen, N.; Zhang, E. Y.; Wang, C. R. *J. Phys. Chem. B.* 2006, 110, 13322-13325.

[132] Cai, T.; Slebodnick, C.; Xu, L.; Harich, K.; Glass, T. E.; Chancellor, C.; Fettinger, J. C.; Olmstead, M. M.; Balch, A. L.; Gibson, H. W.; Dorn, H. C. *J. Am. Chem. Soc.* 2006, 128, 6486-6492.

[133] Cai, T.; Xu, L. S.; Anderson, M. R.; Ge, Z. X.; Zuo, T. M.; Wang, X. L.; Olmstead, M. M.; Balch, A. L.; Gibson, H. W.; Dorn, H. C. *J. Am. Chem. Soc.* 2006, 128, 8581-8589.

[134] Wakahara, T.; Iiduka, Y.; Ikenaga, O.; Nakahodo, T.; Sakuraba, A.; Tsuchiya, T.; Maeda, Y.; Kako, M.; Akasaka, T.; Yoza, K.; Horn, E.; Mizorogi, N.; Nagase, S. *J. Am. Chem. Soc.* 2006, 128, 9919-9925.

[135] Cardona, C. M.; Elliott, B.; Echegoyen, L. *J. Am. Chem. Soc.* 2006, 128, 6480-6485.

[136] Ge, Z. X.; Duchamp, J. C.; Cai, T.; Gibson, H. W.; Dorn, H. C. *J. Am. Chem. Soc.* 2005, 127, 16292-16298.

[137] Stevenson, S.; Harich, K.; Yu, H.; Stephen, R. R.; Heaps, D.; Coumbe, C.; Phillips, J. P. *J. Am. Chem. Soc.* 2006, 128, 8829-8835.

[138] Shiozawa, H.; Rauf, H.; Pichler, T.; Knupfer, M.; Kalbac, M.; Yang, S.; Dunsch, L.; Buechner, B.; Batchelor, D.; Kataura, H. *Phys. Rev. B.* 2006, 73, 20, AR 205411.

[139] Zuo, T.; Beavers, Ch. M.; Duchamp, J. C.; Campbell, A.; Dorn, H. C.; Olmstead, M. M.; Balch, A. L. *J. Am. Chem. Soc.,* 2007, 129, 2035-2043.

[140] Deak, D. S.; Silly, F.; Porfyrakis, K.; Castell, M. R. *Nanotechnology*. 2007, 18, AR 075301.

[141] Diener, M.D.; Alford, J.M.; Kennel, S.J.; Mirzadeh S. *J. Am. Chem. Soc.* 2007, 129. 16, 5131-5138.

[142] Wu, X.; Lu, X. Dimetalloendofullerene $U_2@C_{60}$ has a U-U multiple bond consisting of sixfold one-electron-two-center bonds. *J. Am. Chem. Soc.*, 2007, 129, 2171-2177.

In: Progress in Fullerene Research
Editor: Milton Lang, pp. 235-268
ISBN 978-1-60021-841-5

Chapter 7

ELECTRONIC STRUCTURE OF SPHEROIDAL FULLERENES

Richard Pincak[a,b] ***and Michal Pudlak***[a]
[a]Institute of Experimental Physics, Slovak Academy of Sciences,
Watsonova 47,043 53 Kosice, Slovak Republic
[b]Bogoliubov Laboratory of Theoretical Physics,
Joint Institute for Nuclear Research, Dubna, Russia

Abstract

Both the eigenfunctions and the local density of states (DOS) near the pentagonal defects on the fullerenes was calculated analytically as a numerically. The results shows that the low-energy DOS has a cusp which drops to zero at the Fermi energy for any number of pentagons at the tip except three. For three pentagons, the nonzero DOS across the Fermi level is formed. Graphite is an example of a layered material that can be bent to form fullerenes which promise important applications in electronic nanodevices. The spheroidal geometry of a slightly elliptically deformed sphere was used as a possible approach to fullerenes. We assumed that for a small deformation the eccentricity of the spheroid is much more smaller then one. We are interested in the elliptically deformed fullerenes C_{70} as well as in C_{60} and its spherical generalizations like big C_{240} and C_{540} molecules. The low-lying electronic levels are described by the Dirac equation in (2+1) dimensions. We show how a small deformation of spherical geometry evokes a shift of the electronic spectra compared to the sphere and both the electronic spectrum of spherical and the shift of spheroidal fullerenes were derived. In the next study the expanded field-theory model was proposed to study the electronic states near the Fermi energy in spheroidal fullerenes. The low energy electronic wave functions obey a two-dimensional Dirac equation on a spheroid with two kinds of gauge fluxes taken into account. The first one is so-called K spin flux which describes the exchange of two different Dirac spinors in the presence of a conical singularity. The second flux (included in a form of the Dirac monopole field) is a variant of the effective field approximation for elastic flow due to twelve disclination defects through the surface of a spheroid. We consider the case of a slightly elliptically deformed sphere which allows us to apply the perturbation scheme. We shown exactly how a small deformation of spherical fullerenes provokes an appearance of fine structure in the electronic energy spectrum as compared to the spherical case. In particular, two quasi-zero modes in addition to the true zero mode are predicted to emerge in spheroidal fullerenes. The effect of a weak uniform magnetic field on the electronic structure of slightly deformed fullerene molecules was also studied. It was shown how the existing due to spheroidal deformation fine structure of the electronic energy spectrum splitted in the presence of the magnetic field. We shown that the fine structure of the electronic energy spectrum is very sensitive to the orientation of the magnetic field. We found that the magnetic field pointed in the x direction does not influence the first electronic level whereas it causes a splitting of the second energy level. Exact analytical solutions for zero-energy modes are found. The HOMO (highest occupied molecular orbital)-LUMO (lowest unoccupied molecular orbital) gap was calculated within the continuum field-theory model of fullerenes.

1. Introduction

The purpose of this chapter is to present some examples of using topology and geometry in a study of a new interesting class of carbon materials as fullerenes. Fullerenes are cage-like molecules of carbon atoms. The discovery of these molecules has attracted considerable attention of both experimentalists and theorists due to their unique physical properties. An further interest to carbon nanoparticles originates from the fact that the geometry is accompanied by topological defects. The topologically nontrivial objects play the important role in various physically interesting systems. The 't Hooft-Polyakov monopole in the non-Abelian Higgs model, instantons in quantum chromodynamics, solitons in the Skyrme model, Nielsen-Olesen magnetic vortices in the Abelian Higgs model(see, e.g., [1]) are examples of such topologically nontrivial objects. Notice that the topological origin the elastic flux results in appearance of the disclination-induced Aharanov-Bohm like phase (see [2]). In condensed matter physics for instance, vortices in liquids and liquid crystals, solitons in low-dimensional systems are objects with nontrivial topology.

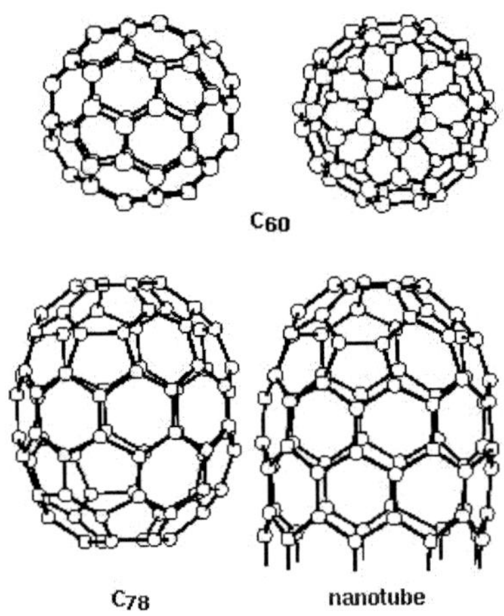

Figure 1. The symmetric structure of fullerene C_{60}, elliptically deformed molecule C_{78} and big deformed structure as carbon nanotube.

The high flexibility of carbon allows producing variously shaped carbon nanostructures: fullerenes, nanotubes, nanohorns, cones, toroids, graphitic onions, etc. Some of

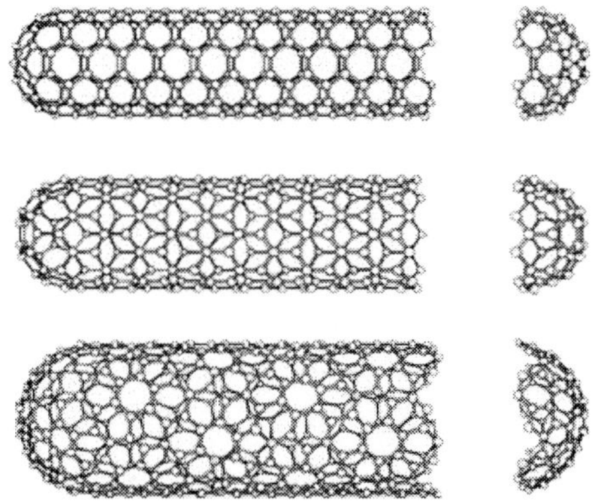

Figure 2. The fullerene C_{60} and spheroidal ones as a cap at the end of some nanotubes.

them as e.g.nanotubes are composition of tubes and fullerenes (as in Fig. 1).Historically, the fullerenes C_{60} (nicknamed also as Buckminsterfullerene or 'bucky ball') were first discovered in 1985 [3]. They are tiny molecular cages of carbon having 60 atoms and making up the mathematical shape called a truncated icosahedron (12 pentagons and 20 hexagons). The amount of C_{60} had been produced in small quantities until 1990 when an efficient method of production was achieved [4]. Since then, there were produced variously shaped fullerene molecules. These molecules are formed when 12 pentagons are introduced in the hexagonal lattice of graphene. Soon after the fullerenes, the carbon nanotubes of different diameters and helicity [5] were produced. We can also image the construction of nanotube with fullerenes as a cap at the end of nanotubes (see Fig. 2). The mechanical, magnetic, and especially electronic properties of carbon nanotubes are found to be very specific (see, e.g., [6]).

The fullerenes are composed of hexagonal and pentagonal carbon rings. However, the structures having heptagonal rings are also possible. A junction which connect carbon nanotubes with different diameters through a region sandwiched by pentagon-heptagon pair has been observed in the transmission electron microscope [7]. A pair of five- and seven-membered rings have to be imposed to connect two different types of carbon nanotubes [8]. Theoretically the closed fullerenes and nanotubes exhibiting high topologies (from genus 5 to genus 21) were suggested in [9]. This follows from the known Euler's theorem that relates the number of vertices, edges and faces of an object. For the hexagonal carbon lattice it can be written in the form [9]

$$...2n_4 + n_5 - n_7 - 2n_8... = \sum(6-x)n_x = \chi = 12(1-g), \qquad (1)$$

where n_x is the number of polygons having x sides, χ is the Euler characteristic which is a geometrical invariant related to the topology of the structure, and g is the genus or a number of handles of an arrangement. According to (1) there is no contribution to the Gaussian curvature for $x = 6$. This means that two-dimensional carbon lattice consisting only of hexagons is flat. By their nature, pentagons (as well as other polygons with $x \neq 6$) in a graphite sheet are topological defects (see Fig. 3).

In particular, fivefold coordinated particles are orientational 60° disclination defects in the sixfold coordinated triangular lattice. Description of the electronic structure re-

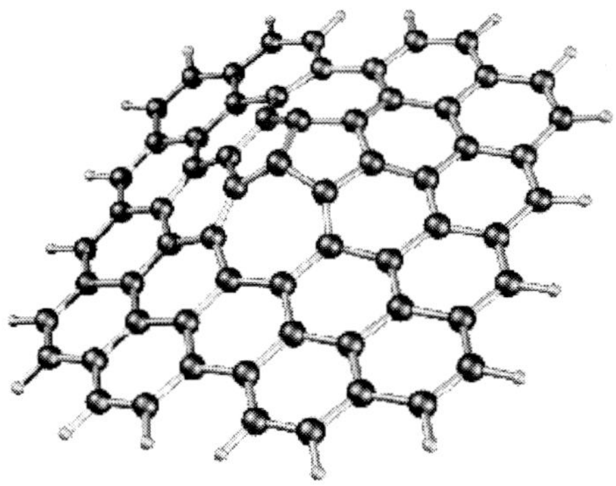

Figure 3. The pentagonal and heptagonal defects in the hexagonal network.

quires formulating a theoretical model describing electrons on arbitrary curved surfaces with disclinations taken into account. The most important fact found in [10, 11, 12] is that the electronic spectrum of a single graphite plane linearized around the corners of the hexagonal Brillouin zone coincides with that of the Dirac equation in (2+1) dimensions. This finding stimulated a formulation of some field-theory models for Dirac fermions on hexatic surfaces to describe electronic structure of variously shaped carbon materials: fullerenes [13, 14] and nanotubes [15]. The field-theory models for Dirac fermions on a plane and on a sphere [16] were invoked to describe variously shaped carbon materials.

2. Electronic Properties

Carbon nanoparticles and their electronic properties have been a subject of an extensive study. Electronic states in nanotubes, fullerenes, nanocones, nanohorns as well as in other carbon configurations are the subject of an increasing number of experimental and theoretical studies. It has been predicted and later observed in experiment that bending or stretching a nanotube change its band structure changing therefore the electrical properties: stretched nanotubes become either more or less conductive. Carbon nanotubes can be either a metal or semiconductor, depending on their diameters and helical arrangement [17]. This finding could allow to build nanotube-based transducers sensitive to tiny forces.

The electronic properties depends on the topological defects. The peculiar electronic states due to topological defects have been observed in different kinds of carbon nanoparticles by scanning tunneling microscopy (STM). For example, STM images with five-fold symmetry (due to pentagons in the hexagonal graphitic network) have been obtained in the C_{60} fullerene molecule [18]. The peculiar electronic properties at the ends of carbon nanotubes (which include several pentagons) have been probed experimentally in [19, 20]. Recently, the electronic structure of a single disclination has been revealed on an atomic scale by STM [21].

The problem of peculiar electronic states near the pentagons in curved graphite nanoparticles was the subject of intensive theoretical studies in fullerenes [14, 13], nanotubes [15], nanohorns [22], and cones [23]. In particular, an analysis within the effective-mass theory has shown that a specific $\sqrt{3} \times \sqrt{3}$ superstructure induced by pentagon defects can appear in nanocones [24]. This prediction has been experimentally verified in [21]. A recent study [25] within both tight-binding and *ab initio* calculations shows a presence of sharp resonant states in the region close to the Fermi energy. The localized cap states in nanotubes have been recently studied in [26].

It is interesting to note that the problem of specific electronic states at the Fermi level due to disclinations is similar to that of the fermion zero modes for planar systems in a magnetic field. Generally, zero modes for fermions in topologically nontrivial manifolds have been of current interest both in the field theory and condensed matter physics. As was revealed, they play a major role in getting some insight into understanding anomalies [27] and charge fractionalization that results in unconventional charge-spin relations (e.g. the paramagnetism of charged fermions) [28] with some important implications for physics of superfluid helium (see, e.g., review [29]). $3D$ space-time Dirac equation for massless fermions in the presence of the magnetic field was found to yield $N-1$ zero modes in the N-vortex background field [30]. As it was shown in [13], the problem of the local electronic structure of fullerene is closely related to Jackiw's analysis [30]. An importance of the fermion zero modes was also discussed in the context of the high-temperature chiral superconductors [31]. Among different theoretical approaches to describe graphene compositions, the continuum model play a special role. It allow us to describe some integral features of similar carbon structures. It should be noted that the formulation of any continuum model begins from description of a graphite sheet (graphene). We briefly report now the effective mass theory for graphene [11, 12]. A single layer of graphite forms a two-dimensional material. Two Bloch functions constructed from atomic orbitals for two inequivalent carbon atoms, A and B, in the unit cell provide the basis function for graphene. The Bloch basis functions Φ_j $(j = A, B)$ are constructed from atomic $2p_z$ orbitals

$$\Phi_A(\vec{k}, \vec{r}) = \sum_n \frac{1}{\sqrt{N}} e^{i\vec{k}\,\vec{r}_n} f(\vec{r} - \vec{r}_n), \tag{2}$$

$$\Phi_B(\vec{k}, \vec{r}) = \sum_n \frac{1}{\sqrt{N}} e^{i\vec{k}\,\vec{r}_n} f(\vec{r} - \vec{d} - \vec{r}_n), \tag{3}$$

$\vec{d}$ is a coordinate of B atom in the unit cell, N is a number of unit cell, $\vec{r}_n$ is a unit cell coordinate, f is $2p_z$ orbital, $\vec{k}$ is the wave vector. Now we want to solve the Schrödinger equation

$$\hat{H}\Phi = E\Phi, \tag{4}$$

where $\Phi = c_A\Phi_A + c_B\Phi_B$ and $\hat{H}$ is the Hamiltonian of the system. We get the secular equations

$$H_{AA}c_A + H_{AB}C_B = EC_A,$$

$$H_{BA}c_A + H_{BB}C_B = EC_B, \tag{5}$$

here

$$H_{ij} = \langle \Phi_i | \hat{H} | \Phi_j \rangle, \quad i, j = A, B. \tag{6}$$

In a tight-binding approximation it have a form

$$\hat{H}_{AA} = \hat{H}_{BB} = \varepsilon_{2p},$$

$$\hat{H}_{AB} = t\left(e^{i\vec{k}\vec{\tau}_1} + e^{i\vec{k}\vec{\tau}_2} + e^{i\vec{k}\vec{\tau}_3}\right), \tag{7}$$

where ε_{2p} is a orbital energy of the $2p_z$ level and $\vec{\tau}_i$ are the coordinates of the three nearest-neighbor B atoms relative to an A atom

$$t = \langle f(\vec{r}) | \hat{H} | f(\vec{r} - \vec{d}\,) \rangle. \tag{8}$$

Solving the secular equation

$$det\,(H - E) = 0, \tag{9}$$

the eigenvalues $E(\hat{k})$ are obtained in the form

$$E(\vec{k}) = \varepsilon_{2p} \pm t\sqrt{1 + 4\cos\frac{ak_x}{2}\cos\frac{\sqrt{3}ak_y}{2} + 4\cos^2\frac{ak_x}{2}}. \tag{10}$$

We have

$$E(\vec{K}) = E(\vec{K}^{'}) = \varepsilon_{2p}, \tag{11}$$

and the $\vec{K}$ and $\vec{K}^{'}$ points are the corners of the Brillouin zone. In $\vec{k}.\vec{p}$ theory we approximate the wave function at wave vector $\hat{k} = \hat{K} + \hat{\kappa}$ by

$$\Phi(\hat{k}, \hat{r}) = c_A(\hat{\kappa}) e^{i\hat{\kappa}.\hat{r}} \Phi_A(\hat{K}, \hat{r}) + c_B(\hat{\kappa}) e^{i\hat{\kappa}.\hat{r}} \Phi_B(\hat{K}, \hat{r}). \tag{12}$$

Inserting Φ into Schrödinger equation, keeping terms of order $\vec{\kappa}$, we get the secular equation ($\hbar = c = 1$ units are used)

$$\frac{\vec{\kappa}}{m}.\begin{pmatrix} \vec{p}_{AA} & \vec{p}_{AB} \\ \vec{p}_{BA} & \vec{p}_{BB} \end{pmatrix}\begin{pmatrix} c_A(\vec{\kappa}) \\ c_B(\vec{\kappa}) \end{pmatrix} = E(\vec{\kappa})\begin{pmatrix} c_A(\vec{\kappa}) \\ c_B(\vec{\kappa}) \end{pmatrix}, \tag{13}$$

where $E(\vec{\kappa}) = E(\vec{k}) - E(\vec{K})$ and

$$\vec{p}_{ij} = \int \Phi_i^*(\hat{K}, \hat{r}) \vec{p} \Phi_j(\hat{K}, \hat{r}) d\vec{r}, \quad i, j = A, B. \tag{14}$$

It can be shown from group theoretical argument (and can be verify directly within the one-orbital tight-binding model) that the secular equation can be write in the form [12, 10]

$$v_F \begin{pmatrix} 0 & \kappa_x - i\kappa_y \\ \kappa_x + i\kappa_y & 0 \end{pmatrix}\begin{pmatrix} c_A(\vec{\kappa}) \\ c_B(\vec{\kappa}) \end{pmatrix} = E(\vec{\kappa})\begin{pmatrix} c_A(\vec{\kappa}) \\ c_B(\vec{\kappa}) \end{pmatrix}, \tag{15}$$

where $v_F = \frac{\sqrt{3}at}{2}$, a is the lattice constant and

$$E(\vec{\kappa}) = \pm v_F |\vec{\kappa}|. \tag{16}$$

The upper half of the energy dispersion curves describes the π^*-energy anti-bonding band, and the lower half is the π energy bonding band. Since there are two π electrons per unit cell, the lower π band is fully occupied. The points K,$K^{'}$ create the Fermi surface of the graphene. When an external gauge fields are imposed, the translation symmetry can be broken and eigenfunction can no longer be labeled by $\vec{\kappa}$. This requires a generalization of the trial wave function

$$\Phi(\hat{r}) = \int c_A(\hat{\kappa})e^{i\hat{\kappa}.\hat{r}}\Phi_A(\hat{K},\hat{r})d\vec{\kappa} + \int c_B(\hat{\kappa})e^{i\hat{\kappa}.\hat{r}}\Phi_B(\hat{K},\hat{r})d\vec{\kappa}. \tag{17}$$

Inserting this function into Schrödinger equation leads to an equation [10, 32]

$$\Big(-iv_F\vec{\sigma}.(\vec{\nabla} - i\vec{W}) - E\Big)\left(\begin{array}{c} \Psi_A(\vec{r}) \\ \Psi_B(\vec{r}) \end{array}\right) = 0, \tag{18}$$

here $\vec{\sigma} = (\sigma_x, \sigma_y)$ are conventional Pauli matrices, $\vec{W}$ is the U(1) external gauge field and $\Psi_A(\vec{r})$, $\Psi_B(\vec{r})$ are envelope functions, smoothly varying in the scale of the lattice constant a, multiplying graphene Bloch function

$$\Phi(\hat{r}) = \Psi_A(\hat{r})\Phi_A(\hat{K},\hat{r}) + \Psi_B(\hat{r})\Phi_B(\hat{K},\hat{r}), \tag{19}$$

$$\Psi_i(\hat{r}) = \int c_i(\hat{\kappa})e^{i\hat{\kappa}.\hat{r}}d\vec{\kappa}, \qquad i = A, B. \tag{20}$$

Eq. (18) is algebraically identical to a two-dimensional Dirac equation, where components of spinor represent localization of electron on the graphene sublattice A and sublattice B, respectively. We can get the similar equation for envelope functions near $K^{'}$ point. Formally Eq. (18) can be obtained from Eq. (15) by exchanging $\kappa_k \rightarrow -i(\partial_k - iW_k)$, $c_{A(B)} \rightarrow \Psi_{A(B)}(\vec{r})$ and $E(\vec{\kappa}) \rightarrow E$.

3. Theory

Now we will formulate a theoretical model describing electrons on arbitrary curved surfaces with disclinations taken into account to investigate the electronic structure of fullerenes. As it was mention in the previous section the electronic spectrum of a single graphite plane linearized around the corners of the hexagonal Brillouin zone coincides with that of the Dirac equation in (2+1) dimensions. This finding stimulated formulation of some field-theory models for Dirac fermions on hexatic surfaces to describe electronic structure of variously shaped carbon materials: fullerenes [13], nanotubes [15, 26], and cones [33, 34].

The effective-mass theory for a graphene sheets can be generalized to graphene surfaces containing five carbon atom rings and creating geometrical structures as a cones or fullerenes. In the continuum description pentagonal ring are described as a localized fictitious gauge fluxes. In such structures there is no globally coherent distinction between K and $K^{'}$ points.

Following the ideas above and describe fermions in a curved background, we need a set of orthonormal frames $\{e_\alpha\}$ which yield the same metric, $g_{\mu\nu}$, related to each other by the local $SO(2)$ rotation,

$$e_\alpha \rightarrow e^{'}_\alpha = \Lambda^\beta_\alpha e_\beta, \quad \Lambda^\beta_\alpha \in SO(2).$$

It then follows that $g_{\mu\nu} = e^{\alpha}_{\mu} e^{\beta}_{\nu} \delta_{\alpha\beta}$ where e^{μ}_{α} is the zweibein, with the orthonormal frame indices being $\alpha, \beta = \{1, 2\}$, and coordinate indices $\mu, \nu = \{1, 2\}$. As usual, to ensure that physical observables are independent of a particular choice of the zweinbein fields, a local $so(2)$ valued gauge field ω_μ must be introduced. The gauge field of the local Lorentz group is known as the spin connection. For the theory to be self-consistent, the zweibein fields must be chosen to be covariantly constant [35]:

$$\mathcal{D}_\mu e^{\alpha}_{\nu} := \partial_\mu e^{\alpha}_{\nu} - \Gamma^{\lambda}_{\mu\nu} e^{\alpha}_{\lambda} + (\omega_\mu)^{\alpha}_{\beta} e^{\beta}_{\nu} = 0,$$

which determines the spin connection coefficients explicitly

$$(\omega_\mu)^{\alpha\beta} = e^{\alpha}_{\nu} D_\mu e^{\beta\nu}. \tag{21}$$

Finally, the Dirac equation on a surface Σ in the presence of the $U(1)$ external gauge field W_μ is written as [36]

$$i\gamma^\alpha e^{\mu}_{\alpha} \left(\nabla_\mu - ia_\mu - iW_\mu\right) \psi(\vec{r}) = E\psi(\vec{r}). \tag{22}$$

where $\nabla_\mu = \partial_\mu + \Omega_\mu$ with

$$\Omega_\mu = \frac{1}{8}\omega^{\alpha\ \beta}_{\ \mu}[\gamma_\alpha, \gamma_\beta], \tag{23}$$

being the spin connection term in the spinor representation and

$$\gamma^\alpha = -I\sigma_\alpha = -\begin{pmatrix} \sigma_\alpha & 0 \\ 0 & \sigma_\alpha \end{pmatrix} \tag{24}$$

$$\oint a_\mu dx^\mu = \frac{2\pi}{4}\tau_2 I. \tag{25}$$

The matrix τ_2 acting in K-spin space. The energy E is measured relative to the Fermi energy, and the spinor $\psi = (\Psi^K_A \Psi^K_B \Psi^{K'}_A \Psi^{K'}_B)^T$ where $\Psi(\vec{r})$ are envelope functions. For our purpose, we need incorporating both a disclination field and a nontrivial background geometry. A possible description of disclinations on arbitrary two-dimensional elastic surfaces is offered by the gauge approach [37]. In accordance with the basic assumption of this approach, disclinations can be incorporated in the elasticity theory Lagrangian by introducing a compensating $U(1)$ gauge fields W_μ. It is important that the gauge model admits exact vortex-like solutions for wedge disclinations [37] thus representing a disclination as a vortex of elastic medium. The physical meaning of the gauge field is that the elastic flux due to rotational defect (that is directly connected with the Frank vector (see the next section)) is completely determined by the circulation of the W_μ field around the disclination line. In the gauge theory context, the disclination field can be straightforwardly incorporated in (22) by the standard substitution $\partial_\mu \to \partial_\mu - iW_\mu$.

Within the linear approximation to gauge theory of disclinations (which amounts to the conventional elasticity theory with linear defects), the basic field equation that describes the $U(1)$ gauge field in a curved background is given by

$$D_\mu F^{\mu k} = 0, \quad F^{\mu k} = \partial^\mu W^k - \partial^k W^\mu, \tag{26}$$

where covariant derivative $D_\mu := \partial_\mu + \Gamma_\mu$ involves the Levi-Civita (torsion-free, metric compatible) connection

$$\Gamma^k_{\mu\lambda} := (\Gamma_\mu)^k_\lambda = \frac{1}{2}g^{kl}\left(\frac{\partial g_{l\lambda}}{\partial x^\mu} + \frac{\partial g_{\mu l}}{\partial x^\lambda} - \frac{\partial g_{\mu\lambda}}{\partial x^l}\right), \tag{27}$$

with $g_{\mu k}$ being the metric tensor on a Riemannian surface Σ with local coordinates $x^\mu = (x^1, x^2)$. For a single disclination on an arbitrary elastic surface, a singular solution to (26) is found to be [37]

$$W^k = -\nu\varepsilon^{k\lambda}D_\lambda G(x, y), \tag{28}$$

where

$$D_\mu D^\mu G(x^1, x^2) = 2\pi\delta^2(x^1, x^2)/\sqrt{g}, \tag{29}$$

with $\varepsilon_{\mu k} = \sqrt{g}\epsilon_{\mu k}$ being the fully antisymmetric tensor on Σ, $\epsilon_{12} = -\epsilon_{21} = 1$. It should be mentioned that eqs. (26-29) self-consistently describe a defect located on an arbitrary surface [37].

The general analytical solution to (22) is known only for chosen geometries. One of them is the cone [33, 34]. For the sphere which are of interest here, there were used some approximations. In particular, asymptotic solutions at small r (which allow us to study electronic states near the disclination line) were considered in [23]. For this reason, also the numerical calculations were performed in [38]. The results of both analytical and numerical studies of spherical geometry will be presented in the next section.

4. Spherical Molecules

According to the Euler's theorem, the fullerene molecule consists of twelve disclinations. Generally, it is difficult to take into account properly all the disclinations. There are two ways to simplify the problem. The first one is considering a situation near a single defect. This approximation we briefly sketch in this chapter. The second possibility is replacing the fields of twelve disclinations by the fields of the magnetic monopole. This approximation will be described in the next section.

4.1. The Model

We start the section with the computation of disclination fields. It will be used as a gauge fields in Dirac equation on the sphere. To describe a sphere, we employ the polar projective coordinates $x^1 = r, x^2 = \varphi$; $0 \leq r < \infty, 0 \leq \varphi < 2\pi$ with R being the radius of the sphere. In these coordinates, the metric tensor becomes

$$g_{rr} = 4R^4/(R^2+r^2)^2,\ g_{\varphi\varphi} = 4R^4r^2/(R^2+r^2)^2,\ g_{r\varphi} = g_{\varphi r} = 0, \tag{30}$$

so that

$$\sqrt{g} := \sqrt{\det||g_{\mu\nu}||} = 4R^4r/(R^2+r^2)^2.$$

Nonvanishing connection coefficients (27) take the form

$$\Gamma^r_{rr} = -\frac{2r}{R^2+r^2}, \quad \Gamma^r_{\varphi\varphi} = -r\frac{R^2-r^2}{R^2+r^2}, \quad \Gamma^\varphi_{r\varphi} = \frac{1}{r}\frac{R^2-r^2}{R^2+r^2},$$

and the general representation for the zweibeins is found to be

$$e^1_{\ r} = e^2_{\ \varphi} = 2R^2 \cos\varphi/(R^2 + r^2),\ e^1_{\ \varphi} = -e^2_{\ r} = -2R^2 \sin\varphi/(R^2 + r^2),$$

which in view of (21) gives

$$\omega_r^{12} = \omega_r^{21} = 0, \quad \omega_\varphi^{12} = -\omega_\varphi^{21} = 2r^2/(R^2 + r^2) =: 2\omega. \tag{31}$$

From the solutions (28) and (29) one can easily found

$$G = \log r; \quad W_r = 0, \quad W_\varphi = \nu, \quad r \neq 0.$$

It describes locally a topological vortex on the Euclidean plane, which confirms the observation that disclinations can be viewed as vortices in elastic media. The elastic flux is characterized by the Frank vector $\vec{\omega}$, $|\vec{\omega}| = 2\pi\nu$ with ν being the Frank index and the elastic flow through a surface on the sphere is given by the circular integral

$$\frac{1}{2\pi} \oint \vec{W} d\vec{r} = \nu.$$

There are no restrictions on the value of the winding number ν apart from $\nu > -1$ for topological reasons. This means that the elastic flux is 'classical' in its origin; i.e., there is no quantization (in contrast to the magnetic vortex). If we take into account the symmetry group of the underlying crystal lattice, the possible values of ν become 'quantized' in accordance with the group structure (e.g., $\nu = 1/6, 1/3, 1/2, ...$ for the hexagonal lattice). We define $\nu = n/6$, where n is the number of pentagons at the apices. It is interesting to note that in some physically interesting applications vortices with the fractional winding number have already been considered. A detailed theory of magnetic vortices on a sphere has been presented in [39].

The Dirac matrices in $2D$ space,can be chosen as the Pauli matrices: $\gamma^1 = -\sigma^2, \gamma^2 = \sigma^1$ and (23) reduces to

$$\Omega_\varphi = i\omega\sigma^3. \tag{32}$$

From the assumptions above the Dirac operator $\hat{\mathcal{D}} := i\gamma^\alpha e_\alpha^{\ \mu}(\nabla_\mu + iW_\mu)$ on the two-sphere becomes

$$\hat{\mathcal{D}} = \hat{\mathcal{D}}^\dagger = \frac{r^2 + R^2}{2R^2} \begin{bmatrix} 0 & e^{-i\varphi}(-\partial_r + \frac{i\partial_\varphi + \nu}{r} + \frac{\omega}{r}) \\ e^{i\varphi}(\partial_r + \frac{i\partial_\varphi + \nu}{r} - \frac{\omega}{r}) & 0 \end{bmatrix}. \tag{33}$$

The generator of the local Lorentz transformations $\Lambda_\alpha^\beta \in SO(2)$ takes the form $-i\partial_\varphi$, whereas the generator of the Dirac spinor transformations $\rho(\Lambda)$ is

$$\Sigma_{12} = \frac{i}{4}[\gamma_1, \gamma_2] = \frac{1}{2}\sigma^3.$$

For massless fermions σ^3 serves as a conjugation matrix, and the energy eigenmodes are symmetric with respect to $E = 0$ $(\sigma^3\psi_E = \psi_{-E})$. The total angular momentum of the $2D$ Dirac system is therefore given by

$$L_z = -i\partial_\varphi + \frac{1}{2}\sigma^3,$$

which commutes with the operator (33). Accordingly, the eigenfunctions are classified with respect to the eigenvalues of $J_z = j + 1/2,\ j = 0, \pm 1, \pm 2, ...$, and are to be taken in the form

$$\psi = \begin{pmatrix} u(r)e^{i\varphi j} \\ v(r)e^{i\varphi(j+1)} \end{pmatrix}. \tag{34}$$

From the (33) is follows that the spin connection term can be taken into account by redefining the wave function as

$$\psi = \tilde{\psi}\sqrt{R^2 + r^2}, \tag{35}$$

which reduces eigenvalue problem (22) to

$$\begin{aligned} \partial_r \tilde{u} - \frac{(j-\nu)}{r}\tilde{u} &= \tilde{E}\tilde{v}, \\ -\partial_r \tilde{v} - \frac{(j+1-\nu)}{r}\tilde{v} &= \tilde{E}\tilde{u}, \end{aligned} \tag{36}$$

where $\tilde{E} = 2R^2E/(R^2 + r^2)$.

4.2. Extended Electron States

We will consider an approximate solution to (36). Because we are interested in electronic states near the disclination line, we can restrict our consideration to the case of small r. In this case, a solution to (36) (with (35) taken into account) is found to be

$$\begin{pmatrix} u \\ v \end{pmatrix} = A \begin{pmatrix} J_\eta(2Er) \\ \pm J_{\bar{\eta}}(2Er) \end{pmatrix} \tag{37}$$

where $\eta = \pm(j - \nu)$, $\bar{\eta} = \pm(j - \nu + 1)$, and A is a normalization factor. Therefore, there are two independent solutions with $\eta(\bar{\eta}) > 0$ and $\eta(\bar{\eta}) < 0$. Notice that respective signs $\pm$ in (37) correspond to states with $E > 0$ and $E < 0$. As already noted, σ^3 serves as the conjugation matrix for massless fermions, and the energy eigenmodes are symmetric with respect to $E = 0$. One can therefore consider either case, for instance, $E > 0$.

The important restrictions come from the normalization condition

$$\int (|u|^2 + |v|^2)\sqrt{g}dx^1dx^2 = 1. \tag{38}$$

From (37), it follows that $A^2 \sim E$. On the other hand, the integrand in (38) must be nonsingular at small Er. This imposes a restriction on possible values of j. Namely, for $\eta, \bar{\eta} > 0$ one obtains $j - \nu > -1/2$, and for $\eta, \bar{\eta} < 0$ one has $j - \nu < -1/2$. As is seen, possible values of j do not overlap at any ν.

In the vicinity of a pentagon, the electron wave function reads

$$\begin{pmatrix} u \\ v \end{pmatrix} \sim \begin{pmatrix} E^{1/2+\eta}r^{\eta} \\ E^{1/2+\bar{\eta}}r^{\bar{\eta}} \end{pmatrix}. \tag{39}$$

In particular, in the leading order, one obtains $\Psi \sim \sqrt{E}$, $\Psi \sim E^{1/3}r^{-1/6}$, and $\Psi \sim E^{1/6}r^{-1/3}$ for $\nu = 0,\ 1/6,\ 1/3$, respectively. Because the local density of states diverges as

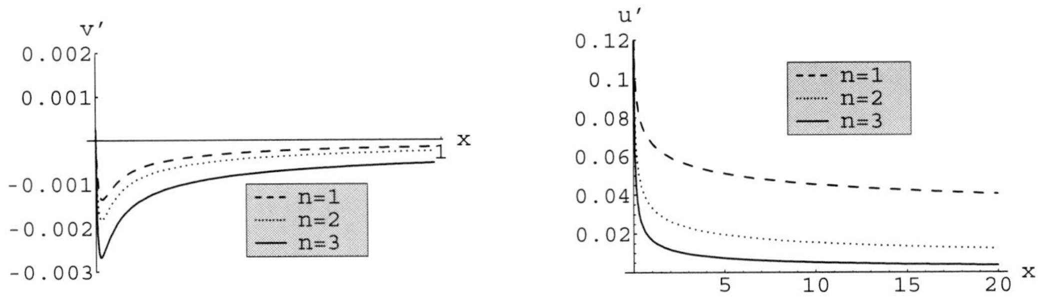

Figure 4. The solutions $v'(x), u'(x)$ for different n.

$r \to 0$, it is more appropriate to consider the total density of states on a patch $0 < r \leq \delta$ for small δ, rather than the local quantities. For this, the electron density should be integrated over a small disk $|r| < \delta$ (recall that r, φ are stereographically projected coordinates on the sphere). The result is

$$D(E,\delta) \propto \begin{cases} (E\delta)\delta, & \nu = 0; \\ (E\delta)^{2/3}\delta, & \nu = 1/6, 5/6; \\ (E\delta)^{1/3}\delta, & \nu = 1/3, 2/3; \\ \delta, & \nu = 1/2; \end{cases} \tag{40}$$

For the defect free case ($\nu = 0$) we obtain the well-known behavior of the total density of electronic states (DOS) in the δ disk given by $D(E,\delta) \sim E\delta^2$ (in accordance with the previous analysis [10]). For $\nu = 1/6, 1/3, 2/3, 5/6$, the low-energy total DOS has a cusp which drops to zero at the Fermi energy. Most intriguing is the case where $\nu = 1/2$ and a region of a nonzero DOS across the Fermi level is formed. This implies local metallization of graphite in the presence of $180°$ disclination. In the fullerene molecule, however, there are twelve $60°$ disclinations, and therefore, the case $\nu = 1/6$ is actually realized.

4.3. Numerical Results

The numerical calculations for the different type of carbon nanostructures as well as for the case of sphere were presented in paper [38]. As a starting point, the analytical asymptotic solutions found in the previous section are considered. The initial value of the parameter r is defined as $r = 10^{-4}$. It is worth noting that the choice of the boundary conditions does not influence the behavior of the calculated wave functions and only the starting point depends on it. A dimensionless substitution $x = Er$ is used. The normalized numerical solutions to (36) are given in Fig. 4. The parameters are chosen to be $E = 0.01$ and $R = 1$.

Notice that here we present the solutions for dotted values $v'(= \tilde{v})$ and $u'(= \tilde{u})$. The local DOS is shown schematically in Figs. 5, 6 for different n. The Fig. 6 describes also the dependence of the local DOS on a position of the maximum value of integrand in (38)(which actually characterizes the numerically calculated localization point of an electron).

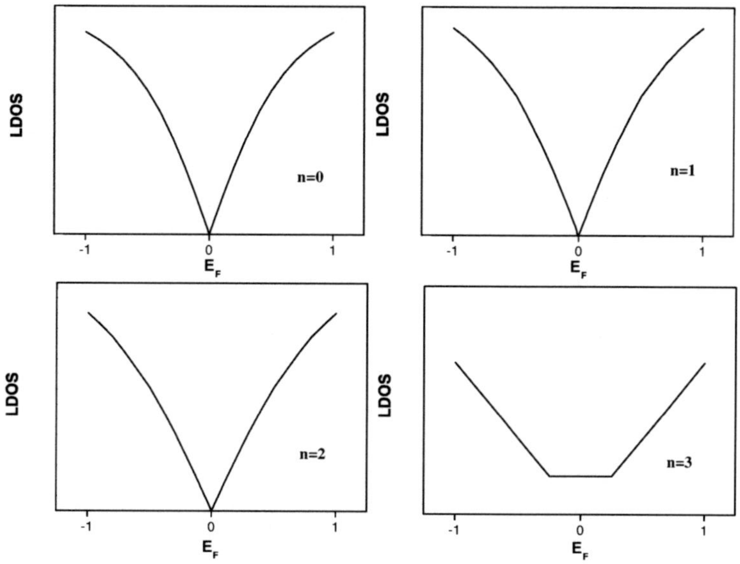

Figure 5. Schematic densities of states near the Fermi energy in the case of sphere.

Here and below $\delta = 0.1$. Notice that in fact the choice of the value of δ does not influence the characteristic behavior of LDOS. As is seen, the DOS has a cusp which drops to zero at the Fermi energy. The case $n = 3$ becomes distinguished. Let us emphasize once more that in the fullerene molecule there are twelve 60° disclinations, so that the case $n = 1$ is actually realized.

5. Spheroidal Geometry Approach to Fullerenes

Geometry, topological defects and the peculiarity of graphene lattice have a pronounced effect on the electronic structure of fullerene molecules. The most extensively studied C_{60} molecule is an example of a spherical fullerene nicknamed a "soccer ball" [40]. Moreover, the spherical fullerenes as C_{60} are stable towards fragmentation than the other bigger fullerenes [41]. The family of icosahedral spherical fullerenes is described by the formula $20(n^2 + nl + l^2)$ with integer n and l. Other fullerenes are either slightly (as C_{70}) or remarkably deformed and their general nickname is a "rugby ball". The electronic structure of C_{70} cluster of D_{5h} geometry has been studied in [42] by using the local-density approximation in the density-functional theory. It was clearly shown that the spheroidal geometry is of a decisive importance for observed peculiarities in electronic states of the C_{70}. What is important, the calculated energy levels were found to be in good agreement with photoemission experiments. Spheroidal fullerenes can be considered as a initially flat hexagonal network which has warped into closed monosurface by twelve disclinations. We are interested here in the slightly elliptically deformed big fullerenes like C_{240} and C_{540} molecules, see Figures 7, 8 [40].

It should be noted that the formulation of any continuum model begins from the description of a graphite sheet (graphene) (see, e.g., [10, 15, 11, 43, 12, 44]). The models which involve the influence of both geometry and topological defects on the electronic structure

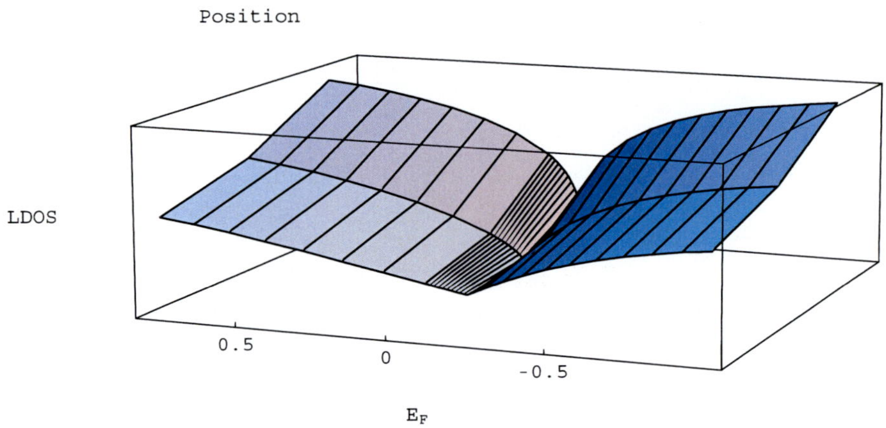

Figure 6. 3d schematic plotting of the DOS near the Fermi energy for $n = 0, 1, 2$ (going from the front side to the back side).

by boundary conditions were developed in Refs. [45, 46]. A different variant of the continuum model within the effective-mass description for fullerene near one pentagonal defect was suggested in previous section. There are attempts to describe electronic structure of the graphene compositions by the Dirac-Weyl equations on the curved surfaces where the structure of the graphite with pentagonal rings taken into account is imposed with the help of the so-called K spin contribution [14, 47]. In the case of spherical geometry, this is realized by introducing an effective field due to magnetic monopole placed at the center of a sphere [14]. In this section we use a similar approach. The difference is that we consider also the elastic contribution. Indeed, pentagonal rings being the disclination defects are the sources of additional strains in the hexagonal lattice. Moreover, the disclinations are topological defects and the elastic flow due to a disclination is determined by the topological Frank index. For this reason, this contribution exists even within the so-called "inextensional" limit which is rather commonly used in description of graphene compositions (see, e.g., [48]).

Recently, in the framework of a continuum approach the exact analytical solution for the low energy electronic states in icosahedral spherical fullerenes has been found [49]. The case of elliptically deformed fullerenes was studied in [50] where some numerical results were presented. In this section, we suggest a similar to [49] model with the Dirac monopole instead of 't Hooft-Polyakov monopole for describing elastic fields. We consider a slightly elliptically deformed sphere with the eccentricity of the spheroid $e \ll 1$. In this case, by analogy with [51] we use the spherical representation for the eigenstates, with the slight asphericity considered as a perturbation. This allows us to find explicitly the low-lying electronic spectrum for spheroidal fullerenes.

The problem of Zeeman splitting and Landau quantization of electrons on a sphere was studied in Ref. [52]. To this end, the Schrödinger equation for a free electron on the surface of a sphere in a uniform magnetic field was formulated and solved.Our studies also cover slightly elliptically deformed fullerenes influent by the weak uniform external magnetic field pointed in the different directions.

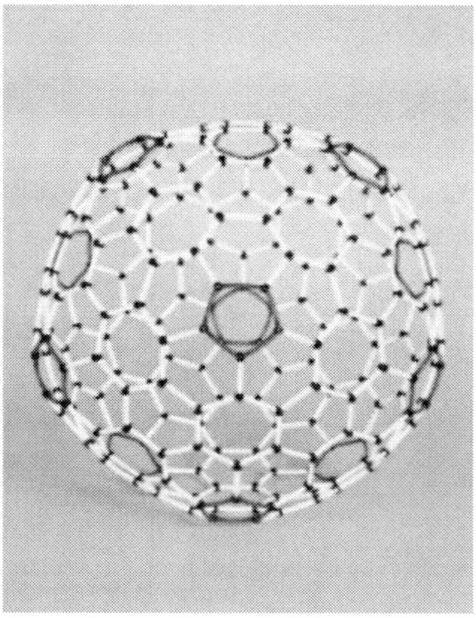

Figure 7. Picture of molecular model of giant spheroidal fullerene C_{240}.

5.1. The Model

Let us start with introducing spheroidal coordinates and writing down the Dirac operator for free massless fermions on the Riemannian spheroid S^2. The Euler's theorem for graphene requires the presence of twelve pentagons to get the closed molecule. In a spirit of a continuum description we will extend the Dirac operator by introducing the Dirac monopole field with charge G inside the spheroid to simulate the elastic vortices due to twelve pentagonal defects. The K spin flux which describes the exchange of two different Dirac spinors in the presence of a conical singularity will also be included in a form of 't Hooft-Polyakov monopole with charge g.

The Dirac equation on a surface Σ in the presence of the abelian magnetic monopole field W_μ and the external magnetic field A_μ is written as [14, 47, 49]

$$i\gamma^\alpha e_\alpha^{\ \mu}[\nabla_\mu - iW_\mu - ia_\mu - iA_\mu]\psi = E\psi. \tag{41}$$

We impose the K spin flux by the gauge field. In Cartesian coordinates

$$a_x = \frac{2gyz}{r(r^2 - z^2)}\tau_2, \; a_y = \frac{-2gxz}{r(r^2 - z^2)}\tau_2, \; a_z = 0. \tag{42}$$

Disclination field we choose in the form

$$W_x = \frac{-Gy}{r(r+z)}, \; W_y = \frac{Gx}{r(r+z)}, \; W_z = 0, \tag{43}$$

in the region R_N (northern hemisphere) and

$$W_x = \frac{Gy}{r(r-z)}, \; W_y = \frac{-Gx}{r(r-z)}, \; W_z = 0, \tag{44}$$

in the region R_S (southern hemisphere). Here $r = \sqrt{x^2 + y^2 + z^2}$.

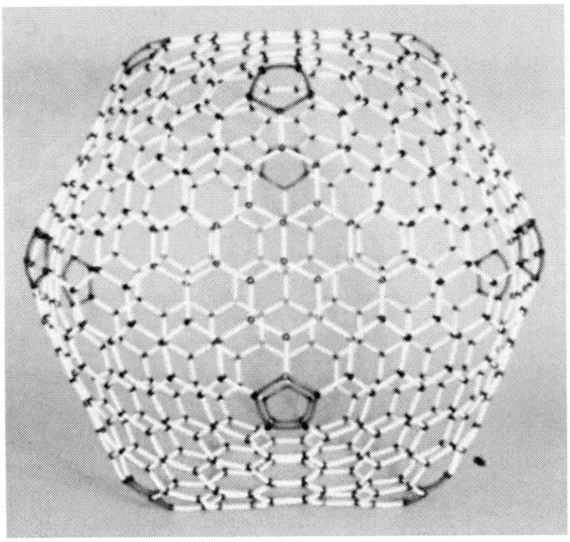

Figure 8. Picture of molecular model of giant spheroidal fullerene C_{540}.

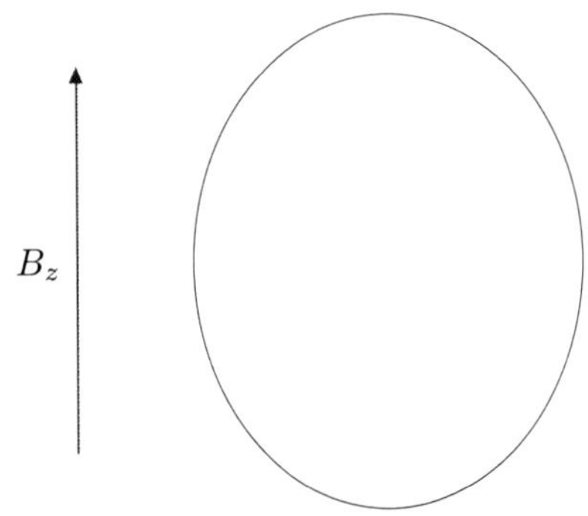

Figure 9. The schematic picture of the spheroidal fullerene in a weak uniform magnetic field pointed in the z direction.

5.2. Electronic States Near the Fermi Energy in Weak Magnetic Field Pointed in the z Direction

Firstly we will study the structure of electron levels near the Fermi energy influent by the weak magnetic field pointed in the z direction (see Fig. 9) so that $\vec{A} = B\left(y, -x, 0\right)/2$. The elliptically deformed sphere or a spheroid

$$\frac{x^2}{a^2} + \frac{y^2}{a^2} + \frac{z^2}{c^2} = 1, \tag{45}$$

may be parameterized by two spherical angles $q^1 = \phi$, $q^2 = \theta$ that are related to the Cartesian coordinates x, y, z as follows

$$x = a\ \sin\theta\cos\phi; \quad y = a\ \sin\theta\sin\phi; \quad z = c\ \cos\theta. \tag{46}$$

The nonzero components of the metric tensor for spheroid are

$$g_{\phi\phi} = a^2 \sin^2\theta; \quad g_{\theta\theta} = a^2\cos^2\theta + c^2\sin^2\theta, \tag{47}$$

where $a, c \geq 0,\ 0 \leq \theta \leq \pi,\ 0 \leq \phi < 2\pi$. Accordingly, orthonormal frame on spheroid is

$$e_1 = \frac{1}{a\,\sin\theta}\,\partial_\phi; \quad e_2 = \frac{1}{\sqrt{a^2\cos^2\theta + c^2\sin^2\theta}}\,\partial_\theta\,, \tag{48}$$

and dual frame reads

$$e^1 = a\,\sin\theta\, d\phi; \quad e^2 = \sqrt{a^2\cos^2\theta + c^2\sin^2\theta}\, d\theta\,. \tag{49}$$

A general representation for zweibeins is found to be

$$e^1{}_\phi = a\,\sin\theta;\ e^1{}_\theta = 0;\ e^2{}_\phi = 0;\ e^2{}_\theta = \sqrt{a^2\cos^2\theta + c^2\sin^2\theta}\,. \tag{50}$$

Notice that $e^\mu{}_\alpha$ is the inverse of $e^\alpha{}_\mu$. The Riemannian connection with respect to the orthonormal frame is written as [35, 53]

$$de^1 = -\omega^1{}_2 \wedge e^2 = \frac{a\cos\theta}{\sqrt{a^2\cos^2\theta + c^2\sin^2\theta}}\, d\phi\wedge e^2, \tag{51}$$

$$de^2 = -\omega^2{}_1 \wedge e^1 = 0. \tag{52}$$

Here $\wedge$ denotes the exterior product and d is the exterior derivative. From Eqs. 51 and 52 we get the Riemannian connection in the form

$$\omega^1_{\phi 2} = -\omega^2_{\phi 1} = \frac{a\cos\theta}{\sqrt{a^2\cos^2\theta + c^2\sin^2\theta}}; \quad \omega^1_{\theta 2} = \omega^2_{\theta 1} = 0. \tag{53}$$

We assume that the eccentricity of the spheroid $e = \sqrt{|1 - (c/a)^2|}$ is small enough. In this case, one can write down $c = a + \delta a$ where δ ($|\ \delta\ | \ll 1$) is a small dimensionless parameter characterizing the spheroidal deformation from the sphere. So, we can follow the perturbation scheme using δ as the perturbation parameter. Within the framework of the perturbation scheme the spin connection coefficients are written as

$$\omega^1_{\phi 2} = -\omega^2_{\phi 1} \approx \cos\theta(1 - \delta\sin^2\theta), \tag{54}$$

where the terms to first order in δ are taken into account. The inclusion of the K-spin connection can be performed by considering two Dirac spinors as the two components of an $SU(2)$ color doublet $\psi = (\psi_\uparrow \psi_\downarrow)^T$ (see Ref. [14] for details). In this case, the interaction with the color magnetic fields in a form of nonabelian magnetic flux ('t Hooft-Polyakov monopole) is responsible for the exchange of the two Dirac spinors which in turn corresponds to the interchange of Fermi points. As was shown in Ref. [16], the Dirac equation for ψ can be reduced to two decoupled equations for $\psi_\uparrow$ and $\psi_\downarrow$, which include an Abelian monopole of opposite charge, $g = \pm 3/2$.

The elastic flow through a surface due to a disclination has a vortex-like structure [13] and can be described by Abelian gauge field W_μ. Similarly to Ref. [14], we replace the fields of twelve disclinations by the effective field of the magnetic monopole of charge G located at the center of the spheroid. The circulation of this field is determined by the Frank index, which is the topological characteristic of a disclination. Trying to avoid an extension of the group we will consider the case of the Dirac monopole. The another possibility is to introduce the non-Abelian 't Hooft-Polyakov monopole (see Ref. [49]). As is known, the vector potential W_μ around the Dirac monopole have singularities. To escape introducing singularities in the coordinate system let us divide the spheroid (similarly as it was done for the sphere in [54]) into two regions, R_N and R_S, and define a vector potential W_μ^N in R_N and a vector potential W_μ^S in R_S. Notice that below W_μ includes both gauge fields W_μ and a_μ defined in Eq. 41.

One has

$$R_N: \ 0 \leq \theta < \frac{\pi}{2} + \Delta; \ 0 \leq \phi < 2\pi,$$

$$R_S: \ \frac{\pi}{2} - \Delta < \theta \leq \pi; \ 0 \leq \phi < 2\pi,$$

$$R_{NS}: \ \frac{\pi}{2} - \Delta < \theta < \frac{\pi}{2} + \Delta; \ 0 \leq \phi < 2\pi, \ (overlap) \tag{55}$$

where Δ is chosen from the interval $0 < \Delta \leq \pi/2$. In spheroidal coordinates (47) the only nonzero components of W_μ are found to be

$$W_\phi^N \approx g\cos\theta(1+\delta\sin^2\theta) + G(1-\cos\theta) - \delta G\sin^2\theta\cos\theta, \tag{56}$$

$$W_\phi^S \approx g\cos\theta(1+\delta\sin^2\theta) - G(1+\cos\theta) - \delta G\sin^2\theta\cos\theta, \tag{57}$$

where the terms of an order of δ^2 and higher are dropped. In the overlapping region the potentials W_ϕ^N and W_ϕ^S are connected by the gauge transformation

$$W_\phi^N = W_\phi^S + iS_{NS}\,\partial_\phi S_{NS}^{-1}, \tag{58}$$

where

$$S_{NS} = e^{2iG\phi}$$

is the phase factor. The wavefunctions in the overlap R_{NS} are connected by

$$\psi_N = S_{NS}\psi_S, \tag{59}$$

where ψ_N and ψ_S are the spinors in R_N and R_S, respectively. Since S_{NS} must be single-valued [55], $2G$ takes integer values. Notice that the total flux due to Dirac monopole in Egs. 56 and 57 is equal to $4\pi G$. What is important, there is no contribution from the terms with δ. In our case, the total flux describes a sum of elastic fluxes due to twelve disclinations, so that the total flux (the modulus of the total Frank vector) is equal to $12 \times \pi/3 = 4\pi$. Therefore, one obtains $G = 1$.

Let us consider the region R_N. In spheroidal coordinates, the only nonzero component of W_μ in region R_N is found to be

$$W_\phi \approx g\cos\theta(1+\delta\sin^2\theta) + G(1-\cos\theta) - \delta G\sin^2\theta\cos\theta, \tag{60}$$

and the external magnetic field pointed in the z-direction reads

$$A_\phi = -\frac{1}{2}Ba^2\sin^2\theta. \quad (61)$$

By using the substitution (93) we obtain the Dirac equation for functions u_j and v_j in the form

$$\left(-i\sigma_1\frac{1}{a}(\partial_\theta + \frac{\cot\theta}{2}) + \frac{\sigma_2}{a\sin\theta}\left(j - m\cos\theta + \frac{1}{2}Ba^2\sin^2\theta\right) + \delta\hat{\mathcal{D}}_1\right)\begin{pmatrix} u_j(\theta) \\ v_j(\theta)\end{pmatrix}$$

$$= E\begin{pmatrix} u_j(\theta) \\ v_j(\theta)\end{pmatrix}, \quad (62)$$

where

$$\hat{\mathcal{D}}_1 = -\frac{\gamma_1}{a}\sin\theta\,(j - 2m\cos\theta) - \gamma_1\frac{Ba}{2}\sin^3\theta. \quad (63)$$

A convenient way to study the eigenvalue problem is to square Eq. (62). For this purpose, let us write the Dirac operator in Eq. (62) as $\hat{\mathcal{D}} = \hat{\mathcal{D}}_0 + \delta\hat{\mathcal{D}}_1$. One can easily obtain that

$$\begin{aligned}\hat{\mathcal{D}}_0^2 &= -\frac{1}{a^2}\left(\partial_\theta^2 + \frac{\cos\theta}{\sin\theta}\partial_\theta - \frac{1}{4} - \frac{1}{4\sin^2\theta}\right) + \frac{(j - m\cos\theta)^2}{a^2\sin^2\theta} + Bj + \frac{B}{2}(\sigma_3 - 2m)\cos\theta \\ &+\sigma_3\frac{m - j\cos\theta}{a^2\sin^2\theta} + \frac{B^2a^2\sin^2\theta}{4}. \end{aligned} \quad (64)$$

Notice that to first order in δ the square of the Dirac operator is written as $\hat{\mathcal{D}}^2 = (\hat{\mathcal{D}}_0^2 + \delta\hat{\Gamma})$, where $\hat{\Gamma} = (\hat{\mathcal{D}}_0\hat{\mathcal{D}}_1 + \hat{\mathcal{D}}_1\hat{\mathcal{D}}_0)$. In an explicit form

$$a^2\hat{\Gamma} = 2j^2 + jx(\sigma_3 - 6m) + 4m(m - \sigma_3)x^2 + 2m\sigma_3 + 3Ba^2(1 - x^2)x(\sigma_3/2 - m)$$

$$+2Bja^2(1 - x^2) + B^2a^4(1 - x^2)^2/2, \quad (65)$$

where the appropriate substitution $x = \cos\theta$ is used. The equation $\hat{\mathcal{D}}^2\psi = E^2\psi$ takes the form

$$[\partial_x(1 - x^2)\partial_x - \frac{(j - mx)^2 - j\sigma_3 x + \frac{1}{4} + \sigma_3 m}{1 - x^2} - a^2BV(x) - \delta a^2\hat{\Gamma}]\begin{pmatrix} u_j(x) \\ v_j(x)\end{pmatrix}$$

$$= -(\lambda^2 - \frac{1}{4})\begin{pmatrix} u_j(x) \\ v_j(x)\end{pmatrix}, \quad (66)$$

where $\lambda = aE, V(x) = j + (\sigma_3 - 2m)x/2$. Since we consider the case of a weak magnetic field the terms with B^2 and δB in Eq. (66) can be neglected. As a result, the energy spectrum for spheroidal fullerenes is found to be

$$(\lambda_{jn}^\delta)^2 = (n + |j| + 1/2)^2 - m^2 + Ba^2(j + A_{jn}) + \delta(2j^2 + A^1 + A^2) \quad (67)$$

where

$$A_{jn} = -\frac{j(m^2 - 1/4)}{p(p+1)},$$

and

$$A^1 = -\frac{j^2(6m^2 - 1/2)}{p(p+1)} + \frac{j^2m^2(4m^2-3)}{p^2(p+1)^2}, \tag{68}$$

$$A^2 = \frac{(4m(m-1)F_n(2\alpha,2\beta) - X)T\, J_n(2\alpha,2\beta) + (4m(m+1)F_n(2\mu,2\nu) + X)\, J_n(2\mu,2\nu)}{T\, J_n(2\alpha,2\beta) + J_n(2\mu,2\nu)} \tag{69}$$

$$X = \frac{2j^2m}{p(p+1)} + \frac{j^2m}{p^2(p+1)^2} - 2m \tag{70}$$

$$F_n(2\alpha,2\beta) = \frac{(n+1)(n+2\alpha+1)(n+2\beta+1)(n+2\alpha+2\beta+1)}{(2p+1)(p+1)^2(2p+3)} + \frac{n(n+2\alpha)(n+2\beta)(n+2\alpha+2\beta)}{(2p-1)p^2(2p+1)} \tag{71}$$

$$J_n(2\alpha,2\beta) = \Gamma(n+2\alpha+1)\Gamma(n+2\beta+1),$$

Here $p = n + \beta + \alpha$ and

$$T = \frac{n + \mid j \mid - m + 1/2}{n + \mid j \mid + m + 1/2} \tag{72}$$

When $\delta = 0$, $B = 0$, one has the case of a sphere with magnetic monopole inside. In this case, the exact solution is known (see, e.g., [49])

$$\left[\partial_x(1-x^2)\partial_x - \frac{(j-mx)^2 - j\sigma_3 x + \frac{1}{4} + \sigma_3 m}{1-x^2}\right]\begin{pmatrix} u^0_{jn}(x) \\ v^0_{jn}(x) \end{pmatrix} = -(\lambda^2_{0n} - \frac{1}{4})\begin{pmatrix} u^0_{jn}(x) \\ v^0_{jn}(x) \end{pmatrix},$$

$$\begin{pmatrix} u^0_{jn} \\ v^0_{jn} \end{pmatrix} = \begin{pmatrix} C_u(1-x)^\alpha(1+x)^\beta P_n^{2\alpha,2\beta}(x) \\ C_v(1-x)^\mu(1+x)^\nu P_n^{2\mu,2\nu}(x) \end{pmatrix}, \tag{73}$$

with the energy spectrum

$$(\lambda^0_{jn})^2 = (n + |j| + 1/2)^2 - m^2. \tag{74}$$

Here

$$\alpha = \frac{1}{2}\left|j - m - \frac{1}{2}\right|, \beta = \frac{1}{2}\left|j + m + \frac{1}{2}\right|,$$
$$\mu = \frac{1}{2}\left|j - m + \frac{1}{2}\right|, \nu = \frac{1}{2}\left|j + m - \frac{1}{2}\right|, \tag{75}$$

$P_n^{2\alpha,2\beta}(x)$ and $P_n^{2\mu,2\nu}(x)$ are Jacobi polynomials of n-th order, and C_u and C_v are the normalization factors. These states are degenerate. For example, the degeneracy of the zero mode is equal to six. Therefore, we have to use the perturbation scheme for the degenerate energy levels [56].

Finally, in the linear in δ approximation, the low energy electronic spectrum of spheroidal fullerenes takes the form

$$E_{jn} = E^0_{jn} + E^{\delta B_z}_{jn} \tag{76}$$

with

$$E^0_{jn} = \pm\sqrt{(2\xi + n)(2\eta + n)},$$
$$E^{\delta B_z}_{jn} = \frac{Ba^2(j + A_{jn}) + \delta(2j^2 + A^1 + A^2)}{2E^0_{jn}}, \qquad (77)$$

where $\xi = \mu$ (ν) and $\eta = \beta$ (α) for $j > 0$ $(j < 0)$, respectively. The parameters α, β, μ, ν are defined above.

Here we came back to the energy variable $E = \lambda/a$ (in units of $\hbar V_F/a$ where V_F is the Fermi velocity). Notice that the non-diagonal matrix element of perturbation $\langle e^{j\phi}|\langle\psi_{jn}|\Gamma|\psi_{-jn}\rangle|e^{-j\phi}\rangle$ turns out to be zero. Therefore, the states with opposite j do not mix and j remains a good quantum number as would be expected. As is seen from Eq. (76), for $\delta = 0$ both eigenstates and eingenvalues are the same as for a sphere (cf. Ref. [49]). At the same time, Eqs. (76) and (77) show that the spheroidal deformation gives rise to an appearance of fine structure in the energy spectrum. The difference in energy between sublevels is found to be linear in δ which resembles the Zeeman effect where the splitting energy is linear in magnetic field. In addition we found a further splitting of the states with opposite j.

Table 1. The structure of the first and higher energy levels for YO-C_{240} fullerene in uniform magnetic field. The hopping integral and other parameters are taken to be $t = 2.5\ eV$ and $V_F = 3t\bar{b}/2\hbar$, $\bar{b} = 1.45$Å, $\overline{R} = 7.03$Å, $SD = 0.17$Å, $\delta = 0.024$. $\bar{b}$ is average bond length, $\overline{R}$ ($\overline{R} = a$) is average radius, SD is standard deviation from a perfect sphere (see Refs. [57, 58]), so that $\delta = SD/\overline{R}$.

$YO - C_{240}$	j	$E^0_{jn}(eV)$	$E^\delta_{jn}(meV)(B = 0)$	$E^{\delta B_z}_{jn}(meV)(Ba^2 = 0.1)$
$n = 0, m = 1/2$	1	1.094	10.5	37.5
	-1	1.094	10.5	-16.5
$n = 1, m = 1/2$	1	1.89	8.8	24.8
	-1	1.89	8.8	-7.2
$n = 0, m = 1/2$	2	1.89	28.4	60.4
	-2	1.89	28.4	-3.6
$n = 0, m = -5/2$	3	1.89	3	27
	-3	1.89	3	-21

As an example, Table 1 shows all contributions (in compliance with Eq. (76)) to the first and second energy levels for YO-C_{240} (YO means a structure given in [57, 58]). Schematically, the structure of the first energy level is shown in Fig. 10. As is clearly seen, the first (double degenerate) level becomes shifted due to spheroidal deformation and splitted due to nontrivial topology. The uniform magnetic field provides the well-known Zeeman splitting. The difference between topological and Zeeman splitting is clearly seen. In the second case, the splitted levels are shifted in opposite directions while for topological splitting the shift is always positive. The more delicate is the structure of the second level which is schematically presented in Fig. 11. In this case, the initial (for $\delta = 0$) degeneracy of

E^0_{jn} is equal to six. The spheroidal deformation provokes an appearance of three shifted double degenerate levels (fine structure) which, in turn, are splitted due to the presence of topological defects. The magnetic field is responsible to Zeeman splitting.

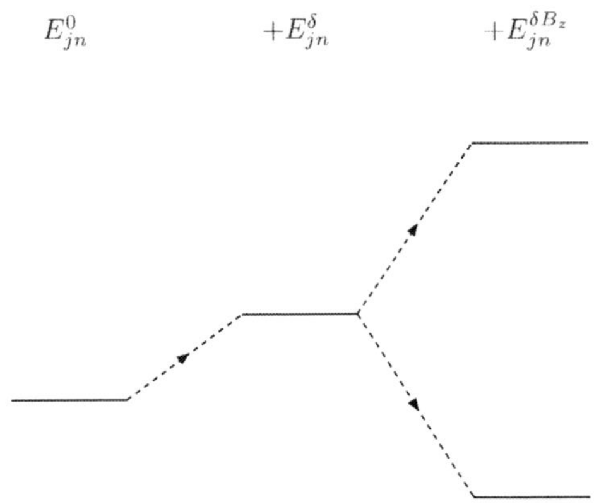

Figure 10. The schematic picture of the first positive electronic level E^{δ}_{jn} for spheroidal fullerenes in a weak uniform magnetic field pointed in the z direction.

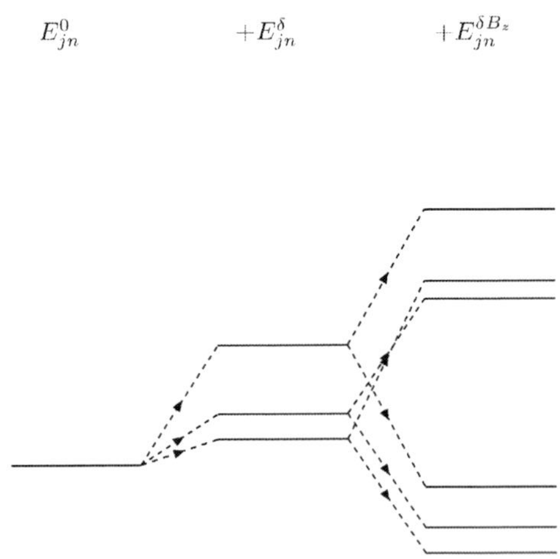

Figure 11. The schematic picture of the second positive electronic level E^{δ}_{jn} of the spheroidal fullerenes in a weak uniform magnetic field pointed in the z direction.

We admit that the obtained values of the splitting energies can deviate from the estimations within some more precise microscopic lattice models and density-functional methods (see, e.g.[59]). Therefore, we focused mostly on the very existence of physically interesting effects. For this reason, the values in both Table and the schematic pictures are presented with accuracy at about one percent of E^0_{jn}.

5.3. Electronic States Near the Fermi Energy in Weak Magnetic Field Pointed in the x Direction

In the second case we introduce that the uniform external magnetic field B is chosen to be pointed in the x direction (Fig. 12) so that $\vec{A} = B\,(0, -z, y)\,/2$.

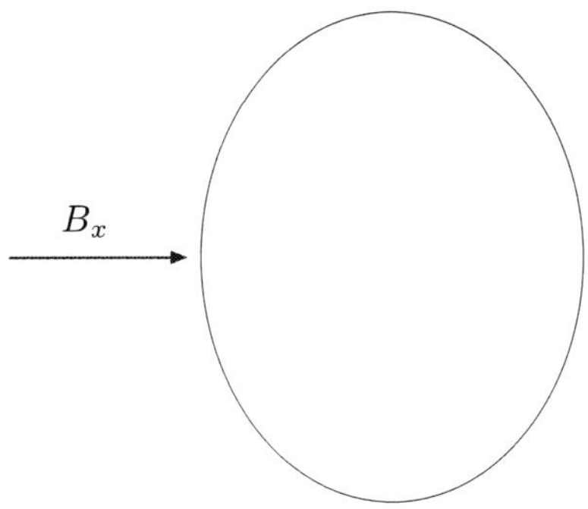

Figure 12. The schematic picture of the spheroidal fullerene in a weak uniform magnetic field pointed in the x direction.

The nonzero component of W_μ in region R_N are the same as in previous case but the external magnetic field are changed

$$A_\phi = -\frac{1}{2}Bac\sin\theta\cos\theta\cos\phi, \tag{78}$$

$$A_0 = -\frac{1}{2}Bac\sin\phi, \tag{79}$$

where $c = a + \delta a$ and only the terms to the first order in δ are taken into account. Following the calculations above with neglecting the terms B^2 and δB we obtain the Dirac equation for functions u_j and v_j in the form

$$\left(-i\sigma_1\frac{1}{a}(\partial_\theta + \frac{\cot\theta}{2} + i\frac{1}{2}Ba^2\sin\phi) + \frac{\sigma_2}{a\sin\theta}\left(j - m\cos\theta + \frac{1}{2}Ba^2\sin\theta\cos\theta\cos\phi\right) + \delta\hat{\mathcal{D}}_1\right)\times$$

$$\times\begin{pmatrix} u_j(\theta) \\ v_j(\theta) \end{pmatrix} = E\begin{pmatrix} u_j(\theta) \\ v_j(\theta) \end{pmatrix}, \tag{80}$$

where

$$\hat{\mathcal{D}}_1 = -\frac{\gamma_1}{a}\sin\theta\,(j - 2m\cos\theta) - \gamma_1\frac{Ba}{2}\sin^2\theta\cos\theta\cos\phi. \tag{81}$$

We find the square of nonperturbative part of Dirac operator above in the form

$$\begin{aligned}\hat{\mathcal{D}}_0^2 &= -\frac{1}{a^2}\left(\partial_\theta^2 + \frac{\cos\theta}{\sin\theta}\partial_\theta - \frac{1}{4} - \frac{1}{4\sin^2\theta}\right) + \frac{(j - m\cos\theta)^2}{a^2\sin^2\theta} \\ &+\sigma_3\frac{m - j\cos\theta}{a^2\sin^2\theta} + BV(\theta,\phi),\end{aligned} \tag{82}$$

where the terms with B^2 and δB was neglecting and

$$V(\theta,\phi) = -i(\cos\phi\partial_\phi - \frac{\sin\phi}{2})\cot\theta - \sigma_3\frac{\cos\phi(1+\sin^2\theta)}{2\sin\theta} - i(\partial_\theta + \frac{1}{2}\cot\theta)\sin\phi$$
$$-\frac{m\cos^2\theta\cos\phi}{\sin\theta}. \tag{83}$$

We denote $V^m_{j,j'} = \langle\psi^m_{jn}|V(\theta,\phi)|\psi^m_{j'n}\rangle$, where ψ^m_{jn} are the eigenfunctions of $\hat{\mathcal{D}}_0$. In the case of first energy mode the terms $V^{1/2}_{-1,-1}$, $V^{1/2}_{1,1}$, $V^{1/2}_{1,-1}$, $V^{1/2}_{-1,1}$ are zero and therefore the weak magnetic field pointed in the x direction do not influent the first energy level. For the second energy mode only nondiagonal terms $V^{1/2}_{-1,-2}$, $V^{1/2}_{1,2}$ were find different from zero. The following numerical values of the nondiagonal terms $V^m_{j,j'}$ was found: $V^{1/2}_{-1,-2} = -2.1$, $V^{1/2}_{1,2} = 2.1$. We use the wave functions for calculations of nondiagonal terms in the case of second energy level in the form:

$$\psi^{1/2}_{0,2}(z,\phi) = \frac{e^{i2\phi}}{\sqrt{2\pi}}\sqrt{\frac{32}{15}}\begin{pmatrix} -i\sqrt{\frac{2}{3}}\sqrt{1-z^2}(1+z) \\ 1-z^2 \end{pmatrix},$$
$$\psi^{1/2}_{0,-2}(z,\phi) = \frac{e^{-i2\phi}}{\sqrt{2\pi}}\sqrt{\frac{32}{15}}\begin{pmatrix} i\sqrt{\frac{2}{3}}\sqrt{1-z^2}(1-z) \\ 1-z^2 \end{pmatrix}, \tag{84}$$

$$\psi^{1/2}_{1,1}(z,\phi) = \frac{e^{i\phi}}{\sqrt{2\pi}}\sqrt{\frac{3}{5}}\begin{pmatrix} -i\sqrt{\frac{2}{3}}(2z^2+z-1) \\ \frac{3}{2}z\sqrt{1-z^2} \end{pmatrix},$$
$$\psi^{1/2}_{1,-1}(z,\phi) = \frac{e^{-i\phi}}{\sqrt{2\pi}}\sqrt{\frac{3}{5}}\begin{pmatrix} i\sqrt{\frac{2}{3}}(-2z^2+z+1) \\ \frac{3}{2}z\sqrt{1-z^2} \end{pmatrix}. \tag{85}$$

The low energy electronic spectrum of spheroidal fullerenes in this case takes the form

$$E_{jn} = E^0_{jn} + E^{\delta B_x}, \tag{86}$$

where

$$E^{\delta B_x} = \frac{\delta\left(\hat{\Gamma}_{22}+\hat{\Gamma}_{11}\right) \pm \sqrt{\delta^2\left(\hat{\Gamma}_{22}-\hat{\Gamma}_{11}\right)^2 + 4|Ba^2V_{1,2}|^2}}{4E^0_{jn}}. \tag{87}$$

We change $1,2$ to $-1,-2$ in the expression above to get the another energy levels in the degenerate energy state which is considered. Where the $\hat{\Gamma}_{ii}$ are the diagonal matrix element of perturbation of Eq. 65 with $B = 0$. Table 2 shows all contributions to the first and second energy levels for YO-C_{240} fullerenes influent by the horizontal weak external magnetic field. Figures 13, 14 present first and second positive energy levels of spheroidal fullerenes in weak magnetic field pointed in the x direction in comparison with Figures 10, 11. As is seen, there is a marked difference between the behavior of the first and second energy levels in magnetic field. Indeed, in both cases the energy levels become shifted due to spheroidal deformation. However, the uniform magnetic field does not influence the first energy level. The splitting takes place only for the second level. We can conclude that there is a possibility to change the structure of the electronic levels in spheroidal fullerenes by altering the direction of magnetic field. It would be interesting to test this prediction in experiment. The values in both Table and the schematic pictures are presented the same way as in the previous case with accuracy at about one percent of E^0_{jn}.

Table 2. The structure of the first and second energy levels for YO-C_{240} fullerene in uniform magnetic field. The hopping integral and other parameters are taken to be $t = 2.5\ eV$ and $V_F = 3t\bar{b}/2\hbar$, $\bar{b} = 1.45$Å, $\overline{R} = 7.03$Å, $SD = 0.17$Å, $\delta = 0.024$. $\bar{b}$ is average bond length, $\overline{R}$ ($\overline{R} = a$) is average radius, SD is standard deviation from a perfect sphere (see Refs. [57, 58]), so that $\delta = SD/\overline{R}$.

$YO - C_{240}$	j	$E^0_{jn}(eV)$	$E^\delta_{jn}(meV)(B=0)$	$E^{\delta B_x}_{jn}(meV)(Ba^2 = 0.1)$
$n = 0, m = 1/2$	1	1.094	10.5	10.5
	-1	1.094	10.5	10.5
$n = 0, m = -5/2$	3	1.89	3	3
	-3	1.89	3	3
$n = 0, m = 1/2$	2	1.89	28.4	
				53/-16/
$n = 1$	1	1.89	8.8	
$n = 1, m = 1/2$	-1	1.89	8.8	
				53/-16/
$n = 0$	-2	1.89	28.4	

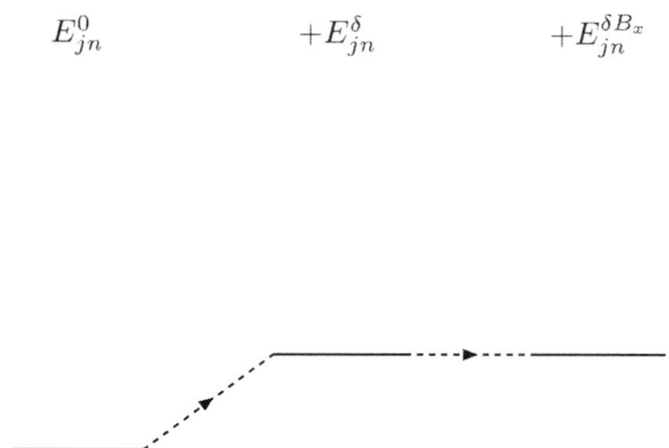

Figure 13. The schematic picture of the first positive electronic level E^δ_{jn} for spheroidal fullerenes in a weak uniform magnetic field pointed in the x direction.

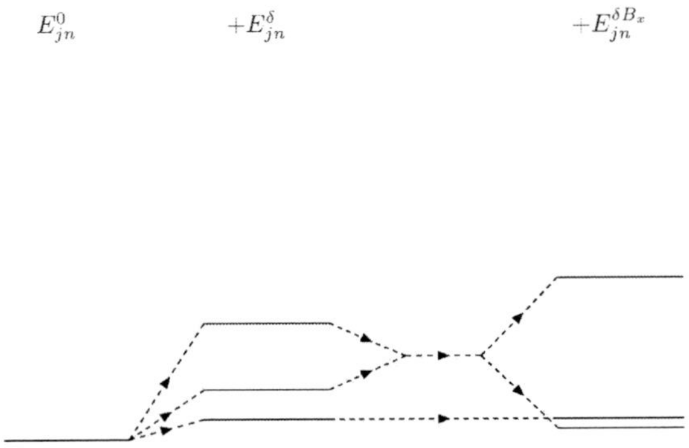

Figure 14. The schematic picture of the second positive electronic level E^δ_{jn} with a mixture of quantum numbers $j = 1, 2$ and $j = -1, -2$ for spheroidal fullerenes in a weak uniform magnetic field pointed in the x direction.

5.4. Zero-Energy Mode

Now we will analyze the special case of an electron state at the Fermi level (so-called zero-energy mode), in this case the right side of equation 41 is equal to zero. We will use the projection coordinates in the form

$$x = \frac{2R^2 r}{R^2 + r^2}\cos\phi; \quad y = \frac{2R^2 r}{R^2 + r^2}\sin\phi; \quad z = -c\frac{R^2 - r^2}{R^2 + r^2}, \tag{88}$$

where R and c are the spheroidal axles. The Riemannian connection with respect to the orthonormal frame is written as [35, 53]

$$-\omega^1_{\phi 2} = \omega^2_{\phi 1} = \frac{R^2 - r^2}{\sqrt{(R^2 - r^2)^2 + 4c^2 r^2}}; \quad \omega^1_{r2} = \omega^2_{r1} = 0. \tag{89}$$

The only nonzero component of the gauge field W_μ in region R_N for spheroidal fullerenes reads

$$W_\phi = G + m\frac{z}{X}, \tag{90}$$

where

$$X = \frac{\sqrt{c^2(r^2 - R^2)^2 + 4R^4 r^2}}{r^2 + R^2}. \tag{91}$$

Notice that the monopole field W_μ in Eq. (90) consists of two parts. The first one comes from the K-spin connection term and implies the charge $g = \pm 3/2$ while the second one is due to elastic flow through a surface. This contribution is topological in its origin and characterized by charge G. For total elastic flux from twelve pentagonal defects one has $G = 1$. The parameter $m = g - G$ is introduced in Eq.(90). The uniform external magnetic field B for simplicity is chosen to be pointed in the z direction. In projection coordinates the only nonzero component of A_μ is written as

$$A_\phi = -\frac{2BR^4 r^2}{(R^2 + r^2)^2}. \tag{92}$$

Let us study Eq.(41) for the electronic states at the Fermi energy ($E = 0$). The Dirac matrices can be chosen to be the Pauli matrices, $\gamma_1 = -\sigma_2, \gamma_2 = -\sigma_1$. By using the substitution

$$\begin{pmatrix} \psi_A \\ \psi_B \end{pmatrix} = \sum_j \frac{e^{i(j+G)\phi}}{\sqrt{2\pi}} \begin{pmatrix} u_j(r) \\ v_j(r) \end{pmatrix}, j = 0, \pm 1, \pm 2, \ldots \tag{93}$$

we obtain

$$\left(\frac{1}{r}(j - m\frac{z}{X} - A_\phi) + \frac{R^2 + r^2}{K}(\partial_r - \frac{r^2 - R^2}{2r(r^2 + R^2)})\right) v(r) = 0, \tag{94}$$

$$\left(\frac{1}{r}(j - m\frac{z}{X} - A_\phi) - \frac{R^2 + r^2}{K}(\partial_r - \frac{r^2 - R^2}{2r(r^2 + R^2)})\right) u(r) = 0, \tag{95}$$

where $K = \sqrt{(R^2 - r^2)^2 + 4c^2r^2}$. We assume that the eccentricity of the spheroid is small enough. In this case, one can write down $c = R + \delta R$ and δ ($| \delta | \ll 1$) is a small dimensionless parameter characterizing the spheroidal deformation. Therefore, one can follow the perturbation scheme using δ as the perturbation parameter. To the leading in δ approximation Eqs. (94) and (95) are written as

$$\partial_r v_j(r) = (-\frac{(R^2 + r^2)^2}{r[(R^2 + r^2)^2 - 4R^2r^2\delta]}[j - m(1 + \delta)\frac{r^2 - R^2}{r^2 + R^2} + m\delta\frac{(r^2 - R^2)^3}{(r^2 + R^2)^3}]$$

$$-\frac{2BR^4r}{[(R^2 + r^2)^2 - 4R^2r^2\delta]} + \frac{r^2 - R^2}{2r(R^2 + r^2)})v_j(r), \tag{96}$$

$$\partial_r u_j(r) = (\frac{(R^2 + r^2)^2}{r[(R^2 + r^2)^2 - 4R^2r^2\delta]}[j - m(1 + \delta)\frac{r^2 - R^2}{r^2 + R^2} + m\delta\frac{(r^2 - R^2)^3}{(r^2 + R^2)^3}]$$

$$+\frac{2BR^4r}{[(R^2 + r^2)^2 - 4R^2r^2\delta]} + \frac{r^2 - R^2}{2r(R^2 + r^2)})u_j(r). \tag{97}$$

The exact solution to Eqs. (96) and (97) is found to be

$$v_j(x) = \frac{(x+1)^{(1/2-m)}}{x^{\frac{1}{2}(j+m+1/2)}} \left(\frac{x + 1 - 2\delta - 2\sqrt{\delta(\delta - 1)}}{x + 1 - 2\delta + 2\sqrt{\delta(\delta - 1)}}\right)^{-\alpha} \left((x+1)^2 - 4\delta x\right)^m, \tag{98}$$

$$u_j(x) = (x+1)^{(m+1/2)} x^{\frac{1}{2}(j+m-1/2)} \left(\frac{x + 1 - 2\delta - 2\sqrt{\delta(\delta - 1)}}{x + 1 - 2\delta + 2\sqrt{\delta(\delta - 1)}}\right)^{\alpha} \left((x+1)^2 - 4\delta x\right)^{-m}, \tag{99}$$

where $x = r^2/R^2$, $\alpha = (j/2)\sqrt{\delta/(\delta - 1)} + BR^2/4\sqrt{\delta(\delta - 1)}$. The function u_j can be normalized if the condition $m - 1/2 < j < 1/2 - m$ is fulfilled. In turn, the normalization condition for v_j has the form $-1/2 - m < j < m + 1/2$. Finally, there are five normalized solutions for u_j with $j = 0, \pm 1, \pm 2$ when $m = -5/2$, and one solution for v_j with $j = 0$ when $m = 1/2$. In the limit $\delta \to 0$ we arrive at the solution for spherical fullerenes

$$v_j(r) = r^{(-1/2-j-m)}(R^2 + r^2)^{(1/2+m)} e^{\frac{BR^4}{R^2+r^2}},$$

$$u_j(r) = r^{(-1/2+j+m)}(R^2 + r^2)^{(1/2-m)} e^{-\frac{BR^4}{R^2+r^2}}. \tag{100}$$

Since normalization conditions do not depend on the parameter δ they are the same as for spherical case. Thus the perturbation does not change the number of zero modes. Notice that Eq. (100) can be obtained from Eqs. (98) and (99) by using $\lim\,[1+(x/a)]^a = e^x, a \to \infty$. Because of the different behaviour of the zero-energy modes there is also possibility to recognize the zero-eigenvalue states in experiment.

5.5. The $SU(2)$ Algebra

The angular-momentum operators for Dirac operator on the sphere S^2 with charge G and total magnetic monopole m looks as follows

$$\hat{L}_z = -i(\partial_\phi \mp iG), \tag{101}$$

$$\hat{L}_x = i\sin\phi\partial_\theta + i\cos\phi\frac{\cos\theta}{\sin\theta}\left(\partial_\phi \mp iG - i\frac{m}{\cos\theta}\right) + \sigma_z\frac{\cos\phi}{2\sin\theta}, \tag{102}$$

$$\hat{L}_y = -i\cos\phi\partial_\theta + i\sin\phi\frac{\cos\theta}{\sin\theta}\left(\partial_\phi \mp iG - i\frac{m}{\cos\theta}\right) + \sigma_z\frac{\sin\phi}{2\sin\theta}, \tag{103}$$

where $-(+)$ sign correspond to north (south) hemisphere. These operators satisfy the standard commutations relations of the $SU(2)$ algebra:

$$\varepsilon_{ijk}\hat{L}_j\hat{L}_k = i\hat{L}_i. \tag{104}$$

The square of Dirac operator and $\hat{L}^2$ may be diagonalized simultaneously

$$\hat{\mathcal{D}}_0^2(B=0) = \hat{L}^2 + \frac{1}{4} - m^2. \tag{105}$$

We can see that the operator $V(\theta,\phi)$ in the case when the magnetic field is pointed in the z direction can be expressed in the form

$$V(\theta,\phi) = \hat{L_z} + \left(\frac{\sigma_3}{2} - m\right)z. \tag{106}$$

In the case when the magnetic field is pointed in the x direction we have

$$V(\theta,\phi) = -\hat{L_x} - \left(\frac{\sigma_3}{2} - m\right)x, \tag{107}$$

where x,z are Cartesian coordinate

$$x = \sin\theta\cos\phi; \quad z = \cos\theta. \tag{108}$$

The square of Dirac operator and operator $\hat{L}^2$ may be diagonalized simultaneously and their eigenvalues are interrelated

$$\langle E, n|\hat{L}^2|E, n\rangle = l(l+1) = E^2 - \frac{1}{4} = (n+|j|)(n+|j|+1), \tag{109}$$

where

$$l = n + |j|. \tag{110}$$

We can use instead of n the angular momentum $l = n + |j|$. We can introduce the eigensteite of $\hat{\mathcal{D}}_0(B = 0)$ in the form

$$\Psi_{l,j}(x, \phi) = \frac{e^{i(j \pm G)\phi}}{\sqrt{2\pi}\Omega} \sqrt{\frac{(l+j)!}{(l-j)!}}$$

$$\times \begin{pmatrix} (1-x)^{-\frac{1}{2}(j-m-1/2)}(1+x)^{-\frac{1}{2}(j+m+1/2)} \frac{d^{l-j}}{dx^{l-j}}(1-x)^{(l-m-1/2)}(1+x)^{(l+m+1/2)} \\ isgn(E)(1-x)^{-\frac{1}{2}(j-m+1/2)}(1+x)^{-\frac{1}{2}(j+m-1/2)} \frac{d^{l-j}}{dx^{l-j}}(1-x)^{(l-m+1/2)}(1+x)^{(l+m-1/2)} \end{pmatrix} \tag{111}$$

where

$$\Omega = 2^l \sqrt{\Gamma(l - m + 1/2)^2 + \Gamma(l + m + 1/2)^2}, \tag{112}$$

and Γ are the Gamma functions (see Ref. [60]). When we act by L_- and L_+ on Ψ we find that

$$\hat{L}_- \Psi_{l,j} = \sqrt{(l+j)(l-j+1)} \Psi_{l,j-1}, \tag{113}$$

$$\hat{L}_+ \Psi_{l,j} = \sqrt{(l+j+1)(l-j)} \Psi_{l,j+1}, \tag{114}$$

where

$$\hat{L}_- = -e^{-i\phi} \left(\partial_\theta - i \frac{\cos\theta}{\sin\theta} \left(\partial_\phi \mp iG - i \frac{m}{\cos\theta} \right) - \frac{\sigma_z}{2\sin\theta} \right), \tag{115}$$

$$\hat{L}_+ = e^{i\phi} \left(\partial_\theta + i \frac{\cos\theta}{\sin\theta} \left(\partial_\phi \mp iG - i \frac{m}{\cos\theta} \right) + \frac{\sigma_z}{2\sin\theta} \right). \tag{116}$$

Now we would like to transform our formulae to the cartesian coordinates. The corresponding transformation rules for spinors are [60]

$$(\Psi)_C = V^\dagger \Psi, \tag{117}$$

and cartesian realization of operator $\hat{L}$ is

$$\hat{L}_C = V^\dagger \hat{L} V, \tag{118}$$

with V-matrices given by

$$V = \exp\left(\frac{i\sigma_y}{2} \theta \right) \exp\left(\frac{i\sigma_z}{2} \phi \right). \tag{119}$$

For example we present the Cartesian realization of $\hat{L}_x$ operator

$$(\hat{L}_x)_C = i \sin\phi \partial_\theta + i \cos\phi \frac{\cos\theta}{\sin\theta} \left(\partial_\phi \mp iG - i \frac{m}{\cos\theta} \right) + \frac{\sigma_x}{2}, \tag{120}$$

that may be expressed in the form ($m = G = 0$)

$$(\hat{L}_x)_C = y\hat{p}_z - z\hat{p}_y + \frac{\sigma_x}{2}. \tag{121}$$

6. Conclusion

Electronic properties of fullerenes molecules have been discussed from a theoretical point of view. The effective-mass approximation is particularly suitable for understanding global and essential features. In this scheme, the motion of electrons in fullerenes is described by Dirac-Weyl's equation. The geometry and topology is found to influence the main physical characteristics of graphite nanoparticles, first of all, their electronic properties. The topological defects (disclinations) appear as generic defects in closed carbon structures.

We have considered the electronic states of spheroidal fullerenes provided the spheroidal deformation from the sphere is small enough. In this case, the spherical representation is used for describing the eigenstates of the Dirac equation, with the slight asphericity considered as a perturbation. The using of the perturbation scheme allows us to find the exact analytical solution of the problem. In particular, the energy spectrum of spheroidal fullerenes is found to possess the fine structure in comparison with the case of the spherical fullerenes. We have shown that this structure is weakly pronounced and entirely dictated by the topological defects, that is it has a topological origin. We found three twofold degenerate modes near the Fermi level with one of them being the true zero mode therefore our finding confirms the results of [42] that there can exist only up to twofold degenerate states in the C_{70}.

Notice that for spherical fullerenes ($\delta = 0$) our results agree with those found in [49]. It is interesting that the predictions of the continuum model for spherical fullerenes are in qualitative agreement with tight-binding calculations [61, 62, 63, 64]. In particular, the energy gap between the highest-occupied and lowest-unoccupied energy levels becomes more narrow as the size of fullerenes becomes larger. It is important to keep in mind, however, that the continuum model itself is correct for the description of the low-lying electronic states. In addition, the validity of the effective field approximation for the description of big fullerenes is also not clear yet. Actually, this approximation allows us to take into account the isotropic part of long-range defect fields. For bigger fullerenes, one has to consider the anisotropic part of the long-range fields, the influence of the short-range fields due to single disclinations as well as the multiple-shell structure.

We have focused also on the structure of low energy electronic states of spheroidal fullerenes in the weak uniform magnetic field. For the states at the Fermi level, we found an exact solution for the wave functions. It is shown that the external magnetic field modifies the density of electronic states and does not change the number of zero modes. For non-zero energy modes, electronic states near the Fermi energy of spheroidal fullerene are found to be splitted in the presence of a weak uniform magnetic field. The case of the x-directed magnetic field was considered and compared with the case of the z-th direction. The z axis is defined as the rotational axis of the spheroid with maximal symmetry. The most important finding is that the splitting of the electronic levels depends on the direction of the magnetic field. Our consideration was based on the using of the eigenfunctions of the Dirac operator on the spheroid, which are also the eigenfunction of $\hat{L}_z$. Let us discuss this important point in more detail.

In the case of a sphere there is no preferable direction in the absence of the magnetic field. The magnetic field sets a vector, so that the z-axis can be oriented along the field. In this case, one has to use such eigenfunctions of the Dirac operator which are also the eigenfunctions of $\hat{L}_z$. For the x-directed magnetic field, the eigenfunction of both the Dirac operator and $\hat{L}_x$ must be used. Evidently, the same results will be obtained in both cases.

The situation differs markedly for a spheroid. The spheroidal symmetry itself assumes the preferential direction which can be chosen as the z-axis. In other words, the external magnetic field does not define the preferable orientation. The symmetry is already broken and, as a result, the case of the magnetic field pointed in the x direction differs from the case of the z-directed field. For instance, there are no eigenfunctions which would be simultaneously the eigenfunctions of both the Dirac operator on the spheroid and $\hat{L}_x$. For this reason, the structure of the electronic levels is found to crucially depend on the direction of the external magnetic field (Figs. 9-14).

It should be mentioned that the zero-energy states in our model corresponds to the HOMO (highest occupied molecular orbital) states in the calculations based on the local-density approximation in the density-functional theory (see, e.g. [42]). In particular, the HOMO-LUMO energy gap is found to be about 1.1 eV for YO-C_{240} fullerene within our model (here LUMO means the lowest unoccupied molecular orbital) [65, 66].

The very big fullerenes like C_{960} and C_{1500} become more deformed, faceted and can no longer form a free-electron model like the electronic shell [67], which was the assumption for this model. For these structures the deviation from the sphericity is larger when the pentagon defects are localized at the opposite poles. In the case when the poles are far away from each other we obtain the structure of nanotubes, and for the exact description some new model related to that proposed here should be used. In our opinion, our predictions are quite general for the fullerene family and are of interest for experimental studies. Finally, we think that the spheroidal geometry approach especially our established generic models could also be related to other physical problems with slightly deformed spherical structures that are common in the nature e.g. in astrophysics for modeling pulsars or kvasars.

Acknowledgment

We would like to thank especially Prof. V. A. Osipov whose is also co-author our results presented in this review for usefully comments and advices during writing this chapter. The work was supported in part by VEGA grant 2/7056/27. of the Slovak Academy of Sciences, by the Science and Technology Assistance Agency under contract No. APVT-51-027904.

References

[1] R. Rajaraman: *Solitons and Instantons*, (North-Holland, Amsterdam 1982)

[2] V.A. Osipov: *Phys. Lett. A* 164 (1992) 327

[3] H.W. Kroto, J.R. Heath, S.C. O'Brien et al.: *Nature* 318 (1985) 162

[4] W. Kratschmer, L.D. Lamb, K. Fostiropoulos et al.: *Nature* 347 (1990) 354

[5] S. Iijima: *Nature* 354 (1991) 56

[6] T.W. Ebbesen: *Physics Today*, (June) (1996) 26

[7] S. Iijima, T.Ichihashi, Y.Ando: *Nature* 356 (1992) 776

[8] R.Saito, G.Dresselhaus, M.S.Dresselhaus: *Phys. Rev. B* 53 (1996) 2044

[9] H. Terrones, M. Terrones: *Carbon* 36 (1998) 725

[10] D.P. DiVincenzo, E.J. Mele: *Phys. Rev. B* 29 (1984) 1685

[11] P.R. Wallace, *Phys. Rev.* 71 (1947) 622

[12] J.C. Slonczewski and P.R. Weiss, *Phys. Rev.* 109 272 (1958) 272

[13] V.A. Osipov, E.A. Kochetov: *JETP Lett.* 72 (2000) 199

[14] J. González, F. Guinea, M.A.H. Vozmediano: *Nucl.Phys. B* 406 (1993) 771

[15] C.L. Kane, E.J. Mele: *Phys. Rev. Lett.* 78 (1997) 1932

[16] J. González, F. Guinea, M.A.H. Vozmediano: *Phys. Rev. Lett.* 69 (1992) 172

[17] T.Ando: *J.Phys.Soc. Jpn.* 74 (2005) 777

[18] J.G. Hou, J. Yang, H. Wang et al.: *Phys. Rev. Lett.* 83 (1999) 3001

[19] D.L. Carroll, P. Redlich, P.M. Ajayan et al.: *Phys. Rev. Lett.* 78 (1997) 2811

[20] P. Kim, T.W. Odom, J.-L. Huang et al.: *Phys. Rev. Lett.* 82 (1999) 1225

[21] B. An, S. Fukuyama, K. Yokogawa et al.: *Appl. Phys. Lett.* 78 (2001) 3696

[22] S. Berber, Y.-K. Kwon, D. Tománek: *Phys. Rev. B* 62 (2000) R2291

[23] V.A. Osipov, E.A. Kochetov, M. Pudlak: *JETP* 96 (2003) 140

[24] K. Kobayashi: *Phys. Rev. B* 61 (2000) 8496

[25] J.-C. Charlier, G.-M. Rignanese: *Phys. Rev. Lett.* 86 (2001) 5970

[26] T. Yaguchi, T. Ando: *J.Phys.Soc.Jpn.* 71 (2002) 2224

[27] R. Jackiw, C. Rebbi: *Phys. Rev. D* 16 (1977) 1052

[28] R. Jackiw, J.R. Schrieffer: *Nucl. Phys. B* 190 (1981) 253

[29] M.M. Salomaa, G.E. Volovik: *Rev. Mod. Phys.* 59 (1987) 533

[30] R. Jackiw: *Phys. Rev. D* 29 (1984) 2375

[31] G.E. Volovik: *JETP Lett.* 70 (1999) 792

[32] J.M.Luttinger, *Phys.Rev.*84 (1951) 814

[33] P.E. Lammert, V.H. Crespi: *Phys. Rev. Lett.* 85 (2000) 5190

[34] V.A. Osipov, E.A. Kochetov: *JETP Lett.* 73 (2001) 631

[35] M. Nakahara, *Geometry, Topology and Physics*, (Institute of Physics Publishing Bristol 1998)

[36] N.D. Birrell and P.C.W. Davies, *Quantum Fields in Curved Space*, (Cambridge 1982)

[37] E.A. Kochetov, V.A. Osipov: *J.Phys. A: Math.Gen.* 32 (1999) 1961

[38] R. Pincak, V.A. Osipov: *Phys. Lett. A* 314 (2003) 315

[39] B.A. Ovrut, S. Tomas: *Phys.Rev. D* 43 (1991) 1314

[40] H. Kroto, *Reviews of Modern Physics*, 69 (1997) 703

[41] R.T. Chancey et al., *Phys. Rev. A* 67 (2003) 043203

[42] S. Saito and A. Oshiyama, *Phys. Rev. B* 44 (1991) 11532

[43] W.H. Lomer, *Proc. Roy. Soc. (London)* A227 (1955) 330

[44] J.M. Luttinger and Kohn, *Phys. Rev.* 97 (1955) 869

[45] H. Matsumura and T. Ando, *J. Phys. Soc. Jpn.* 67 (1998) 3542

[46] R. Saito, G. Dresselhaus and M.S. Dresselhaus, *Physical Properties of Carbon Nanotubes* (Imperial College Press, London, 2003)

[47] P.E. Lammert and V.H. Crespi, *Phys. Rev. B* 69 (2004) 035406

[48] J. Tersoff, *Phys. Rev. B* 46 (1992) 15546

[49] D.V. Kolesnikov and V.A. Osipov, *Eur.Phys.Journ. B* 49 (2006) 465

[50] R. Pincak, *Phys. Lett. A* 340 (2005) 267

[51] K. Clemenger, *Phys. Rev. B* 32 (1985) 1359

[52] H. Aoki and H. Suezawa, *Phys. Rev. A* 46 (1992) R1163

[53] M.Göckeler and T.Schücker, *Differential geometry, gauge theories, and gravity* (Cambridge University Press 1989)

[54] T.T.Wu and G.N.Yang, *Nucl. Phys. B* 107 (1976) 365

[55] T.T.Wu and G.N.Yang, *Phys. Rev. D* 12 (1975) 3845

[56] L.D. Landau and E.M. Lifshitz, *Quantum Mechanics* (Elsevier Science, Oxford, 2003)

[57] M. Yoshida and E. Osawa, *Fullerene Sci. Tech.* 1 (1993) 55

[58] J. P. Lu and W. Yang, Phys. Rev. B 49 (1994) 11421

[59] R.A. Broglia, G. Colò, G. Onida, H.E. Roman, *Solid State Physics of Finite Systems*, (Springer 2004)

[60] A.A. Abrikosov,jr., *Int. Journ. of Mod. Phys. A* 17 (2002) 885

[61] A.Ch. Tang and F.Q. Huang, *Phys. Rev. B* 51 (1995) 13830

[62] E. Manousakis, *Phys. Rev. B* 44 (1991) 10991

[63] A. Perez-Garrido, F. Alhama and J.D. Catala, *Chem. Phys.* 278 (2002) 71

[64] Y.L. Lin and F. Nori, *Phys. Rev. B* 49 (1994) 5200

[65] M. Pudlak, R. Pincak and V.A. Osipov, *Phys. Rev. B* 74 (2006) 235435

[66] M. Pudlak, R. Pincak and V.A. Osipov, *Phys. Rev. A* 75 (2007) 025201

[67] N. Mizorogi et al., *Chem. Phys. Lett.* 378 (2003) 598

In: Progress in Fullerene Research
Editor: Milton Lang, pp. 269-297
ISBN: 978-1-60021-841-5

Chapter 8

THEORETICAL STUDIES OF FULLERENE FORMATION IN PLASMA

G. N. Churilov*, A. S. Fedorov and P. V. Novikov
L.V. Kirensky Institute of Physics SB RAS, Krasnoyarsk, 660036 Russia

Abstract

Three aspects of fullerene formation were studied: the influence of electron concentration in ionized carbon vapor on C_{60} the processes of fullerene and metallofullerene formation, influence of metal atom properties on metallofullerene formation, and influence of buffer gas on these processes. Total rates of formation of C_{60} fullerene and $Me@C_{84}$ (*Me*=Sc, Fe, Pt) endohedral metallofullerene were calculated using classical collision theory and quantum chemical calculations of ionization potentials of reacting particles. It was shown that fullerene and metallofullerene formation rates essentially depend on electron concentration and temperature of carbon containing plasma. The temperature dependence of optimal electron concentration corresponded to maximum formation rate was determined. Also the estimations made showed that waves of electron concentration in plasma can increase the rate of C_{60} fullerene assembling. For metallofullerenes it was shown that the metal atom *Me* adheres to the growing carbon cluster at the last stage of $Me@C_{84}$ molecule formation and the probability of the formation is sensitive to the metal-cluster chemical bonding and metal potential ionization value. Further, the heating and cooling of carbon clusters by buffer gas during the fullerene formation process was considered. Reduced description scheme allowed us to calculate kinetics of a huge number of reactions: coagulation, isomerization and heating/cooling by buffer gas. The kinetic calculations gave the qualitatively correct correlation between the fullerene yields in helium and argon.

Introduction

In 1985, the discovery of buckminsterfullerene prompted a flurry of activity as scientists sought to characterize this new form of carbon and produce it in bulk quantities. When bulk synthesis was achieved by Krätschmer et al. [1] in 1990, scientific investigation of fullerene intensified and diversified as various research groups produced a wide array of both

* E-mail address: churilov@iph.krasn.ru

endohedral and exohedral derivatives. Though there are many works devoted to the formation of fullerenes and their derivatives in carbon containing plasma, the mechanism of their formation has not been fully clarified yet[2],[3].

Up to now the majority of articles concerning fullerene formation were devoted to the reactions of intermediate neutral carbon clusters and processes of their transformation. The most perspective models consider these precursor clusters as polycyclic rings [4], smaller fullerenes of 30-40 atoms [5] and nonclassical fullerenes with heptagons [6]. With few exceptions [7],[8], the role of charges of clusters in the process of their coagulation was generally outside the research scope. In Ref. [7] it was demonstrated experimentally that formation of fullerenes in ionized carbon plasma is more efficient than the synthesis in neutral carbon vapor. In Ref. [8] the authors considered theoretically the reactions between charged and neutral carbon polycyclic structures.

In the first section of this chapter we present a theoretical model of fullerene formation in arc discharge plasma, which elucidates the very important role of electron concentration in ionized carbon vapor [9]. Our model is based on the classical theory of collisions. The model takes into account the influence of charges of clusters on the cross section of their collisions. We consider the ways of fullerene formation depending on the value of carbon cluster charges, electron concentration and temperature. Also we have shown that waves of electron concentration in plasma may influence the rate of C_{60} fullerene formation. In our previous work [10] we demonstrated that efficient synthesis by the Krätschmer's method [1] was due to the existence of spontaneous ionization waves in plasma. The fullerene synthesis at atmospheric pressure in the carbon-helium plasma induced by the arc discharge of kilohertz range frequency was also shown to be efficient [11]. This was ascribed to the existence of forced ionization waves registered by us. Thus, we consider in this chapter the influence of waves of electron concentration on the rate of C_{60} formation.

After the discovery of the C_{60} molecule by Kroto et al. [12] and the detail determination of its properties, great interest was attracted to the synthesis of carbon cage with some extra-atoms encapsulated inside. For the first time such molecule was detected in the same year, as C_{60} [13].

After its experimental discovery a lot of works were devoted to their physical-chemical properties determination, and macroscopic amounts of endohedral metallofullerenes [14]-[17] were successively produced and isolated. While researchers have defined experimentally many interesting properties of such substances, so far the metallofullerene synthesis remains the process of the extremely low yield of generation. At that the method of arc discharge in carbon-containing gas, proposed by W. Kratschmer [1] still remains the major method of fullerenes and metallofullerenes synthesis.

Though there are many theoretical works devoted to the fullerenes formation in carbon containing plasma, the mechanism of the metallofullerene formation in the arc discharge plasma is still unclear. Up to now the majority of articles have performed molecular dynamics (MD) simulations of the clusterization process [18]. Though chemical properties of the metal atoms and the temperature influence on the metallofullerene yield were taken into account in MD simulations, the role of another plasma parameter – electron concentration n_e was usually omitted.

The aim of the second section of this chapter is to elucidate the roles of metal and carbon cluster chemical bonding, ionization potential of metal, the temperature T and the electron concentration n_e to the metallofullerene yield in the arc discharge synthesis [19]. Therefore, in this section we also have not considered detailed carbon cluster transformations in the plasma during the metallofullerene coagulation.

According to experimental studies, the metals such as Sc, Ca, Sr, Va, Y and lanthanides can be encapsulated inside the fullerene cage quite easily. At this it is observed experimentally that the major scandium fullerenes produced are Sc@C_{82} or Sc_2@C_{84} [20]. The endohedral nature of Y@C_{82}, Sc@C_{82}, Sc_2@C_{82} metallofullerenes has been confirmed by synchrotron X-ray powder diffraction [21]. On the other hand, Fe, Ni or Co facilitate the carbon nanotubes (CNTs) formation [22],[23]. These metals are catalysts of CNT formation because they easily form MeC compounds. At the same time their possibility to be encapsulated in the fullerene cage is not clear. However, there is the experimental evidence that Fe can be encapsulated in the fullerene [24]. In this work the contact-arc vaporization of graphite in an atmosphere of $Fe(CO)_5$ has been carried out. After product analysis, the presence of FeC_{60} has been confirmed. Analyzing Fe-C distances by means of Fe-K EXAFS measurements, the conclusion that atom Fe is located inside the fullerene cage has been drawn. At last, due to weak chemical activity, Pt can form only weak combined exohedral metallofullerenes or complex compounds with fullerenes by means of intermediate atoms [26]. Consequently, the choice of the metal atoms (Sc, Fe, Pt), which potentially can form metallofullerene Me@C_n was inspired by their different ionization potential and chemical bonding with carbon clusters. At that the covalent radiuses of these atoms do not differ significantly (1.44Å, 1.3Å, 1.17Å), but their first ionization potentials do differ substantially (6.56, 7.90, 8.96 eV correspondingly). The choice of the fullerene cage C_{84} was made because this fullerene together with C_{82} form metallofullerenes with transition metals most frequently.

In the first and second sections the role of cluster inner temperature, which is connected with phonon excitation in the cluster (vibration temperature) was out of scope. But obviously that formation or destruction of chemical bonds in the cluster would strongly change this temperature because of the high energy of the C-C bond. So the cluster vibration temperature variation would change all processes of carbon clusters transformation during fullerenes formation. Also the role of buffer gas molecules in balancing of vibration and translational cluster temperatures was not taken into consideration. Moreover, all carbon clusters were differed only by number of atoms, but not the number of dangling (radical) bonds in the cluster.

In the third section of the chapter we present the further development of the theoretical model of fullerene formation in carbon containing plasma [25]. In this model unaccounted earlier factors are taking into account. In the model all carbon clusters are described by the reduced description scheme, firstly proposed in [27],[28]. In this scheme all carbon clusters are distinguished by both the number of cluster atoms n and number of free radical bonds l in the cluster, moreover, cluster concentrations now include information about cluster vibration temperature T_{vib}. So, the variation of vibration temperature as a result of chemical reactions and buffer gas cooling/heating is taken into account. In the third section we consider the evolution of carbon vapor at constant temperature near the most probable temperature 3000K of fullerene formation [29]. So, the required cluster concentrations $C(n,l,T_{vib},t)$ in the plasma depend on n, l, T_{vib} and time t.

1. Calculations of Fullerene C_{60} Formation in a Carbon Plasma

In this section we considered a low temperature plasma of carbon clusters with different numbers of atoms beginning from carbon dimer C_2. Also we considered the plasma in the state of partial local thermodynamics equilibrium. On that ground one can use Saha equations [30] to determine particle charge distribution in the plasma.

In our calculations we studied the processes of C_{60} molecule formation in one or two stages. In the one-stage mechanism, we considered the assembly of C_{60} molecule as a result of merging of any pair of carbon clusters, with the total mass of the pair being equal to the mass of C_{60}:

$$C_k + C_m \to C_{60}, \qquad k + m = 60, \qquad k = 2, 4, \ldots, 30,\ m = 60 - k; \tag{1}$$

In the two-stages mechanism the previous stage of formation of C_k and C_m clusters was similarly considered:

$$1^{st}\text{ stage: } \begin{cases} C_i + C_j \to C_k, & i + j = k, \quad i = 2, \ldots, k/2 \\ C_l + C_n \to C_m, & l + n = m, \quad l = 2, \ldots, m/2 \end{cases}, \tag{2}$$

$$2^{nd}\text{ stage: } C_k + C_m \to C_{60}, \qquad k + m = 60, \qquad k = 2, 4, \ldots, 30.$$

In the calculations, even carbon clusters were taken into account only because the mass-spectrometric analysis of clusters during the fullerene formation in the experiments [32] showed that the quantity of even clusters was larger than that of odd clusters by some orders.

For the calculations, we needed to know the mass distribution of carbon clusters during the process of fullerene formation. As the basic distribution, we used the smoothed mass distribution between C_2 and C_{60} presented in paper [31] (Fig.1, A). This distribution was theoretically defined from solving of kinetic equations for different carbon clusters coagulation and dissociation during the process of fullerene synthesis. Similar parabolic-like mass distribution was observed in the experiment [32] where fullerenes were produced by laser ablation. To check sensitivity of the results from the shape of carbon cluster mass distribution, we used two other distributions (Fig. 1, B, C). As we found, the calculated coagulation rates of C_{60} and *Me*@C_{84} are not very sensitive to the shape of the cluster distribution. This fact allows us to expect that a similar coagulation rate would be realized at any carbon cluster distribution, even at a slightly non-equilibrium distribution.

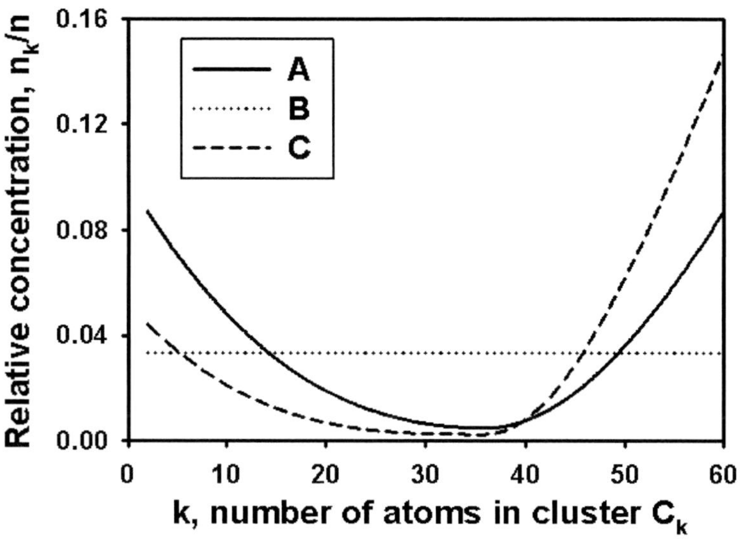

Figure 1. Distribution of carbon clusters by masses from C_2 to C_{60}.

The foundation of our model is the determination of rates of the merging reactions between carbon clusters and between carbon cluster and metal atoms. Due to the large values of hitting particle masses (~$10^4 \div 10^5$ electron mass), it is natural to use classical scattering theory to determine the collision process rates. In the classical theory of gases there is a well-known formula for the collision rate R_{km} of two clusters C_k and C_m: $R_{km} = n_m n_k \mathrm{v}_{km} \sigma_{km}$ [[33]]. Hereinafter in the expressions for collision rate instead of any cluster C_m, C_k may be used metal atom *Me*. Then, if clusters C_k and C_m have charges q_k and q_m respectively, the rate of their collision process:

$$R_{km}(q_k, q_m) = n_m(q_m) \cdot n_k(q_k) \cdot \mathrm{v}_{km} \sigma_{km}(q_k, q_m), \quad (3)$$

where $n_m(q_m)$, $n_k(q_k)$ are the concentrations of clusters C_k and C_m with charges q_k and q_m, $\mathrm{v}_{km} = \sqrt{8k_B T / \pi m_{km}}$ is the average relative velocity of the clusters, $m_{km} = m_k m_m / (m_k + m_m)$ is the reduced mass of C_k and C_m, $\sigma_{km}(q_k, q_m)$ is the cross-section of the cluster collisions depending on the cluster charges.

The effective cross-section of the collision of these clusters $\sigma_{km}(q_k, q_m)$ can be found by formula [34]:

$$\sigma_{km}(q_k, q_m) = \sigma_0 \left(1 - \frac{q_k q_m e^2}{r_{km} \varepsilon} \right), \quad (4)$$

where $\sigma_0 = \pi r_{km}^2$ is the cross-section of the neutral clusters C_k and C_m, $q_k q_m e^2 / r_{km}\varepsilon$ is the ratio of the electrostatic energy and the average kinetic energy of the cluster collision. By integrating over the Maxwell distribution of velocities we obtain the average kinetic energy $\varepsilon = 3/2 \cdot k_B T$. The minimal distance r_{km} between the clusters during the collision have been determined as $r_{km} = r_k + r_m$, where the effective radius of a cluster C_k was approximated as $r_k = 1/2 \cdot (d_x d_y d_z)^{1/3}$, and d_x, d_y and d_z were the dimensions of the cluster C_k.

For the case of charges of opposite sign ($q_1 q_2 < 0$) there is an additional attraction between clusters: $\sigma_{12} > \sigma_0$, but for the case $q_1 q_2 > 0$: $\sigma_{12} < \sigma_0$. If the coulomb repulsion of two clusters is larger than the average kinetic energy of both clusters ($E_{km} / \varepsilon > 1$) the clusters can not touch each other, thus $\sigma_{km} = 0$.

For the total rate of formation of C_{k+m} cluster from clusters C_m and C_k it is necessary to sum up all the concentrations of charged components of both clusters C_k and C_m:

$$R_{km} = \sum_{q_m=-2}^{+1} \sum_{q_k=-2}^{+1} n_k(q_k) \cdot n_m(q_m) \cdot \mathrm{v}_{km}\, \sigma_{km}(q_k, q_m), \qquad (5)$$

Because of merging cluster diversity, zero potential barrier approach has been used for merging reactions of all C_k and MeC_k clusters. In principle, any constant barrier of merging for all clusters could not change the results of the calculations, except an overall rate norm coefficient, which was omitted. Therefore, in this model stage any collision of two clusters C_k and C_m (or MeC_k and C_m pair) led to the joint C_{k+m} (or MeC_{k+m}) cluster formation.

In the calculations the charges $q_i = \{-2, -1, 0, +1\}$ have been used for all C_k clusters. The probability of existence of clusters with charges $q_k > +1$ or $q_k < -2$ in the arc discharge plasma is negligible, because the third electron affinity (EA3) is negative and the second ionization potential (IP2) is very large (IP2 >> kT) for all analyzed clusters. For example, calculated IP2 were equal 19.9, 19.6 and 17.1 eV for the ring-shape cluster C_{10}, coranulene-shape cluster C_{20} and fullerene C_{60} respectively. The same charges $q_i = \{-2, -1, 0, +1\}$ have been used for all metal atoms Me and MeC_k clusters as well.

To define the average cross-section for all clusters the concentrations n_k^q of C_k cluster with charge q were defined by Saha equations [30]:

$$\frac{n_k^{q+1} \cdot n_e}{n_k^q} = 2 \frac{Z_k^{+1}}{Z_k^0} \cdot \left(\frac{m_e k_B T}{2\pi\hbar^2} \right)^{3/2} \cdot \exp\left(-IP_k^q / k_B T\right),\ q = -2, -1, 0, +1 \qquad (6)$$

where IP_k^q is the ionization potential of cluster C_k^q, $Z_k^q(T) = \sum_{n=1}^{n_{\max}} g_n \exp[-(\varepsilon_n - \varepsilon_1)/k_B T]$ is the statistical sum for cluster C_k with charge q at temperature T, n_e is the electron concentration in plasma and g_n is the degeneracy of the electron level ε_n in cluster C_k^q.

Concentrations n_k^0, n_k^{+1}, n_k^{-1}, n_k^{-2} were found from the fact of mass conservation:

$$n_k^0 + n_k^{+1} + n_k^{-1} + n_k^{-2} = n_k . \tag{7}$$

The average charge of cluster C_k in plasma with electron concentration n_e and temperature T is:

$$\langle q_k \rangle = \frac{1}{n_k} \sum_{q=-1}^{+2} q \cdot n_k^q . \tag{8}$$

So, the sum rate of C_{60} creation (5) can be expressed as $R_{km} = R_{km}^0 \cdot P_{km}$, where $R_{km}^0 = n_m n_k \mathrm{v}_{km} \sigma_0$ is the rate of C_{k+m} cluster formation from neutral clusters. Factor P_{km} is

$$P_{km} = \sum_{q_m=-2}^{+1} \sum_{q_k=-2}^{+1} p_m(q_m) p_k(q_k) \left(1 - \frac{q_k q_m e^2}{r_{km} \varepsilon}\right), \tag{9}$$

where $p_i(q) = n_i^q / n_i$ is the fraction of C_k cluster with the charge q.

Then the average rate of C_{k+m} cluster formation during some period τ is

$$\overline{R}_{km} = \frac{1}{\tau} \int_0^{\tau} R_{km} dt = R_{km}^0 \frac{1}{\tau} \int_0^{\tau} P_{km} dt . \tag{10}$$

From the above formulas it is possible to calculate the rate of C_{60} molecule assembly. There are only unknown values of ionization potentials IP_k^0 and electron affinities IP_k^{-1} and IP_k^{-2} for every kind of carbon cluster. The values of IP_k^q for a cluster C_k were calculated as the differences between binding energies of this cluster with different charges: $E_b(C_k^0)$, $E_b(C_k^{+1})$, $E_b(C_k^{-1})$ and $E_b(C_k^{-2})$. The calculations of E_b were made using VASP package [35]. This package allows *ab initio* calculations basing on the plane wave decomposition and the Vanderbilt type pseudopotential for every atom [36] in the framework of DFT (density functional theory) approach.

The values of IP_k^q were calculated for clusters C_2, C_4, C_8, C_{10}, C_{14}, C_{18}, C_{20}, C_{40}, C_{60}, C_{70} and C_{80} of different shapes. We used the cluster shapes consisting of hexagons only or pentagons and hexagons in agreement with our previous work [10]. Moreover, we used the clusters C_{20}, C_{40}, C_{60}, containing one heptagon according to the ref. [6], where the important role of nonclassical fullerenes, containing one heptagon, in fullerene growth was shown. If there were used some isomers of a cluster C_k, the average values of IP_k^q for this k were calculated. After that, the values of IP_k^q were interpolated for all ($k = 2 - 84$). Also the ionization potentials and the electron affinities for the same clusters which included metal atoms above the cluster surface were calculated. The results of IP_k^q calculations for C_k and different MeC_k clusters are shown on Figure 1.

The rate of C_{60} formation in the one stage process was defined as a sum of rates R_{km} defined by (5) with the condition $k + m = 60$:

$$\overline{R_{60}^{I}} = \sum_{k=2}^{30} \overline{R}_{k,60-k} \quad . \tag{11}$$

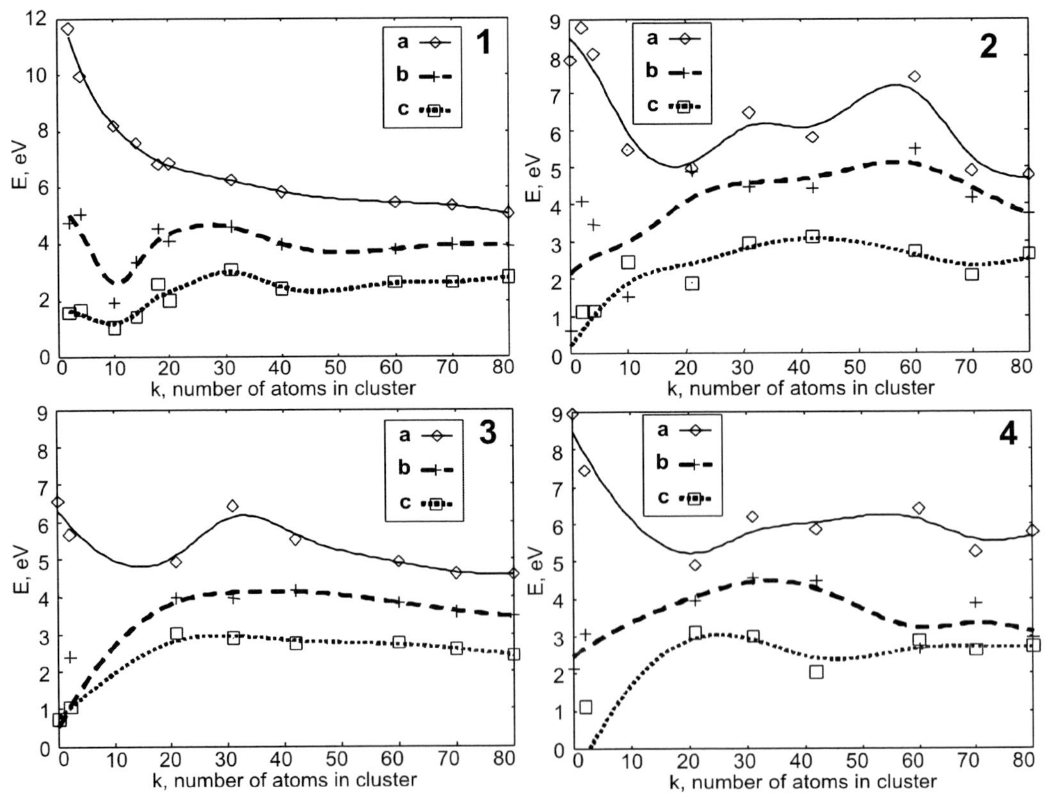

Figure 2. Interpolated average ionization potential IP (a), electron affinity EA (b) and anion electron affinity EA2 (c) for clusters C_k (1), for clusters FeC_k(2), ScC_k (3), PtC_k (4).

The dependence of average (integrated over the time interval $0 < \tau < 10^{-5}$ sec) rate $\overline{R_{60}^{I}}$ of C_{60} formation on electron concentration and temperature is shown in figure 3 (surface 1). In surface 2 of this figure the same dependence of rate of C_{60} formation from neutral clusters is shown. This rate was computed similarly to eq.(11), but for surface 2 we used $R_{km} = R_{km}^{0}$. There is a drastic difference between the rates of C_{60} formation from neutral and charged clusters.

In figure 4 the dependence of the total rate of C_{60} formation in two stages (2) integrated over the above described interval of τ is shown. In the process of C_{60} formation all the intermediate reactions of cluster merging were summarized. The concentrations of different clusters C_m were accumulated over the whole range of τ: $\Delta n_m = \sum_{n+k=m} \overline{R}_{nk} \cdot \tau$. Taking into account all intermediate changes of cluster concentration we calculated the total rate $\overline{R_{60}^{II}}$ of C_{60} formation in two stages in the first order by Δn_m:

$$\overline{R_{60}^{II}} = \overline{R_{60}^{I}} + \sum_{m+k=60} (\Delta n_m n_k + n_m \Delta n_k) \mathrm{v}_{km} \sigma_{km}^{0} \cdot \frac{1}{\tau} \int_0^{\tau} P_{km}\, dt \,. \quad (12)$$

As seen, the surfaces in figures 3 and 4 are similar. Thus, one may expect that the dependence of C_{60} assembling rate on n_e and T in several (more than two) stages of cluster merging will be similar to that for the two-stage process.

From figures 3 and 4 it is possible to determine the optimal electron concentration for every temperature for C_{60} formation. For example for temperature ~2500K the optimal electron concentration is $\sim 10^{16}$ $1/m^3$.

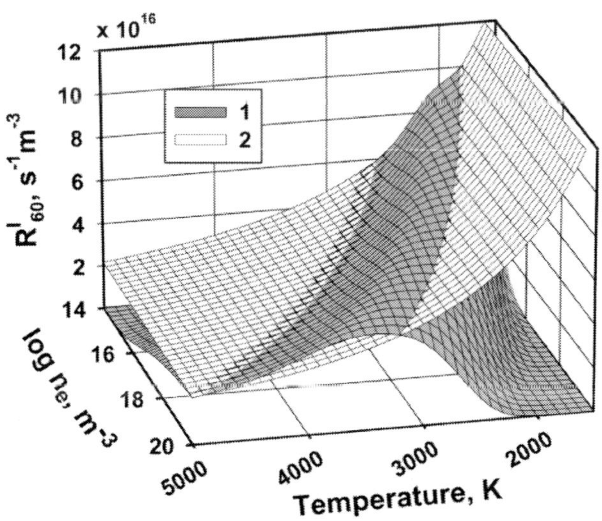

Figure 3. Rate of the C_{60} molecule formation in one stage from charged clusters (1) and neutral clusters (2).

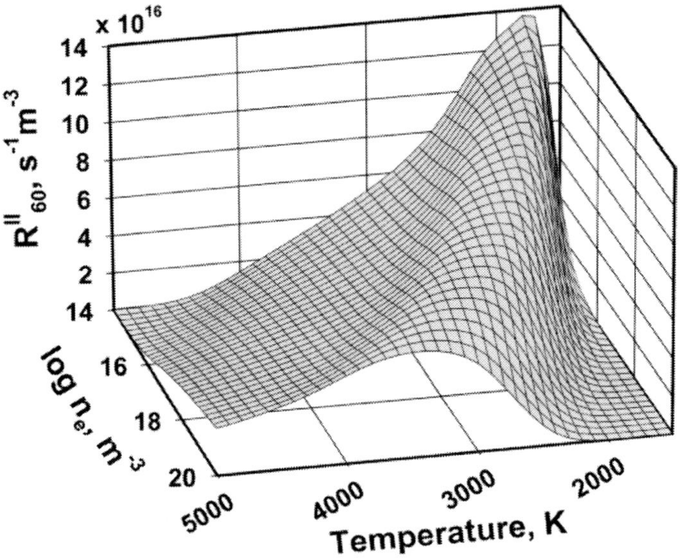

Figure 4. Rate of the C_{60} molecule formation in two stages.

The existence of the optimal parameters for C_{60} assembly can be explained by the strong dependence of the charges of carbon clusters on temperature. This dependence for clusters C_2 and C_{58} is shown in figure 5. The charge difference is the result of difference between IP_k^q for corresponding clusters. From the calculations made we can conclude that the existence of the opposite charges of clusters C_2 and C_{58} (-0.5 and 0.8 consequently) at optimal for C_{60} assembly plasma parameters (T ~2500K, $n_e \sim 10^{16}$ 1/m^3) stimulates essential increase (~ 30 times) in the cross-section of these clusters $\sigma_{km}(q_k, q_m)$ and, therefore, accelerates C_{60} assembling.

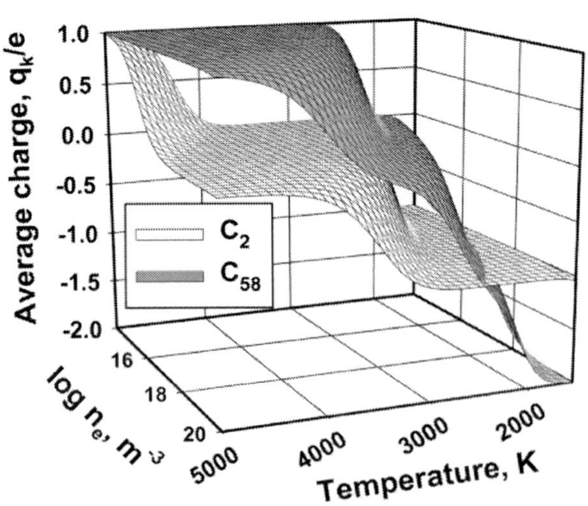

Figure 5. Dependence of charges of the clusters C_2 and C_{58} from temperature and electron concentration in charges of the electron.

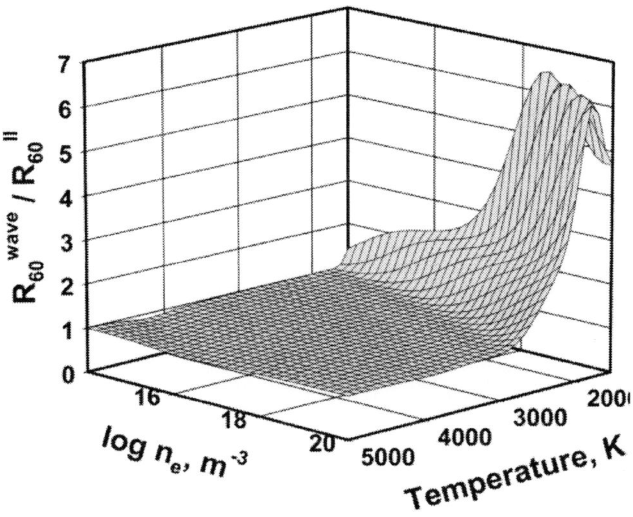

Figure 6. Ratio $\overline{R_{60}^{wave}}/\overline{R_{60}^{II}}$ of the fullerene formation rate with the waves of electron concentration influence $\overline{R_{60}^{wave}}$ and without them $\overline{R_{60}^{II}}$.

In our previous work [10] we proved the existence of electron concentration waves in plasma at atmospheric pressure in our arc discharge reactor. The waves were excited with an arc fed by alternating current of high frequency ~10^5 sec^{-1}. Therefore, we consider here the wave influence on the C_{60} assembling in the framework of a simple model. For the conditions inside our setup we estimate the average time between two cluster collisions Δt_c (~10^{-6} sec) at 2500K and the period of oscillations of the feeding current Δt_f (~10^{-5} sec). Therefore oscillations of the electron concentration in the wave would affect the average charges of carbon clusters and, consequently, the cross-section collisions of these clusters. As the rate of any cluster pair merging R_{km} from eq. (5) is a complicated function of the electron concentration, the total rate of C_{60} molecule formation would be a function of variation of the electron concentration. To find the average rate of C_{60} formation one period of the current oscillation $\tau = 1/\nu$ was chosen as an integration time interval. We have considered oscillations of electron concentration as $n_e(t) = n_{e0}(1 + \alpha \sin(2\pi t/\tau))$. The factor α was chosen to get the ratio $\max(n_e(t))/\min(n_e(t)) \approx 10$.

In figure 6 the ratio $\overline{R_{60}^{wave}}/\overline{R_{60}^{II}}$ of the fullerene formation rate in the conditions of the existence of the waves of electron concentration $\overline{R_{60}^{wave}}$ and without the waves $\overline{R_{60}^{II}}$ is shown. Both rates were calculated by eq. (12). The electron concentration was equal to the constant $n_e = n_{e0}$ without the waves and that was $n_e = n_e(t)$ when the waves exist. In figure 6 the electron concentration logarithm corresponding to an average value of the electron concentration $< n_e(t) >$ is shown at abscissa axis for the case of wave existence.

One can see that the waves of electron concentration can change the rate of C_{60} formation significantly (~7 times as a maximum). Thus, the electron concentration waves can increase the C_{60} formation rate in the conditions of high electron concentration.

2. Calculations of Endohedral Metallofullerene *Me*@C_{84} Formation in a Carbon Plasma

Basic ideas of this section are similar with the model, proposed for pure fullerene formation. The plasma is considered in a state of partial local thermodynamics equilibrium, which is the general base for consideration of synthesis in the arc discharge plasma.

As thorough theoretical consideration of the metallofullerene synthesis in the carbon containing plasma is unrealistic at present, we introduce some simplifications in the model of metallofullerene formation.

First, the process of *Me*@C_{84} coagulation in plasma we consider as the stationary process, so concentrations of any kind of carbon cluster are kept constant. At that the plasma parameters – electron concentration n_e and temperature T are considered as independent parameters, so the metallofullerene yield is a function of them: $R_{Me@C84} = R_{Me@C84}(n_e, T)$. Though these parameters in the plasma are self-consistent, one can vary n_e significantly at constant temperature by inputting of electron donors (alkali metals) or acceptors (halides).

Second, we assume the *Me*@C_{84} yield is proportional to the rate of the *Me*@C_{84} molecule coagulation. It may be proved by the relative chemical stability of the metallofullerene molecules, so the process of their coagulation prevails over the destruction process. At the same time the *Me*@C_{84} coagulation rate is calculated in two successive stages. The *Me*@C_{84} molecule is considered as a result of merging of some carbon clusters C_m with some metal containing cluster MeC_k:

$$1^{st}\text{ stage: } Me + C_k \rightarrow MeC_k \qquad (13)$$

$$2^{nd}\text{ stage: } MeC_k + C_m \rightarrow C_{Me@84},\ k + m = 84,\ k = 2, 4, \ldots, 82.$$

Using (3 ,4 ,11), it is possible to calculate the rate of cluster C_{k+m} (MeC_{k+m}) formation from clusters C_k (or MeC_k) and C_m. Thus the total rate of *Me*@C_{84} molecule coagulation $R^{total}_{Me@C_{84}}$ in the two-stage process was defined as a sum of all possible channels of metallofullerene formation $\overline{R}_{Mek,m}$ defined by (11), on conditions that at the second stage the equality $k + m = 84$ was kept:

$$R^{total}_{Me@C_{84}} = \sum_{k=2}^{82} \overline{R}_{Mek,m}\ ,\ m = 84 - k. \qquad (14)$$

For calculations of the rates, used in (14), it is necessary to know the average concentrations n_{Mek} of a clusters MeC_k which were formed during the first stage. These concentrations were calculated as:

$$n_{Mek} = \alpha_k \tau_{Me} v_{Me,k} \cdot n_k\ , \qquad (15)$$

where n_k is the concentration of a cluster C_k, τ_{Me} – the average life time of MeC_k cluster before its dissociation to *Me* and C_k cluster, $\nu_{Me,k}$ is the average frequency of cluster C_k and atom *Me* collisions. α_k is the form factor, dependent from the cluster C_k shape.

The average life time of MeC_k cluster has been determined as $\tau_{Me} = \nu_{Me}^{-1}$, where the frequency ν_{Me} of the *Me* atom desorption from carbon cluster surface can be calculated by Arrenius type formula [37]:

$$\nu_{Me} = \nu_1 \exp(-E_c/kT), \qquad (16)$$

where ν_1 is the rate constant and E_c is cohesive energy of metal atom at carbon cluster surface. Zeiri et al. [38] have developed a theoretical description which connects the rate constant ν_1 and the surface-absorbate vibration frequency ν_{vib}

$$\nu_1 = \nu_{vib} f(T). \qquad (17)$$

The factor $f(T)$ is unity for single atoms and typically $\nu_{vib} \sim 10^{11}$–10^{13} s^{-1}. The E_c values for all MeC_k clusters have been approximately calculated as interaction energy of the atom *Me* and an infinite neutral carbon plane. We assume the shape of carbon cluster is not so important for E_c evaluation, because the metal atom forms chemical bonds with nearest carbon atoms only. Moreover, actually we calculated the cohesion energy of the metal atom and the form-averaged carbon cluster C_k, so the flat carbon plate would be good choice for the E_c determination. At that Me-carbon plane system was modeled by repetitive 18-atom flat carbon cluster C_{18} and metal atom *Me* above it.

The cohesion energy was defined as $E_c = E(\mathbf{R}_0) - E(\mathbf{R} = \infty)$, where R_0 is the distance between atom *Me* and the carbon plate in equilibrium. This equilibrium distance R_0 was determined by geometry optimization calculations. After that the potential profile E(R) of the atom *Me* being close to the equilibrium position R_0 above the carbon plate was approximated by a parabola, and then the harmonic oscillation frequency ν_{vib} had been calculated for all metal atoms.

At that the E_c and ν_{vib} values were calculated for the neutral and charged joint system $C_{18}Me$ (q=0, ±1). The calculations have shown that the charge of the system does not change E_c and ν_{vib} values essentially. Small influence of the charge on these values may be explained as the distributed among the cluster atoms additional charge does not modify the chemical bonding carbon cluster-metal essentially. Thus in the calculations the averaged by charge E_c and ν_{vib} values have been used (Table 1).

Table 1. The average values of $<\nu_{vib}>$, $<E_c>$ and life time $<\tau_{Me}>$ for different metals and charges.

System	$<\nu_{vib}>$, 10^{13} s^{-1}	$<E_c>$, eV	$<\tau_{Me}>$, 10^{-8} s at T = 3000 K
$C_{18}Fe$	1.04	3.81	4.52
$C_{18}Sc$	0.93	3.04	1.30
$C_{18}Pt$	0.47	2.47	0.29

The form factor α_k was used to take into account deviation of the globular cluster MeC_k life time τ from that of the cluster composed from the atom *Me* above the carbon flat cluster C_k. The factor α_k can lead to significant increasing of the τ value for the cluster $Me@C_k$ in which the metal atom locate inside carbon cage even this cage is not fully closed. From simple geometry one can approximately calculate $\alpha_k = S_{84}/(S_{84} - S_k)$, where S_{84} is the surface area of C_{84} fullerene molecule and S_k is the surface area of C_k cluster. At that $(S_{84} - S_k)$ is the surface of the hole in the incomplete C_{84} cluster. Obviously, when the $Me@C_k$ cluster shape is closed to that for C_{84} fullerene, the life time of $Me@C_k$ would be very large in comparison with that for any flat cluster. In the calculations $\alpha_k = 1$ has been used for all small clusters C_k with k<20.

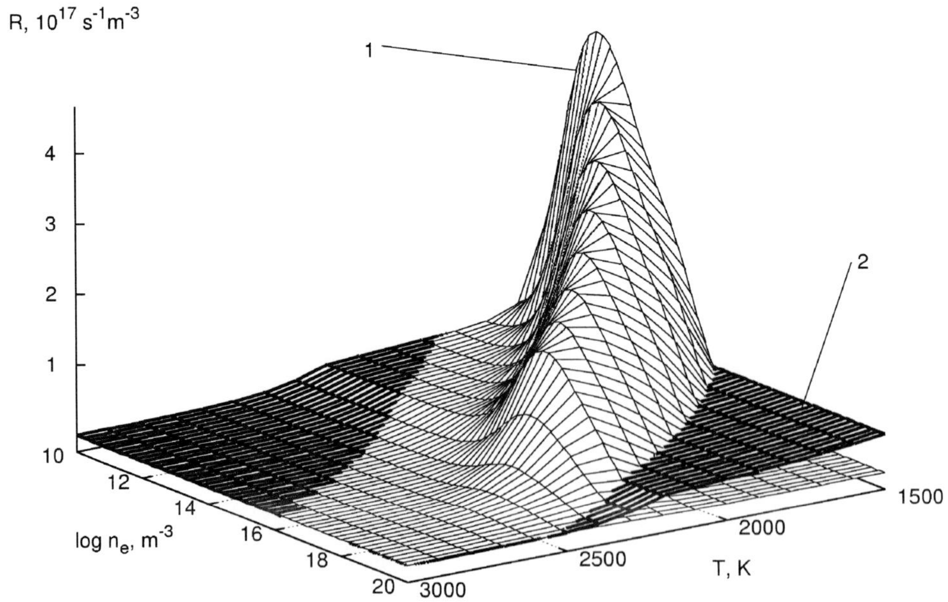

Figure 7. Rate of the $Fe@C_{84}$ molecule formation in one stage from the charged (1) and neutral clusters (2).

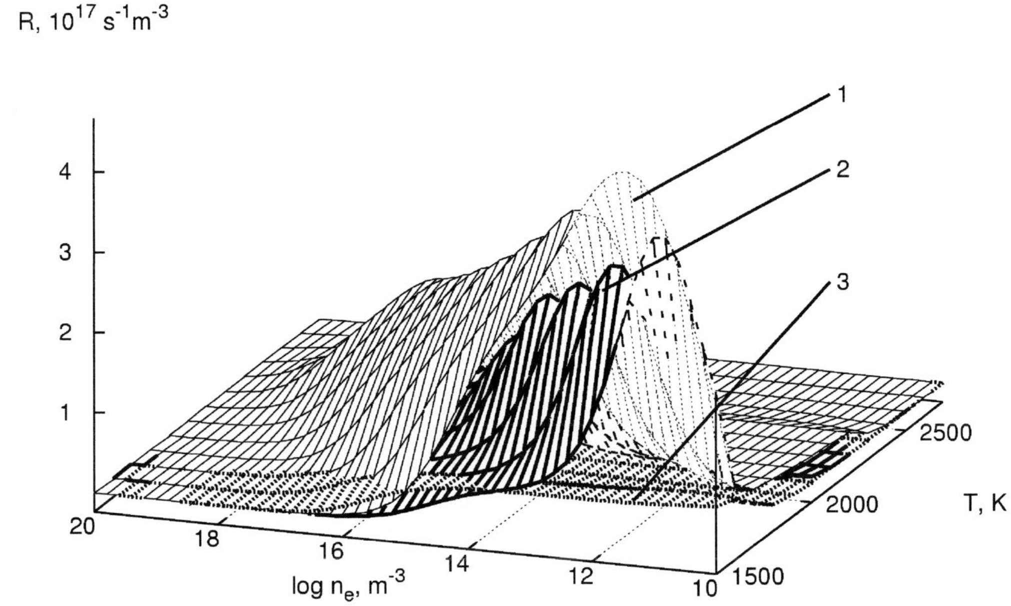

Figure 8. Rates of $Me@C_{84}$ formations: (1) $Fe@C_{84}$, (2) $Sc@C_{84}$, (3) $Pt@C_{84}$.

From the comparison of the average life time $<\tau_{Me}>$ of $Me@C_k$ cluster existence (10^{-8} s) and average time between cluster collision $\tau = 1/R_{km} \approx 10^{-6}$ s important conclusion follows that metal atoms stick to the growing carbon cluster at the last stages of metallofullerene formation.

The dependence of the coagulation rate R_{total} ($Me=Fe$) on the electron concentration n_e and the temperature T is shown in Figure 7 (surface 1). The surface 2 presents the rate of $Fe@C_{84}$ formation from neutral clusters. This rate was calculated by (14), but there neutral cluster collision cross-sections σ^0_{km} were used instead of $\sigma_{km}(q_k, q_m)$. There is a drastic difference between the rates of $Fe@C_{84}$ formation from neutral and charged clusters. Rapid decreasing of the rate at temperatures greater then 3000K (Fig. 7) is connected with exponential growth of the metal atom desorption frequency $\nu_{k,Me}$ with the temperature, see (15 , 16 , 17).

In Figure 8 the dependences $R_{total} = R_{total}$ (n_e,T) of $Me@C_{84}$ coagulation rate are shown for the metals Me={Fe, Sc, Pt}. From the data shown in the Figures 7 and 8 it is possible to determine for the electron concentration optimal for the metallofullerene $Me@C_{84}$ formation at every temperature (for example T^{opt} ~2500K and n_e^{opt} ~10^{16} m^{-3}). From the calculations we have determined the main contribution into the rate of $Me@C_{84}$ formation is given by the C_2 and $Me@C_{82}$ cluster coagulation rate. It occurs because these clusters have large charges of opposite sign ($q_2 \approx -0.5$, $q_{82,Me} \approx 0.4$) at the optimal for the $Me@C_{84}$ synthesis plasma parameters T^{opt} and n_e^{opt} as in the case of pure clusters assembling (see Fig. 5). The difference of the cluster charges stimulates essential increasing of these cluster cross-sections $\sigma_{km}(q_k, q_m)$ according (4) and, therefore, accelerates *Me*@C_{84} assembling.

Also a self-consistent electron concentration $n_e^{scf}(T)$ in the carbon-helium plasma with slight (~8% of carbon molar content) metal (Fe, Sc, Pt) impurity has been calculated using electro neutrality condition:

$$n_e^{scf}(T) = n_C^{ion}(T) + n_{Me}^{ion}(T) = \sum_k \sum_{q_k=-2}^{+1} q_k n_k(q_k) + \sum_{q=-2}^{+1} q n_{Me}(q). \quad (18)$$

Due to low metal content and relatively high IP of the metals, the $n_e^{scf}(T)$ values practically coincide for all the metal-containing plasmas and for the pure carbon-helium plasma. So, the electron concentration in the pure carbon-helium plasma is shown by thick solid line in Fig. 9 (line No.1). The ridges of the maximum formation rate surfaces of C_{84} and *Me*@C_{84} are shown by other lines. For example, the dotted line (No.3) corresponds to the ridge of the maximum rate of the Fe@C_{84} formation (see Fig. 7). One can see all the lines are close to the $n_e^{scf}(T)$ line at the temperature T < 3700 K. Also the contour line of the C_{84} formation rate is shown in Fig. 9 by a thin solid line.

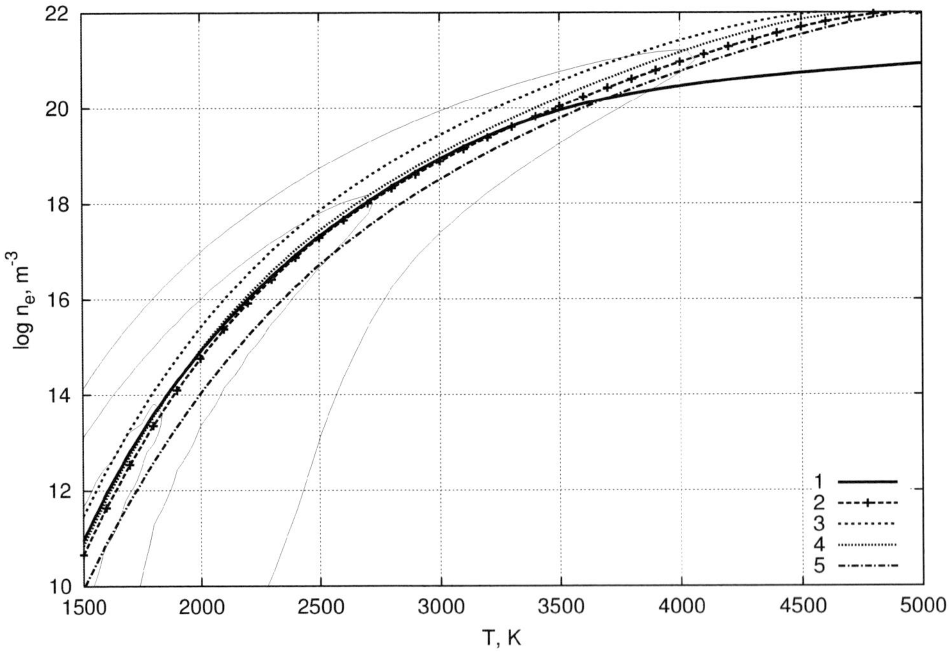

Figure 9. (1) T-dependence of the self-consistent electron concentration $n_e^{scf}(T)$, (2)- ridge of the C_{84} formation rate surface $R_{C_{84}}^{total}(n_e, T)$, (3-5)- the ridges of the Me@C_{84} formation rate surfaces (Me=Fe, Sc, Pt correspondently). Also a few contour lines of the C_{84} formation rate are shown by thin solid lines.

The dependences of the C_{84} and *Me*@C_{84} maximum formation rate (correspond to the ridges on Fig. 9) on the temperature are shown in Fig. 10. One can see the Fe@C_{84} formation rate dominates in comparison with the Sc@C_{84} and Pt@C_{84} formation rates. It can be explained by the dependencies of the metal-cluster chemical bonding, ionization potential and electron affinities on the cluster size (see Fig. 2). The union of these dependencies for pure carbon and FeC_k clusters gives a combination of the metal-carbon cluster life time τ_{Me} and cluster charges, which yields the maximum rate of the metallofullerene formation.

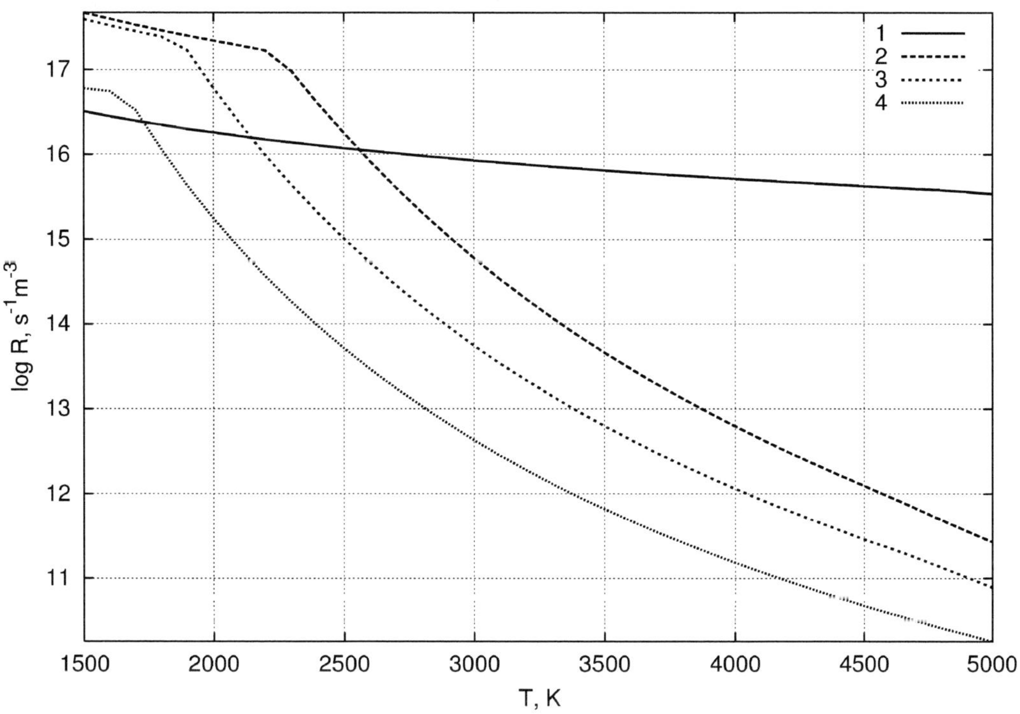

Figure 10. Dependence of the maximum formation rates of the C_{84} and *Me*@C_{84} on the temperature: (1)– C_{84}, (2) – Fe@C_{84}, (3) – Sc@C_{84}, (4) – Pt@C_{84}.

Thus, in this section it was shown that the rate of the metallofullerene *Me*@C_{84} (*Me*= Sc, Fe, Pt) formation essentially depends on the temperature and electron concentration in plasma. It was shown that the metal atom *Me* adhere to growing carbon cluster at the last stage of *Me*@C_{84} molecule formation. At the same time the formation probability is sensitive to the metal-cluster chemical bonding and the metal ionization potential value. In the row (Me=Fe,Sc, Pt) the metallofullerene yield decreases because the low ionization potential value for Sc is veiled by the low value of the MeC_n life time $< \tau_{Me} >$. At that the Pt@C_{84} yield is negligible because the Pt ionization potential is very high and the metal atom is kept near carbon clusters a very short time $< \tau_{Me} >$ only (see Table 1).

From Fig. 10 one can see the Me@C_{84} yield is increased at the tempetatures T<2500K. But the temperature about 2500K is the lower bound of temperature range in which the fullerene formation process is possible [29]. Thus metallofullerenes are formed at the last stages of pure fullerenes formation in a cooling carbon vapor. Therefore, one can make the prognosis that metallofullerene yield will be very low at direct evaporation of carbon and metal in carbon-helium plasma. Thus, the possible way of metallofullerene yield increase lies in development of special plasma sources different from ordinary arc discharge.

It is necessary to note that the calculations above may be used for exohedral metallofullerene formation too, except roles of the form-factor α_k (at that it should be equal one).

3. Influence of Buffer Gas on Fullerene Formation in a Carbon Plasma

As described in the introduction to this chapter we consider kinetics of fullerene formation on the base of the reduced description scheme, firstly proposed in [27],[28]. In this scheme all carbon clusters are distinguished by both the number of cluster atoms n and number of free radical bonds l in the cluster, so concentration $C(n,l)$ of every kind of cluster in the plasma depend on n,l. As more stable form of small carbon clusters condensed from plasma are the clusters with sp or sp2 hybridization (chains, rings, sections of grapheme planes and their combinations), in the model we consider carbon atoms with 0, 1 or 2 radical bonds per atom only. At that fullerene molecules would not have radical bonds ($l = 0$); monocyclic clusters have one radical per atom ($l = n$); chain clusters have one radical bond per atom, except terminal atoms ($l = n + 2$). All other clusters (bi-, tricyclic and unclosed fullerene shells) would contain intermediate number of radicals per atom. At this stage of model development we keep in mind the fullerenes C60 and C70 only because they are effectively obtained in experiments. Therefore in the calculations only the clusters with n = 60 and 70 could have zero number of radicals ($l = 0$).

This reduced scheme allows is to see distinct clusters of different shape and to trace the cluster transformation which usually reduces the number of free radical bonds in the cluster during annealing. At that this scheme is quite simple to use in intensive calculations of cluster kinetics.

Second, in the model all cluster concentrations $C(n,l)$ include now information about cluster vibration temperature T: $C(n,l) \rightarrow C(n,l,T)$. The variation of carbon cluster temperature is the result of new carbon bonds formation/destruction in clusters. The cluster kinetic temperature is considered to be equal to the plasma temperature.

Third, the model takes into account the influence of the buffer gas on the temperature of all clusters. If the buffer gas temperature is T_{gas}, all carbon clusters having temperature $T<T_{gas}$ would be heated and clusters with $T>T_{gas}$ would be cooled.

At last, allow non-stationary process of fullerenes formation, the model observes evolution of every cluster concentration C during time t, so the cluster concentrations are dependent on the time: $C(n,l,T) \rightarrow C(n,l,T,t)$. At that cluster distribution in space considered to be uniform, so the concentrations $C(n,l,T,t)$ do not depend on space coordinates r.

According to the model, processes of clusters collision involve information about three cluster concentrations:

$$\sum_{n_1,l_1,T_1}\sum_{n_2,l_2,T_2}(C(n_1,l_1,T_1,t)+C(n_2,l_2,T_2,t)\rightarrow C(n_1+n_2,l_1+l_2-2,T_3,t)) \qquad (19)$$

Solving of such dynamic equations for cluster concentration require long-duration calculations. So in the calculations we fixed the plasma temperature T_{gas}=3000K and electron concentration $n_e=10^{17}$ m^{-3}, which were close to the plasma parameters during fullerene synthesis.

The model describes the process of fullerene formation in the carbon contained plasma which consist from the buffer gas and different carbon clusters with concentration *C(n,l,T,t)*. To trace dependence of fullerene synthesis on buffer gas sort, in the calculations we have used He and Ar. At that the plasma is considered in the thermal equilibrium. As discussed in above sections, we have taken into account the even carbon clusters only. To model graphite evaporation from graphite electrodes or plates which is the source of carbon clusters in the plasma-chemical synthesis, at the beginning of the modeling (t=0) the dimer molecules C_2 were lived only, so $C(n,l,T,t) \neq 0$ at n=2, l=4. Here were used experimental facts, that fullerenes are formed from very small clusters [39].

According to the model, the kinetic equations for all cluster concentrations *C(n,l,T,t)* were integrated over the time t. The concentrations $C(n,l_n,T_n,t)$ (19) were changed due to several processes. The main process was the process of cluster coagulation: concentration of a cluster C_n increased in reactions of coagulation of smaller clusters $C_k+C_m\rightarrow C_n$ ($k+m=n$) and decreased in reactions of coagulation of this cluster with others $C_n+C_p\rightarrow C_{n+p}$ ($n+p\leq n_{\max}$). These processes were described by reaction constants $K_{k,m}^{l_k,l_m}$ and $K_{n,p}^{l_n,l_p}$, respectively.

The second process was the process of cluster structure transformation with rate constant $M_n^{l_n}$. We considered this transformation as step-by-step association/dissociation of inner cluster bonds with respective decreasing/increasing of radical number l in the cluster. The third process was the process of cluster heating/cooling by buffer gas atoms with rate constant $N_n^{l_n}$. At that the processes of cluster dissociation were not taken into account due to high activation energy of cluster break-down. So, the total set of dynamic equations for all cluster concentrations was:

$$\begin{aligned}\frac{d}{dt}C(n,l_n,T_n,t) = & \sum_{\substack{k+m=n\\ l_k+l_m-2=l_n}} K_{k,m}^{l_k,l_m}C(k,l_k,T_k,t)C(m,l_m,T_m,t)\\ & -\sum_{n+p\leq 120} K_{n,p}^{l_n,l_p}C(n,l_n,T_n,t)C(p,l_p,T_p,t)\\ & -M_n^{l_n}C(n,l_n-2,T_n^{+},t)-N_n^{l_n}C(n,l_n,T_n^{*},t)\end{aligned} \qquad (20)$$

From the classical collision theory used for description of the process of two cluster (C_k, C_m) coagulation: $C(k,l_k,T_k,t)+C(m,l_m,T_m,t)\rightarrow C(k+m,l_k+l_m-2,T_{km},t)$ one can get the reaction constant $K_{k,m}^{l_k,l_m}$ which is equal to $K_{k,m}^{l_k,l_m} = \sigma_{k,m}^{l_k,l_m} P_{k,m} \text{v}_{k,m} P_{bond}$; where $\text{v}_{k,m} = \sqrt{8T/\pi\mu_{\text{k,m}}}$ is the average relative velocity of both clusters, $\mu_{km} = m_k m_m/(m_k + m_m)$ is the reduced mass of C_k and C_m ; $\sigma_{k,m}^{l_k,l_m} = \sigma_{k,m}^{l_k,l_m}(q_k, q_m)$ is the cross-section of two cluster collision depending on the cluster charges q_k, q_m. We assume here that the cross-section $\sigma_{k,m}^{l_k,l_m} = \pi r_b^2 l_k l_m$ of two cluster coagulation is proportional to the numbers of free radical bonds l_k и l_m in the both clusters. It means that both clusters can be associated together in the places of dangling radical bonds only. r_b here is the affected zone of single free radical bond influence. In all calculations r_b =1.7Å have been used. After coagulation the number of radical bonds in joined cluster was equal $l = l_k + l_m - 2$. Further decreasing of l would happen at subsequent internal transformation of the cluster. At that just after coagulation the joined cluster temperature is $T_{k+m} = (kT_k + mT_m)/(k+m)$, where k and m are the number of atoms in clusters C_k and C_m.

Energy barrier E_a for both clusters association lead to modification of the rate of coagulation on Arrenius factor $P_{bond} = e^{-E_a/T_{gas}}$, where T_{gas} is the plasma temperature. Due to extreme complexity of E_a calculations for all clusters, it was used an average value E_a= 2.0 eV, received from [40].

According to the collision theory of charged particles, if the clusters C_k and C_m have charges q_k and q_m respectively, the cross-section of the cluster collisions $\sigma_{k,m}^{l_k,l_m}$ is multiplied on factor $P_{k,m}$ (9).

The values of IP_k^q were calculated analogously with previous sections and were interpolated for all $k \in \{2–120\}$.

For the processes of inner cluster transformation the constant M_n^l, could be calculated in the way similar with association processes:

$$M_n^l = \nu_{phon} \exp(-E_{bar} / k_B T_{n,l}); \tag{21}$$

Here $T_{n,l}$ is the vibration cluster temperature, which is defined from the energy of cluster phonon vibrations. Because of difficulty of all phonon frequency calculations for all clusters, we used the average value $\nu_{phon} = 10^{11}$ s^{-1} for all clusters $C(n,l,T,t)$. This value was calculated by averaging of phonon frequencies for set of different carbon clusters.

The processes of cluster heating or cooling take place if the cluster vibration temperature $T_{n,l}$ is different from the buffer gas temperature T_{gas} (plasma temperature). Because of complexity of such cluster cooling/heating rate calculation, we applied a simple approach for such calculations. This approach uses the fact that an average velocity $< \text{v}_{gas} >$ of buffer gas molecules at $T_{gas} \sim 3000\,\mathrm{K}$ are $< \mathrm{v}_{gas} > \sim 5 \cdot 10^3$ m/s and interaction radius between gas molecule and a carbon atom of the interacting cluster is ~1Å. At that it is easy to estimate that time of both particle interaction $\tau_{\mathrm{int}} \le \nu_{phon}^{-1}$. So the collision of the buffer gas molecule and a carbon cluster can be considered as a two-stage process. At first the elastic collision of the gas molecule and a carbon atom in the cluster structure occurs. An increment or decrement of carbon atom kinetic energy occurs. Because this atom is chemically bonded with other atoms in the cluster, at the next stage the redistribution of vibration energy in the cluster occurs. Therefore, on the average, buffer gas collisions lead to cooling of hot clusters ($T_{n,l} > T_{gas}$) and heating of cold ones ($T_{n,l} < T_{gas}$).

It was considered that the vibration temperature of a cluster C_k is decreased by the rate

$$dT_k \,/\, dt = \Delta E_{He} \cdot \nu_{He} \,. \tag{22}$$

Here

$$\nu_{He} = n_{He} V_{k,He} \sigma_{k,He} \tag{23}$$

is the collision frequency of a cluster C_k with helium atoms, n_{He} is the buffer gas concentration; $V_{k,He}$ is the velocity of relative movement of a He atom and a cluster C_k; and $\sigma_{k,He} = \pi (r_k + R_{He})^2$ is their collision cross section;

$$\Delta E_{He} = m_{He} \Delta V_{He}^{\;2} \,/\, 2 \tag{24}$$

is the change of the helium atom kinetic energy upon the collision with a carbon cluster; $\Delta V_{He}^{\;2} = < V_{He}^2 - (V'_{He})^2 >$ is the change of the squared velocity of a helium atom before and after the collision averaged over the Maxwell distribution of helium atom velocities, where

$$V'_{He} = \frac{(m_{k,He} - m_k) \cdot V_{He} + 2 m_k V_{eff}}{m_k + m_{He}} \tag{25}$$

is the velocity of a helium atom after the elastic collision with a carbon atom of the cluster contacted with it. $V_{eff} = \pm V_{He} \pm V_C$ is the relative velocity of a helium atom and a carbon atom of the cluster contacted with it. Upon the collision, the velocity of this carbon atom may be estimated as the total velocity $V_C = V_T \pm V_{vib}$, where $V_T = \sqrt{k_B T / m_k}$ is the average translation velocity of the cluster, and $V_{vib} = \sqrt{1/3 \cdot E_{vib} / m_C}$ is the characteristic velocity of oscillations of the cluster walls and

$$E_{vib} = \frac{3k-6}{2} k_B T \cdot \frac{1}{k} \tag{26 }$$

is the average oscillation energy of the cluster per atom, where $(3k-6)$ is the degrees of freedom.

Also we estimated the probabilities of inelastic processes of cooling/heating with involving of ionized or exited states of buffer gas atoms (He, Ar). The values of these probabilities are proportional to Arrenius-type factor $p \sim \exp(-E / k_B T_{gas})$ where E is ionization potential IP or first excitation energy E* of the buffer gas atom. For argon and helium these values are: $IP_{Ar} = 15.8\,\text{eV}$, $E^*_{Ar} \sim 10\,\text{eV}$, $IP_{He} = 24.6\,\text{eV}$, $E^*_{He} \sim 14\,\text{eV}$ correspondently. Even for the lowest of these energies $p \sim 10^{-15}$ at $T_{gas} = 3000\,\text{K}$, so at temperature under consideration accounting of buffer gas ions or excited states can be omitted.

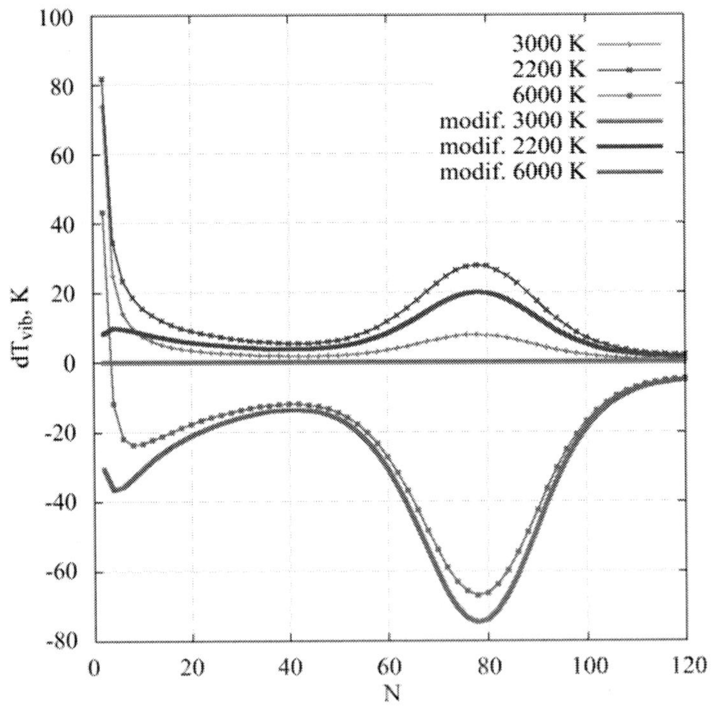

Figure 11. Dependence of the rate of cluster vibration temperature change on number of atoms N in it due to collisions with He atoms for different cluster temperatures. The original (thin lines with markers) and corrected (thick lines) dependences are shown. dT_{vib} is average change of vibration temperature during one time step $\Delta t = 10^{-9}$ s.

Then an average change of the cluster vibration temperature due to the collisions with buffer gas molecules was calculated on the base of the proposed model. This change of cluster temperature $\tilde{f}(n,T_{n,l},T_{gas})$ during one time step (10^{-9} s) is shown in Figure 11. In Fig. 11 one can see (thin lines with markers) that even this simple model describes well the cooling/heating processes, which means that any cluster having the vibration temperature higher (colder) than the plasma (buffer gas) temperature would be cooled (heated) at collisions. For more exact accordance, function $\tilde{f}(n,T_{n,l},T_{gas})$ was modified as: $f(n,T_{n,l},T_{gas})=\tilde{f}(n,T_{n,l},T_{gas})-\tilde{f}(n,T_{gas},T_{gas})$. This function is shown in the Fig. 11 by thick lines for T_{gas}=3000K.

While as we have defined response rate for all processes of cluster concentration modification, the dynamical equations for all cluster concentrations have been integrated over time. In the calculations for integration over time, the step $\Delta t = 10^{-9}$ s and the total time duration $t_{final} = 2\cdot 10^{-4}$ s were used.

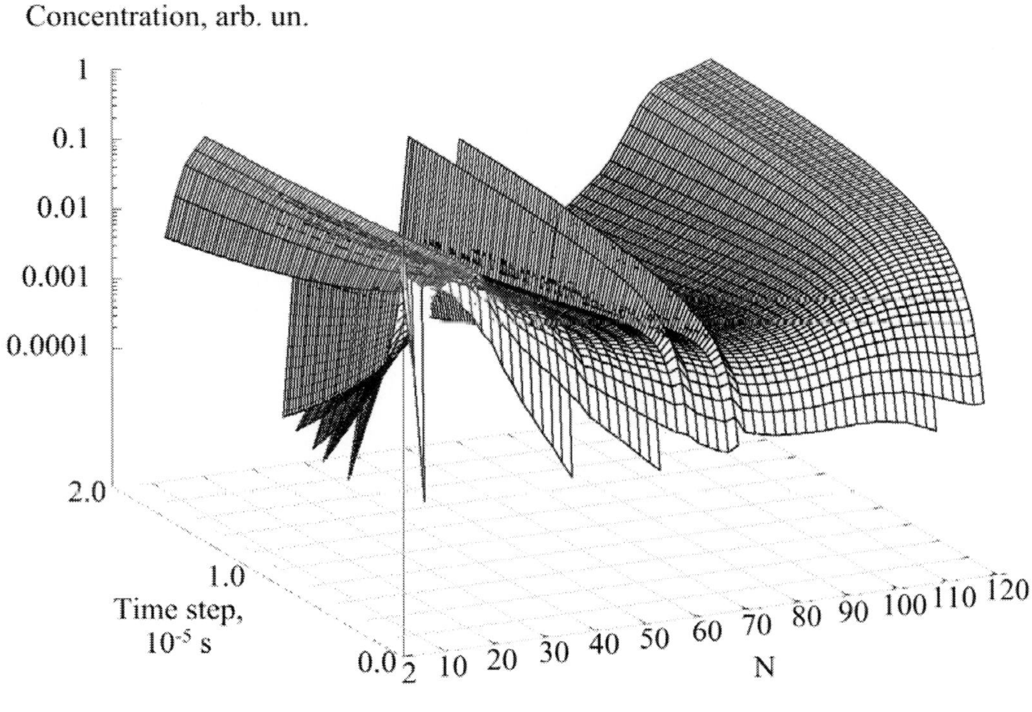

Figure 12. Change of the integrated cluster distribution $C(N,t)$ during the time. N is the number of atoms in carbon cluster.

In Fig. 12 the change of the cluster distribution $C(N,t)=\sum_{l,T} C(N,l,T,t)$ integrated over radicals and vibration temperatures is shown. One can see the narrow peaks of C_{60} and C_{70} concentration growth. The growth of C_{60} and C_{70} clusters concentration occurs due to the presence of C_{60} and C_{70} fullerenes which have no radicals ($l = 0$). These peaks are explained by the absence of free radicals in fullerene molecules which suppress further reactions of fullerene coagulation with other clusters. All other clusters having free radicals are coagulated into bigger clusters (with final N=100-120 atoms in our calculations) which form soot particles in experiments of fullerene synthesis. In our calculations we associate the growth of big clusters with N=100-120 with soot particles,too.

The second fact leading to fullerene accumulation is a very low probability of bond break in fullerenes in comparison with other clusters, having radicals. Fullerene molecules have compact and stable geometrical structure. Therefore, after bond breaking between two atoms due to geometry steadiness then atoms immediately form the bond again, i.e. the time of radical existence in fullerene is smaller than average time between collisions of this molecule with other carbon clusters. So radical formation in fullerenes is negligible in the model.

Because of low yield of higher fullerenes (C_{76}, C_{80}, C_{82}, etc.) in experiments, we neglect such molecules which have no radicals ($l = 0$) in the current stage of the model. But such molecules having radicals are taken into account. Despite this approximation the calculation results show the basic features of the fullerene formation process.

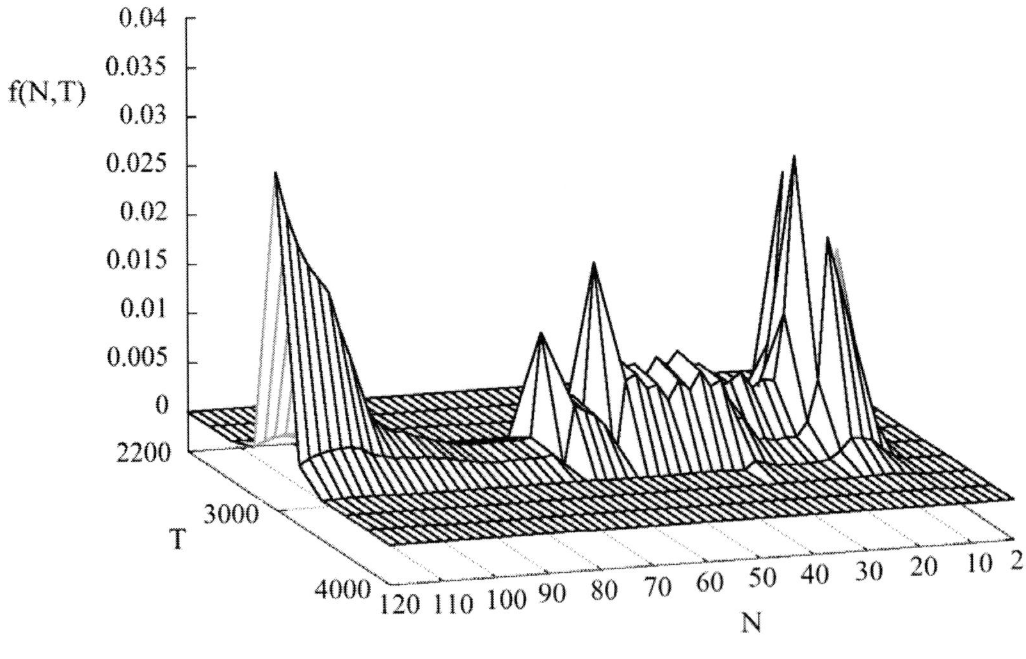

Figure 13. Integrated over the radical number cluster distribution $C(N,T)$ at the last time point. N is the number of atoms in carbon cluster. T is vibration temperature of the cluster.

In Fig. 13 the cluster distribution $C(N,T) = \sum_l C(N,l,T,t_{final})$ integrated over radical number at the last time step t_{final} is shown. One can see that the concentration of small clusters is big. The distributions along cluster temperature for every N have a quite smeared peak near the gas temperature T_{gas}=3000K. It proves that any model of fullerene formation should take into account significant temperature variation of clusters.

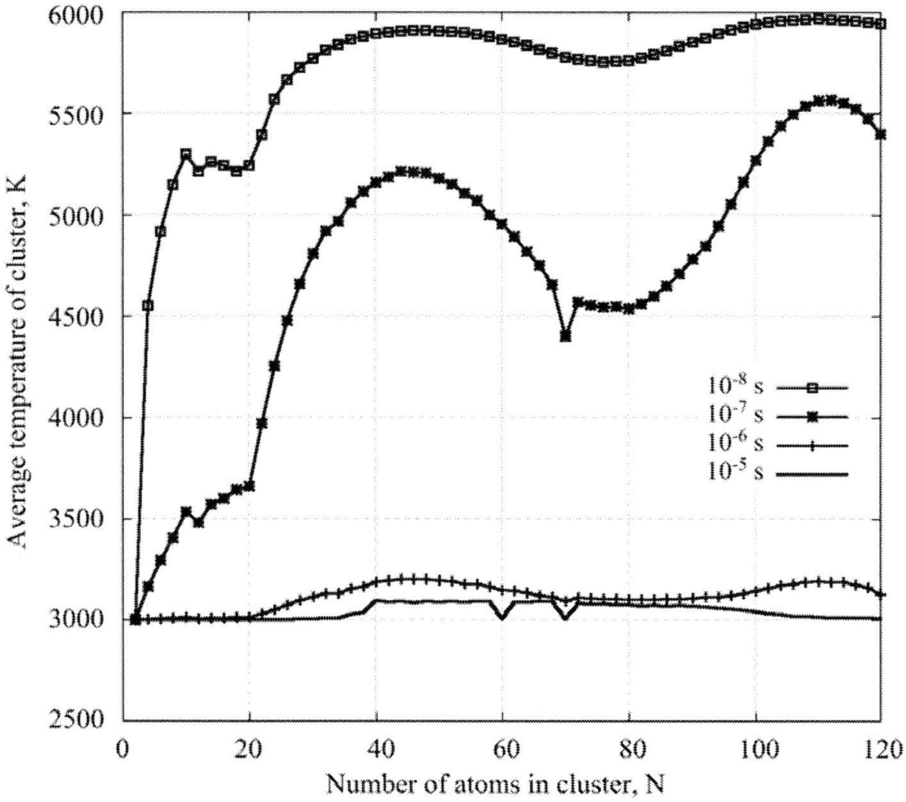

Figure 14. Average cluster temperatures at different time moments in helium. T_{gas}=3000K.

In Fig. 14, dependences of cluster average temperatures on number of atoms in it are shown for different time moments in helium. At the beginning of coagulation, carbon clusters are heated by the formation of new chemical bonds. Here their cooling rate by buffer gas occur rather quickly. At the time of $10^{-6} - 10^{-5}$ s all average temperatures become close to the gas temperature. The peculiarities at N=60,70 arise because of fullerenes accumulation. Fullerenes have no free radicals therefore they can only be cooled by buffer gas and can not be heated by forming new bonds.

In Fig. 15, the full distribution of C_{60} clusters over the cluster temperature T_{vib} and the number of radicals is shown at the final time step. The concentration of fullerene C_{60} with no radicals (L = 0) having vibration temperature T_{vib} equal to gas temperature (T_{vib} = 3000K) prevails over other concentrations by order because of its stability. During the processes of carbon coagulation the C_{60} clusters with high number of radicals (monocyclic rings etc.) are formed from smaller clusters. They have average T_{vib} near the T_{gas} because they have the most intensive interaction with buffer gas due to their relatively big cross section. During decreasing of radical number in the cluster, its vibration temperature is increased due to formation of new chemical bonds. Therefore, the most C_{60} clusters with $L \leq 30$ have high vibration temperatures (about 6000K). But substantial part of all fullerene molecules C_{60} (with $L = 0$) have the average T_{vib} equal to T_{gas} again because it cooled by buffer gas only and does not get heated due to new bonds formation.

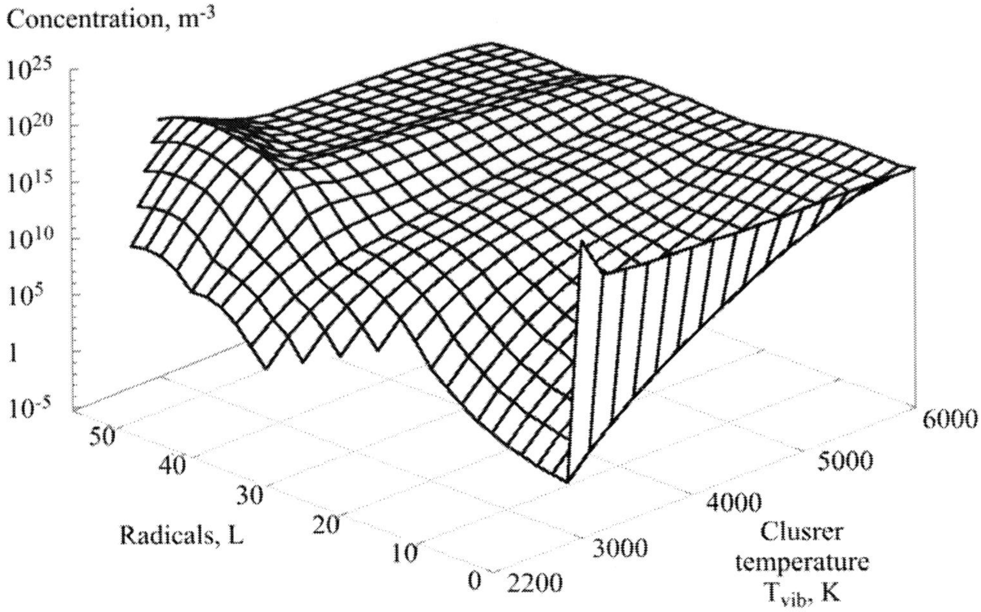

Figure 15. Distribution of C_{60} cluster $C(60, L, T, t_{final})$ over the radical number L and the cluster temperature T_{vib} at the last time point . T_{gas}=3000K.

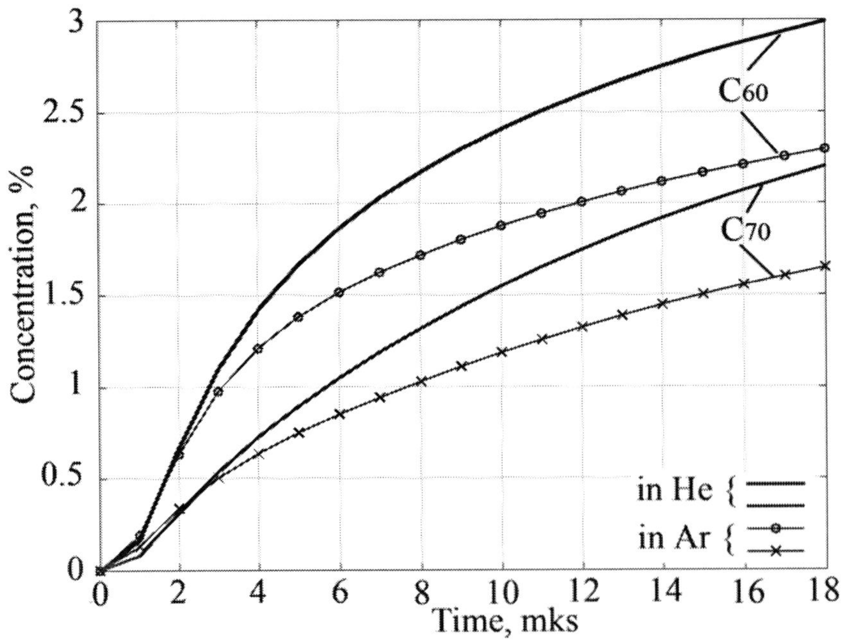

Figure 16. Change of the C_{60} and C_{70} fullerene concentration in argon and helium during the time.

Figure 16 shows the main result of the calculations. The curves show the fullerenes C_{60} and C_{70} growth in different buffer gases. The total fullerene concentration in helium (5.2%) is about 30% higher than in argon (4.0%). This result qualitatively coincides with experimental data. The difference in fullerene yield in argon and helium is lower then in experiments. It can be explained by variations of plasma parameters in space and in time in experimental conditions, which were not included in the model. Due to these variations the mass and heat transfer in the plasma can play an important role. But in the suggested model the processes of heat conductivity are not considered because in the model the cluster distribution depends on time but it is uniform in space.

However, the model shows that the mass of buffer gas molecules plays the essential role in the carbon clusters cooling/heating processes. Here the obtained results qualitatively coincide with experimental data. Further development of the model taking into account cluster dissociation and higher fullerenes with detailed consideration can give a more exact difference between fullerenes yield in argon and helium and a more reliable C_{60}:C_{70}:$C_{N>70}$ ratio.

Conclusion

The dependence of the C_{60} fullerene and Me@C_{84}endofullerene formation rates on the electron concentration and the temperature in partially ionized carbon vapor was theoretically investigated. Both C_{60} and Me@C_{84} assembling was treated as a result of carbon cluster merging in one or two stages. It was shown that taking the carbon cluster charges into account drastically changed the rate of molecule C_{60} or Me@C_{84} formation because the charges of the clusters change the cross section of collisions of these clusters.

On the basis of these calculations the optimal electron concentrations of carbon contained plasma can be determined for the temperatures realized in experimental plasma-chemical setups.

In the framework of our model it was shown that the waves of electron concentration in plasma can significantly increase the rate of fullerene assembling. Under the waves influence the additional changes of different cluster concentration are accumulated during the wave period and the rate of molecule C_{60} assembly is increased.

It was shown that the metallofullerenes are formed at the last stages of pure fullerenes formation in a cooling carbon vapor because of the short lifetime of the metal atom on the cluster surface. At the same time the formation probability is higher for metals with lower ionization potential. Based on these facts, one can make the prognosis that metallofullerene yield will be very low at direct evaporation of carbon and metal in carbon-helium plasma. Thus, the possible way of increasing metallofullerene yield lies in the development of special plasma sources different from ordinary arc discharge.

The kinetic model of carbon cluster coagulation in a carbon contained plasma was developed taking into account the processes of cluster heating/cooling in chemical bond formation and in collisions with buffer gas. The kinetic calculations have shown that the efficiency of fullerene formation depends on carbon cluster cooling during coagulation processes. The calculations qualitatively reproduced the fact that the fullerene yield in helium is higher than in argon. Further development of the model taking into account cluster dissociation and higher fullerenes with detailed consideration can give a more exact difference between fullerenes yield in argon and helium and a more reliable ratio of fullerene concentrations.

References

[1] Kratschmer, W.; Fostiropoulos, K.; Huffman, D.R. *Chem. Phys. Lett.* 1990, 170, 167.

[2] Lozovik, Yu.E; Popov, A.M. *Physics Uspekhi* 1997, 40, 7, 717–737.

[3] Curl, R.F. *Nature* 1993, 363, 14–15.

[4] von Helden G; Gotts NG; Bowers MT. *Nature* 1993, 363 , 60-33.

[5] Heath, J.R. in Fullerenes: Synthesis, Properties and Chemistry of Large Carbon Clusters; Hammond, G.S.; Kuck, V.J.; *Am. Chem. Soc.*: Washington, DC, 1991;pp 1–23.

[6] Hernandez, E.; Ordejon, P.; Terrones, H. *Phys. Rev. B* 2001, 63, 193403(4).

[7] Afanas'ev, D.V; Dyuzhev, G.A.; Karataev, V.I. *Tech. Phys. Lett.* 1999, 25, 3, 182–184.

[8] Alekseev, N.I.; Dyuzhev, G.A. *Tech. Phys.* 1999, 44, 1435–1439.

[9] Churilov, G.N.; Fedorov, A.S.; Novikov, P.V. *Carbon* 2003, 41, 173.

[10] Churilov, G.N.; Novikov, P.V.; Taraban'ko, V.E.; Lopatin, V.A.; Vnukova, N.G.; Bulina, N.V. *Carbon* 2002, 40, 891-896.

[11] Churilov, G.N.; Solovyov, L.A.; Churilova, Y.N.; Chupina, O.V.; Malcieva, S.S. *Carbon* 1999, 37, 421-431.

[12] Kroto, H.W.; Heath, J.R.; O'Brien, S.C.; Curl, R.F.; Smalley, R.E. *Nature* 1985, 318, 162.

[13] Heath, J.R.; O'Brien, S.C.; Zhang, Q.; Liu, Y.; Curl, R.F.; Kroto, H.W.; Tittel, F.K.; Smalley, R.E. *J. Am. Chem. Soc.* 1985, 107, 7779.

[14] Chai, Y.; Guo, T.; Jin, C.; Haufler, R.E.; Chibante, L.P.F.; Fure, J.; Wang, L.; Alford, J.M.; Smalley, R.E. *J. Phys. Chem.* 1991, 95, 7564.

[15] Shinohara, H.; Sato, H.; Saito, Y.; Ohkohchi, M.; Ando, Y. *J. Phys. Chem.* 1992, 96, 3571.

[16] Kikuchi, K.; Suzuki, S.; Nakao, Y.; Nakahara, N.; Wakabayashi, T.; Shiromaru, H.; Saito, K.; Ikemoto, I.; Achiba, Y. *Chem. Phys. Lett.* 1993, 216, 23.

[17] Takata, M.; Umeda, B.; Nishibori, E.; Sakata, M.; Saito, Y.; Ohno, M.; Shinohara, H. *Nature* 1995, 377, 46.

[18] Yamaguchi, Y.; Maruyama, S. *Eur. Phys. J. D* 1999, 9, 385.

[19] Fedorov, A.S.; Novikov, P.V.; Churilov, G.N. *Chem. Phys.* 2003, 293, 253.

[20] Shinohara, H.; Sato, H.; Ohchochi, M.; Ando, Y. et al. *Nature* 1992, 357, 52.

[21] Nishibori, E.; Takata, M.; Sakata, M. et al. *Chem. Phys. Lett.* 1998, 298, 79.

[22] Iijima, S.; Ichihara, T. *Nature*, 1993, 603, 363.

[23] Thess, A. et al., *Science* 1996, 483, 273.

[24] Pradeep, T.; Kulkarrii, G.U.; Kannan, K.R. et al., *J. Am. Chem. Soc.* 1992, 114, 2272.

[25] Fedorov, A.S.; Novikov, P.V.; Martinez, Yu.S.; Churilov, G.N. *J. Nanosci. Nanotech.* 2007, 7, 1-6.

[26] Balch, A.L.; Olmstead, M.M. *Chem. Reviews* 1998, 98, 2149.

[27] Alekseev, N.I.; Dyuzhev, G.A. *Tech. Phys.* 1999, 69, 104.

[28] Alekseev, N.I.; Dyuzhev, G.A. *Tech. Phys.* 2001, 71, 67.

[29] Yamaguchi, Y.; Maruyama, S. *Chem. Phys. Lett.*, 1998, 286, 336.

[30] Mitchner, M.; Kruger, C.H. Partially Ionized Gases; John Willey & Sons: New York, 1973.

[31] Nerushev, O.A.; Sukhinin, G.I. *Tech. Phys.* 1997, 42, 160–167.

[32] Cox, D.M.; Reichmann, K.C.; Kaldor, A. *J. Chem. Phys.* 1988, 88, 1588–1597.

[33] Vincenti, V.; Kruger, C.H. Introduction to Physical Gas Dynamics; Wiley, 1965; pp 52.
[34] Landau, L.D.; Lifshitz, E.M. Mekhanika (Mechanics). Nauka: Moscow, 1973. Translated into English Pergamon Press: Oxford, 1980.
[35] Kresse, G.; Furthmuller, *J. Phys. Rev. B* 1996, 54, 11169.
[36] Vanderbilt, D. *Phys. Rev. B* 1990, 41, 7892.
[37] Redhead, P.A. *Vacuum* 1962, 12, 203.
[38] Zeiri, Y.; Redondo, A.; Goddard, W.A. *Surf. Sci.* 1983, 131, 221.
[39] Ebbesen, T.W.; Tabuchi, J.; Tanigaki, K. *Chem. Phys. Lett.* 1992, 191, 336.
[40] Strout, D.L.; Scuceria, G.E. *J. Phys. Chem.* 1996, 100, 6492.

In: Progress in Fullerene Research
Editor: Milton Lang, pp. 299-317

ISBN: 978-1-60021-841-5

Chapter 9

MEMBRANES BASED ON POLYPHENYLENE OXIDE MODIFIED BY FULLERENE C_{60}

***G. A. Polotskaya*[1] *and A. M. Toikka*[2]**
[1]Russian Academy of Sciences, Institute of Macromolecular Compounds, Bolshoy pr.31, Saint-Petersburg, 199004, Russia
[2]Saint-Petersburg State University, Department of Chemical Thermodynamics & Kinetics, Universitetsky pr., 26, Petrodvoretz, Saint-Petersburg, 198504, Russia

Abstract

Polymer membrane efficiency depends on polymer structure and specific properties. Polymer modification by nano-size particles is a way for the development of new membrane materials. Polyphenylene oxide (PPO) modification by fullerene incorporation can be made by two modes. One of them is dispersing fullerene C_{60} and PPO that leads to composition with donor-acceptor binding. Another mode is chemical modification by a three-step synthesis that leads to obtaining covalently bonded PPO-C_{60}. Transport properties of dispersed and bonded PPO-C_{60} are compared exclusively in gas separation. PPO modified by dispersing fullerene is successfully used for preparation of homogeneous (dense) and asymmetric (porous) membranes. The effect of fullerene addition on structure and transport properties of these membranes is considered in gas separation, pervaporation, and ultrafiltration processes. Homogeneous membranes based on PPO-C_{60} compositions are effective in pervaporation of the reacting mixture: ethanol – acetic acid – water – ethyl acetate in binary and quaternary combination of components. The use of membranes modified by fullerene leads to shifting of the reaction equilibrium and maximizing ethyl acetate yield in the hybrid process. The asymmetric porous membranes containing fullerene can be applied for the removing estrogenic pollutants from water. The membrane efficiency to estrogens removal is estimated in adsorption and ultrafiltration processes. Transport properties of membranes under study are improved by fullerene incorporation.

1. Introduction

Membrane technologies have been more and more frequently applied in industrial and ecological processes due to its operational simplicity, low power consumption, modular and compact equipment comparing to physical and chemical analogues. Synthetic membranes that are mainly used for the industrial application could be divided in two base classes: organic and inorganic ones. The majority of organic membranes is made on the base of polymer materials and presents the most important type of membrane for separation and purification technology [1]. One of the main advantages of polymer membranes for separation of complex mixtures is connected with the possibility to use a wide variety of polymer materials for different tasks. The selectivity of membrane separation processes depends on a numerous factors: the chemical nature of macromolecules, physical structure of dense or composite membrane, properties of separated mixture, etc. Currently a great number of polymer membranes were developed for separation, purification and concentration of different liquid and gas mixtures. Unfortunately, most of these membranes are used for laboratory work only and have no significant application for the industrial purposes. Thus in the case of nonporous isotropic membranes the transmembrane fluxes are too low for the separation or purification in industrial practice. As a result the assortment of industrially produced polymer membranes is limited. The search of new membrane materials both polymeric and composite with high productivity and selectivity is an important task of chemical engineering [1, 2].

Development of new membranes based on hybrid polymer materials is a promising trend of membrane technology. Modification of polymer materials by incorporation of ultra- or nano-size particles leads to a pronounced effect on material properties [3-7] and accordingly on membrane separation efficiency [8-13]. Fullerenes have attracted much attention due to its unique chemical and physical properties [14, 15]. Recent research shows that fullerenes exhibit great potential for improving polymer properties. In the case of fullerene addition to polymer and their possible interaction, the π-electron system of the fullerene molecule is subjected to minimum transformations. Fullerene retains its unique properties in the polymer matrix, but the properties of the polymer itself are transformed [16-19].

In the last few years, some research on polymer membranes modified by fullerenes came in to literature [20-34]. Among the first polymers modified by fullerene was polyphenylene oxide which gas separation properties enhances with increasing fullerene content [22]. In the subsequent research on polystyrene modified by dispersing 1-2 % fullerene, the prospects for polymer modification by fullerene to improve gas transport properties have been suggested [23, 24]. This effect is directly related to structure peculiarities of new membrane materials. Membranes consisting of Matrimid and the dispersed benzylamine-modified fullerene have been studied in [25]. Matrimid and the fullerene derivative interact strongly, exhibit enhanced glass transition temperature and density. Increase in C_{60} content leads to increasing He/N_2 selectivity and decreasing gas permeability due to the presence of the benzylamine-modified fullerene that serves as impenetrable volumes and rigidifying elements within the polymer matrix.

The most recent research concerning the effect of fullerene on the properties of membrane materials is devoted to polyphenylene oxide – fullerene C_{60} compositions [22, 26-34]. In this chapter the effect of C_{60} inclusion in polymer membranes is analyzed in details by the example of poly-(2,6-dimethyl-1,4-phenylene oxide) (PPO). PPO being a commercially available polymer has been investigated for various membrane applications [35-39]. PPO is one of the most gas permeable glassy polymers with sufficiently high free volume [1, 37-44]. However, relatively low selectivity limits PPO application in gas separation. To improve gas transport properties, attempts have already been made to modify PPO by harmful chemical methods (bromination, sulphonation, silylation, etc) that are not enough efficient for promoting [45-49]. The modern approach of PPO modification by fullerene C_{60} inclusion has been made by two different modes. Chemical modification by a three-step synthesis leads to obtaining PPO-C_{60} bonded covalently to methyl group of PPO [28, 29]. Dispersion of fullerene C_{60} and PPO in solution or powder leads to PPO-C_{60} compositions with nonvalent binding [29, 32]. The gas separation properties of bonded and dispersed PPO-C_{60} differ significantly: this feature is discussed in the presented chapter. Unfortunately, information on PPO-C_{60} bonded is limited by gas separation only. Therefore, much attention is given to membranes based on PPO-C_{60} dispersed. Physical properties of PPO-C_{60} compositions, membranes preparation and their structure will be a subject of consideration. We will pay a peculiar attention on transport properties of membranes based on PPO modified by fullerene C_{60} in the processes of gas separation, pervaporation, and ultrafiltration.

2. PPO- C_{60} Dispersed Composition

PPO-C_{60} compositions containing up to 2-4 wt% C_{60} for gas separation and pervaporation can be prepared by the mixing toluene solutions of PPO and fullerene in the required quantity. These compositions are used for the obtaining homogeneous membranes. PPO-C_{60} compositions containing 4 and above wt% C_{60} are prepared by the mixing powders of PPO and fullerene in the required quantity and can be used for asymmetric membranes preparation.

The mixing of poly(2,6-dimethyl-1,4-phenylene oxide) (PPO) and fullerene C_{60} results in the forming donor-acceptor complex due to the interaction between the π-electron system of C_{60} molecule with phenyl ring and unshared electron pair of oxygen atom in PPO. Moreover, C_{60} being in complex with PPO molecule is capable of additional intermolecular interaction that leads to physical cross-linking of PPO chains by C_{60} molecules.

Hydrodynamic and electrooptic studies show that these complexes are not decomposed in solution [50]. This indicates that the PPO-C_{60} nonvalent bonds are relatively strong. Hydrodynamic studies showed that intrinsic viscosity and the asymmetry of PPO molecule decrease. C_{60} molecules exert a great orienting effect on PPO chain segments. It should be noted that the hydrodynamic radius of an equivalent sphere of PPO chain in isolated state is ~ 50 times greater then the radius of the C_{60} molecule.

The PPO-C_{60} complex does not destroyed after transition from liquid solution to solid state. Photoluminiscent spectra show the existence of molecular complexes in PPO-C_{60} films containing up to 2 wt% C_{60}, whereas the major part of C_{60} molecules are not bonded to PPO in PPO-C_{60} films containing 4 wt% C_{60} [51].

Inclusion of heterogeneous particles in polymer systems could be followed by compacting or loosening the polymer chain structure. The character of structural changes could be partly estimated on the base of the data on the density of PPO samples (films) with different C_{60} contents. In particular the molecular packing coefficients and free volume values were calculated (Table 1).

Table 1 shows that density of fullerene-containing films increases with the addition of fullerene, that is, polymer coils become more compact. Such transformation manifests itself in the increasing molecular packing coefficient and the decreasing free volume of C_{60}-PPO compositions [32].

Table 1. Physical properties and parameters of PPO-C_{60} films.

Polymer	Density, g/cm^3	Molecular packing coefficient	Free volume, cm^3/g	ρ_v 10^{15}, Ω cm
PPO	1.057	0.623	0.181	1.76
PPO+1%C_{60}	1.093	0.639	0.150	-
PPO+2%C_{60}	1.120	0.649	0.128	68.1

ρ_v –specific electric resistance

Method of dielectric losses and polarization was used for studying relaxation properties and C_{60} molecules distribution in PPO matrix. PPO is a slightly polar glassy polymer. The most of C_{60} molecules are distributed more uniformly in PPO-C_{60} composition then in fullerene-containing polystyrene [24]. Specific electric resistance increases with the rising of C_{60} concentration in films (Table 1).

Contact angle is the most widespread characteristic of solid surface tension. Several angles formed by the liquid drop on solid surface are common observed. The maximum (advancing, θ_a) and minimum (receding, θ_r) angles were determined by the Wilhelmy plate technique (tensiometric method) from the values of force exerted as the sample was immersed into a liquid or withdrawn from it [52]. PPO is a hydrophobic polymer since water contact angle on its surface is greater than 90^0 (Table 2). C_{60} inclusion increases contact angle and hydrophobicity. In contrast, PPO surface is well moistened in ethanol. C_{60} addition to PPO decreases contact angles due to enhance affinity of ethanol. These results are important for the explanation of C_{60} effect on PPO sorption and pervaporation properties.

Table 2. Contact angles (degree) for PPO-C_{60} membranes.

Membrane	Water		Ethanol	
	θ_a (°)	θ_r (°)	θ_a (°)	θ_r (°)
PPO	91.2	42.6	20.2	16.1
PPO+2%C_{60}	94.4	51.5	17.9	15.1
PPO+10%C_{60}	102.1	58.8	13.9	12.5

3. Homogeneous Membranes

In this section we present main results concerning properties of homogeneous fullerene-containing PPO membranes and separation processes with their use.

Homogeneous membranes with dense nonporous structure of ~60 μm thickness are obtained from 5 wt% polymer solution in toluene by casting on a cellophane surface and drying at 40°C. Fullerene-containing PPO membranes are of brown color; the intensity of coloration depends on the quantity of C_{60} in composition.

Separation through homogeneous nonporous membranes is a result of different permeation rate of species that are components of the mixture to be separated. Gas separation and pervaporation processes differ only by aggregate state of feed mixture (gas and liquid accordingly). The gradient of the pressure is a driving force in both processes. The main approach which is commonly used for the consideration of the mechanism of these processes is so-called "solution–diffusion model" [see, e.g. 2]. According to this model the permeation of species through membrane includes following stages: sorption, solution, diffusion, and desorption. The separation of species could be determined by following main factors: *i.* the difference in solubility of species in the material of membrane; *ii.* the difference of species diffusion rates. Permeability in gas separation is mainly controlled by diffusion factor depending on the molecular size of the penetrant (any component or few species of separated mixture) and also on the value and distribution of free volume in the polymer membrane. The permeability of liquids in pervaporation is significantly controlled by the solubility of penetrant in the polymer membrane and, thus, characterizes the polymer sorption activity. The combination of gas separation and pervaporation tests provides exhaustive data on the transport of small molecules in the polymer. The membrane performance depends on two basic factors – *permeability* and *selectivity*. The absolute rate of permeation is known as permeability (or productivity). Selectivity is usually determined by the ratio of concentrations of components in permeate (a product of membrane separation) and feed mixture (see, e.g. [1, 2, 53]). For example in the case of pervaporation of binary mixture selectivity coefficient (or separation factor) α_{pv}, could be defined as:

$$\alpha_{\text{pv}} = \frac{y_i / y_k}{x_i / x_k},$$

where y_i, y_k и x_i, x_k – molar (or weight) fraction of component i or k in permeate and feed mixture accordingly. Selectivity could be also determined as the ratio of permeate rates of two different species in transmembrane process [54]. The high-performance membranes should combine high permeability and high selectivity. Unfortunately, the increasing permeability is often accompanied by the decreasing membrane selectivity. E.g. membranes with low effective thickness reveal a high permeability but the reduction of the thickness also reduces the role of diffusion stage; accordingly the selectivity could fall significantly. The search of high-performance membranes and new membrane materials continues to be one of the main trends in membrane science and technology. In next sections we will consider some membrane processes that were realized with the use of new fulleren-containing membrane materials.

3.1. Gas Separation

Gas transport properties of membranes based on PPO modified by dispersing fullerene C_{60} have been studied by two scientific groups [26, 28-29, 32]. The results presented in cited references are in agreement.

PPO membranes modified by dispersing C_{60} were estimated by measuring the permeability of H_2, O_2, N_2, CH_4, and CO_2 in [26, 32]. The data listed in Table 3 show the decrease in the values of permeability for all gases with the increasing C_{60} content in samples. The selectivity of PPO-C_{60} membranes is higher than PPO membranes. As a rule, gas transport properties are determined by diffusion factor that significantly depends on polymer free volume. According to Table 1, the addition of C_{60} to PPO leads to the decrease of polymer free volume. Such changes could be the reason for decreasing gas permeability through PPO membranes modified by dispersing C_{60}.

Table 3. Gas transport properties [26].

Membrane	Permeability, barrers					Selectivity			
	H_2	O_2	N_2	CH_4	CO_2	H_2/N_2	O_2/N_2	CO_2/CH_4	CO_2/N_2
PPO	112.98	19.08	4.65	6.43	90.68	24.3	4.1	14.1	19.5
PPO+1%C_{60}	103.91	18.13	4.05	5.57	80.90	25.6	4.5	14.5	20.0
PPO+2%C_{60}	92.64	15.12	3.29	4.59	67.11	28.2	4.6	14.6	20.4

1 barrer= 10^{-10} cm^3 (STP)cm /(cm^2 s cmHg) = $0.33 \cdot 10^{-15}$ mol·m/(m^2·s·Pa)

In works [28, 29] gas permeability properties of PPO modified by C_{60} dispersed are compared with that of PPO modified by C_{60} bonded covalently to methyl group of PPO. The gas permeability of PPO+2%C_{60} bonded increases up to 80 % in comparison with pure PPO, whereas selectivity of gas separation retains the same. Probably such effect is due to the stiffening of polymer chain by C_{60} inclusion that leads to increase of polymer free volume [29].

One of the reason of decreasing permeability PPO-C_{60} dispersed with increasing C_{60} content is the clustering of the dispersed C_{60} and the higher crystalline gas impermeable fraction (due to clustering) than in pure PPO or PPO-C_{60} bonded. The clusters seem to form lamellas oriented parallel to the membrane plane.

The results of the work [29] show that glass transition temperature (T_g) of PPO-C_{60} dispersed decreases with C_{60} content from 214^{0}C for pure PPO to 208^{0}C for PPO+2%C_{60}. T_g of PPO-C_{60} bonded increases with C_{60} content from 214^{0}C for pure PPO to 235^{0}C for PPO+2%C_{60}. This interesting result indicates that PPO-C_{60} bonded become more stiffened polymer, and there is suppressed interchain packing and some possible conformational change in the backbone that leads to the increasing membrane free volume and gas permeability.

3.2. Pervaporation

Pervaporation as a membrane separation process has a wide range of application [1, 2, 13, 53, 55]. One of advantages and important applications of pervaporation is the use of this process as a part of hybrid reactive–separation processes (e.g. see [56-61]). Bellow we consider the system with ester synthesis reaction as an example.

The effect of fullerene addition on PPO transport properties was studied in the pervaporation of the reacting mixture: ethanol – acetic acid – water – ethyl acetate in binary and quaternary combination of components. These compounds form the reaction mixture in ethyl acetate synthesis reaction:

$$C_2H_5OH + CH_3COOH = H_2O + CH_3COOC_2H_5$$

The shifting of the esterification equilibrium in hybrid process (reaction + pervaporation) is one of the main ways for the maximizing yield of ethyl acetate.

3.2.1. Polymer – Liquid Interaction

In pervaporation the polymer acts as "a molecular sieve" with sorption active centers that are able to interact with components of liquid mixture by van der Waals, dipole-dipole, and ion-dipole forces, or hydrogen- bonding [62]. For binary and multicomponent mixtures, the mass transport significantly depends on both permeant–membrane and permeant–permeant interactions. Quantitatively, the permeant–membrane interaction (or affinity of membrane material to the permeating species) can be expressed in terms of the Flory-Huggins parameter χ. It is known that χ directly characterizes the solubility of a polymer in a given liquid [63]. According to Flory-Huggins theory [63] the Gibbs energy of mixing ΔG_m is given by equation:

$$\frac{\Delta G_m}{RT} = \ln \varphi_i + \varphi_p + \chi \varphi_p^2$$

where φ_i is the volume fraction of the solvent in the polymer and φ_p is the volume fraction of the polymer. The value of the dimensionless binary interaction parameter χ could be calculated from the following expression:

$$\chi = \frac{-\ln \varphi_i - \varphi_p + \dfrac{\Delta G_m}{RT}}{\varphi_p^2} = \frac{-\ln \varphi_i - \varphi_p}{\varphi_p^2} + \frac{\Delta G_m}{RT\varphi_p^2}$$

For practical purposes, in the case of equilibrium sorption of the liquid in the polymer, the dimensionless binary interaction parameter can be estimated by equation (see e.g. [32, 36]):

$$\chi = \frac{-\ln \varphi_i - \varphi_p}{\varphi_p^2}$$

that would be valid if ΔG_m is zero. More correctly the left side of last equation corresponds to the value

$$\chi' = \chi - \frac{\Delta G_m}{RT\varphi_p^2}$$

that should be valid in general case of $\Delta G_m \neq 0$. If the interaction between the polymer and the penetrant increased, the amount of penetrant inside the polymer also increases and χ decreases. In contrast, χ increases with decreasing affinity between the polymer and the penetrant. Therefore, a lower value of χ implies higher sorption.

To estimate the binary interaction parameters, equilibrium vapour sorption of ethanol, acetic acid, water, and ethyl acetate in PPO and PPO-C_{60} dispersed homogeneous membranes was studied. Table 4 lists data that show a better sorption capacity of fullerene-containing membranes for ester, acid, and alcohol than that of PPO membranes. Ethyl acetate exhibits the highest sorption (the lowest value of χ), whereas water appears to be absolutely inert to polymers under study.

Table 4. Sorption capacities and interaction parameters χ' for polymer membranes and components of esterification reaction.

Component	Sorption, wt.%		Parameter χ'	
	PPO	PPO+2% C_{60}	PPO	PPO+2% C_{60}
Ethanol	13.0	15.0	1.45	1.34
Acetic acid	23.6	24.2	1.29	1.24
Water	0	0	-	-
Ethyl acetate	23.4	26.4	1.22	1.13

Absence of water sorption and high sorption capacity for ethanol and ethyl acetate play an important role in pervaporation of systems involving these components, especially, in binary systems: ethyl acetate – water and ethanol – water.

3.2.2. Pervaporation of Ethyl Acetate – Water Mixture

The pervaporation of ethyl acetate – water mixtures has been studied in dilute aqueous solutions containing < 7 wt% ethyl acetate [32]. The reason is the phase separation (liquid phase splitting) of ethyl acetate – water system over a wide composition range. This mixture is homogeneous at ethyl acetate concentration below ~7 and above ~97 wt%.

Figure 1 shows high selectivity with respect to ethyl acetate that is caused by hydrophobic properties of membranes. After membrane swelling in ethyl acetate and formation of transport canals, water permeation is difficult because of low water solubility in ester and, correspondingly, high positive activity coefficients. Membrane selectivity increases due to C_{60} addition to PPO membranes. At the same time, C_{60} addition only slightly changes the flux across the membranes: from 0.48 kg m^{-2} h^{-1} (PPO membrane) to 0.52 kg m^{-2} h^{-1} (PPO+2% C_{60} membrane) in pervaporation of 7 wt% ethyl acetate aqueous. This may be a result of two opposite factors: decreasing free volume of fullerene-containing membranes, on the one hand, and increasing their swelling in ethyl acetate, on the other hand.

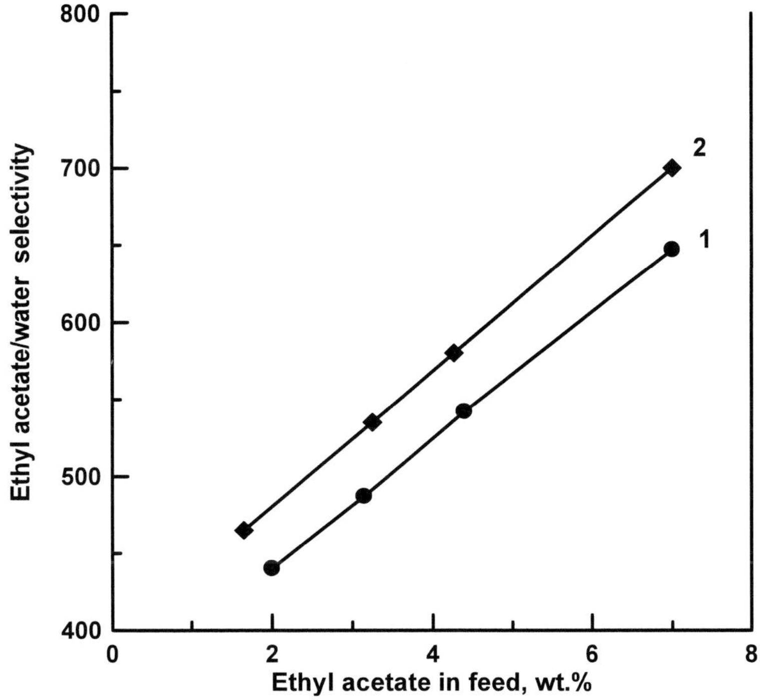

Figure 1. Dependence of selectivity on ethyl acetate concentration in the feed for the pervaporation of ethyl acetate - water mixture through (1) PPO, (2) PPO+2% C_{60} dispersed membranes.

3.2.3. Pervaporation of Ethanol – Water Mixture

The pervaporation of ethanol – water mixture through membranes based on PPO is determined by sorption properties of membranes with respect to ethanol and water. The mechanism of mass transfer in the pervaporation of ethanol – water mixture is presented in [32].The mass transfer starts from the sorption where ethanol alone takes part because of water is inert to PPO. Ethanol molecules interact with PPO backbone by hydrogen-bonding and create transport channels. Owing to high affinity to ethanol (unlimited solubility in water–ethanol system), water penetrates these channels whose size is commensurable with the kinetic diameter of water molecules. Water is the more permeable component of the mixture due to the smaller size of its molecules than those of ethanol. Increasing concentration of ethanol (swelling agent) in feed leads to the weakening intra- and intermolecular bonds in PPO membrane and to its plasticization. This facilitates the membrane permeability.

The pervaporation of ethanol – water mixtures was studied over a wide composition range. Figure 2 shows that the course of the permeate-feed concentration dependence for pervaporation differs essentially from vapour–liquid equilibrium diagram for the ethanol – water system. In pervaporation, the permeate is enriched with water. For each composition of the feed mixture, ethanol content in the permeate decreases with increasing C_{60} amount in membranes. Hence, the addition of C_{60} to PPO leads to the increase in selectivity of water removal process.

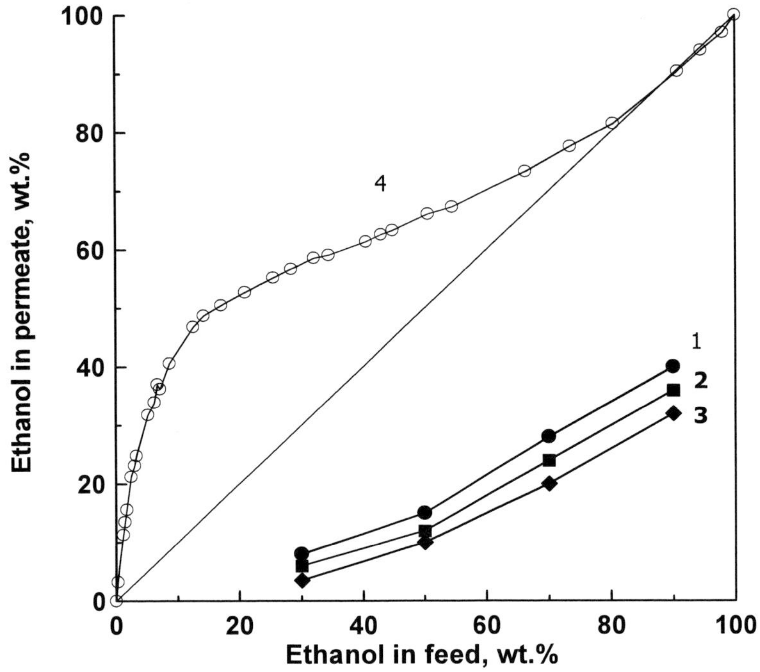

Figure 2. Dependence of ethanol concentration in permeate on ethanol concentration in feed for the pervaporation of ethanol - water mixture through (1) PPO, (2) PPO+1% C_{60}, and (3) PPO+2% C_{60} membranes. (4) vapour–liquid equilibrium curve.

In the pervaporation of feed mixture containing 10wt% ethanol, the flux across the PPO+2% C_{60} membrane is nearly twice then flux across PPO membrane. Furthermore, membranes permeability and selectivity increase with the increasing ethanol concentration in the feed, although ethanol is less permeable component of the mixture. This fact may be connected with sorption capacity of membranes with respect to ethanol that increases with C_{60} additives. Transport properties are higher in PPO membranes modified by dispersing C_{60}.

3.2.4. Pervaporation of Ethanol – Acetic Acid – Water – Ethyl Acetate Mixture

The selectivity in pervaporation of quaternary reacting mixture depends not only on binary interactions between component and the membrane polymer, but also on more complex competitive interplay between all species of the reacting system. Pervaporation of the quaternary system was studied for the feed composition (16.7 mol % acetic acid, 16.7 mol % ethanol, 33.3 mol % ethyl acetate, 33.3 mol % water) that is close to the chemically equilibrium state. Figure 3 shows that initial reagents penetrate through the membrane only slightly. Ethanol penetrates partly through membranes; acetic acid almost completely remains in the reaction mixture in spite of high value of acetic acid sorption. Similar small selectivity with respect to acetic acid has been noted in pervaporation of quaternary mixture during hybrid process of butyl acetate synthesis [60].

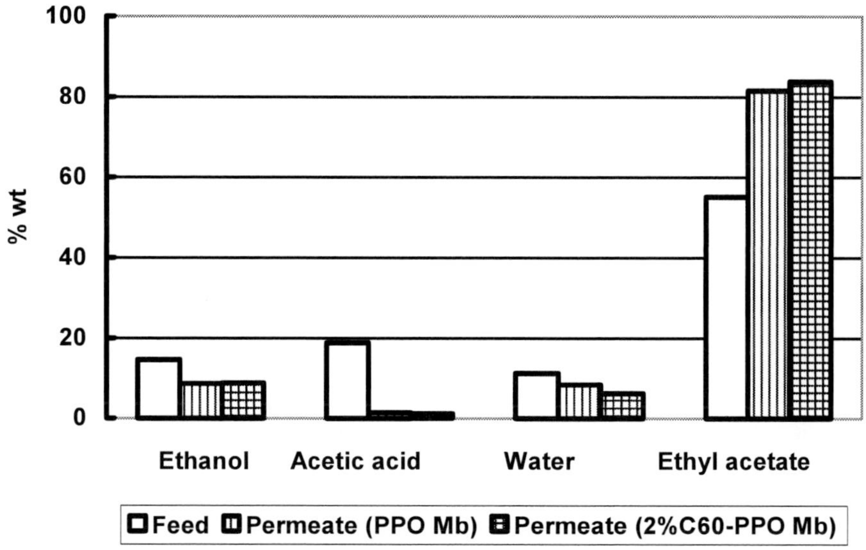

Figure 3. Content of quaternary mixture in feed and permeate in pervaporation through PPO and PPO+2 wt% C_{60} membranes.

Figure 3 shows that reaction products penetrate by different modes through membranes based on PPO. Water penetrates to a little extent, furthermore, its amount decreases in comparison with the initial amount. Hence, the main component of permeate becomes ethyl acetate. The ethyl acetate removal promotes shifting of the reaction equilibrium and the subsequent accelerating esterification. The use of membranes modified by fullerene C_{60} leads to decreasing water content and increasing ethyl acetate content in permeate.

4. Asymmetric Porous Membranes

Asymmetric membranes are widely used in nano-, ultra-, and microfiltration processes for the purification, concentration and fractionation of dissolved substances and for solving environmental problems. Separation through porous membranes has filtration mechanism that is based on different molecular sizes of mixture components. Membrane separation properties are characterized by the *rejection coefficient* (φ):

$$\varphi = 1 - c_p/c_c ,$$

where c_p is concentration of dissolved substance in permeate, c_c is concentration of dissolved substance in concentrate. *Flux* through the membrane is the amount of substance that penetrates through membrane area in unit time. Asymmetric membrane structure and properties depend on the formation technique and membrane material.

4.1. Preparation and Structure

PPO-C_{60} compositions containing 2 and 10 wt% C_{60} are obtained by mixing fullerene C_{60} and PPO powders. Asymmetric membranes are prepared by phase inversion technique, that is, by immersion precipitation. Solution of PPO-C_{60} composition in mixed chloroform-butanol solvent is cast on a glass plate which is immersed in a coagulation bath with a nonsolvent - ethanol. The membrane structure is formed as a result of two interconnected processes that take place in the coagulation bath: 1) diffusion of solvent and nonsolvent and 2) phase separation [27].

PPO	PPO+10wt%C_{60}

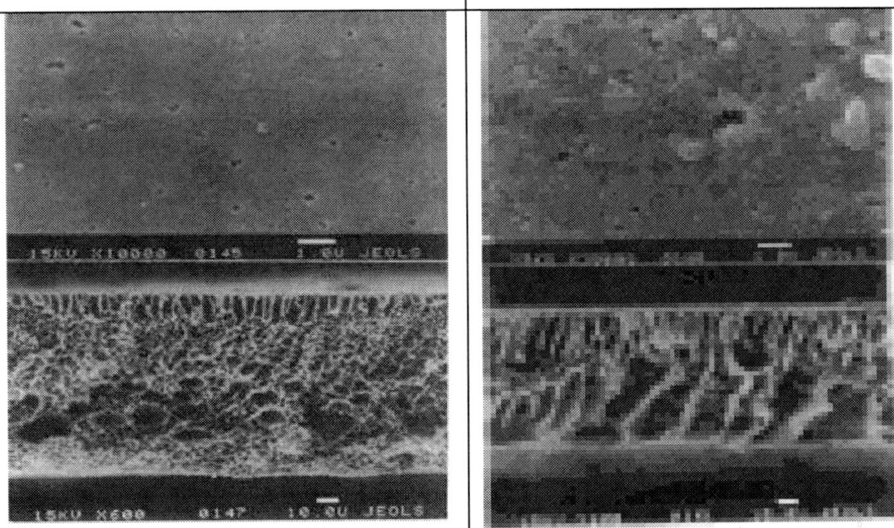

Figure 4. SEM micrographs of top surface and cross-section of PPO and PPO+10%C_{60} membranes

Figure 4 demonstrates asymmetric membranes morphology. All membranes consist of a porous support and a dense top layer that also contains pores. Porous structures of PPO and PPO-C_{60} membranes differ greatly. Cross-section of PPO membrane has a sponge-like structure. Membranes containing 2 or 10% C_{60} exhibited finger-like structure. Furthermore, pore size and porosity of top layer are higher in the case of PPO- C_{60}. This structure is a result of physical cross-linking of several PPO chains by C_{60} molecules that leads to the increasing of the microheterogeneity of the system in solution and favours the formation of more developed porous structure of fullerene-containing membranes after precipitation [27].

4.2. Application for Estrogenic Pollutants Removal

Endocrine-disrupting compounds commonly known as environmental estrogens are usually present in surface water in the range of 1-30 ng/l. Estrogenic compounds could exert undesirable effect on the developmental and reproductive systems of wildlife and human beings [64-67]. To solve this important environmental problem, membrane techniques may be used [68-73]. It has been reported that hydrophobic polymers are a promising membrane material for natural organic compounds removal from water [70]. Since estrogens are hydrophobic compounds of low volatility, sorption plays an important role in the removal of these compounds. It has been shown that the application of adsorption ability of activated carbon in conjunction with microfiltration and ultrafiltration membranes is the emerging technology for removal of organic compounds in surface water [71]. Research of fullerene adsorption properties has been presented [74]. Fullerenes as adsorbents of organic compounds in water are much more effective than soot or activated carbon. Adsorption ability of fullerenes is realized mainly through dispersion interactions. Efficiency of fullerene-containing membranes to estrogenic pollutants removal was estimated in adsorption and ultrafiltration experiments [33].

4.2.1. Adsorption

The adsorption capacity was measured by the shaking membrane pieces in aqueous solution of 5 ppb estrone for 24 h. Figure 5 shows high adsorption activity of both PPO and PPO-C_{60} membranes with respect to estrone. At the first contacts of membranes with aqueous solution of estrone, half of estrone is immediately sorbed on the membrane surface. Individual properties of membranes are manifested in the following stage where PPO + 2 %C_{60} membrane exhibits the fastest adsorption rate. Finally, all membranes reach ultimate adsorption with a similar value of adsorption capacity [33].

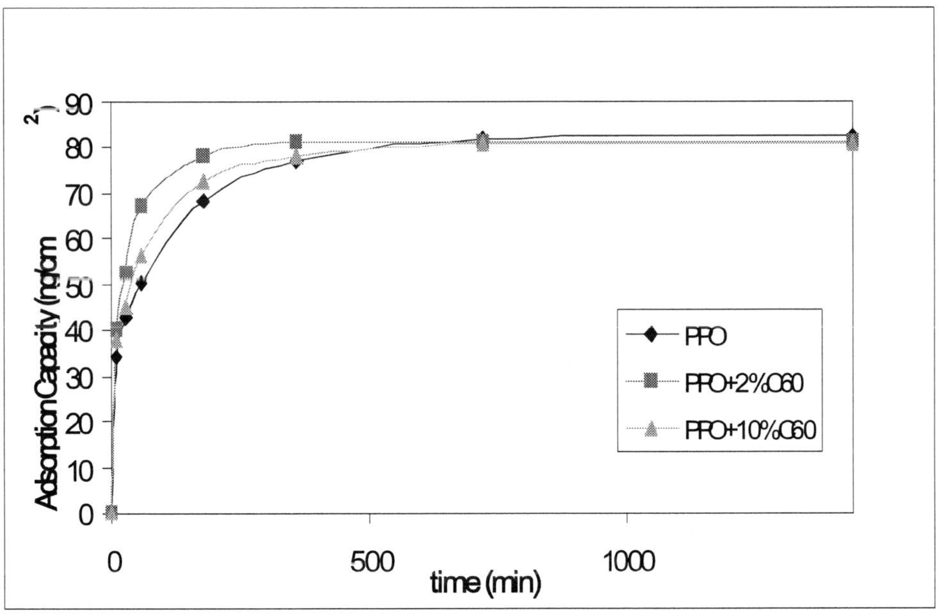

Figure 5. Adsorption capacity of membranes to estrone.

4.2.2. Ultrafiltration

Table 5 lists the data on fluxes through membranes and estrone rejection in the dead-end filtration tests. All membranes demonstrated very good rejection (more than 96%). Fullerene-containing membranes exhibit higher flux and rejection than pure PPO. The best rejection is in the case of PPO+2 %C_{60} membrane.

Table 5. Flux and estrone rejection in the ultrafiltration of 2 ppb estrone in water at 1 bar.

Membrane	Flux, cm^3/cm^2 h	Rejection, %
PPO	0.71	96.1
PPO + 2 % C_{60}	1.74	98.2
PPO + 10 % C_{60}	5.41	96.6

Thus, independent experiments on both adsorption and filtration show enhanced efficiency of PPO+2 % C_{60} membrane. The attempt to supply an explanation of the observed results is given in [33]. The adsorption of estrone molecules on the membrane surface depends on following factors: size of contact area in membrane and the presence of C_{60} molecules that is known as a lipophilic agent. Accordingly fullerene C_{60} being lipophilic agent can give additional sites in the membrane for estrone adsorption. It is evident that the amount of adsorption sites depends on the contact area. Porous structure of PPO and PPO-C_{60} membranes differs greatly (Fig. 4). Different porosity implies different inside area that is available for adsorption. Moreover, the PPO membrane exhibits the largest inside area due to its sponge-like porous structure. On the other hand, membranes containing 2 and 10wt% C_{60} have additional adsorption sites due to the fact of C_{60} molecules are active adsorbents but finger-like inside surfaces of PPO-C_{60} possess smaller area than PPO. As a result of the opposition of these two factors, the optimal variant is realized by PPO + 2 % C_{60} membrane.

5. Conclusion

The PPO modification by fullerene C_{60} sufficiently changes physical and transport properties of membranes. Effect of C_{60} inclusion into PPO depends on the way of modification. In the case of chemical modification with obtaining PPO-C_{60} bonded covalently, gas permeability increases, whereas selectivity of gas separation retains the same. PPO-C_{60} compositions with donor-acceptor binding exhibit lower gas permeability and slightly higher selectivity. Unfortunately, PPO-C_{60} bonded membranes have been studied for gas separation only.

Membranes based on PPO-C_{60} compositions are effective in pervaporation of the reacting mixture: ethanol – acetic acid – water – ethyl acetate; the permeate containing essentially ethyl acetate is obtained. The ethyl acetate removal promotes the shifting of the reaction equilibrium and the subsequent accelerating of esterification. PPO modification by dispersing fullerene C_{60} provides for a better sorption capacity for ester, acid, and alcohol than that of PPO membranes and better selectivity in pervaporation of binary and quaternary systems.

Asymmetric membranes based on PPO-C_{60} compositions can be used for the removing estrogenic pollutants from water. Membrane containing 2 wt% C_{60} exhibits the higher rejection and the fastest adsorption rate to estrone than PPO and PPO + 10 % C_{60}. Fullerene-containing membranes exhibit higher flux than pure PPO membranes.

Thus, for the most studied membrane processes, PPO modification by fullerene C_{60} has a favorable effect on transport properties and contributes to the efficiency of these processes.

Acknowledgement

Thanks are due to our co-workers listed in the references on our work for their most valuable contribution to this research. This work was supported by Russian Foundation for Basic Research (grant 06-03-32493).

References

[1] Mulder, M. Basic Principles of Membrane Technology, Kluwer Academic Publishers, Dodrecht/Boston/London, 1995.

[2] Baker, R.W. *Membrane Technology and Applications*. McGraw-Hill, N.Y., 2000.

[3] Pomogailo, A D. Hybrid polymer-inorganic nanocomposites, *Russian Chemical Reviews* 2000, 69, 53-80.

[4] Ray, S.S.; Okamoto, M. Polymer/layered silicate nanocomposites: a review from preparation to processing, *Progress in Polymer Sci.* 2003, 28, 1539-1641.

[5] Ray, S.S.; Yamada, K.; Okamoto, M.; Ueda, K. New polylactide-layered silicate nanocomposites. 2. Concurrent improvements of material properties, biodegradability and melt rheology, *Polymer* 2003, 44, 857–866.

[6] Förster, S.; Konrad, M. From self-organizing polymers to nano- and biomaterials, *J. Mater. Chem.* 2003, 13, 2671–2688.

[7] Andrews, R.; Weisenberger, M.C. Carbon nanotube polymer composites, *Current Opinion in Solid State & Materials Sci.* 2004, 8, 31-37.

[8] Yang, H.H.; Park, H.C.; Kang, Y.S.; Won, J.; Kim, W.N. Zeolite-filled polyimide membrane containing 2,4,6-triaminopyrimidine, *J. Membrane Sci.* 2001, 188, 151–163.

[9] Yang, H.; Nguyen, Q.T.; Ping, Z.; Long, Y.; Hirata, Y. Desorption and pervaporation properties of zeolite-filled poly(dimethylsiloxane) membranes, *Materials Research Innovations* 2001, 5, 101-106.

[10] Liu, Q.-L.; Xiao, J. Silicalite-filled poly(siloxane imide) membranes for removal of VOCs from water by pervaporation, *J. Membrane Sci.* 2004, 230, 121-129.

[11] Naidu, B.V.K.; Sairam, M.; Raju, K.V.S.N.; Aminabhavi, T.M. Pervaporation separation of water+isopropanol mixtures using novel nanocomposite membranes of poly(vinyl alcohol) and polyaniline, *J. Membrane Sci.* 2005, 260, 142-155.

[12] Sairam, M.; Vijaya, B.; Naidu, K.; Nataraj, S.K.; Sreedhar, B.; Aminabhavi, T.M. Poly(vinyl alcohol)-iron oxide nanocomposite membranes for pervaporation

dehydration of isopropanol, 1,4-dioxane and tetrahydrofuran, *J. Membrane Sci.* 2006, 283, 65-73.

[13] Shao, P.; Huang, R.Y.M. Polymeric membrane pervaporation, *J. Membrane Sci.* 2007, 287, 162-179.

[14] Hammond, G.S.; Kuck, V.J. (Eds), *Fullerenes: Synthesis, Properties, and Chemistry of Large Carbon Clasters*, ACS, Washington, DS, 1992.

[15] Hirsch, A. *The Chemistry of the Fullerenes;* Georg Thieme Verlag, Stuttgart, 1994.

[16] Sapurina, I.; Mokeev, M.; Lavrentev, V.; Zgonnik, V.; Trchova, M.; Hlavata, D.; Stejskal, J. Polyaniline complex with fullerene C_{60}, *Europ. Polymer J.* 2000, 36 (11), 2321-2326.

[17] Weber, V.; Duval, M.; Ederle, Y.; Mathis, C. Physico-chemical behavior in solution of star-shaped polystyrene with a C_{60} core, *Carbon* 1998, 36 (5-6), 839-842.

[18] Lee, T-W.; Park, O.O.; Kim, J.; Kim, Y.C. Application of a novel fullerene-containing copolymer to electroluminescent devices, *Chem. Mater.* 2002 14, 4281-4285.

[19] Yevlampieva, N.P.; Vinogradova, L.V.; Ryumtsev, E.I. Effect of fullerene C_{60} as a branching point on molecular and polarization properties of star-shaped polystyrenes, *Polymer Sci. (Russ)* 2006, 48, 106-113.

[20] Garaud, J.L.; Janot, J.M.; Miquel, J.; Seta, P. Photoinduced electron transfer properties of porous polymer membranes doped with fullerene C_{60} associated with phospholipids, *J.Membrane Sci.* 1994, 91, 259-264.

[21] Higuchi, A.; Yoshida, T.; Imizu, T.; Mizoguchi, K.; He, Z.; Pinnau, I.; Nagai, K.; Freeman, B.D. Gas Permeation of fullerene-dispersed poly(1-trimethylsilyl-1-propine) membranes, *J.Polym.Sci.: part B: Polym.Phys.* 2000, 38, 1749-1755.

[22] Polotskaya, G.A.; Andreeva, D.V.; Elyashevich, G.K. Investigation of gas diffusion through films of fullerene-containing poly(phenyleneoxide), *Technical Physics Letters.* 1999, 25, 555-557.

[23] Higuchi, A.; Agatsuma, T.; Uemiya, S.; Kojima, T.; Mizoguchi, K.; Pinnau, I.; Nagai, K.; Freeman, B.D. Preparation and gas permeation of immobilized fullerene membranes, *J.Appl. Polymer Sci.* 2000, 77, 529-537.

[24] Polotskaya, G.A.; Gladchenko, S.V.; Zgonnik, V.N. Gas diffusion and dielectric studies of polystyrene-fullerene compositions. *J.Appl.Polymer Sci.* 2002, 85(14), 2946-2951.

[25] Chung, T.-C.; Chan, S.S.; Wang, R.; Li, Z.; He, C. Characterization of permeability and sorption in Matrimid/C_{60} mixed matrix membranes, *J.Membrane Sci.* 2003, 211, 91-99.

[26] Polotskaya, G.; Biryulin, Yu.; Pientka, Z.; Brozova, L.; Bleha, M. Transport properties of fullerene - polyphenylene oxide homogeneous membranes. *Fullerenes, Nanotubes, and Carbon Nanostructures* 2004, 12 (1), 387-391.

[27] Polotskaya, G.A.; Biryulin, Yu.F.; Rozanov, V.V. Asymmetric membranes with porous skin based on fullerene-containing polyphenylene oxide. *Fullerenes, Nanotubes, and Carbon Nanostructures* 2004, 12 (1), 392-397.

[28] Sterescu, D.M.; Bolhuis-Versteeg, L.; van der Vegt, N.F.A.; Stamatialis, D.F.; Wessling, M. Novel gas separation membranes containing covalently bonded fullerenes, *Macromolecular Rapid Communications* 2004, 25, 1674-1678.

[29] Sterescu, D.M.; Stamatialis, D.F.; Mendes, E.; Wubbenhorst, M.; Wessling M. Fullerene-modified poly(2,6-dimethyl-1,4-polyphenylene oxide) gas separation membranes: Why binding is better than dispersing, *Macromolecules* 2006, 39, 9234-9242.

[30] Polotskaya, G.A.; Penkova, A.V.; Toikka, A.M. Fullerene-containing polyphenylene oxide membranes for pervaporation, *Desalination* 2006, 200 (1-3), 400-402.
[31] Jin, X.; Hu, J.Y.; Tint, M.L.; Ong, S.L.; Biryulin, Y.; Polotskaya, G. Estrogenic compounds removal by fullerene containing membranes, *Desalination* 2007, 214 (1-3), 83-90.
[32] Polotskaya, G.A.; Penkova, A.V.; Toikka, A.M.; Pientka, Z.; Brozova, L.; Bleha, M. Transport of small molecules through polyphenylene oxide membranes modified by fullerene, *Separation Science and Technology*, 2007, 42 (2), 333–347.
[33] Ong, S.L.; Hu, J.Y.; Biryulin, Yu.F.; Polotskaya, G.A. Fullerene-containing polymer membranes for rejection of estrogenic compounds in water. *Fullerenes, Nanotubes, and Carbon Nanostructures* 2006, 14 (2-3), 463-466.
[34] Wang, H.; De Sousa, R.; Gasa, J.; Tasaki, K.; Stucky, G.; Jousselme, B.; Wudl F. Fabrication of new fullerene composite membranes and their application in proton exchange membrane fuel cells, *J. Membrane Sci.* 2007, 289, 277-283.
[35] Doghieri, F.; Nardella, A.; Sarti, G.C.; Valentini, C. Pervaporation of methanol-MTBE mixtures through modified poly(phenylene oxide) membranes, *J. Membrane Sci.* 1994, 91, 283-291.
[36] Khayet, M.; Villaluenga, J.P.G.; Godino, M.P.; Mengual, J.I.; Seoane, B.; Khulbe, K.C.; Matsuura, T. Preparation and application of dense poly(phenylene oxide) membranes in pervaporation, *J. Colloid and Interface Sci.* 2004, 278, 410-422.
[37] Semenova, S.I. Polymer membranes for hydrocarbon separation and removal, *J. Membrane Sci.* 2004, 231, 189-207.
[38] Toi, K.; Morel, G.; Paul, D.R. Gas sorption and transport in polyphenylene oxide in compare with other glassy polymers, *J.Appl. Polymer Sci.* 1982, 27, 2997-3006.
[39] Polotskaya, G.A.; Agranova, S.A.; Gazdina, N.V.; Kuznetsov, Yu.P.; Nesterov, V.V. Effect of molecular weight parameters on gas transport properties of poly(2,6-dimethyl-1,4-phenylene oxide*), J.Appl. Polymer Sci. 1996*, 62, 2215-2218.
[40] Lapkin, A.A.; Roschupkina, O.P.; Ilinitch, O.M. Transport of C1-C3 hydrocarbons in poly(phenylene oxides) membranes, *J.Membrane Sci.* 1998, 141, 223-229.
[41] *Polyphenylene Oxide and Modified Polyphenylene Oxide Membranes: Gas, Vapor and Liquid Separation,* Chowdhury, G.; Kruczek, B.; Matsuura, T. (Eds.), Kluwer Academic Publishers, 2001.
[42] Khulbe, K.C.; Matsuura, T.; Lamarche, G.; Kim, H.J. The morphology characterisation and perfomance of dense PPO membranes for gas separation, *J. Membrane Sci.* 1997, 135, 211-223.
[43] Ilinitch, O.M.,; Fenelonov, V.B.; Lapkin, A.A.; Okkel, L.G.; Terskikh, V.V.; Zamaraev, K.I. Intrinsic microporosity and gas transport in polyphenylene oxide polymers, *J. Microporous and Mesoporous Materials*. 1999, 31, 97-110.
[44] Ilinitch, O.M.; Lapkin, A.A. Sorbate-induced structural rearrangements and permeation of gases in the polyphenylene oxide co-polymer, *Polymer* 2002, 43, 3209-3215.
[45] Chern, R.T.; Sheu, F.R.; Jia, L.; Stannett, V.T.; Hopfenberg, H.B. Transport of gases in unmodified and arylbrominated 2,6-dimethyl- 1,4-poly (phenylene oxide)*, J. Membrane Sci.* 1987, 35, 103-115
[46] Assogna, A.; Perego, G.; Roggero, A.; Sisto, R.; Valentini, C. Structure and gas permeability of silylated polyphenylene oxide, *J.Membrane Sci.* 1992, 71, 97-103.

[47] Perego, G.; Roggero, A.; Sisto, R.; Valentini C. Membranes for gas separation based on silylated polyphenylene oxide, *J. Membrane Sci.* 1991, 55, 325-331.

[48] Le Roux, J.D.; Paul, D.R.; Kampa, J.; Lagow, R.J. Surface fluorination of poly (phenylene oxide) composite membranes. Transport properties, *J.Membrane Sci.* 1994, 90, 21-35.

[49] Polotskaya, G.A.; Agranova, S.A.; Antonova, T.A.; Elyashevich, G.K. Gas transport and structural features of sulfonated poly(phenylene oxide), *J. Appl. Polymer Sci.* 1997, 66, 1439-1443.

[50] Lavrenko, P.N.; Evlampieva, N.P.; Volokhova, D.M.; Vinigradova, L.V.; Melenevskaya, E.Yu.; Zgonnik, V.N. Hydrodynamic and electrooptic properties of C_{60}-poly(phenylene oxide) complex in solution, *Polymer Sci. (Russ)* 2002, A44, 289-296.

[51] Biryulin, Yu.F.; Melenevskaya, E.Yu.; Mikov, S.N.; Orlov, S.E.; Petrikov, V.D.; Syckmanov, D.A.; Zgonnik, V.N. Optical properties of fullerenecontaining polydimethylphenylene oxide free films, *Physics & Techniques of Semiconductors (Russ)* 2003, 37, 110-116.

[52] Rusanov, A.I., Prokhorov, V.A. *Interfacial Tensiometry,* Elsevier, Amsterdam, 1996.

[53] Feng, X.; Huang, R.Y.M. Liquid separation by membrane pervaporation: A Review, *Ind. Eng. Chem. Res.* 1997, 36, 1048-1066.

[54] R.E.Kesting, *Synthetic Polymer Membranes.* A Wiley-Interscience publication, NY, 1985.

[55] Semenova, S.I.; Ohya, H.; Soontarapa, K. Hydrophilic membranes for pervaporation: An analytical review, *Desalination* 1997, 110, 251-286.

[56] Aiouache, F.; Goto, S. Reactive distillation-pervaporation hybrid column for tert-amyl alcohol etherification with ethanol, *Chem. Eng. Sci.* 2003, 58, 2465-2477.

[57] Assabumrungrat, S.; Kiatkittipong, W.; Praserthdam, P.; Goto, S. Simulation of pervaporation membrane reactors for liquid phase synthesis of ethyl tert-butyl ether from tert-butyl alcohol and ethanol, *Catalysis Today.* 2003, 79-80, 249-257.

[58] Benedict, D.J.; Parulekar, S.J.; Tsai, S.-P. Esterification of lactic acid and ethanol with/without pervaporation, *Ind. Eng. Chem. Res.* 2003, 42, 2282-2291.

[59] Steinigeweg, S., Gmehling, J.. Transesterification processes by combination of reactive distillation and Pervaporation, *Chem. Eng. Proc.* 2004, 43, 447–456.

[60] Liu, K.; Tong, Z.; Liu, L.; Feng, X. Separation of organic compounds from water by pervaporation in the production of n-butyl acetate via esterification by reactive distillation, *J. Membrane Sci.* 2005, 256, 193-201.

[61] Peters, T.A.; Benes, N.E.; Keurentjes, J.T.F. Zeolite-coated ceramic pervaporation membranes; pervaporation-esterification coupling and reactor evaluation, *Ind. Eng. Chem. Res.* 2005, 44, 9490-9496.

[62] Prausnitz, J.M.; Lichtenthaler, R.N.; Azevedo, E.G. Molecular Thermodynamics of Fluid Phase Equilibria: 3d edition. Prentice-Hall PTR: Upper Saddle River, NY, 1999.

[63] Flory, P.J. Principles of Polymer Chemistry, Cornell Univ. Press, New York, 1953.

[64] Johnson, A.C. & Sumpter, J. P. Removal of endocrine-disrupting chemicals in activeated sludge treatment works. *Environmental Science and Technology* 2001, 35, 4697-4703.

[65] Hu, J.Y.; Yuan, T.; Ong, S.L.; Song, L.F.; Ng, W.J. Identification and quantification of bisphenol A by gas chromatography and mass spectrometry in lab-scale dual membrane system. *J. Environ. Monit.* 2003, 5, 141-144.

[66] Ternes, T.A.; Stumpf, M.; Mueller, J.; Haberer, K.; Wilken, R.D.; Servos, M. Behavior and occurrence of estrogens in municipal sewage treatment plants-I. Investigations in Germany, Canada and Brazil. *The Science of the Total Environment* 1999, 35, 81-90.

[67] Meng, Z.; Chen, W.; Mulchandani, A. Removal of Estrogenic Pollutants from Contaminated Water Using Molecularly Imprinted Polymers, *Environmental Science and Technology* 2005, 39, 8958-8962.

[68] Kiso, Y.; Nishimura, Y.; Kitao, T.; Nishimura, K. Rejection properties of non-phenylic pesticides with nanofiltration membranes. *J. Membrane Sci.* 2000, 171, 229-237.

[69] Kiso, Y.; Kon, T.; Kitao, T.; Nishimura, K. Rejection properties of alkyl phthalates with nanofiltration membranes, *J. Membrane Sci.* 2001, 182, 205-214.

[70] Chang, S.; Waite, T.D.; Schafer, A.I.; Fane, A.G. Adsorption of trace steroid estrogens to hydrophobic hollow fibre membranes, *Desalination* 2002, 146, 381-386.

[71] Konieczny, K.; Klomfas, G. Using activated carbon to improve natural water treatment by porous membranes, *Desalination* 2002, 147, 109-116.

[72] Schäfer, A.I.; Nghiem, L.D.; Waite, T.D. Removal of the natural hormone estrone from aqueous solutions using nanofiltration and reverse osmosis, *Environmental Science and Technology* 2003, 37, 182-188.

[73] Nghiem, L.D.; Schafer, A.I. Adsorption and transport of trace contaminant estrone in NF/RO membranes, *Environ. Eng. Sci.* 2002, 19, 441-451.

[74] Berezkin, V.I.; Viktorovskii, I.V.; Vul', A.Ya.; Golubev, L.V.; Petrova, V.N.; Khoroshko, L.O. Fullerene microcrystals as adsorbents of organic compounds, *Semiconductors* 2003, 37 (7), 802-810.

In: Progress in Fullerene Research
Editor: Milton Lang, pp. 319-330
ISBN: 978-1-60021-841-5

Chapter 10

FULLERENOÏD OXIDES

S. Boudin, R. Retoux and M. Hervieu
Laboratoire CRISMAT, ENSICAEN, Bd Maréchal Juin, 14050 Caen Cedex, FRANCE

Abstract

A great interest has been taken, for a long time, in tetrahedral oxides like silicates and aluminates, due to their numerous applications in various fields ranging from catalysis, ion exchange, molecular sieves, to thermo and photoluminescence. These compounds often exhibit complex frameworks characterized by the presence of hexagonal and pentagonal windows, but no structural relationships had been established with the different forms of carbon and especially with fullerenes. For the first time, three years ago, has been reported a complex oxide, which exhibits a 3D-framework of AlO_4 tetrahedra forming huge "Al_{84}" spheres similar to those of the D2d isomer of the C_{84} fullerenes, labeled by the way *"fullerenoïd"*[1]. The detailed analysis of its structure and non-stoichiometry mechanisms has opened a promising route for the generation of a large family of fullerenoïd oxides.

1. Routes Towards Fullerenoïd Compounds: State of the Art

After the discovery, by Krätschner *et al*[2] in 1990, of carbon fullerenes that exhibit various particular properties, amounts of work have been devoted to the feasibilities for isolating similar but carbon-free nano-particles.

The first route for building so-called *inorganic fullerenes* (IF) was to curl layered compounds, starting from hexagonal atomic arrangements similar to graphite. Using this approach, nested boron-nitride cages were obtained by irradiation of crystalline, amorphous and nanotubes of BN under the electron beam of a transmission electron microscope (F. Banhart *et al*[3]. and Stéphan et al[4].). Onion-like particles, with a diameter up to 700 Å, were observed. Similarly after considering the layered metal sulphides (2H-WS_2 and 2H-MoS_2), whose layer surfaces offer a buckling hexagonal atomic arrangement, nested WS_2 and MoS_2 IF's were synthesized by sulphurizing tungsten and molybdenum oxides under H_2S flux [5,6,7,8].

A second route for building IF's was to start from 3D compounds, whose structures are flexible enough to be accommodated into balls. Thus C. Ducati *et al.*[9] created thin films of $(TiO_2)_n$ fullerenoïds. The existence of TiO_2 nanotubes and the possible non-stoichiometry of titanium oxide were favourable for the formation of a curled TiO_2 network. Using an ablation method, a Ti source was oxidized by a He / O plasma, leading to the formation of particles with edges-sharing TiO_6 octahedra.

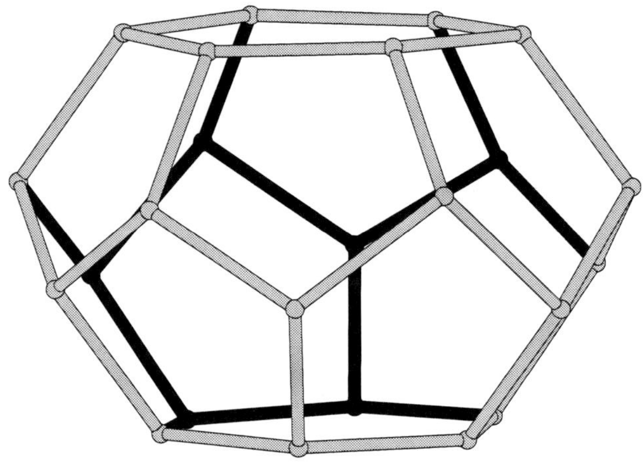

Figure 1. Cage, with 12 pentagonal and 2 hexagonal windows, in α-melanophlogite.(for clarity, only the Si atoms are represented).

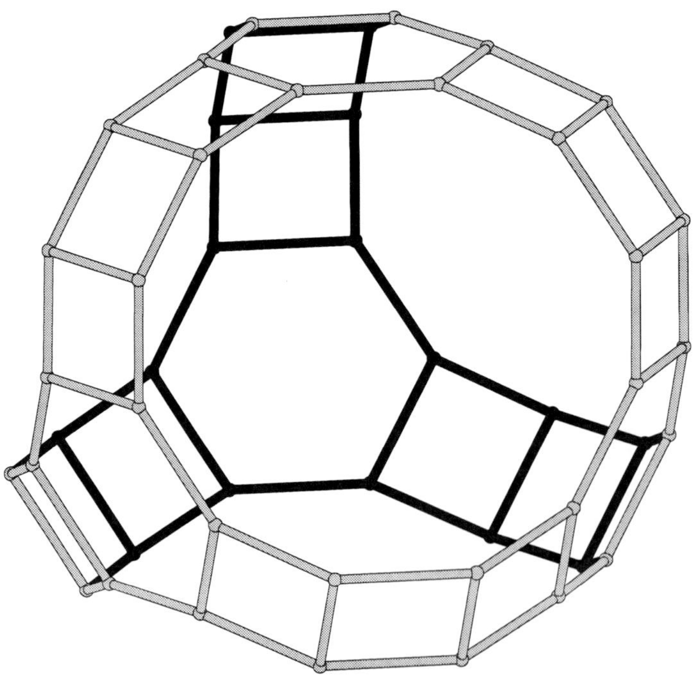

Figure 2. Cage, with 8 square, 4 rectangular, 4 hexagonal and 4 dodecagonal windows, in X type zeolithe. (For clarity only the Si and Al atoms are represented).

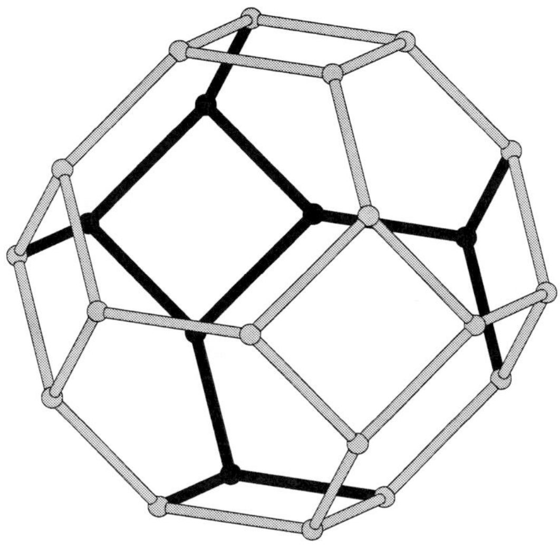

Figure 3. Sodalite cage, with 6 square and 8 hexagonal windows. (for clarity, only Al atoms are represented).

Other possible 3D structures can be imagined from systems forming open frameworks, exhibiting hexagonal and pentagonal cycles of corner sharing polyhedra, as exemplified by natural silicates. This is the case of the melanophlogite described by Nakagawa *et al*[10], which contains a $(SiO_2)_n$ framework : corner sharing SiO_4 tetrahedra form large cages with 12 pentagonal and 2 hexagonal windows (see drawing of one cage in Fig. 1) and the inner cages are filled up with neutral molecules of CO_2, CH_4 and N_2. Other examples are the natural silicates and alumino-silicates: the zeolites contain $(Al/Si)O_4$ tetrahedra forming huge cages (see, for example in Fig 2., drawing of a cage delimited by 8 square, 4 rectangular, 4 hexagonal and 4 dodecagonal windows, in X zeolite[11]). In zeolites, the $(MO_2)_n$ framework (M = Al, Si) is negative and has therefore to be filled up with alkaline or alkaline earth cations. In the 'pure' aluminate system, the sodalite-type structure exhibits cages built up from 6 square and 8 hexagonal windows (see drawing of cage in sodalite, in Fig. 3). Due to the higher negativity of the aluminate frameworks than the alumino-silicate ones, the filling of the cages is rather ensured by alkaline earth cations and even by trivalent cations. Among aluminates, some sodalites[12, 13] and $Sr_6(Bi_2O_3)(AlO_2)_{12}$[14] are particularly interesting because the stuffing of one cage is ensured by more complex species : either a $(MO_6)^{2-}$ polyanion with M= Cr, Mo, W (in sodalites[12]) surrounded by alkaline earth cations or by a $(PbO)_4$ neutral molecule surrounded by lanthanides cations (in sodalites[13]) or by a Bi_2O_3 neutral molecule surrounded by alkaline earth cations (in $Sr_6(Bi_2O_3)(AlO_2)_{12}$[14]). Such arrangements of nested arrays of molecules or cations are favourable for the formation of large cages.

2. Strategy and Discovery of the first C_{84} Related Fullerenoïd Oxides

Our strategy was firstly guided by the nature of the natural silicates and alumino-silicates, keeping in mind that we are looking for large M_n spheres exhibiting an arrangement of hexagonal and pentagonal windows similar to those of the C fullerenes and that these M atoms should adopt a tetrahedral coordination, involving very large cages. It means that the negative $M_nO_{5n/2}$ spheres should be stuffed with voluminous species and therefore could be stabilized by large cations. In that way, due to their non-bonded electrons, which tend to repulse, assembling several lone pair cations would favour the formation of a cationic shell around it. Based on these observations, we focussed our researches on aluminates and post-transition metals systems. We have also privileged a second direction of research, which should allow working under normal pressure. Starting from this observation, we have explored the Ca-Bi-Al-O and Sr-Bi-Al-O systems, working by solid-state reaction with mixtures of CaO or SrO_2, Bi_2O_3 and Al_2O_3 wrapped in a gold tube, heated in an evacuated quartz ampoule at 850°C (to avoid Bi oxide sublimation) for 48 hours, and cooled down to room temperature for 18 hours. The step-by-step characterization of the phase compositions and of the structural characteristics (cell parameters and possible space groups) has been carried out by electron diffraction and Energy dispersive Spectroscopy (EDS); the synthesis process was modified after each observation. The detailed methods for synthesizing the first fullerenoïds, $Ca_{33}Bi_{22+x}Al_{48}O_{138+1.5x}$ and $Sr_{33}Bi_{24+x}Al_{48}O_{141+1.5x}$, crystal growth and structure determination are described elsewhere[15,16]. Both crystal structures were solved using single crystal X ray diffraction (see for details ref [15, 16]). The fullerenoïd oxides crystallise in a $F\bar{4}3m$ cubic cell with a= 24.423(3) Å and a = 25.090(6) Å for Ca and Sr oxide respectively.

3. Structural Description of $Ca_{33}Bi_{22+x}Al_{48}O_{138+1.5x}$ and $Sr_{33}Bi_{24+x}Al_{48}O_{141+1.5x}$

The Aluminate Framework

The framework is composed of corner-sharing AlO_4 tetrahedra. The polyhedra are connected into huge $Al_{84}O_{210}$ cages (Fig 4), where the aluminium atoms are disposed like carbon atoms in the D2d isomer of the C_{84} fullerene (Fig 5). The 'Al_{84}' ball is delimited by 12 pentagonal AlO_5 and by 32 hexagonal AlO_6 windows. The balls are connected into the aluminate framework in a compact face centred cubic (fcc) arrangement, so that one ball is connected, by hexagonal windows, to 12 other balls (see Fig 6, one plane of 'Al_{84}' balls). Note that, in these oxides, the fullerene moieties are not isolated like in the aforementioned chalcogenides and the titanium oxide, but they are connected to each other. At the connection of 6 balls (i.e. on a octahedral interstitial site of the fcc array), the AlO_4 tetrahedra form a large $Al_{48}O_{106}$ cubic unit. Inside, three types of smaller cages (labelled A, B, C) are observed (Fig 7a,b,c). A C cage, lying at the centre of the $Al_{48}O_{106}$ unit, is occupied by one calcium (resp. strontium) cation. At the connection of 4 'Al_{84}' balls (i.e. on a tetrahedral interstitial site of the fcc array), AlO_4 tetrahedra delimit a fourth type of small cages (labelled D) (fig 7d). Except C cages, all the small cages are empty. The other alkaline earth and bismuth cations are located inside the huge fullerenoïd cages. The aluminate framework formulates $[Al_{48}O_{94}]^{-44n}{}_{n}$.

Figure 4. $Al_{84}O_{210}$ cages in $Ca_{33}Bi_{22+x}Al_{48}O_{138+1.5x}$ and $Sr_{33}Bi_{24+x}Al_{48}O_{141+1.5x}$

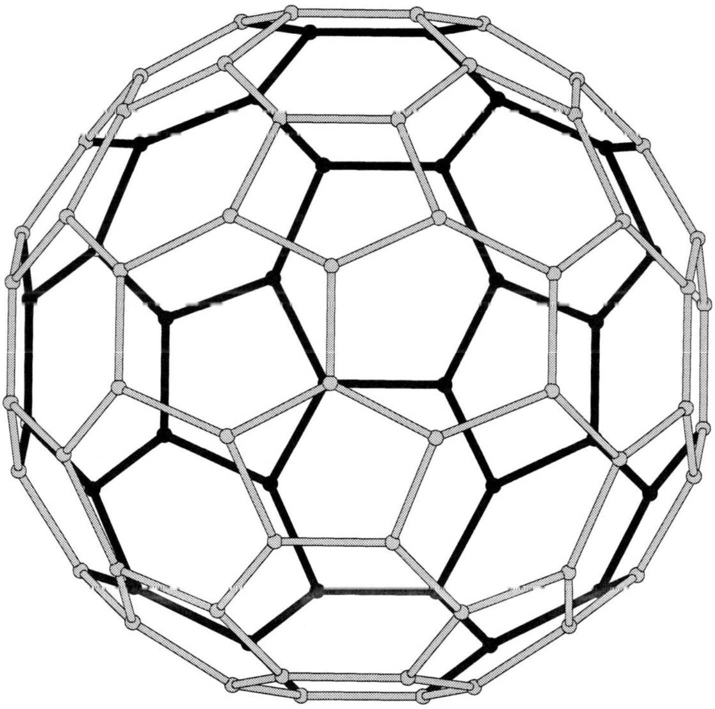

Figure 5. 'Al_{84}' ball in $Ca_{33}Bi_{22+x}Al_{48}O_{138+1.5x}$ and $Sr_{33}Bi_{24+x}Al_{48}O_{141+1.5x}$. (for clarity, only Al atoms are represented).

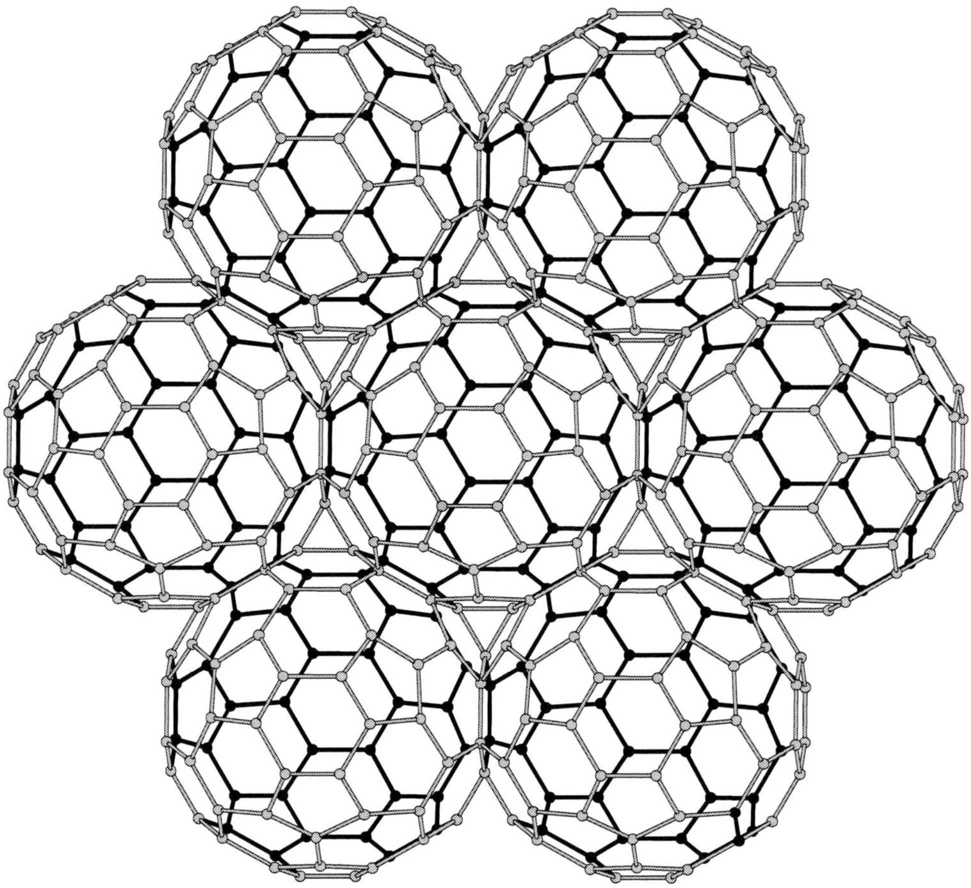

Figure 6. One plane of the face centered cubic array of 'Al_{84}' balls.

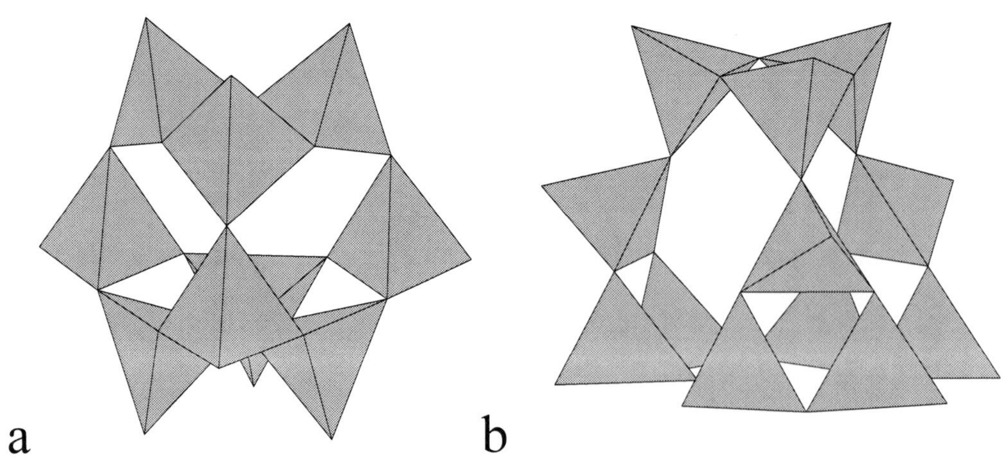

Figure 7. Continued on next page

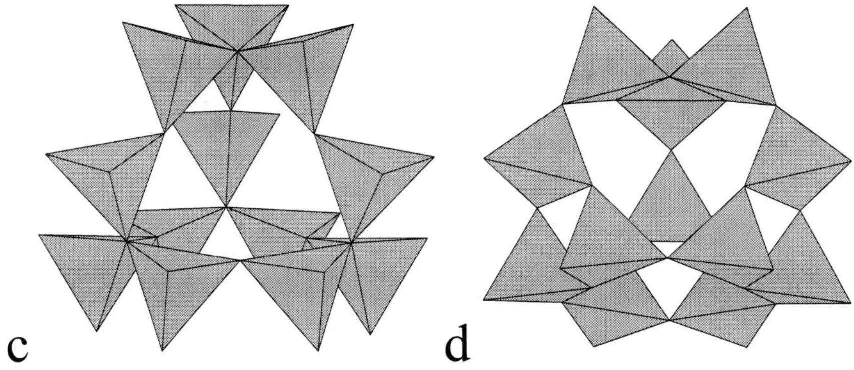

Figure 7. Small (a) A, (b) B and (c) C type cages in $Ca_{33}Bi_{22+x}Al_{48}O_{138+1.5x}$ and $Sr_{33}Bi_{24+x}Al_{48}O_{141+1.5x}$.

Filling of the Fullerene Cages

The centre of this cage is filled by a polyanion composed of bismuth and oxygen ions. Due to the great ability of Bi^{3+} cation to show various configurations, the geometry and the oxygen content of the polyanion differs from Ca to Sr oxide. In the Ca oxide, the $(Bi_{16}O_{45.05})^{-y}$ anion (Fig 8a) is ideally composed of 4 BiO_3 trigonal pyramids (in dark grey in Fig 8a) and of 8 BiO_4 distorted tetrahedra (in light grey in Fig 8a). In such polyanion, the oxygen (in white circles) shared between 3 BiO_4 tetrahedra, is partially occupied leading to a $(Bi_{16}O_{45.05})^{-y}$ formula for the polyanion. In the Sr oxide, the $(Bi_{16}O_{47.375})^{-z}$ anion (Fig 8b) is ideally composed of 4 BiO_6 octahedra (in dark grey in Fig 8b) and of 8 BiO_5 square pyramids (in light grey in Fig 8b). In such polyanion, the oxygen (in white circles) shared between one BiO_6 octahedron and 2 BiO_5 square pyramids is partially occupied leading to a $(Bi_{16}O_{47.375})^{-z}$ formula for the polyanion.

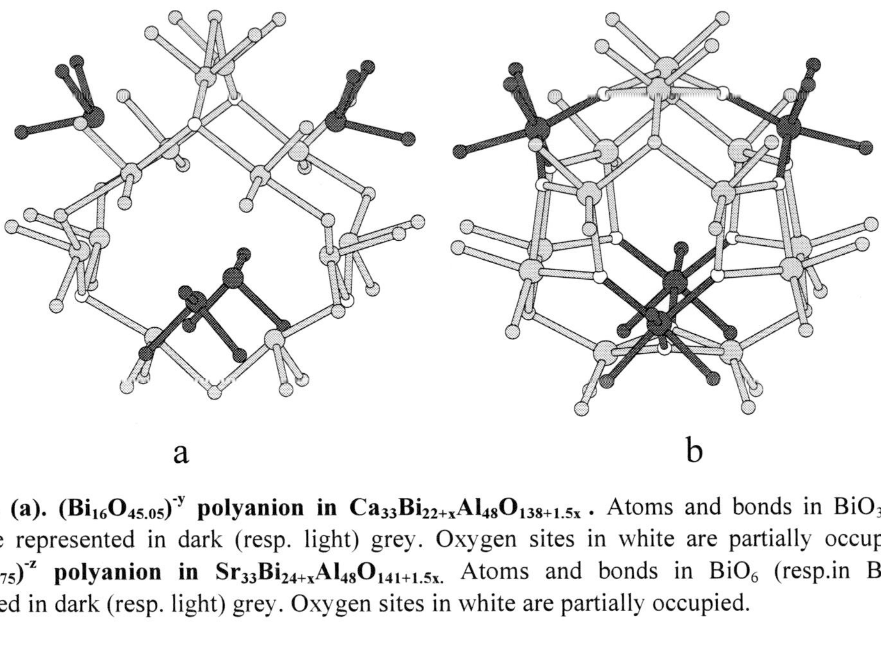

Figure 8. **(a). $(Bi_{16}O_{45.05})^{-y}$ polyanion in $Ca_{33}Bi_{22+x}Al_{48}O_{138+1.5x}$.** Atoms and bonds in BiO_3 (resp.in BiO_4) are represented in dark (resp. light) grey. Oxygen sites in white are partially occupied. **(b). $(Bi_{16}O_{47.375})^{-z}$ polyanion in $Sr_{33}Bi_{24+x}Al_{48}O_{141+1.5x}$.** Atoms and bonds in BiO_6 (resp.in BiO_5) are represented in dark (resp. light) grey. Oxygen sites in white are partially occupied.

In the Ca (resp. Sr) oxides, the bismuth polyanion is surrounded by 32 alkaline earth cations and by 6.7 (resp 8.25) Bi^{3+} cations, whose sites are partially occupied. Thus the resulting cationic content inside one fullerene cage is equal to $(Ca_{32}Bi_{22.7}O_{45.05})^{+42}$ (resp. $(Sr_{32}Bi_{24.25}O_{47.375})^{+42}$). It ensures with the Ca^{2+} (resp Sr^{2+}) cation, located inside a $[Al_{48}O_{94}]^{-44}$ unit of the framework, the charges balance. Among the 32 alkaline earth cations, 12 ones are located in front of the pentagonal windows of the 'Al_{84}' cages, whereas the 20 others and the Bi^{3+} cations, surrounding the polyanion, are located in front of hexagonal windows.

4. Non-stoichiometry and Micro Structural Mechanisms

Associated to X ray characterisations, a systematic investigation of the fullerenoïd type phases has been undertaken using Transmission Electron Microscopy[15-18]. In a first step Electron Diffraction performed, on powder samples, allowed us to confirm the cell parameters and the space group used in the X ray diffraction study.

High Resolution Electron Microscopy of Regular Structure

HREM observations attested the good crystallinity of the Ca and Sr fullerenoïd oxides. A good correlation has been established between the observed contrasts in the through focus series and the simulated images, calculated using the X ray atomic parameters. The original arrangement of the Al, Sr/Ca and Bi atoms in these fullerenoïd type compounds involves very typical contrasts on HREM images whatever the zone axis investigated. For example, Figure 9 presents typical [001] experimental and calculated (inset) HREM images for pure Sr (Fig. 9a) and pure Ca (Fig. 9b) based materials (focus values close to -350 and -650Å respectively). Similarly Figure 10 presents typical [111] experimental and calculated (inset) HREM images for pure Sr (Fig. 10a) and pure Ca (Fig. 10b) compounds (focus values are close to -650 and -1050Å respectively).

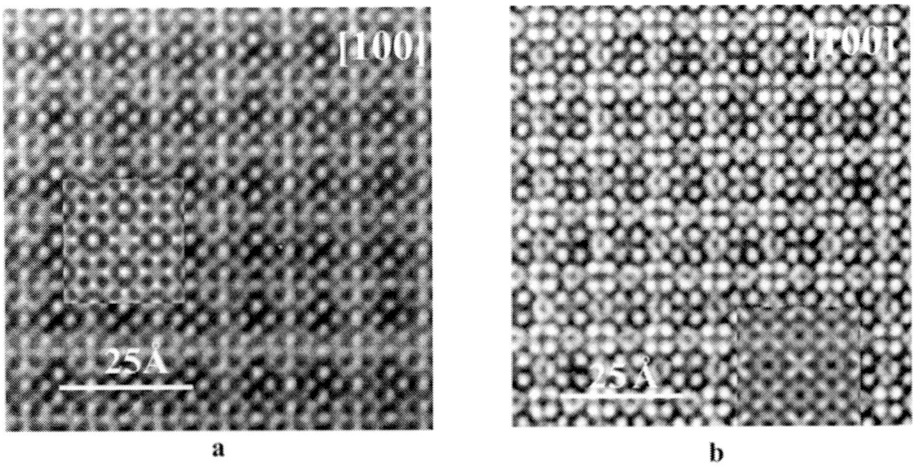

a b

Figure 9. HREM image recorded along [001] at focus values close to -350Å (a) and -650Å (b) respectively for $Sr_{33}Bi_{24+x}Al_{48}O_{141+1.5x}$ and $Ca_{33}Bi_{22+x}Al_{48}O_{138+1.5x}$.

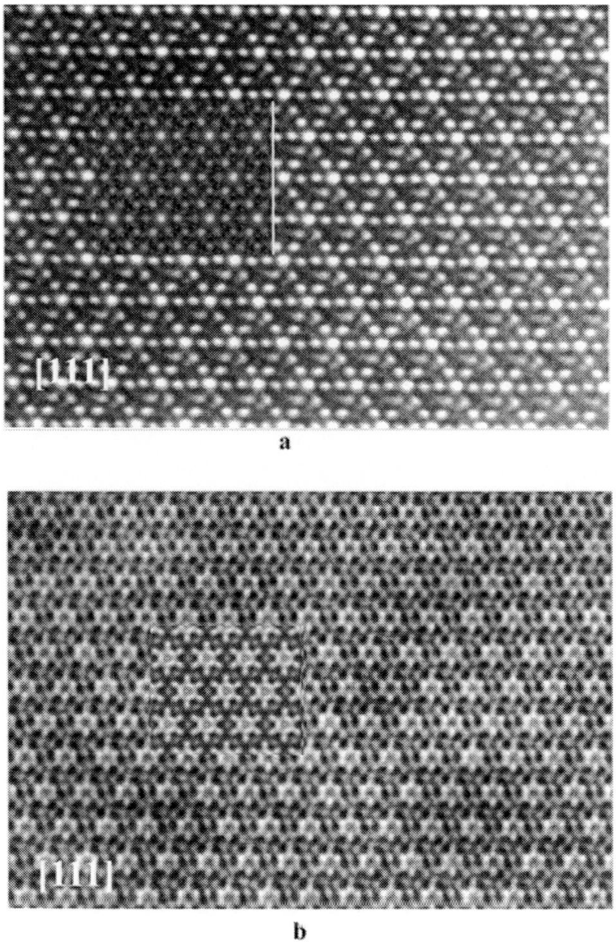

Figure 10. HREM image recorded along [111] at focus values close to -650Å (a) and -1050Å (b) respectively for Sr and Ca based fullerenoïdes.

Local Variations of the Contrasts

Energy Dispersive Spectroscopy analyses of these series showed significant variations in the cationic content of the crystals. These phenomena suggest that large domains of non-stoichoimetry exist around the cationic compositions determined from the X ray calculations. Higher and lowest cationic content can be explained by variable occupations of the cationic and anionic sites, located inside the fullerenoïd cages, and by the filling of the empty A, B and D cages, located in the aluminate framework. Using simulated images, calculated by taking into account these variations, the comparison with the HREM images evidenced this phenomenon[17]. Thus local variations of the contrasts are observed and pointed by white arrows, in a [001] HREM image of the Sr compound (Fig 11).

Similarly, HREM observations on a mixed $Sr_{0.5}/Ca_{0.5}$ fullerenoïd oxide ($Sr_{16.5}Ca_{16.5}Bi_{22+\delta}Al_{48}O_{138+3\delta/2}$, Figure 12) reveal inhomogeneities in contrasts. According to series of calculated images, it can correlated to different cationic contents and to different occupancies of the A, B, C, D and fullerene cages[18]. In addition few other extended defects, as antiphase boundaries or stacking faults, have been evidenced in this mixed Sr/Ca fullerenoïd oxides[18].

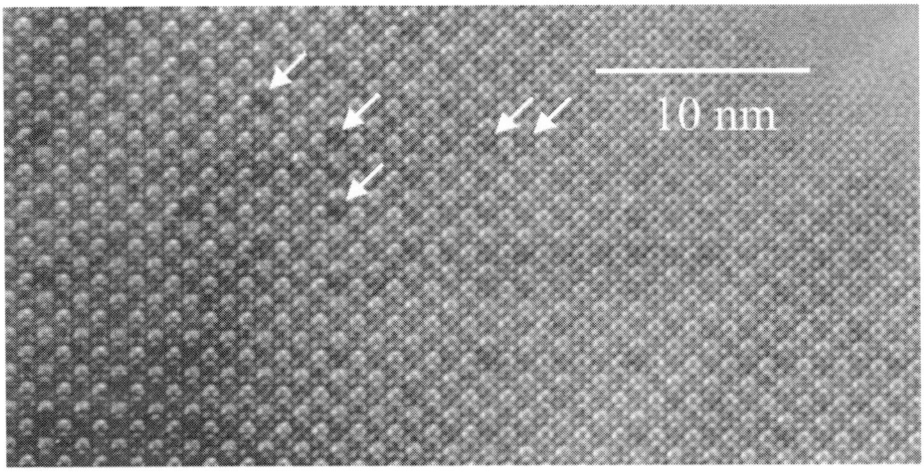

Figure 11. HREM image recorded along [001] on $Sr_{33}Bi_{24+x}Al_{48}O_{141+1.5x}$. Here white arrows point local variations of the contrasts randomly dispersed all over the crystal.

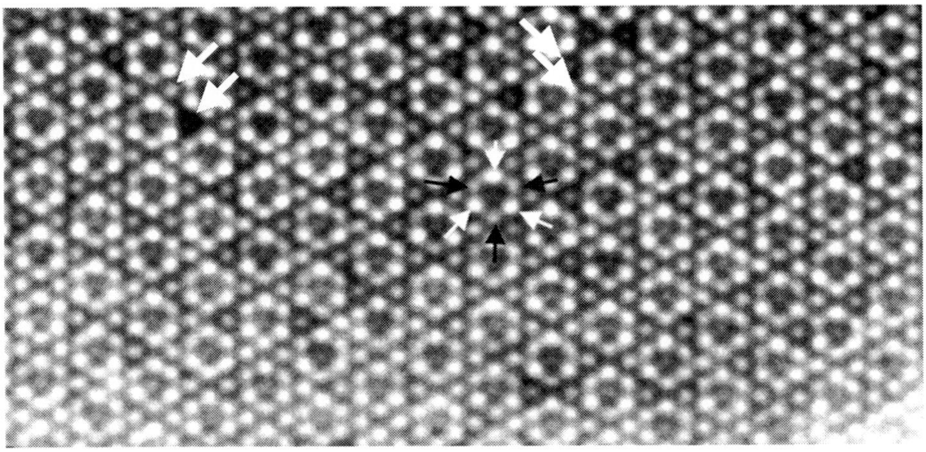

Figure 12. HREM image recorded along [111] at focus values close to 0Å for the $Sr_{0.5}/Ca_{0.5}$ fullerenoïd oxide : $Sr_{16.5}Ca_{16.5}Bi_{22+\delta}Al_{48}O_{138+3\delta/2}$.

5. Perspectives

The great similarity between the architecture of the spheres in these fullerenoïds with those of the D2d isomer of C_{84} fullerenes, opens the route to the exploration of other systems for the discovery of new members. The goals are to work along two directions. The first one concerns the framework, by varying the size of the "Al_n" spheres and their mode of connection. It means that one has to consider the possibility to introduce other tetrahedral species on the aluminium sites, such as Si, Ge, Ga, Fe… The second deals with the nature of the species filling the center of the sphere and balancing the charges, firstly selecting systems including lone pair cations (like Pb^{2+}, Sn^{2+}, Sb^{3+}, Tl^{+}...).

Another working route results directly from the observation of the micro-structural mechanisms, which have been detected by high resolution electron microscopy. The observation of complex mechanisms such as the formation of dislocation, twinning domains and anti-phase boundaries which obey exactly the same laws that the metal atoms in fcc is really fascinating as soon as they engage $Al_{84}O_{210}$ spheres. From the fundamental point of view, these mechanisms combined with the possibility to fill the empty cages of the mother structure show the paramount richness of these new families.

Lastly, these Al based materials can indeed be regarded as potential materials for applications in photoluminescence or thermoluminescence up to magnetism, frequency doubling or non linear optical properties. The techniques classically used by solid state chemists for improving these types of properties have to be applied.

References

[1] Hervieu M., Mellenne B., Retoux R., Boudin S., Raveau B. (2004). *Nat. Mat.,* 3(4), 269-273.

[2] Krätschner W., Lamb L. D., Fostiropoulos K. & Huffman D. R. (1990). *Nature,* 347, 354-358.

[3] Banhart F., Zwanger M., Muhr H. J. (1994). *Chem. Phys. Lett,* 231, 98-104.

[4] Stéphan O., Bando, Y., Loiseau A., Willaime F., Shramchenko N., Tamiya T., Sato T. (1998). *Appl. Phys., A*67, 107-111.

[5] Tenne R., Margulis L., Genut M., Hodes G. (1992). *Nature,* 360, 444-446.

[6] Margulis L., Salitra G., Tenne R., Talianker M. (1995). *Nature,* 365, 113-114.

[7] Feldman Y., Tenne R. (1995). *Science,* 267, 222-225.

[8] Feldman Y., Frey G. L., Homyonfer M., Lyakhovitskaya V., Margulis L., Cohen H., Hodes G., Hutchison J. L., Tenne R. (1996). *J. Am. Chem.. Soc.,* 118, 5362-5367.

[9] Ducati C., Barborini E., Bongiomo G., Vinati S., Milani P., Midgley P. A. (2005). *Appl. Phys. Lett.,* 87, 201906.

[10] Nakagawa T., Kihara K., Harada K. (2001). *Am. Mineral.,* 86, 1506-1512.

[11] Zhu L., Seff K. (2000). *J. Phys. Chem. B,* 104, 8646-8951.

[12] Depmeier W., Schmid H., Setter N., Werk M. L. (1987). *Acta Cryst C*43, 2251-2255. Van Smaalen S., Dinnebier R. E., Katzke H., Depmeier W. (1997). *J. Solid State Chem.,* 129, 130-143. Depmeier W., Bührer W., (1991). *Acta Cryst. B*47, 197-206. Depmeier W. (1984). *Acta Cryst. C*40, 226-231. R. Kondo (1965). *J. of the Ceram. Ass. of Japan,* 73, 1-8. Hu X., Depmeier W., Wildermuth G., Doyle S. (1996). *Z. für Kristallographie,* 211, 679-688.

[13] Scheikowski M., Müller-Buschbaum H. (1993). *Z. für Anorg. und all. Chemie,* 619, 1755-1758. Müller-Buschbaum H., Werner J. P. (1996). *Z. für Naturforschung Teil B Anorg. Chemie, Orga Chemie,* 51,883-887.

[14] Bakakin V. V. et al. (1994). *Zh. Strukt. Khim.,* 35, 92-99.

[15] Boudin S., Mellenne B., Retoux R., Hervieu M. Raveau B. (2004). *Inorg. Chem.,* 43, 5954-5960.

[16] Mellenne B., Boudin S., Retoux R., Hervieu M., Raveau B. (2005). *Mater. Res. Bull.,* 40, 1388-1401.

[17] Retoux R., Mellenne B., Boudin S., Hervieu M., Raveau B. (2005). *J. Solid State Sciences,* 7, 736-744.

[18] Lebedev O. I., Bals S., VanTendeloo G., Snoeck G. E., Retoux R., Boudin S., Hervieu M., (2006). *Int. J. of Mat. Res,* 97 (7), 978-984.

In: Progress in Fullerene Research
Editor: Milton Lang, pp. 331-367

ISBN: 978-1-60021-841-5

Chapter 11

1,3-DIPOLAR CYCLOADDITION REACTIONS OF SOME 1-ARYL-3-OXIDOPYRIDINIUM BETAINES WITH C_{60} –A THEORETICAL STUDY

Lemi Türker* and Selçuk Gümüş
Middle East Technical University, Department of Chemistry,
06531, Ankara, Turkey

Abstract

In the present study, the molecular orbital properties of 1,3-dipolar cycloaddition products of certain substituted pyridinum betaines with C_{60} have been investigated theoretically at the level of PM3 (RHF) type semi empirical quantum chemical calculations. The betaines act as 4π-components across their 2,6-positions and C=C bond of C_{60} present either between the two hexagonal rings (66-type) or the pentagonal and hexagonal rings (56-type) acts as 2π-component in the 1,3-dipolar cycloadditions considered presently. The results indicate that in all the cases, the cycloadducts formed with 66-type dienophile are more stable and less endothermic than adducts formed with the 56-type. Some structural, spectral and physicochemical properties of these cycloadducts have been studied.

Key words: Fullerene, C_{60}, 1,3-dipolar cycloaddition, pyridinium betaines, PM3.

1. Introduction

Every organic reaction can be classified into one of three, more or less exclusive categories; ionic, radicalic and pericyclic. Ionic reactions involve pairs of electrons moving in one direction. In a unimolecular reaction, like the ionization of a tertiary alkyl halide, the carbon-halogen bond cleaves with both electrons reside on the chloride ion, leaving an electron deficiency behind forming a carbocation. In a bimolecular ionic reaction, one component, called the nucleophile, provides both electrons for a new bond, and the other, called the electrophile, receives them. Ionic reactions are by far the most numerous of organic reactions.

Radical reactions involve the correlated movement of unpaired electrons. In a unimolecular reaction like the photolytic cleavage of chlorine, one electron moves to one atom and the other electron moves to the other. In bimolecular reactions, a new bond is formed when one electron from one component pairs up with an electron from the other component. In radicalic reactions both components have equal contribution for the formation of a new bond, in contrast to ionic reactions, where both electrons are provided by the nucleophile.

Pericyclic reactions constitute the third distinct class. They are characterized with cyclic transition structures in which all bond-forming and bond breaking takes place synchronously, without the formation of an intermediate. The Diels-Alder reactions stand for one of the examples, that there is no absolute sense in which the electrons flow from one component to the other. In other words, neither component can be responsible alone with the supply of electrons to any of the new bonds.

It is convenient to divide pericyclic reactions into four main classes; cycloadditions, electrocyclic reactions, sigmatropic rearrangements, and group transfer reactions.

Cycloadditions are characterized by two components coming together to form two new σ-bonds, at expense of some π-bonds, at the termini of both components, joining them together to form a ring. Cycloadditions are by far the most abundant, featureful, and useful of all pericyclic reactions [1].

Generally, cycloaddition reactions are regarded as proceeding in either a concerted or in a nonconcerted fashion involving biradicalic or dipolar intermediates. In the past, all these suggested mechanisms were discussed critically on the basis of experimental data by Sauer and Huisgen [2,3]. According to Sauer [2], a concerted addition, which can also be described in terms of molecular orbitals, in which two new σ-bonds are formed simultaneously, offers the best explanation of the experimental facts. Huisgen [4-6] also strongly suggested that cycloaddition reactions should take place in a concerted fashion.

A five membered instead of a six membered ring should result from a $[4\pi+2\pi]$ cycloaddition if the 4π-electrons in the diene component is compressed into only a three-atom skeleton. This has been most commonly seen in the 1,3-dipolar cycloadditions. Note that the term dipolar here refers to the necessary formal charges in the structure of 4π-electron component. Thus, combination of such a zwitterionic molecule (the dipole) with a multiple bond system (dipolarophile) is referred to as 1,3-dipolar cycloaddition reaction (See Scheme 1). The zwitterionic component can be either allyl (bent) of propargyl-allenyl type (linear) [4]. Since its discovery, the 1,3-dipolar cycloaddition has found significant application in organic synthesis and has been a valuable tool for the construction of many types of molecules.

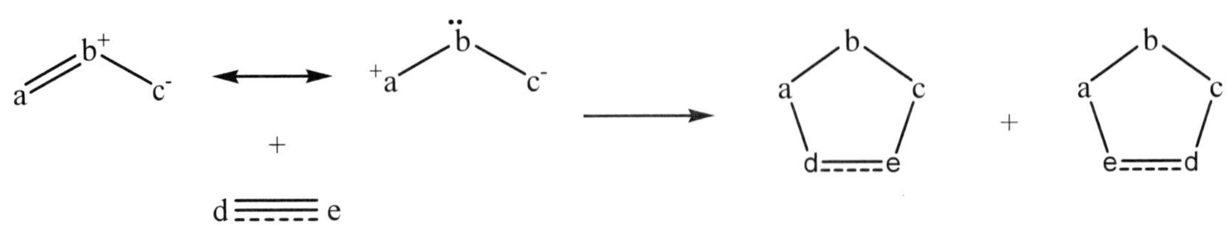

Scheme 1. Simple representation of a 1,3-dipolar cycloaddition reaction.

In 2001, Kolb and Sharpless proposed a new approach to lead discovery and optimization, which they termed 'click chemistry'. The click chemistry process makes use of a limited number of robust reactions, of which the 1,3-dipolar cycloaddition is the most prominent [7,8]. Utilizing the efficiency, mild character, and high-yielding properties of the cycloadditions, azides and alkynes which form 1,2,3-triazoles, attracted attention (beside the others) and several laboratories were able to explore the application of this chemistry to biological problems. For instance, Huisgen cycloaddition reaction can be used as a powerful tool to identify unique and potent enzyme inhibitors that can be easily translated into drugs. A number of labs have been able to successfully incorporate the triazole formation to develop inhibitors of diverse enzyme active sites. In the literature, the synthesis and characterization of inhibitors that target acetyl cholinesterase (AChE) [9,10], carbonic anhydrase II (CAII) [11], HIV-1 protease (HIV-PR) [12], and fucosyl transferase VI (Fuc-T) [13] have been described. An *in situ* approach was employed in the synthesis of the AChE and CAII inhibitors; that is, the enzymes were used to accelerate the cycloaddition reaction as well as to design the best-fit-inhibitor. In the absence of enzyme the azide-alkyne condensation is extremely slow, even at elevated temperatures. To overcome this time constraint, Rostovtsev *et. al* developed catalytic triazole synthesis [14]. In addition to improving the rate of the cycloaddition, the copper(I)-mediated reaction was found to be remarkably regioselective. This improvement allowed for the application of the Huisgen process to high throughput screening. The utility of the reaction is apparent in the discovery of inhibitors of HIV-1 PR and Fuc-T [12,13].

As mentioned above, the versatility, mild conditions, regioselectivity, and high yield of the reaction are good indications of its potential application to more complex and diverse biological systems. Click chemistry provides a novel and potentially general way of uniting two compounds to produce new generations of enzyme inhibitors.

C_{60} is an electron-deficient spherically shaped polyene and is known to undergo a variety of reactions, viz., radical, nucleophilic addition, mono- and polycyclic additions, etc. Fullerene reactions have been studied extensively both experimentally and theoretically by several workers [15-56], and the reactivity of fullerene toward various cycloadditions has been well documented by Hirsch [57-60] and Silwa [61] and specifically, 1,3-dipolar cycloadditions involving C_{60} have been surveyed by Yurovskaya and Ovcharenko [62] in 1998. In 2002, Yurovskaya and Trushkov [63] have reviewed further progress on the C_{60} cycloadditions. 1,3-Dipolar reactions have been used to form fullerene-fused heterocycles and some of these reactions have been modeled [64]. Among various dipoles that have been employed to derivatize fullerene, diazomethane, nitrile oxide and nitrone are very important. Diazoalkanes add to C_{60}, first to form fullerene-fused pyrazolines and this on subsequent thermolysis loses N_2 to produce methanofullerenes [63]. The pyrazoline intermediate has been found to be stable and in certain instances, it has been isolated and characterized [63]. Nitrile oxides and nitrones add to C_{60} to form fullerene-fused isoxazolines. The ability of C_{60} to act as a dipolarophile was first reported in 1992 [65]. Wudl and co-workers have performed a systematic study of the reactions of C_{60} with substituted diazomethanes and proved that C_{60} is a reactive dipolarophile [66].

In the present study, the 1,3-dipolar cycloaddition reaction products of certain substituted pyridinum betaines with C_{60} have been investigated theoretically at the level of PM3 (RHF) type semi empirical quantum chemical calculations. The aim of the present work is, firstly, to study the regiochemistry of the 1,3-dipolar cycloaddition reaction by reacting the dipole across the termini of 6,6 or 5,6 junctions of C_{60} (See Figure 1). Secondly, to investigate the effect of the type and the position of the substituent (X: H, F, Cl, Br, I, OMe, NO_2) on the 4π-component (1-Aryl-3-oxidopyridinium betaines) (Figure 2).

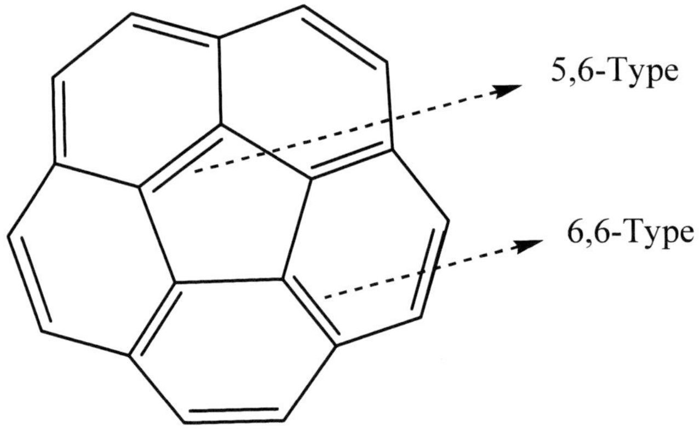

Figure 1. Illustration of the 56-type and 66-type 2π-components of the C_{60} system (just a part of C_{60} is shown).

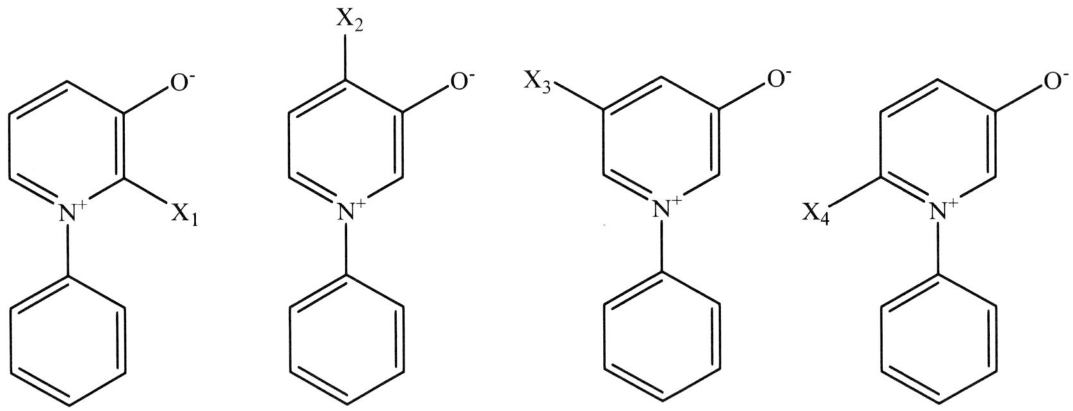

Figure 2. The substituted 1-aryl-3-oxidopyridinium betaines under consideration, which act as 4π-components in the 1,3-dipolar cycloaddition reactions. The substituents are indicated as X_1 through X_4.

2. Method

The geometry optimizations of all the structures leading to energy minima were achieved by using first MM+ and then, PM3 self-consistent fields molecular orbital (SCFMO) [67,68] method at the restricted Hartree-Fock (RHF) level [69]. For the optimizations the steepest-descent method followed by conjugate gradient methods, Fletcher-Reeves and Polak-Ribiere were employed, consecutively (convergence limit of 4.18×10^{-4} kJ/mol (0.0001 kcal/mol) and RMS gradient of 4.18×10^{7} kJ/Mmol (0.001 kcal/A mol)).

All the optimized structures were subjected to vibrational analysis (using the same level of calculation employed in the geometry optimization of them), in order to judge that the optimized geometry at least stands for a local energy minimum. All these computations were performed by using Hyperchem (release 7.5) [70] and Gaussian 03 package programs [71].

3. Result and Discussion

In the present study, the structural and molecular orbital properties of the cycloadducts obtained by the reaction of certain 1-aryl-3-oxidopyridinium betaines with C_{60} [(5,6)-fullerene-60-I_h] have been studied theoretically at the level of semi-empirical PM3 type quantum chemical calculations. 3-Oxidopyridinium betaines contain 8π-electrons. Such a number of π-electron array shows more than one mode of addition. Depending on the frontier molecular orbital (FMO) characteristics, these 8π-electrons participate in cycloadditions within the Woodward-Hoffman rules. These betaines react thermally with 4π-addends across their 2,4-positions (2π/6π-component) and with 2π-addends across their 2,6 positions (4π-component). The magnitude of the interaction energy between the dipoles and dipolarophiles in a cycloaddition reactions dictates the regioselectivity and periselectivity in course of the reaction. During the cycloaddition reactions presently considered, the betaines and fullerene (C_{60}) act as 4π- and 2π- components, respectively. The FMO coefficients, as well as the energies, of the dipoles and the dipolarophiles depend on the type and number of substituents attached to the system. The underlying principle in these is that the bonding interactions occur between lobes of the FMO of the same symmetry. Then the signs of the HOMO-LUMO coefficients in both the betaine and the dipolarophile determine the possible mode of addition. Examining the nodal properties of the FMO's of the betaine and the dipolarophile, it is an easy task to detect the matching nodal properties in both participants. Then, which opposite pair of the FMO interaction is the predominant, is determined by the energy gap separating the interacting orbitals. Presently, the product of 1,3-dipolar cycloaddition reactions of 1-aryl-3-oxidopyridinium betaines with C_{60} mainly depends on four factors;

(i) In C_{60} molecule there are two distinct types of junctions, namely; 56-type and 66-type (Figure 1). The former one is in between a five and a six membered ring whereas the later one is on the border of two adjacent six-membered rings of C_{60}. In the π-skeleton of C_{60} these junctions may act as a 2π-component (dipolarophile) in 1,3-dipolar cycloaddition reactions. Therefore, the structure of cycloadduct formed will depend on whether the reaction takes place through 56-type or 66-type 2π-system.

(ii) As shown in Figure 2, 1-aryl-3-oxidopyridinium betaines have four different positions for substitution (X_1-X_4). Thus, the cycloaddition reaction of a betaine with C_{60} will yield different isomeric cycloadducts (regio and stereoisomers) depending on the position of the substituent in parent dienophile system.

(iii) In cycloaddition reactions, the type of the substituent is another dictating factor. In the present study, the authors have considered six different substituents. These are F, Cl, Br, I, NO_2 and OMe. Of these substituents four different halogens considered will reveal information about various effects (the increase in the volume of the heteroatom, while decrease in electro negativity, thus influencing inductive and mesomeric effects) on the stability and product distribution of the cycloadducts. The effect of electron withdrawing and donating character of the substituents on the cycloadduct stability has also been investigated by considering NO_2 and OMe substituents.

(iv) As illustrated in Scheme 1, the termini (ac) of 4π-component(dipole) may interact with the 2π-system as -de- or -ed- fashion. This is also true for the present case. Although, the reaction through 66-type junction yield the same cycloadducts for either attack, the interactions through 56-type 2π-systems result two different regio isomers. These isomers are named as X-56-6 or X-56-5 in this study (X: substituent, the right most number, 5 or 6, indicates that the enone system is next to a 5-or 6-membered ring of C_{60}).

The possible cycloadducts of the 1,3-dipolar cycloadditions of a 1-aryl-3-oxdopyridinium betaine with C_{60} are shown in Figures 3 and 4.

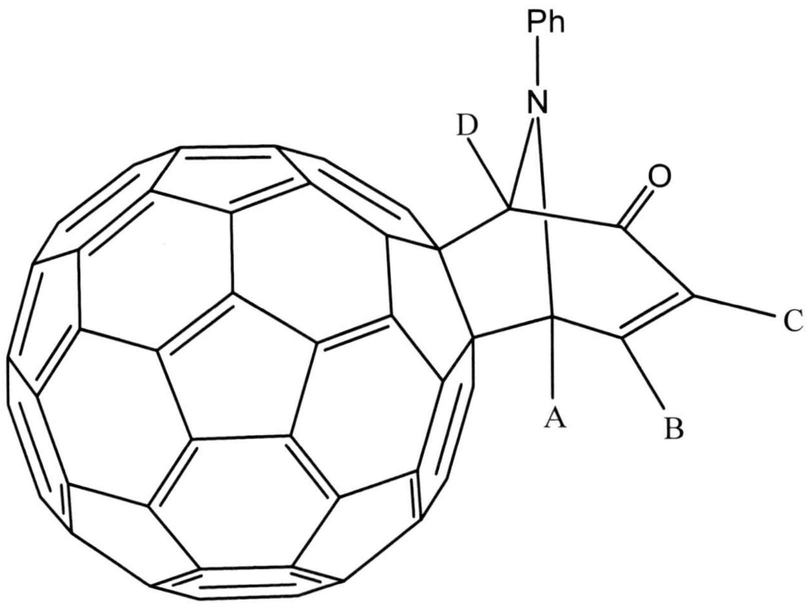

Figure 3. One of the possible cycloadducts of the 1,3-dipolar cycloaddition reaction of pyridinum betaines with C_{60}. Position of the substituents are shown by letters A-D.

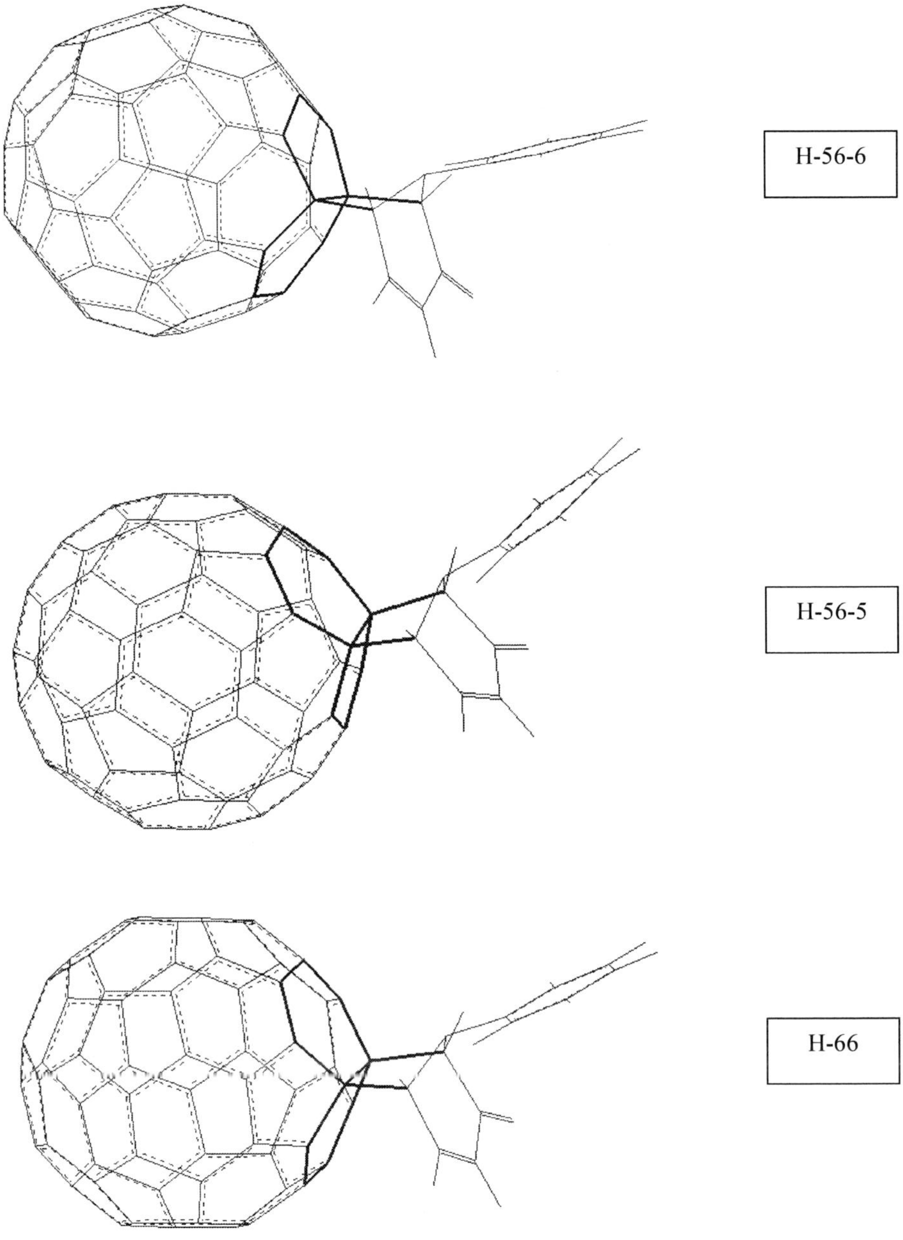

Figure 4. Geometry optimized structures of the cycloadducts obtained by 1,3-Dipolar cycloaddition reactions of the unsubstituted betaines with C_{60}.

A) The Geometries and Stabilities

a) Unsubstituted cycloadduct

The geometry optimized structures of the cycloadducts obtained by the reaction of the unsubstituted betaine and C_{60} can be seen in Figure 4. As seen in this figure, the perfect ball character of C_{60} has been distorted after the reaction. One can judge the stabilities of the cycloadducts of the unsubstituted betaine and C_{60} by looking at the total and binding energies in Table 1. The most stable and the least endothermic among the isomers in this series is H-66. This result is in parallel with the findings of Kavitha and Venuvanalingam [72], and Türker [44]. According to Kavitha and Venuvanalingam C_{60} will react with nitrile oxide or nitrone through 66-junction, since the cycloadducts of that pathway were found to be the most stable. Türker studied the 1,3-dipolar cycloaddition reaction of 1-phenyl-phthalazinium-1-olate with C_{60}. He also found the product obtained through the 66-junction to be more stable than adducts obtained through 56-junction. On the other hand, the stabilities of the present H-56-6 and H-56-5 are almost equal, being 56-6 more favorable. Probably, a kind of favorable interaction occurs between the enone system present and the 6-membered aromatic ring of C_{60}. Whereas, 5-membered rings of C_{60} which are radialene type in character should not interact favorably with the enone moiety. Note that the enone system is inclined to C_{60} surface so that the lone-pair of the bridge nitrogen atom necessarily points to either a 6- or 5-membered ring of C_{60}. Then, the lone-pair next to the electron deficient 5-membered ring presents a situation which is less repulsive than comparatively electron rich 6-membered ring.

A broad variety of methano-bridged fullerenes are accessible by the reaction of C_{60} with diazomethanes [58,59]. In these reactions C_{60} behaves as an 1,3-dipolarophile. One of its double bounds, commonly 66-type, is affected. However, such kind of cycloadditions can be accompanied or followed by thermal disrotatory electrocyclic ring opening reactions of ${}_{\sigma}2_s+{}_{\pi}4_s$ type leading to a larger ring formation [44]. In the present study, such kind of electrocyclic ring opening reactions are not considered. Based on calculations at the MNDO level and substituted diphenyldiazomethanes, a simple rule was formulated stating that "for C_{60} as well as for C_{60} derivatives it is energetically unfavorable to introduce double bonds at 56-ring junctions or to reduce the greatest possible number of [5]-radialene rings" [46].

Table 1. Various energies of the unsubstituted (X1-X4:H) cycloadducts (Energies in kJ/mol).

	Type		
Energy	**56-6**	**56-5**	**66**
Total Energy	-866448.12	-866446.76	-866527.01
Binding Energy	-49978.73	-49977.37	-50057.62
Isolated Atomic Energy	-816469	-816469	-816469
Electronic Energy	-13629873	-13630550	-13633899
Core-Core Interaction	12763425	12764103	12767372
Heat of Formation	3470.47	3471.83	3391.58

b) F, Cl, Br, I, NO_2 and OMe substituted cycloadducts

Figures 5-10 show the geometry optimized structures for the F, Cl, Br, I, NO_2, and OMe substituted isomers, respectively. The structures differ from each other by the orientation of the enone system.

Various energies of the series have been given in Tables 2-7. As in the H (unsubstituted) case, the cycloadducts of the 66-type have been found to be the most stable and least endothermic for all substituents. The energy difference between the 66-type and 56-type isomers is about 80 kJ/mol. For all series, the effect of the position of the substituent on the stability of the cycloadducts have also been investigated. All the cycloadducts irrespective of the kind and position of substituent are endothermic in character. In the case of unsubstituted series of cycloadducts, the least endothermic structure is from 66-type geometry. It is the most stable adduct (the total and binding energies). The fluorine substituted adducts exhibit stability order of B>C>A>D for 56-6 and 66-types but B=D>C>A for 56-5 type. The least endothermic adduct occurs when the F atom is at B-position. The B and D positions for 56-5 type yields equivalent values. The chlorine atom at position B results in the most stable and least endothermic adduct. The stability order of B>C>A>D is exhibited in this case. On the other hand, the presence of bromine changes the stability order and A-type adduct becomes the most stable and the least endothermic no matter whether 56- or 66- type is considered. Whereas, in the case of iodine substituted adducts, the B-types are the most stable and the least endothermic ones with the exception of 56-6 series in which D is the least endothermic isomer. In the case of nitro substituent, in all cases, the stability shines on C-type and the order is C>B>A>D. The C-type is also the least endothermic adduct within the series. Again in all the cases, the methoxide at B-position results in the most stable and the least endothermic adduct formation and the stability order is B>C>A>D as it is in most of the cases of different substituents.

The energetics (total energy, binding energy, etc.) of all the possible substituted isomers are given in Tables 2-7

Table 2. Various energies of the F substituted cycloadducts (Energies in kJ/mol).

Energy	56-6 A	56-6B	56-6 C	56-6 D
Total Energy	-907427.19	-907440.74	-907436.14	-907425.97
Binding Energy	-50004.37	-50017.93	-50013.33	-50003.16
Isolated Atomic Energy	-857422.81	-857422.81	-857422.81	-857422.81
Electronic Energy	-14125318	-14086296	-14066554	-14126742
Core-Core Interaction	13217891	13178855	13159118	13219316
Heat of Formation	3305.87	3292.32	3296.91	3307.09
Energy	**56-5A**	**56-5 B**	**56-5 C**	**56-5D**
Total Energy	-907427.87	-907439.81	-907434.94	-907425.88
Binding Energy	-50005.06	-50016.99	-50012.12	-50003.07
Isolated Atomic Energy	-857423	-857423	-857423	-857423
Electronic Energy	-14121559	-14085698	-14065669	-14121559
Core-Core Interaction	13214131	13178258	13158234	13119131
Heat of Formation	3305.18	3293.24	3298.12	3307.18
Energy	**66 A**	**66 B**	**66 C**	**66 D**
Total Energy	-907507.69	-907519.56	-907514.88	-907502.64
Binding Energy	-50084.88	-50096.75	-50092.07	-50079.83
Isolated Atomic Energy	-857423	-857423	-857423	-857423
Electronic Energy	-14126493	-14090324	-14070266	-14130792
Core-Core Interaction	13218986	13182805	13162752	13223289
Heat of Formation	3225.36	3213.49	3218.18	3230.42

Table 3. Various energies of the Cl substituted cycloadducts (Energies in kJ/mol).

Energy	56-6 A	56-6 B	56-6 C	56-6 D
Total Energy	-895510.39	-895526.61	-895523.83	-895504.46
Binding Energy	-49890.08	-49906.31	-49903.52	-49884.16
Isolated Atomic Energy	-845620	-845620	-845620	-845620
Electronic Energy	-14098540	-14061484	-14039471	-14103798
Core-Core Interaction	13203030	13165957	13143947	13208293
Heat of Formation	3462.42	3446.19	3448.97	3468.34
Energy	**56-5 A**	**56-5 B**	**56-5 C**	**56-5 D**
Total Energy	-895508.72	-895525.51	-895522.53	-895502.72
Binding Energy	-49888.42	-49905.21	-49902.23	-49882.42
Isolated Atomic Energy	-845620	-845620	-845620	-845620
Electronic Energy	-14098469	-14060692	-14038457	-14103631
Core-Core Interaction	13202960	13165167	13142935	13208128
Heat of Formation	3464.08	3447.29	3450.27	3470.08
Energy	**66 A**	**66 B**	**66 C**	**66 D**
Total Energy	-895588.98	-895605.46	-895602.69	-895582.44
Binding Energy	-49968.68	-49985.16	-49982.39	-49962.13
Isolated Atomic Energy	-845620	-845620	-845620	-845620
Electronic Energy	-14101988	-14065481	-14043239	-14108522
Core-Core Interaction	13206399	13169876	13147637	13212939
Heat of Formation	3383.82	3367.34	3370.11	3390.36

Table 4. Various energies of the Br substituted cycloadducts (Energies in kJ/mol).

Energy	56-6 A	56-6 B	56-6 C	56-6 D
Total Energy	-899058.99	-899057.4	-899053.2	-899050.9
Binding Energy	-49835.38	-49833.78	-49829.58	-49827.29
Isolated Atomic Energy	-849224	-849224	-849224	-849224
Electronic Energy	-14097044	-14055113	-14034614	-14104153
Core-Core Interaction	13197985	13156056	13135560	13205102
Heat of Formation	3507.71	3509.3	3513.5	3515.8
Energy	**56-5 A**	**56-5 B**	**56-5 C**	**56-5 D**
Total Energy	-899060.85	-899056.52	-899051.91	-899052.97
Binding Energy	-49837.23	-49832.90	-49828.29	-49829.35
Isolated Atomic Energy	-849224	-849224	-849224	-849224
Electronic Energy	-14098965	-14054584	-14033682	-14105613
Core-Core Interaction	13199904	13155528	13134630	13206560
Heat of Formation	3505.85	3510.19	3514.79	3513.73
Energy	**66 A**	**66 B**	**66 C**	**66 D**
Total Energy	-899139.87	-899136.31	-899131.85	-899131.99
Binding Energy	-49916.26	-49912.69	-49908.23	-49908.38
Isolated Atomic Energy	-849224	-849224	-849224	-849224
Electronic Energy	-14103562	-14059314	-14038291	-14109586
Core-Core Interaction	13204422	13160178	13139159	13210454
Heat of Formation	3426.83	3430.39	3434.85	3434.71

Table 5. Various energies of the I substituted cycloadducts (Energies in kJ/mol).

Energy	56-6 A	56-6 B	56-6 C	56-6 D
Total Energy	-892734.46	-892783.89	-892781.39	-892796.24
Binding Energy	-49707.64	-49757.07	-49754.56	-49769.41
Isolated Atomic Energy	-843027	-843027	-843027	-843027
Electronic Energy	-14073413	-14033823	-14017546	-14084617
Core-Core Interaction	13180678	13141039	13124764	13191821
Heat of Formation	3630.33	3580.89	3583.4	3568.55
Energy	**56-5 A**	**56-5 B**	**56-5 C**	**56-5 D**
Total Energy	-892736.98	-892783.79	-892780.20	-892719.59
Binding Energy	-49710.15	-49756.97	-49753.38	-49692.77
Isolated Atomic Energy	-843027	-843027	-843027	-843027
Electronic Energy	-14075963	-14034321	-14017093	-14080064
Core-Core Interaction	13183226	13141538	13124313	13187344
Heat of Formation	3627.81	3581.00	3584.59	3645.20
Energy	**66 A**	**66 B**	**66 C**	**66 D**
Total Energy	-892813.36	-892862.83	-892860.15	-892796.24
Binding Energy	-49786.54	-49836.01	-49833.32	-49769.41
Isolated Atomic Energy	-843027	-843027	-843027	-843027
Electronic Energy	-14080653	-14038116	-14021362	-14084619
Core-Core Interaction	13187839	13145253	13128502	13191823
Heat of Formation	3551.43	3501.96	3504.64	3568.55

Table 6. Various energies of the NO_2 substituted cycloadducts (Energies in kJ/mol).

Energy	56-6 A	56-6 B	56-6 C	56-6 D
Total Energy	-936954.23	-936998.21	-936998.8	-936940.82
Binding Energy	-50702.87	-50746.85	-50747.44	-50689.45
Isolated Atomic Energy	-886251	-886251	-886251	-886251
Electronic Energy	-14900582	-14804887	-14753990	-14914843
Core-Core Interaction	13963628	13867889	13816991	13977902
Heat of Formation	3499.51	3455.54	3454.95	3512.93
Energy	**56-5 A**	**56-5 B**	**56-5 C**	**56-5 D**
Total Energy	-936952.30	-936997.06	-936997.79	-936937.89
Binding Energy	-50700.93	-50745.70	-50746.43	-50686.53
Isolated Atomic Energy	-886251	-886251	-886251	-886251
Electronic Energy	-14901120	-14802716	-14750019	-14915681
Core-Core Interaction	13964167	13865719	13813021	13978743
Heat of Formation	3501.45	3456.69	3455.96	3515.85
Energy	**66 A**	**66 B**	**66 C**	**66 D**
Total Energy	-937030.61	-937076.44	-937077.34	-937014.03
Binding Energy	-50779.25	-50825.08	-50825.98	-50762.67
Isolated Atomic Energy	-886251	-886251	-886251	-886251
Electronic Energy	-14904641	-14808762	-14756843	-14919004
Core-Core Interaction	13967610	13871686	13819766	13981990
Heat of Formation	3423.14	3377.308435	3376.41	3439.71

Table 7. Various energies of the MeO substituted cycloadducts (Energies in kJ/mol).

Energy	56-6 A	56-6 B	56-6 C	56-6 D
Total Energy	-909150.55	-909189.35	-909175.12	-909139.01
Binding Energy	-51508.33	-51547.12	-51532.89	-51496.79
Isolated Atomic Energy	-857642	-857642	-857642	-857642
Electronic Energy	-14571610	-14484157	-14455610	-14583398
Core-Core Interaction	13662460	13574968	13546435	13674259
Heat of Formation	3341.06	3302.27	3316.49	3352.6
Energy	**56-5 A**	**56-5B**	**56-5 C**	**56-5 D**
Total Energy	-909141.57	-909195.60	-909167.42	-909131.19
Binding Energy	-51499.35	-51553.38	-51525.20	-51488.96
Isolated Atomic Energy	-857642	-857642	-857642	-857642
Electronic Energy	-14552075	-14607651	-14472773	-14560755
Core-Core Interaction	13642933	13698456	13563606	13651624
Heat of Formation	3350.05	3296.01	3324.19	3360.43
Energy	**66 A**	**66 B**	**66 C**	**66 D**
Total Energy	-909226.54	-909260.01	-909254.06	-909216.15
Binding Energy	-51584.31	-51617.78	-51611.84	-51573.92
Isolated Atomic Energy	-857642	-857642	-857642	-857642
Electronic Energy	-14572948	-14524795	-14458990	-14586553
Core-Core Interaction	13663722	13615535	13549736	13677337
Heat of Formation	3265.07	3231.6	3237.55	3275.46

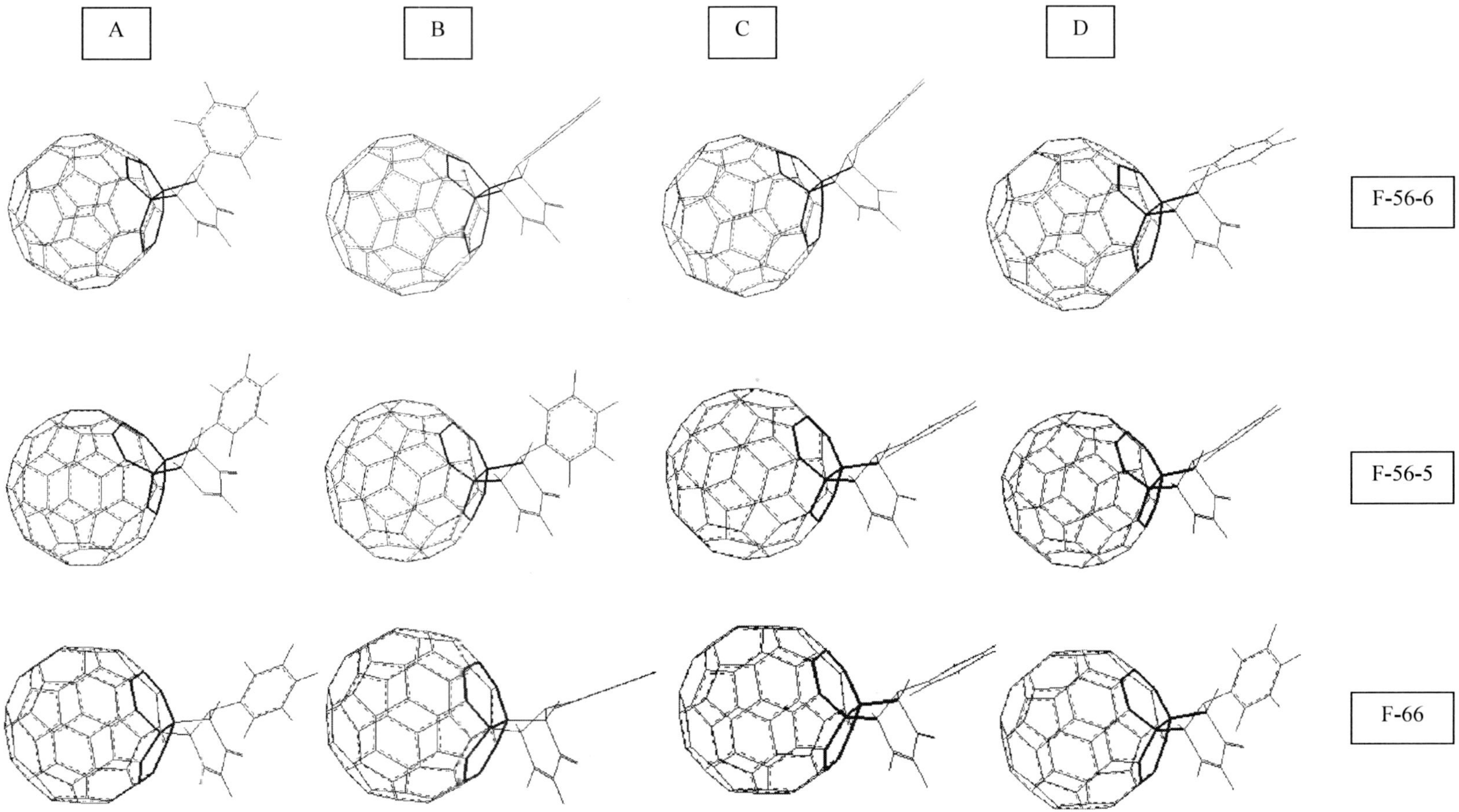

Figure 5. Geometry optimized structures of the cycloadducts obtained by 1,3-Dipolar cycloaddition reactions of the F-substituted betaines with C_{60}.

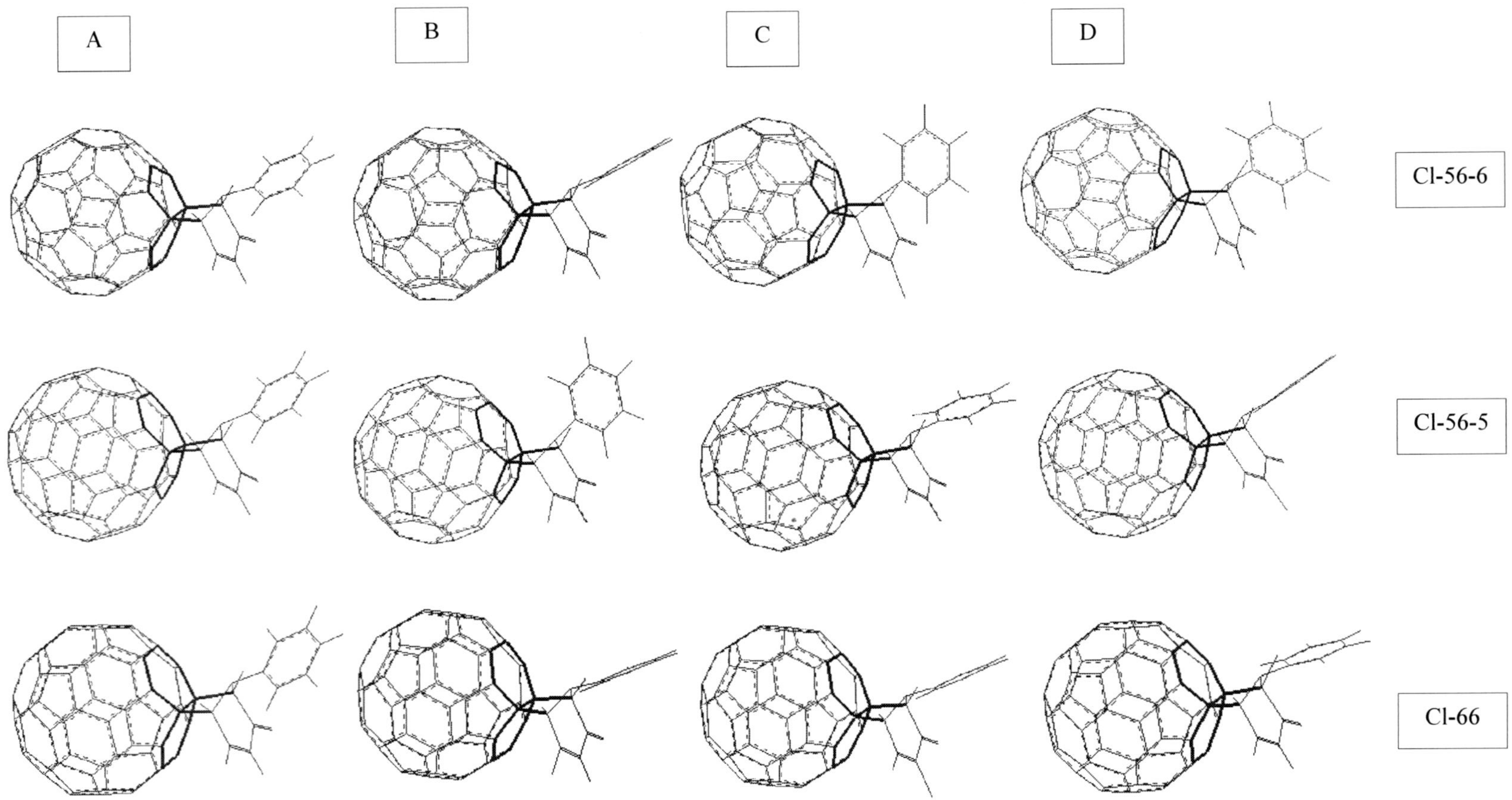

Figure 6. Geometry optimized structures of the cycloadducts obtained by 1,3-Dipolar cycloaddition reactions of the Cl-substituted betaines with C_{60}.

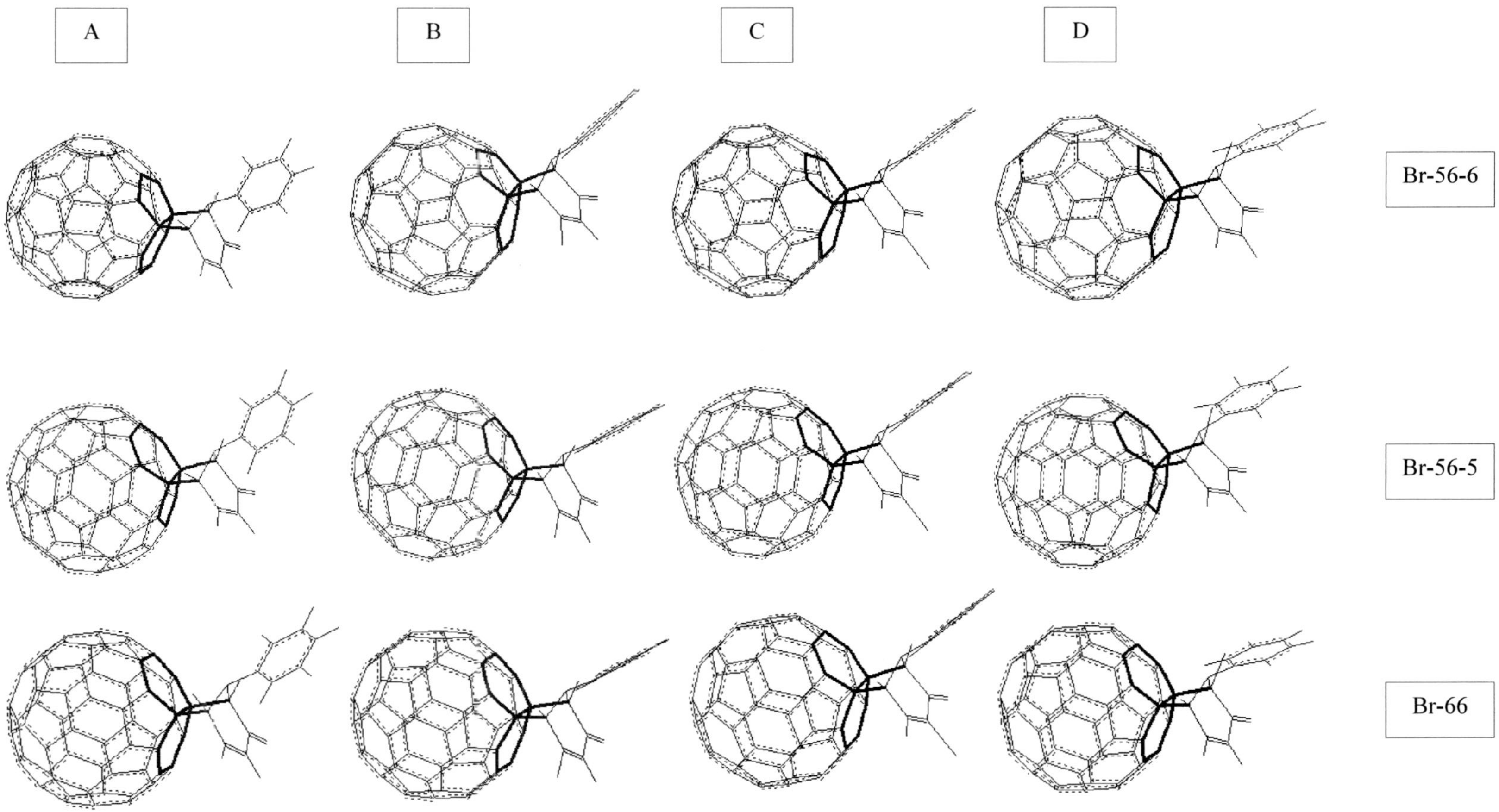

Figure 7. Geometry optimized structures of the cycloadducts obtained by 1,3-Dipolar cycloaddition reactions of the Br-substituted betaines with C_{60}.

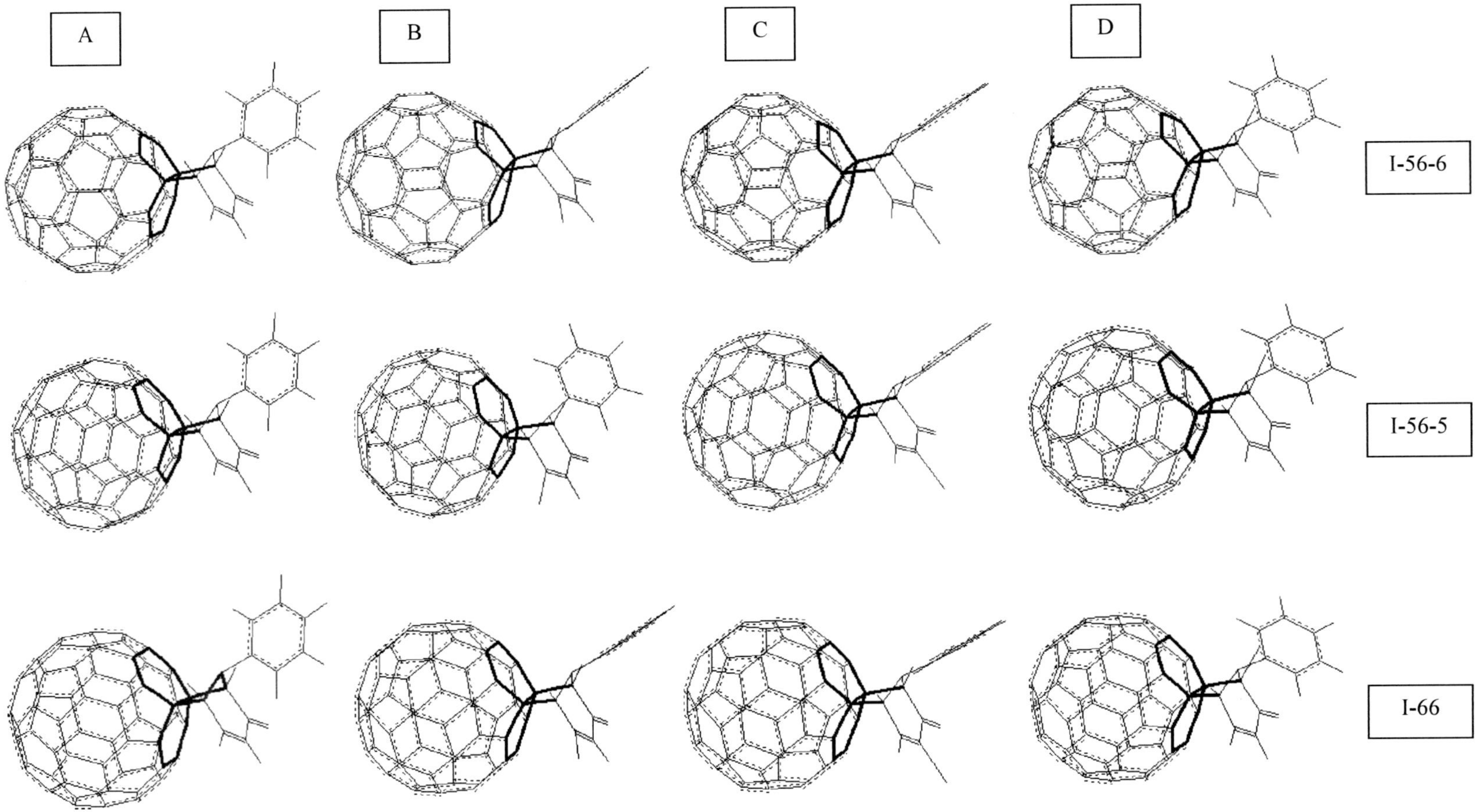

Figure 8. Geometry optimized structures of the cycloadducts obtained by 1,3-Dipolar cycloaddition reactions of the I-substituted betaines with

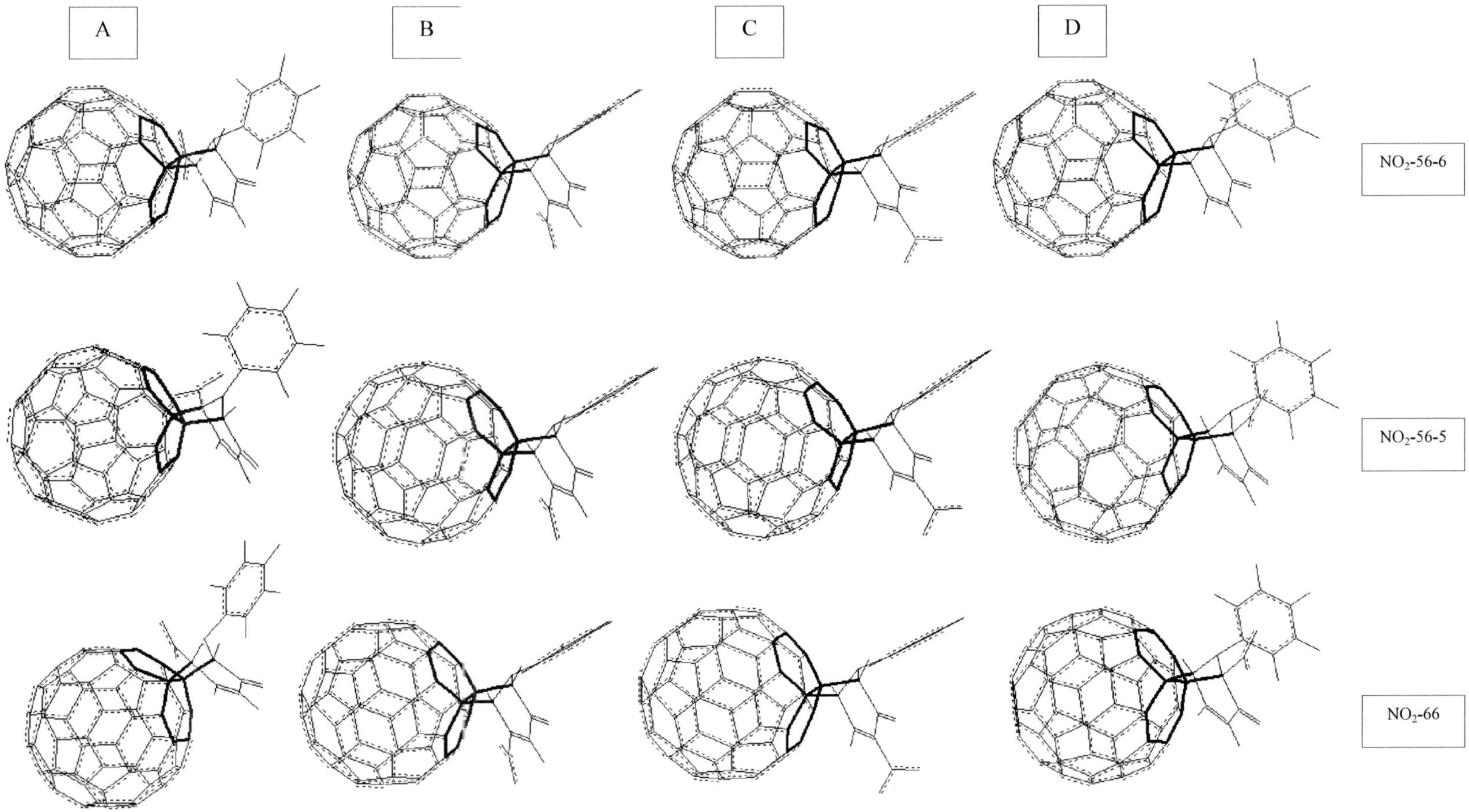

Figure 9. Geometry optimized structures of the cycloadducts obtained by 1,3-Dipolar cycloaddition reactions of the NO_2-substituted betaines

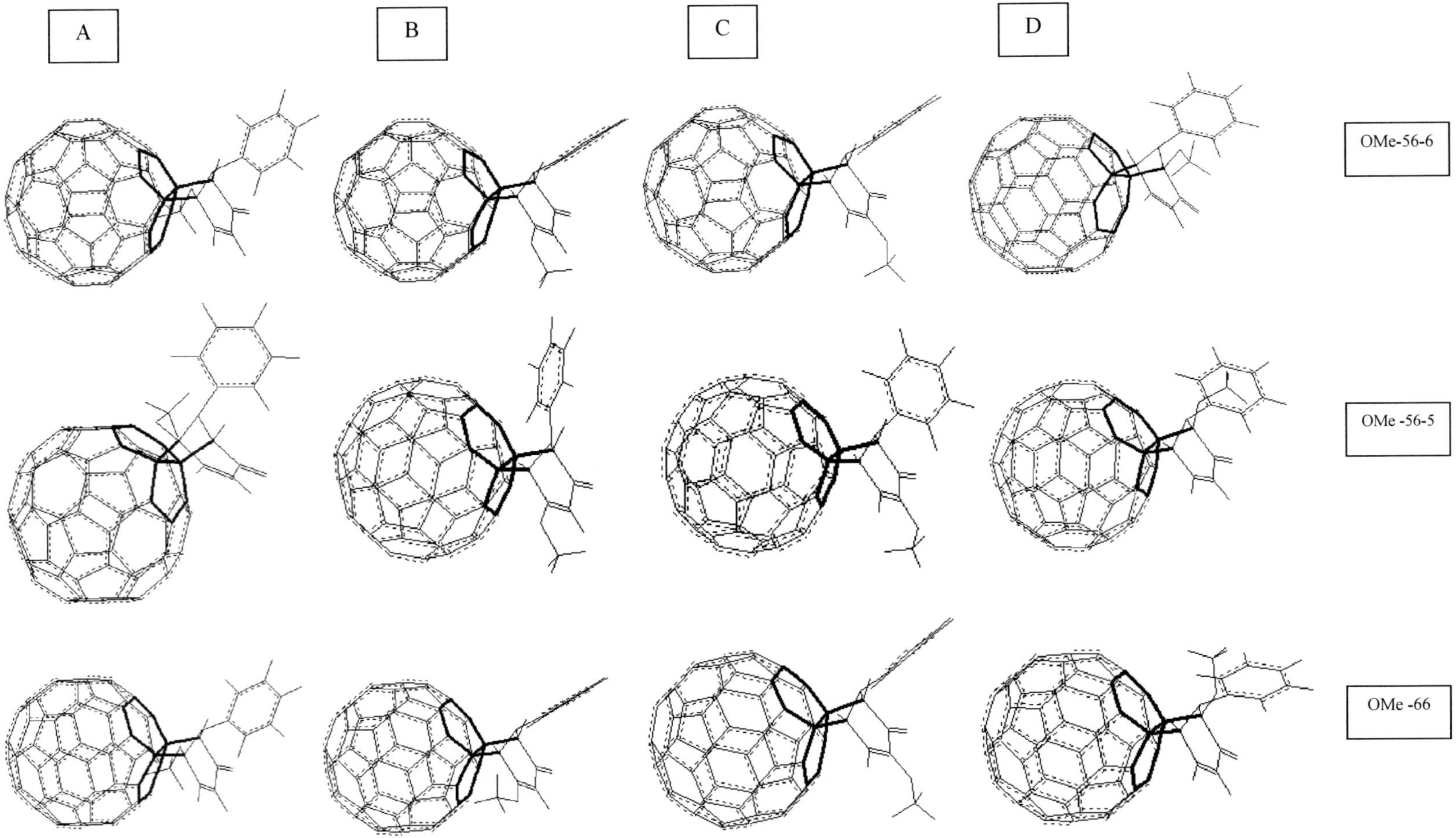

Figure 10. Geometry optimized structures of the cycloadducts obtained by 1,3-Dipolar cycloaddition reactions of the MeO-substituted betaines with C_{60}.

B) Conformation of the Phenyl Ring

As seen in figures for the geometry optimized structures (Figures 4-10) the cycloadducts prefer a regio- and stereo-isomerism in such a way that the enone moiety and phenyl ring are on the same side of a virtual plane comprising the bridge-head carbons and the bridge-nitrogen atom. This is the case exhibited by the cycloadducts irrespective of the kind and position of the substituents X_1-X_4, originating from the 3-oxidopyridinium moiety. In most of the cases, possibility of some sort of attractive interaction between the HOMO involving the phenyl part and the LUMO spread over the enone system makes phenyl ring to be inclined towards the enone system (However, in the case of A- or D-type adducts, steric repulsions between the substituent and the phenyl ring may overcome the attractive forces operating). On the other hand, possibility of some repulsive interactions between the phenyl ring and π-electron cloud of the fullerene body could be claimed responsible for the fact. Although, this repulsive interaction could be the cause of the emerged calculated position of the phenyl group, otherwise it would be face to face with a hexagonal ring of C_{60} in the case of 66- and 56-5, it cannot be a valid effect when it would be face to a pentagonal ring, because the pentagonal rings are radialenic in character having 5 π-electrons. In other words, they are somewhat electron deficient and willingly accept electrons (thus C_{60} can accommodate up to 12 electrons in reductions [51,58,74]). Hence, the interaction with the N-phenyl ring would not be repulsive.

On the other hand, rapid inversion of nitrogen at the bridge position should also make the phenyl ring to incline the other side (opposite to the enone side). However, it seems it is not favored, besides the interaction of nitrogen lone-pair and pentagonal π-system should be favorable. To have a deeper look to such kind of effects we have also investigated the rotation of the N-phenyl ring in 56-6 A type cycloadducts.

A rotational analysis at PM3 level has been performed, in order to investigate the effect of conformation of the phenyl ring (attached to N atom) on the total energy of the system. For this purpose, 56-6A substituted systems have been considered. The phenyl ring has been rotated around the N-Ph bond starting from 0° to 360° by 10 degrees increment in each step. The variations of total energy by the rotation of the phenyl ring around the N-Ph bond can be seen in Figures 11-17.

The effect of rotation of the phenyl ring on unsubstituted system (H-56-6) can be seen in Figure 11. The calculations resulted in two maxima at 60° and 230°, and two minima at 150° and 330°. The global minimum of the rotational analysis has been found to be at 330° with the corresponding total energy of -330.0060718 Hartree (-866448.122 kJ/mol). This result completely reflects what has been found by the geometry optimization calculations. The maxima have been obtained when the phenyl ring is perpendicular to the enone moiety. This conformation may possibly increase the repulsion between the phenyl and enone moieties, thus increasing the energy of the system.

The minimum and maximum energies in the conformational analysis of F-56-6 A (Figure 12) system have been found to be at 30° and 160°, respectively. In contrast to the unsubstituted system, the maximum energy conformation has been obtained when the phenyl and enone moieties are face to face with each other.

In the rotational analysis of Cl-56-6 A system the minimum energy is observed with 350° of torsional angle (Figure 13), which is equal to the angle obtained by the geometry optimization at PM3 level. When the torsional angle is 260° the phenyl and enone moieties are perpendicular, which stands for the least favorable structure.

In the case of Br substituted derivative, the structure having 0° torsional angle is the most stable one. The least stable structure, on the other hand, has the torsional angle of 140° (see Figure 14).

As for the results of the rotational analysis of the I- substituted (I-56-6 A) derivative shown in Figure 15, the most favorable structure has the torsional angle of 40° with phenyl and enone moieties perpendicular to each other. Whereas, at the angle of 150° the system has the highest energy, corresponding to geometry in which the phenyl and the enone moieties are face to face, as in the other cases except the unsubstituted and methoxy substituted systems. Similar results are obtained for the NO_2 case (56-6 A) (see Figure 16).

In the case of MeO-56-6 A derivative, the most stable structure is associated with the torsional angle of 10° with -346.2702222 Hartree (-909150.4957 kJ/mol) of total energy. In this case the lowest energy system possesses the phenyl and enone moieties in face to face arrangement, in contrast to the other cases, except the unsubstituted derivative. When the torsional angle is 270° and the phenyl and enone are perpendicular in the structure, the system has the highest total energy, being in the least favorable conformation (see Figure 17).

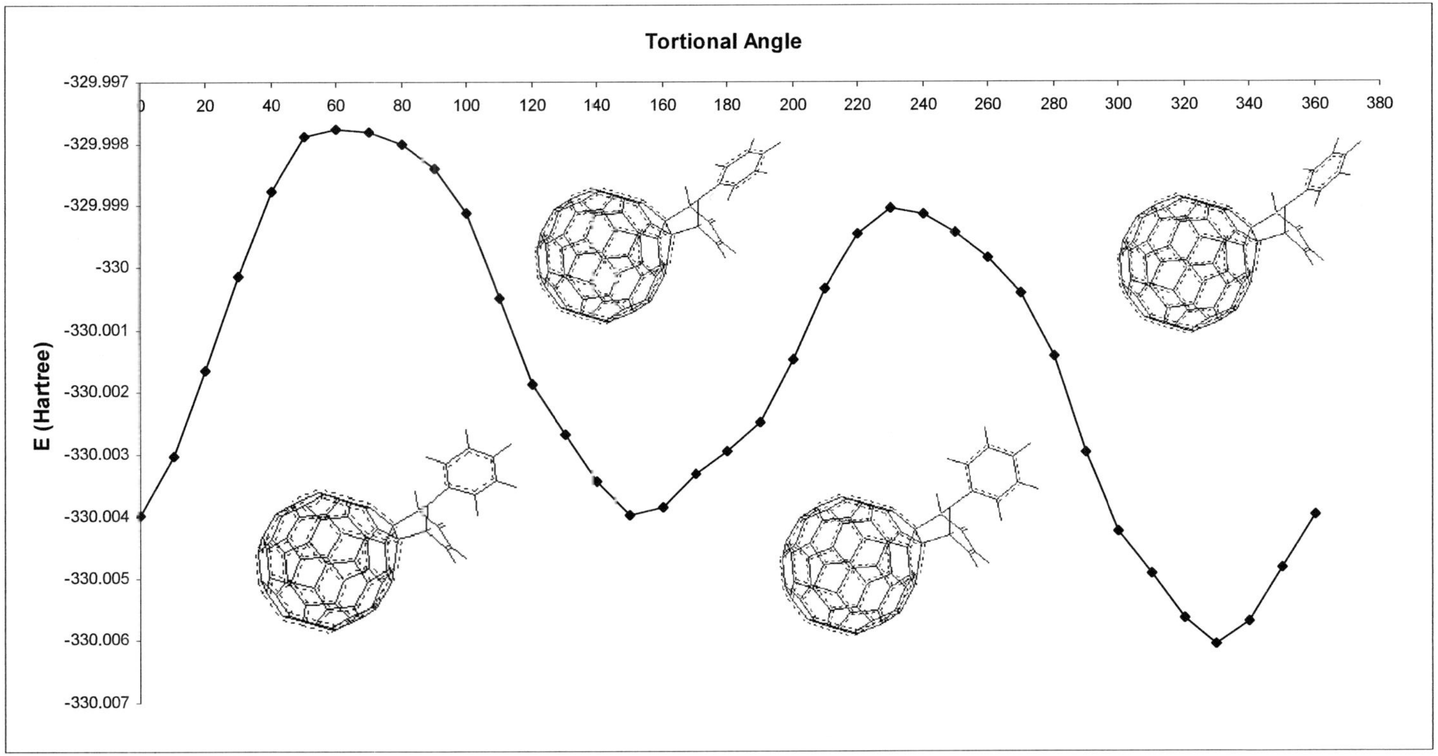

Figure 11. Variation of energy as a function of torsional angle of the phenyl ring in unsubstituted cycloadduct (56-6 A type).

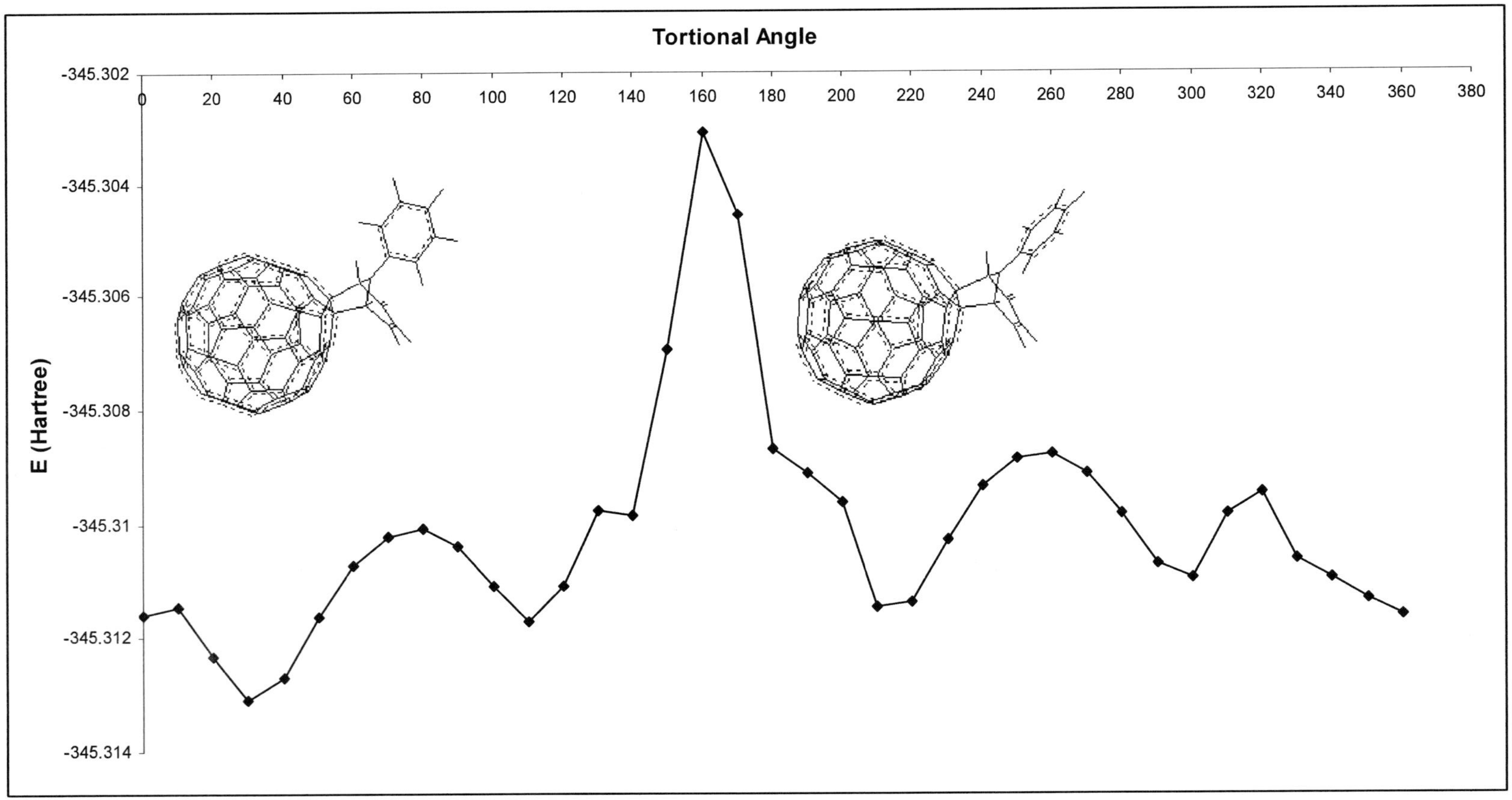

Figure 12. Variation of energy as a function of torsional angle of the phenyl ring in fluorine substituted cycloadduct (56-6 A type).

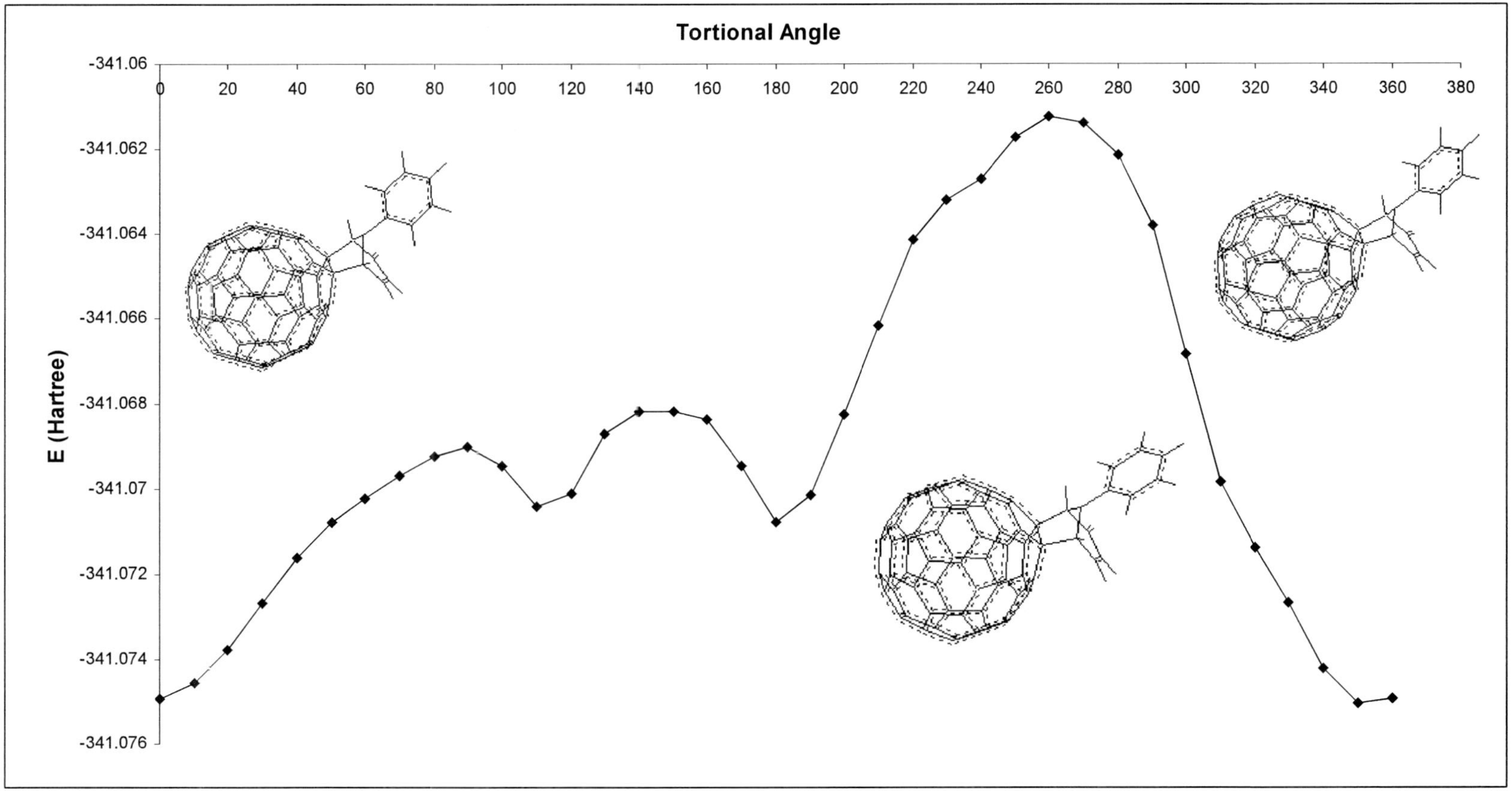

Figure 13. Variation of energy as a function of torsional angle of the phenyl ring in chlorine substituted cycloadduct (56-6 A type).

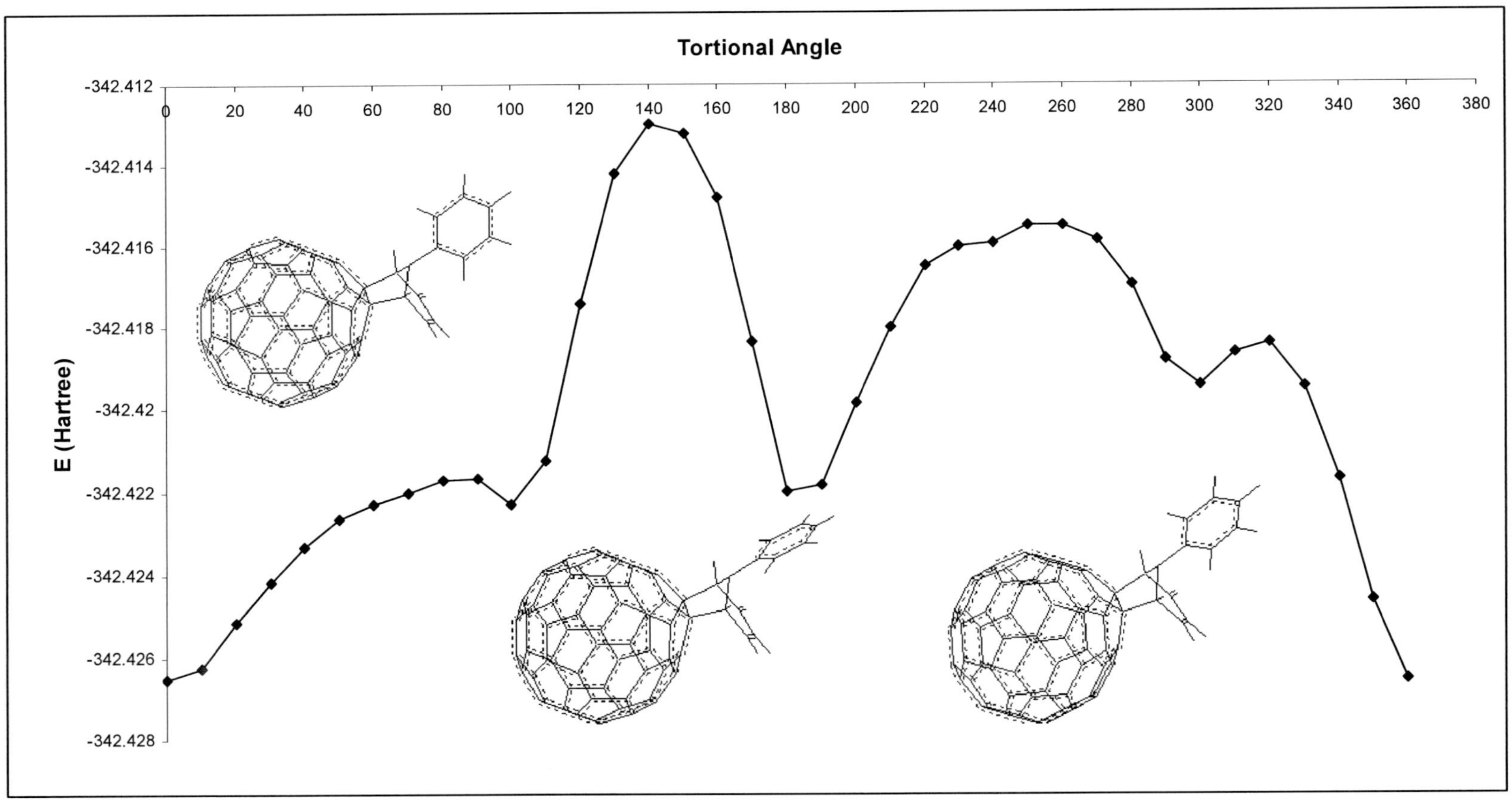

Figure 14. Variation of energy as a function of torsional angle of the phenyl ring in bromine substituted cycloadduct (56-6 A type).

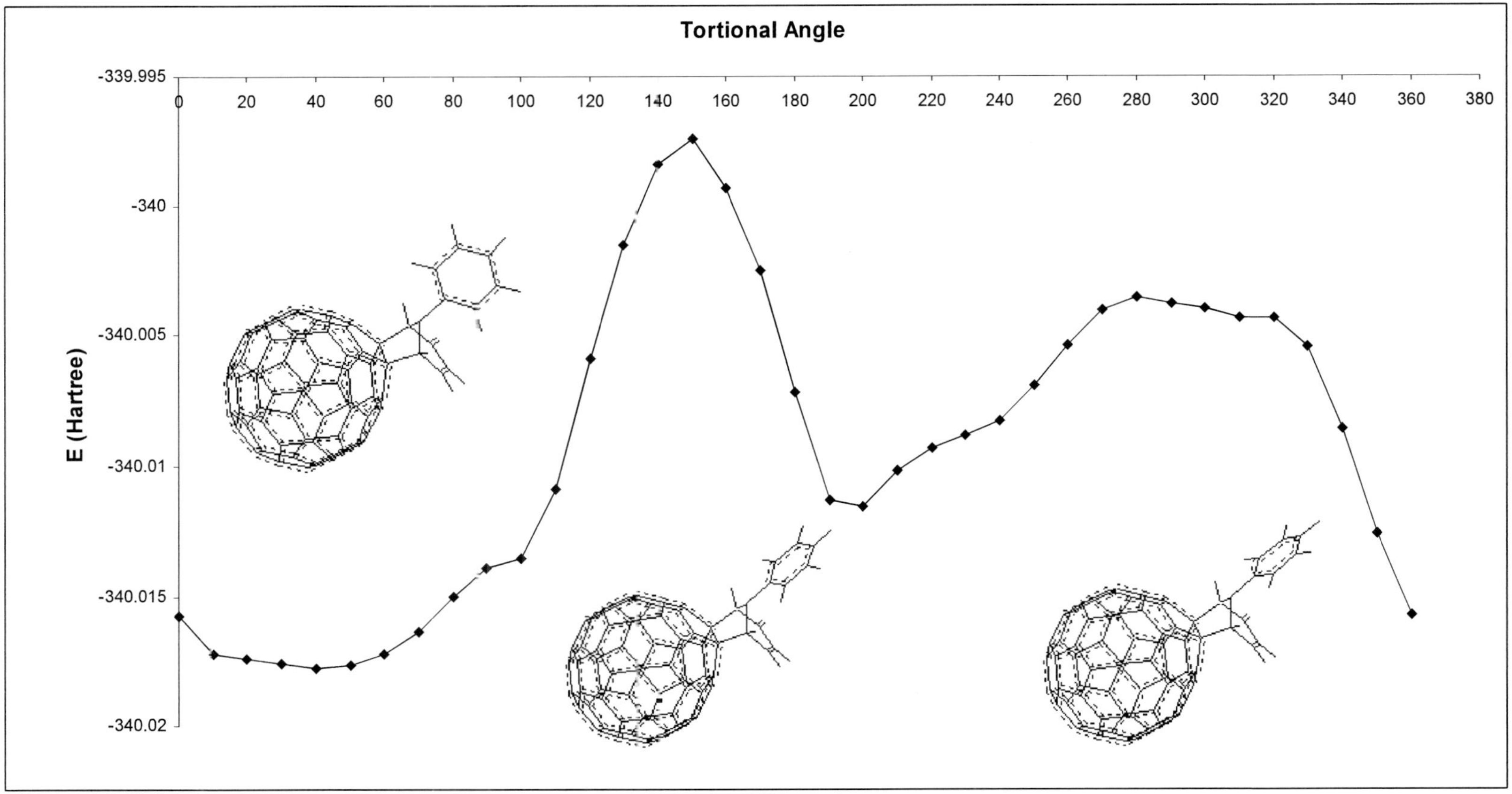

Figure 15. Variation of energy as a function of torsional angle of the phenyl ring in iodine substituted cycloadduct (56-6 A type).

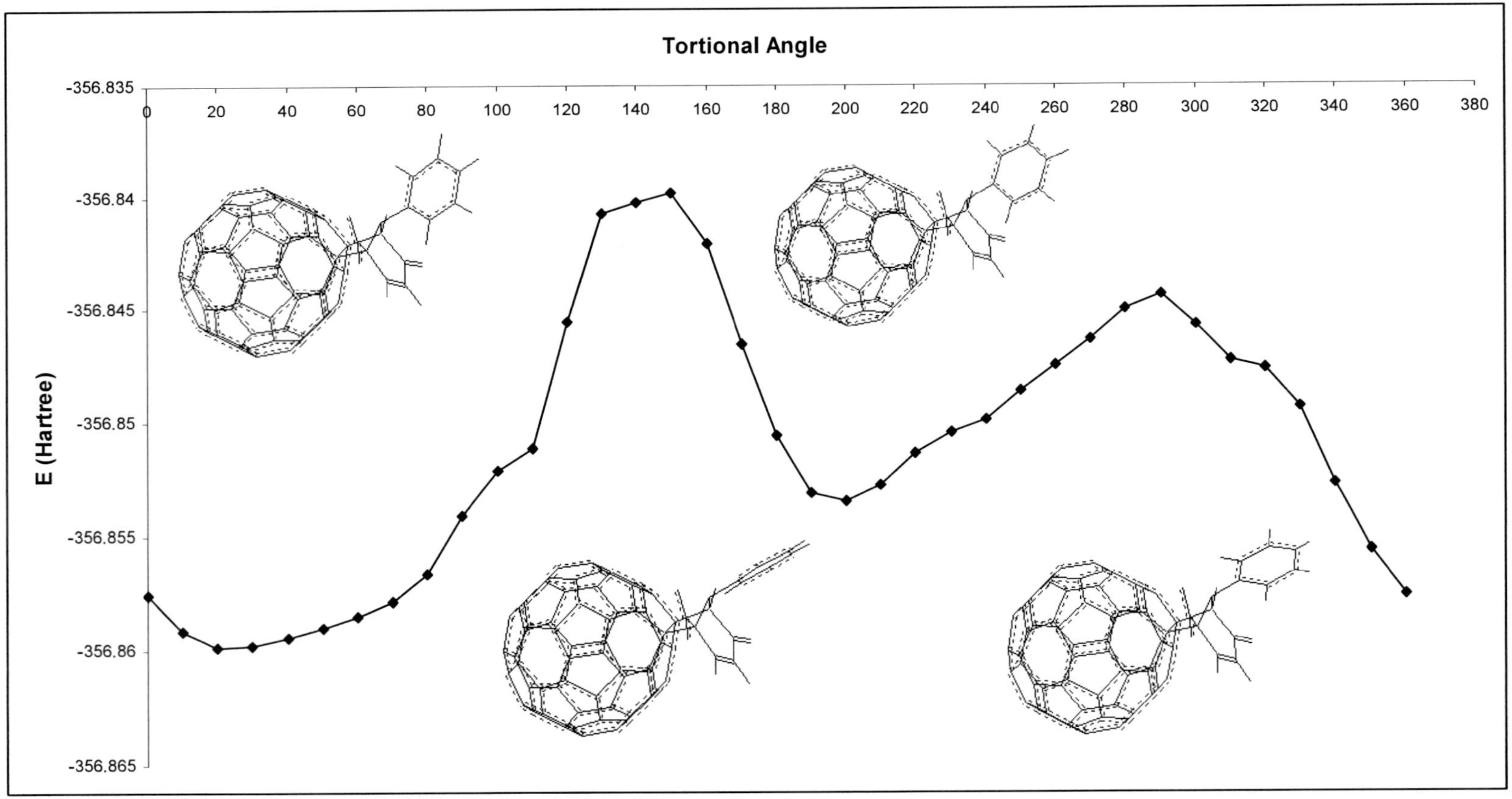

Figure 16. Variation of energy as a function of torsional angle of the phenyl ring in nitro substituted cycloadduct (56-6 A type).

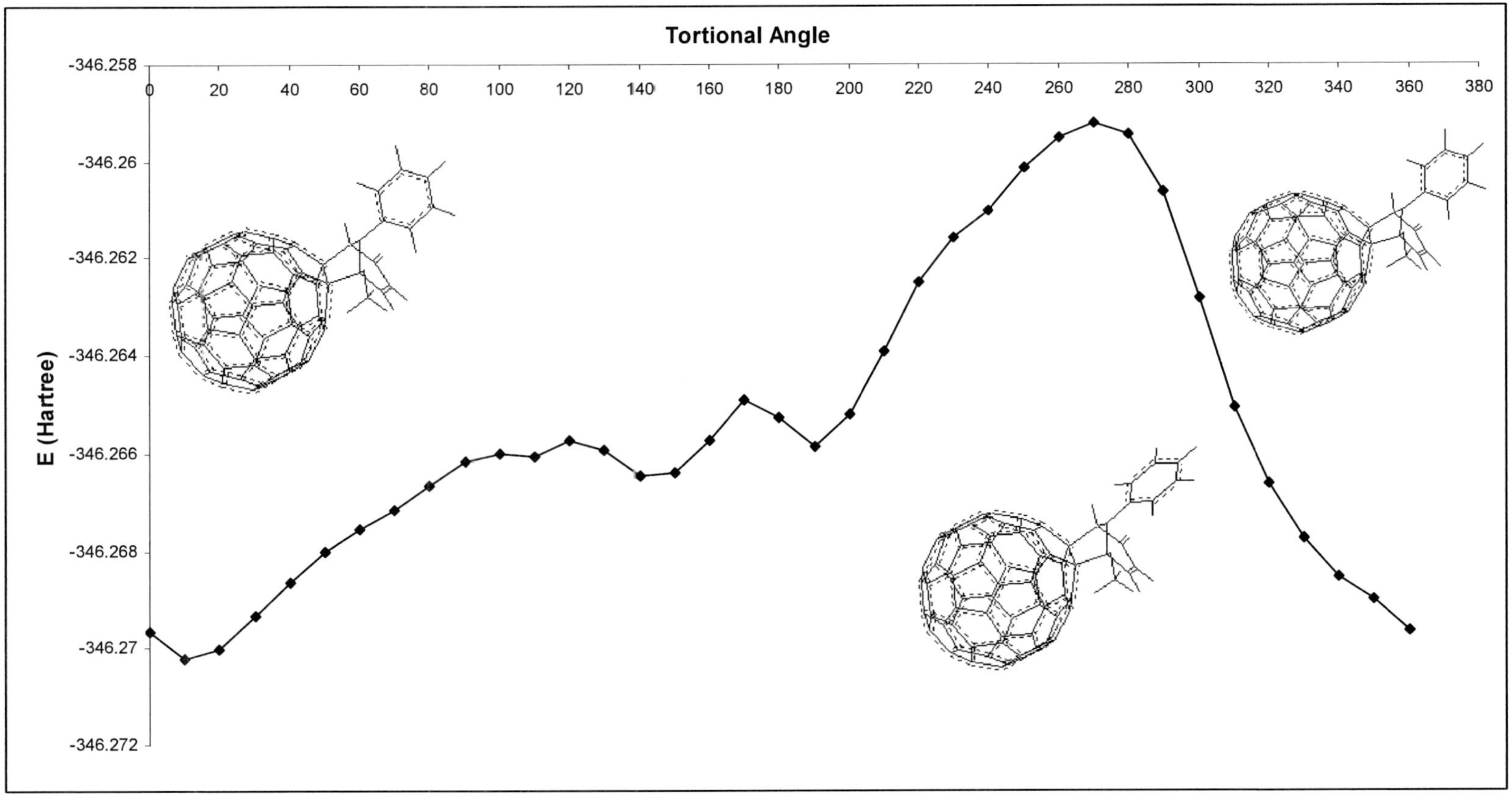

Figure 17. Variation of energy as a function of torsional angle of the phenyl ring in Methoxy substituted cycloadduct (56-6 A type).

C) The Molecular Orbital Properties

a) Unsubstituted cycloadduct

In Table 8 HUMO (Highest Unoccupied Molecular Orbital), LUMO (Lowest Unoccupied Molecular Orbital), HOMO (Highest Occupied Molecular Orbital), LOMO (Lowest Occupied Molecular Orbital) and $\Delta\varepsilon$ (ε_{LUMO}-ε_{HOMO}) have been given for the series. The LOMO and HUMO values are almost the same for all, but H-66 has a lower HOMO and higher LUMO and thus, a greater interfrontier energy gap, than the other two isomers (H-56-6 and H-56-5).

Table 8. Frontier molecular orbital energies of the unsubstituted cycloadducts (Energies in eV).

Energy	56-6	56-5	66
HUMO	6.49	6.49	6.46
LUMO	-3.08	-3.06	-2.85
HOMO	-8.95	-8.92	-9.03
LOMO	-48.32	-48.32	-48.33
$\Delta\varepsilon$	5.87	5.86	6.18

b) F, Cl, Br, I, NO_2 and OMe substituted cycloadducts

As for the molecular orbital energies for the series of substituted cycloadducts (Table 9-14), 66-type systems have the lowest HOMO and the highest LUMO energies and hence, the greatest $\Delta\varepsilon$ value for all the series. Therefore, the 66-type systems should be less susceptible to oxidation and reduction reactions as compared to the others. The LOMO values are almost equal for all cases except for the Cl case, which possesses the lowest energy values.

The HOMO and LUMO molecular orbital energy schemes, as well as the molecular orbital energy diagram of F-56-6 B can be seen in Figure 18. Both HOMO and LUMO are mostly in fullerene character.

3D-electrostatic potential map for F-56-6 B can be seen in Figure 19. The surface is plotted with 0.05 a.u. isosurface potential. The dark and light loops represent negative and positive isosurface potential, respectively. The dark loops are located over carbonyl bond, nitrogen and around the phenylene moiety. The dashed lines indicate the dipole moment of the system. The head of the dipole moment vector in all the cases points to the origin of the axes of inertia which shows that C_{60} moiety has the tendency to attract electrons. It is known that C_{60} shows an extremely facile reduction profile [73] and it is predicted that high electron affinity of C_{60} should allow accommodation of even 12 electrons under suitable conditions [51,58,74]. Thus, the cycloadducts of present concern should also have such kind of reduction profile. Additionally, enone system of the cycloadducts can undergo not only reductions but it is possible to get some carbonyl derivatives of them as well.

Table 9. Frontier molecular orbital energies of the F substituted cycloadducts (Energies in eV).

Energy	56-6 A	56-6 B	56-6 C	56-6D	56-5 A	56-5 B	56-5 C	56-5 D	66 A	66 B	66 C	66 D
HUMO	6.28	6.37	6.19	6.24	6.33	6.37	6.19	6.37	6.29	6.33	6.16	6.26
LUMO	-3.13	-3.14	-3.14	-3.14	-3.09	-3.11	-3.12	-3.11	-2.88	-2.89	-2.90	-2.87
HOMO	-8.98	-9.01	-9.01	-9.00	-8.96	-8.99	-8.99	-8.99	-9.24	-9.19	-9.16	-9.23
LOMO	-48.33	-48.36	-48.37	-48.35	-48.34	-48.36	-48.37	-48.36	-48.35	-48.37	-48.38	-48.34
Δε	5.85	5.87	5.87	5.86	5.87	5.88	5.87	5.87	6.36	6.30	6.26	6.36

Table 10. Frontier molecular orbital energies of the Cl substituted cycloadducts (Energies in eV).

Energy	56-6 A	56-6 B	56-6 C	56-6 D	56-5 A	56-5 B	56-5 C	56-5 D	66 A	66 B	66 C	Cl -66 D
HUMO	6.36	6.43	6.30	6.30	6.37	6.43	6.30	6.30	6.35	6.39	6.27	6.27
LUMO	-3.12	-3.10	-3.11	-3.12	-3.08	-3.08	-3.08	-3.08	-2.86	-2.87	-2.87	-2.87
HOMO	-8.98	-8.97	-8.97	-8.98	-8.96	-8.95	-8.95	-8.95	-9.23	-9.10	-9.09	-9.20
LOMO	-55.63	-56.00	-55.55	-55.33	-55.60	-56.00	-55.55	-55.30	-55.60	-56.03	-55.57	-55.35
Δε	5.86	5.87	5.86	5.86	5.88	5.87	5.87	5.87	6.37	6.23	6.22	6.33

Table 11. Frontier molecular orbital energies of the Br substituted cycloadducts (Energies in eV).

Energy	56-6 A	56-6 B	56-6 C	56-6 D	56-5 A	56-5 B	56-5 C	56-5 D	66 A	66 B	66 C	Br-66 D
HUMO	6.34	6.32	6.348	6.31	6.34	6.32	6.34	6.31	6.30	6.29	6.31	6.27
LUMO	-3.11	-3.10	-3.11	-3.11	-3.06	-3.08	-3.09	-3.05	-2.85	-2.86	-2.87	-2.86
HOMO	-8.97	-8.97	-8.98	-8.97	-8.93	-8.94	-8.96	-8.92	-9.21	-9.12	-9.10	-9.21
LOMO	-48.33	-48.33	-48.35	-48.33	-48.32	-48.33	-48.34	-48.31	-48.33	-48.34	-48.36	-48.33
Δε	5.86	5.87	5.87	5.86	5.87	5.86	5.87	5.87	6.36	6.26	6.23	6.35

Table 12. Frontier molecular orbital energies of the I substituted cycloadducts (Energies in eV).

Energy	56-6 A	56-6 B	56-6 C	56-6 D	56-5 A	56-5 B	56-5 C	56-5 D	66 A	66 B	66 C	I-66 D
HUMO	6.40	6.39	6.43	6.40	6.40	6.39	6.42	6.40	6.36	6.36	6.40	6.36
LUMO	-3.08	-3.08	-3.09	-3.07	-3.03	-3.06	-3.06	-3.02	-2.83	-2.84	-2.86	-2.83
HOMO	-8.94	-8.94	-8.95	-8.93	-8.89	-8.92	-8.93	-8.86	-9.18	-9.05	-9.05	-9.15
LOMO	-48.29	-48.31	-48.32	-48.28	-48.28	-48.30	-48.32	-48.27	-48.29	-48.32	-48.33	-48.28
$\Delta\varepsilon$	5.86	5.86	5.86	5.86	5.86	5.86	5.87	5.84	6.35	6.21	6.19	6.32

Table 13. Frontier molecular orbital energies of the NO_2 substituted cycloadducts (Energies in eV).

Energy	56-6 A	56-6 B	56-6 C	56-6 D	56-5 A	56-5 B	56-5 C	56-5 D	NO_2-66 A	NO_2-66 B	NO_2-66 C	NO_2-66 D
HUMO	6.62	6.551	6.81	6.79	6.66	6.53	6.81	6.86	6.62	6.52	6.788	6.82
LUMO	-3.25	-3.25	-3.25	-3.19	-3.22	-3.19	-3.22	-3.19	-2.97	-2.97	-2.98	-2.95
HOMO	-9.12	-9.12	-9.12	-9.07	-9.11	-9.07	-9.10	-9.07	-9.36	-9.32	-9.35	-9.33
LOMO	-48.45	-48.46	-48.48	-48.40	-48.45	-48.45	-48.47	-48.42	-48.46	-48.47	-48.49	-48.43
$\Delta\varepsilon$	5.87	5.87	5.87	5.88	5.89	5.88	5.88	5.88	6.39	6.35	6.37	6.38

Table 14. Frontier molecular orbital energies of the MeO substituted cycloadducts (Energies in eV).

Energy	56-6 A	56-6 B	56-6 C	56-6 D	56-5 A	56-5 B	56-5 C	56-5 D	MeO -66 A	MeO -66 B	MeO -66 C	MeO -66 D
HUMO	6.45	6.70	6.46	6.39	6.41	6.86	6.46	6.44	6.42	6.68	6.43	6.51
LUMO	-3.09	-3.06	-3.06	-3.10	-2.96	-3.09	-3.09	-2.95	-2.85	-2.89	-2.84	-2.84
HOMO	-8.94	-8.92	-8.92	-8.95	-8.82	-8.96	-8.95	-8.81	-9.18	-9.08	-9.01	-9.17
LOMO	-48.32	-48.3	-48.30	-48.33	-48.24	-48.34	-48.35	-48.22	-48.33	-48.38	-48.31	-48.31
$\Delta\varepsilon$	5.85	5.86	5.86	5.85	5.86	5.87	5.86	5.86	6.33	6.19	6.17	6.33

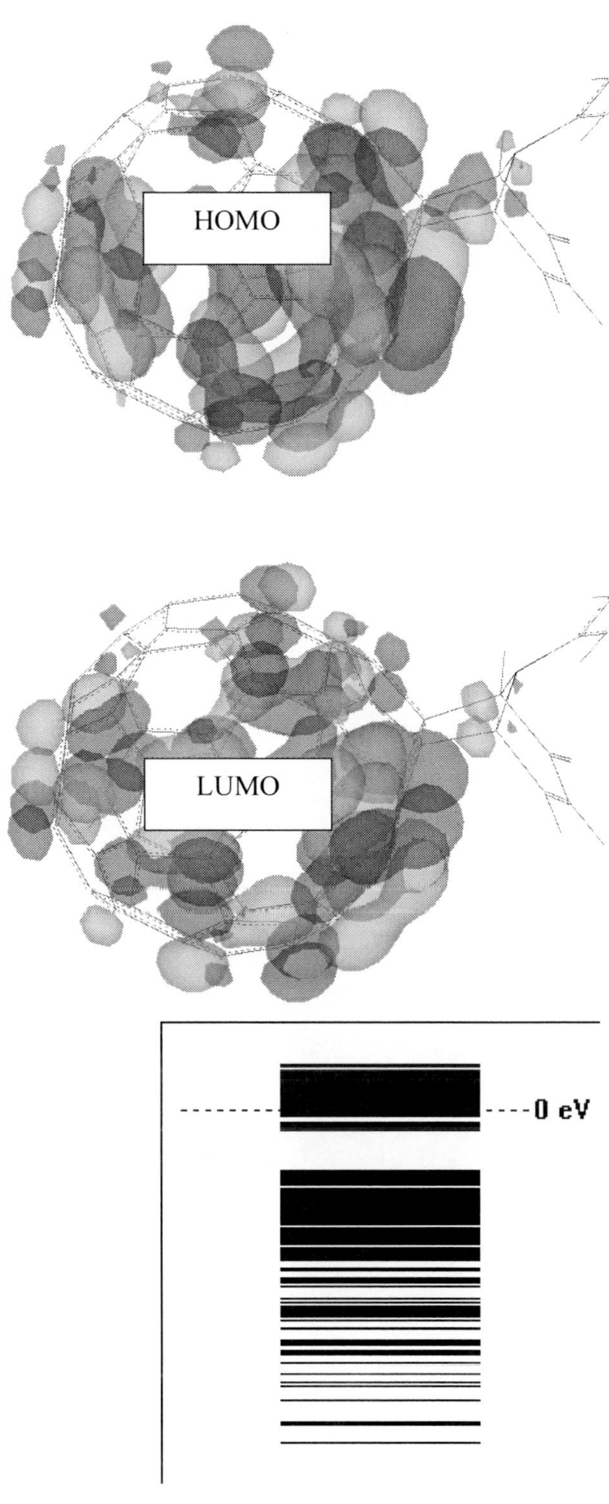

Figure 18. The HOMO, LUMO schemes and the molecular orbital eigenvalue spectra of the F-56-6 B system.

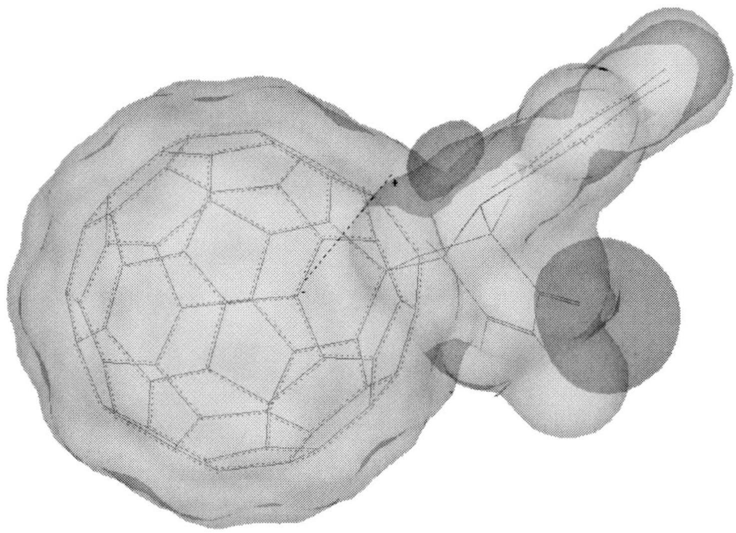

Figure 19. 3D electrostatic potential map of F-56-6 B system. Dashed lines indicate the direction of the dipole moment.

D) Vibrational Analysis

The vibrational analysis of all the structures have been performed to judge their stabilities. All of the vibrational frequencies have been found to be positive indicating that the geometry optimized structures are at least a local minimum. A representative IR-spectrum, of F-56-6 B (which will be informative for these still nonexistent systems), is given in Figure 20. The values should be corrected by scaling factor of 0.86 for the PM3 type calculations. Comparing with experimental frequencies of similar stretching of other compounds, it seems a correction factor of ca. 0.86 is necessary for PM3 (RHF) vibrational frequencies. However, one should note that the values of vibrational frequencies calculated by AM1 (RHF) must be corrected by a scaling factor of 0.95 [75].

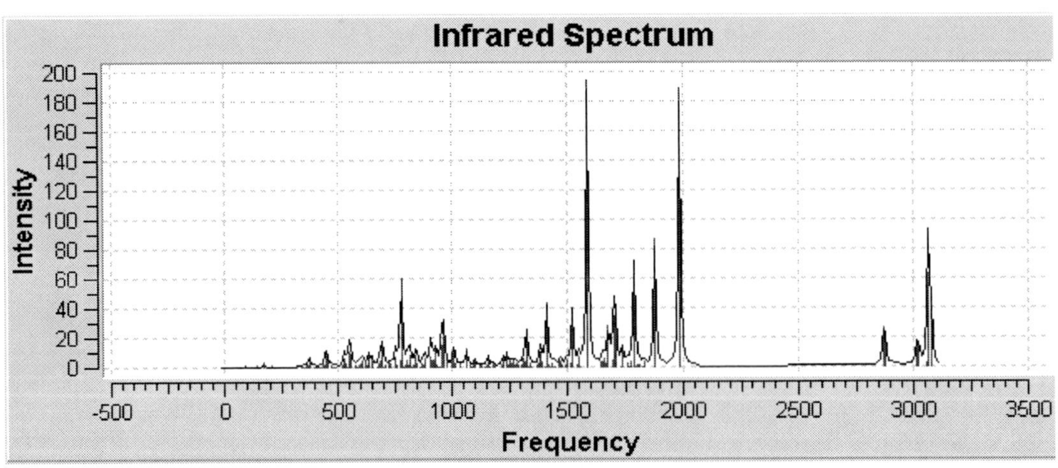

Figure 20. Vibrational spectrum of the F-56-6 B system.

Thus, the peak at 2000 cm^{-1} turns out to be 1720 cm^{-1} and it stands for the carbonyl stretching frequency. However, characteristic αβ-unsaturated carbonyl stretchings occur at 1680-1700 cm^{-1}. This means that the fluorine at the β-position strongly attracts electrons via inductive effect and the enone carbonyl system behaves as if it were a saturated carbonyl group. The substituents presently considered greatly affect CO strechings when they are located at the β-position.

4. Conclusion

The present theoretical study on 1,3-dipolar cycloaddition reactions of various 1-aryl-3-oxidopyridinium betaines with C_{60} has revealed that the cycloadducts obtained across the terminal positions of a 66-type junction (the junction between two hexagonal rings) are more stable. The present calculations also indicated that among the 56-type cycloadducts a more stable structure occurs when the bridge-nitrogen atom is next to a pentagon and the enone system faces to a hexagon of C_{60} moiety (presently 56-6 type adducts). All the cycloadducts were found to be stable but endothermic in nature, thus cycloadditions of C_{60} with various 1-aryl-3-oxidopyridinium betaines should open new synthetic routes to derivatization of C_{60}.

References

[1] Ian Fleming, Pericyclic Reactions, Oxford Science Publications, Cambridge, 1998.

[2] J. Sauer, Diels-Alder reactions II: The Reaction Mechanism, *J. Angew. Chem. Int. Ed. Engl.* 6, 16, 1967.

[3] Huisgen, R., Cycloadditions with Polar Intermediates, *Angew. Chem. Int. Ed. Engl.* 7, 312, 1968.

[4] Huisgen, R., 1,3-Dipolar cycloadditions. 76. Concerted nature of 1,3-dipolar cycloadditions and the question of diradical intermediates, *J. Org. Chem.* 41, 403, 1976.

[5] Huisgen, R., Kinetics and Mechanism of 1,3-Dipolar Cycloadditions, *Angew. Chem.* Int. *Ed. Engl.* 2, 633, 1963.

[6] Huisgen, R., Mechanism of 1,3-dipolar cycloadditions. Reply, *J. Org. Chem.* 33, 2291, 1968.

[7] Kolb, H.C.; Sharpless, K. B., The growing impact of click chemistry on drug discovery, *Drug Discovery Today*, 2003, 8, 1128 - 1136.

[8] Kolb, H.C., M.G. Finn, K.B., Sharpless Click Chemistry: Diverse Chemical Function from a Few Good Reactions, *Angew. Chem. Int. Ed.* 2001, 40, 2004 - 2021.

[9] W.G. Lewis, L.G. Green, F. Grynszpan, Z. Radic, P.R. Carlier, P. Taylor, M.G. Finn, K.B. Sharpless, Click Chemistry In Situ: Acetyl cholinesterase as a Reaction Vessel for the Selective Assembly of a Femtomolar Inhibitor from an Array of Building Blocks, *Angew. Chem. Int. Ed.* 2002, 42, 1053 – 1057.

[10] Manetsch, R.; Krasinski, A.; Radic, Z.; Raushel, J.; Taylor, P.; Sharpless, K. B.; Kolb, H. C., In Situ Click Chemistry: Enzyme Inhibitors Made to Their Own Specifications, *J. Am. Chem. Soc.* 2004, 126, 12809 – 12818.

[11] V.P. Mocharla, B. Colasson, L.V. Lee, S. Röper, K.B. Sharpless, C. Wong, H.C. Kolb, In Situ Click Chemistry: Enzyme-Generated Inhibitors of Carbonic Anhydrase II, *Angew. Chem. Int. Ed.* 2005, 44, 116 – 120.

[12] Brik, J. Muldoon, Y.-C. Lin, J.H. Elder, D.S. Goodsell, A.J. Olson, V.V. Fokin, K.B. Sharpless, C.-H. Wong, Rapid Diversity-Oriented Synthesis in Microtiter Plates for In Situ Screening of HIV Protease Inhibitors, *Chem. BioChem*, 2003, 4, 1246-1248.

[13] Lee, L. V.; Mitchell, M. L.; Huang, S.-J.; Foci, V. V.; Sharpness, K. B.; Wong, C.-H., A Potent and Highly Selective Inhibitor of Human α-1,3-Fucosyltransferase via Click Chemistry, *J. Am. Chem. Soc.* 2003, 125, 9588-9589.

[14] Rostovtsev, V.V. Enantioselective Catalytic Hydroamination of Alkenes, *Angew. Chem.* 2002, 114, 2708 – 2711.

[15] Haddon, R.C., Chemistry of the Fullerenes: The Manifestation of Strain in a Class of Continuous Aromatic Molecules, *Science* 1993, 261, 1545.

[16] Fowler, P. W.; Ceulemans, A., Electron Deficiency of the Fullerenes, *J. Phys. Chem.* 1995, 99, 508.

[17] Sola, M.; Mestres, J.; Duran, M., Molecular Size and Pyramidalization: Two Keys for Understanding the Reactivity of Fullerenes, *J. Phys. Chem.* 1995, 99, 10752.

[18] Haddon, R. C., C_{60}: Sphere or Polyhedron?, *J. Am. Chem. Soc.*, 1997, 119, 1797.

[19] Haddon, R. C.; Elser, V., Icosahedral C_{60} revisited: An aromatic molecule with a vanishingly small ring current magnetic susceptibility, *Chem. Phys. Lett.* 1990,169, 362.

[20] Fowler, P. W.; Lazzaeretti, P.; Zanasi, R., Electric and magnetic properties of the aromatic sixty-carbon cage, *Chem. Phys. Lett.* 1990, 165, 79.

[21] Haddon, R. C.; Schneemeyer, L. F.; Waszczak, J. V.; Glarum, S. H.; Tycko, R.; Dabbagh, G.; Kortan, A. R.; Mu¨ ller, A. J.; Mujsce, A. M.; Rosseinsky, M. J.; Zahurak, S. M.; Makhija, A. V.; Thiel, F. A.; Raghavachari, K.; Cockayne, E.; Elser, V. *Nature* 1991, 350, 46.

[22] Fowler, P. W.; Lazzaeretti, P.; Malagoli, M.; Zanasi, R., Magnetic properties of C_{60} and C_{70}, *Chem. Phys. Lett.* 1991, 179, 174.

[23] Pasquarello, A.; Schlutter, M.; Haddon, R. C., Ring currents in topologically complex molecules: Application to C_{60}, C_{70}, and their hexa-anions, *Phys. Rev. A* 1993, 47, 1783.

[24] Fowler, P. W.; Collins, D. J.; Austin, S. J. *J. Chem. Soc., Perkin Trans.* 2 1993, 275. [25] Zanasi, P.; Fowler, P. W., Ring currents and magnetisability in C_{60}, *Chem. Phys. Lett.* 1995, 238, 270.

[25] Zanasi, R.; Lazzaeretti, P.; Fowler, P. W., Magnetic properties of C_{60} calculated by continuous transformation of the origin of the current density, *Chem. Phys. Lett.* 1997, 278, 251.

[26] Ruud, K.; Agren, H.; Helgaker, T.; Dhle, P.; Kock, H.; Taylor, P.R., The Hartree–Fock magnetizability of C_{60}, *Chem. Phys. Lett.* 1998, 285, 205.

[27] Haddon, R. C., Magnetism of the carbon allotropes, *Nature* 1995, 378, 249.

[28] Eckert, J. F.; Nicoud, J. F.; Nierengarten, J. F.; Liu, S. G.; Echegoyen, L.; Barigelletti, F.; Armaroli, N.; Ouali, L.; Krasnikov, V.; Hadziioannou, G., Fullerene-Oligophenylenevinylene Hybrids: Synthesis, Electronic Properties, and Incorporation in Photovoltaic Devices, *J. Am. Chem. Soc.* 2000, 122, 7467-7479.

[29] Martin, N.; Sanchez, L.; Herranz, M. A.; Guldi, D. M., Evidence for Two Separate One-Electron Transfer Events in Excited Fulleropyrrolidine Dyads Containing Tetrathiafulvalene (TTF), *J. Phys. Chem. A* 2000, 104, 4648-4657.

[30] Prato, M.; Maggini, M., Fulleropyrrolidines: A Family of Full-Fledged Fullerene Derivatives, *Acc. Chem. Res*. 1998, 31, 519-526.

[31] Yin, G.; Xu, D. P.; Mei, Y. H.; Xu, Z. *New Carbon Mater*. 2002, 17, 18-21.

[32] Espildora, E.; Delgado, J. L.; de la Cruz, P.; de la Hoz, A.; Lopez-Arza, V.; Langa, F., Synthesis and properties of pyrazolino[60]fullerene-donor systems, *Tetrahedron* 2002, 58, 5821-5826.

[33] Kawamura, Y.; Akaishi, K.; Nishiuchi, M.; Tsukayama, M., Structures and Electrical Properties of the Layered Tin-Iodide Perovskite Compounds with Various Organic Cations, *Mol. Cryst. Liq. Cryst*. 2002, 376, 219-224.

[34] Silwa, A.M.S.; Cavaleiro, J.A.S., Synthesis of novel [60]fullerene–flavonoid dyads, *Tetrahedron Lett*. 2002, 43, 1689-1691.

[35] Langa, F.; de la Cruz, P.; Espildora, E.; de la Hoz, A.; Bourdelande, J. L.; Sanchez, L.; Martin, N., C_{60}-Based Triads with Improved Electron-Acceptor Properties: Pyrazolylpyrazolino[60]fullerenes, *J. Org. Chem*. 2001, 66, 5033-5041.

[36] Campidelii, S.; Deschenaux, R., Liquid-Crystalline Fulleropyrrolidines, Preliminary Communication, *Helv. Chim. Acta* 2001, 84, 589-593.

[37] Herranz, M.A.; Illescas, B.; Martin, N.; Luo, C.P.; Guldi, D.M., Donor/Acceptor Fulleropyrrolidine Triads, *J. Org. Chem*. 2000, 65, 5728-5738.

[38] Mikami, K.; Matsumoto, S.; Okubo, Y.; Fujitsuka, M.; Ito, O.; Suenobu, T.; Fukuzumi, S., Stepwise Bond Formation in Photochemical and Thermal Diels-Alder Reactions of C_{60} with Danishefsky's Dienes, *J. Am. Chem. Soc*. 2000, 122, 2236-2243.

[39] Curl, R.E., Dawn of Fullerenes: Conjecture and Experiment (Nobel Lecture), Angew. Chem., Int. Ed. Engl. 1997, 36, 1566-1576.

[40] Kroto, H., Symmetry, Space, Stars, and C60 (Nobel Lecture), *Angew. Chem., Int. Ed. Engl*. 1997, 36, 1578-1593

[41] Vaissiere, B.; Sandall, J.P. B.; Fowler, P. W.; Oliveira, P.; Bensasson, R. V., Regioselectivity in radical reactions of C_{60} derivatives, *J. Chem. Soc., Perkin Trans*. 2 2001, 821-823.

[42] Chen, Z.; Thiel, W., Performance of semi empirical methods in fullerene chemistry: relative energies and nucleus-independent chemical shifts, *Chem. Phys. Lett*., 2003, 367, 15-25.

[43] Türker, L., 1,3-Dipolar cycloaddition reactions of C_{60} with 3-phenyl-phthalazinium-1-olate. A theoretical study, *J. Mol. Struct. (Theochem)* 2002, 588,165-174.

[44] Okamoto, Ab Initio Investigation of Hydrogenation of C_{60}, Y. *J. Phys. Chem. A* 2001, 105, 7634-7637.

[45] Prato, M.; Suzuki, T.; Foroudian, H.; Li, Q.; Khemani, K.; Wudl, F., [3 + 2] and [4 + 2] Cycloadditions of fullerene C_{60}, *J. Am. Chem. Soc*., 1993, 115, 1594-1595.

[46] Becker, H.; Javahery, G.; Petrie, S.; Bohme, D.K., [4+2+] Diels-Alder Cycloaddition to C_{60} and C_{70} Radical Cations in the Gas Phase: A Comparison with Solution, *J. Phys. Chem* 1994, 98, 5591-5592.

[47] Sandall, J.P. B.; Fowler, P. W.; Taylor, R., Some cycloadditions of dienes with $C_{60}F_{18}$: structures and relative stabilities derived from theoretical calculation, *J. Chem. Soc., Perkin Trans*. 2 2002, 1718-1721.

[48] Ito, A.; Monobe, T.; Yoshii, T.; Tanaka, K., Thermodynamic stability in [36-D_{6h}]fullerene, *Chem. Phys. Lett*. 2000, 330, 281-286.

[49] Maurizio, P.; Maggini, M., Fulleropyrrolidines: A Family of Full-Fledged Fullerene Derivatives, *Acc. Chem. Res.* 1998, 31, 519-526.

[50] Sternfeld, T.; Thilgen, C.; Hoffman, R. E.; Heras, M.R. C.;Diederich, F.; Wudl, F.; Scott, L.T.; Mack, J.; Rabinovitz, M., An Insight into the Aromaticity of Fullerene Anions: Experimental Evidence for Diamagnetic Ring Currents in the Five-Membered Rings of C_{60}^{6-} and C_{70}^{6-}, *J. Am. Chem. Soc.* 2002, 124, 5734-5738.

[51] Qian, W.; Rubin, Y., Complete Control over Addend Permutation at All Six Pseudooctahedral Positions of Fullerene C_{60}, J. Am. Chem. Soc. 2000, 122, 9564-9565.

[52] Laszlo, I.; Rassat, A., Toroidal and spherical fullerene-like molecules with only pentagonal and heptagonal faces, *Int. J. Quantum Chem.* 2001, 84, 136-139.

[53] Aihara, J.; Takata, S., Aromatic character of typical C_{60} derivatives, J. Chem. Soc., Perkin Trans. 2 1994, 65-69.

[54] Braun, T.; Schubert, A. P.; Kostoff, R. N., Growth and Trends of Fullerene Research as Reflected in Its Journal Literature, *Chem. Rev.* 2000, 100, 23-37.

[55] Wang, G.; Saunders: M.; Cross. R. J., Reversible Diels-Alder Addition to Fullerenes: A Study of Equilibria Using ^3He NMR Spectroscopy, *J. Am. Chem. Soc.* 2001, 123, 256-259.

[56] Hirsch, A. Synthesis 1995, 895.

[57] Hirsch, A. *The Chemistry of Fullerenes;* Thieme: Stuttgart, Germany, 1994.

[58] Hirsch, A., *Fullerenes and Related Structures*, Top. Curr. Chem. 1999, 1, 199.

[59] Silwa, W. *Fullerene Sci. Technol.* 1995, 3, 243.

[60] Silwa, W. *Fullerene Sci. Technol.* 1997, 5, 1133-1146.

[61] Yurovskaya, M. A.; Ovcharenko, A. A. *Chem. Heterocycl. Compd.* 1998, 291.

[62] Yurovskaya, M. A.; Trushkov, I. V., Cycloaddition to buckminsterfullerene C_{60}: advancements and future prospects, *Russ. Chem. Bull., Int. Ed.* 2002, 51, 3, 367-443.

[63] Cases, M.; Duran, M.; Mestres, J.; Martin, N.; Sola, M., Mechanism of the Addition Reaction of Alkyl Azides to [60]Fullerene and the Subsequent N_2 Extrusion to Form Monoimino-[60]fullerenes, *J. Org. Chem.* 2001, 66, 433-442.

[64] Sola, M.; Mestres, J.; Marti, J.; Duran, M., An AM1 study of the reactivity of buckminsterfullerene (C_{60}) in a Diels-Alder model reaction, *Chem. Phys. Lett.*, 1994, 231, 325-333.

[65] Suzuki, T.; Li, Q.; Khemani, K. C.; Wudl, F., Dihydrofulleroid H3C61: synthesis and properties of the parent fulleroid, *J. Am. Chem. Soc.* 1992, 114, 7301-7302.

[66] Stewart, J.J.P., Optimization of parameters for semi empirical methods I. Method, *J. Comput. Chem.* 10 (1989) 209-220.

[67] Stewart, J.J.P., Optimization of parameters for semi empirical methods II. Applications, *J. Comput. Chem.* 10 (1989) 221-230.

[68] Leach, A.R., *Molecular modeling*. Essex, Longman; 1997.

[69] *Hyperchem*. Gainesville, FL, USA: Hypercube Inc. 2002.

[70] M.J. Frisch, G.W. Trucks, H.B. Schlegel, G.E. Scuseria, M.A. Robb, J.R. Cheeseman, J.A. Montgomery Jr., T. Vreven, K.N. Kudin, J.C. Burant, J.M. Millam, S. S. Iyengar, J. Tomasi, V. Barone, B. Mennucci, M. Cossi, G. Scalmani, N. Rega,; G.A. Petersson, H. Nakatsuji, M. Hada, M. Ehara, K. Toyota, R. Fukuda, J. Hasegawa, M. Ishida, T. Nakajima, Y. Honda, O. Kitao, H. Nakai, M. Klene, X. Li, J.E. Knox, H.P. Hratchian, J.B. Cross, C. Adamo, J. Jaramillo, R. Gomperts, R.E. Stratmann, O. Yazyev, A.J. Austin, R. Cammi, C. Pomelli, J.W. Ochterski, P.Y. Ayala, K. Morokuma, G.A. Voth, P.

Salvador, J.J Dannenberg, V.G. Zakrzewski, S. Dapprich, A.D. Daniels, M.C. Strain, O. Farkas, D.K. Malick, A.D. Rabuck, K. Raghavachari, J.B. Foresman, J. V. Ortiz, Q. Cui, A.G. Baboul, S. Clifford, J. Cioslowski, B.B. Stefanov, G. Liu, A. Liashenko, P. Piskorz, I. Komaromi, R.L. Martin, D.J. Fox, T. Keith, M.A. Al-Laham, C.Y. Peng, A. Nanayakkara, M. Challacombe, P.M.W. Gill, B. Johnson, W. Chen, M. W. Wong, C. Gonzalez and J.A. Pople, *Gaussian* 03, Revision C. 02, Gaussian, Inc.: Wallingford CT, 2004.

[71] Kavitha K.; Venuvanalingam, P., Open versus Closed 1,3-Dipolar Additions of C_{60}: A Theoretical Investigation on Their Mechanism and Regioselectivity, *J. Org Chem*. 2005, 70, 5426-5435

[72] Kroto, H.W.; Walton, D.R.M., *The Fullerenes*, Cambridge Press, Cambridge, 1993.

[73] Haddon, R.C.; Brus, L.E.; Raghavachari, K., Electronic structure and bonding in icosahedral C_{60}, *Chem. Phys. Lett*. 125 (1986) 459.

[74] D.C. Young, Computational Chemistry, John Wiley and Sons, New York, 2001.

In: Progress in Fullerene Research
Editor: Milton Lang, pp. 369-382

ISBN: 978-1-60021-841-5

Chapter 12

SURFACE MODIFICATION OF FULLERENE THIN FILMS BY DIFFERENT MULTIPLE CHARGED IONS

B. Todorović-Marković[1*], Z. Marković[1], I. Draganić[1], D. Vasiljević-Radović[2], N. Romčević[3], M. Romčević[3] and M. Dramićanin[1]

[1]„Vinča" Institute of Nuclear Sciences, P.O.B. 522, 11001 Belgrade, Serbia
[2]Institute of Microelectronic Technologies and Single Crystals, Njegoševa 12, 11000 Belgrade, Serbia
[3]Institute of Physics, P.O.B. 68, 11001 Belgrade, Serbia

Abstract

In this chapter, the results of structural modification of fullerene thin films induced by multiple charged ion bombardment have been presented. The fullerene thin films have been bombarded by different multiple charged ions: N^{5+}, B^{3+}, Fe^{6+}. The applied energies were in the range of 45 to 90 keV. As-deposited and irradiated films have been characterized by atomic force microscope, Raman and Fourier transform infrared spectroscopy and UV/Vis spectrophotometry.

The results of Raman spectroscopy indicate the formation of highly disordered structure with sp^3 bonds spread throughout sp^2 structures during irradiation by multiple charged ions. The morphology of bombarded films has been changed significantly. Results of FTIR spectroscopy indicate the formation of new C-N and B-C bonds while Fe-C bonds could not be identified. The optical band gap was found to be varied from 1.7 to 0.7 eV depending on the type of used ions.

After irradiation of fullerene thin films by multiple charged nitrogen ions, amorphous carbon nitride phase was formed inside the bombarded films. During bombardment by miltiple charged boron ions, amorphous boron carbide films could be observed. New Fe-C bonds could not be observed during bombardment of fullerene thin films by multiple charged iron ions. Results of Raman measurements indicate that the percentage of sp^3 bonds is raised after bombardment by multiple charged iron ions.

PASC numbers. 61.48.+c, 78.30.-j, 79.20.Rf

---

[*] E-mail address: biljatod@vin.bg.ac.yu. Phone. +38111 2455451, Fax.+381113440100 (Corresponding author)

1. Introduction

Ion implantation is a useful technique for modifying surface properties of different materials. Many surface properties can be improved with ion implantation: these include enhanced hardness, wear resistance and resistance to chemical attack and reduced friction [1,2]. By applying this technique, it is possible to synthesize different carbon based materials: carbon nitride, boron carbide etc [3-7].

A lot of reports investigating ion implantation of fullerene films have been published up to now [8-15]. Based on results of the bombardment of fullerene thin films with different single charged ions (He, Ar, B, N, Si, Ag, Ne, etc.), one could conclude that bombardment of fullerenes by energetic particles was accompanied by disintegration of fullerene molecules which resulted in the formation of amorphous carbon [8]. Ion irradiation transforms fullerene with the density of 1.65 to 1.72 $g{\cdot}cm^{-1}$ to a material with the density of amorphous carbon (around 2.0 to 2.2 $g{\cdot}cm^{-1}$) or with even higher density [16]. Palmetshofer and Kastner observed polymerization and amorphization of C_{60} films during bombardment of these films with H^+, He^+, C^+ and Ar^+ at energies from 60-600 keV and at fluences from $1x10^{12}$ to $1x10^{16}$ cm^{-2} [10]. The final structure of fullerene thin films bombarded by single charged ions can be described by a highly disordered amorphous carbon network with high sp^2 carbon bond distortion giving origin to six- and non-six-membered rings and poor graphitic in-plane order [17].

There are only a few studies investigating the effect of multiple charged ions on fullerene thin films [16,18]. Some studies on the impact of low fluence highly charged highly energetic ions (72.3 MeV $^{127}I^{13+...25+}$) onto fullerite have showed a pronounced increase in C^+ sputtering yield and a weak one for sputtered C_{60}^+ molecules with increasing charge state [18].

Highly charged ions can unlock a large class of new phenomena in various areas of atomic, plasma, nuclear and chemical physics [19,20]. When a highly charged ion (HCI) approaches a solid surface, multiple surface electrons move to the unoccupied levels of the HCI through certain neutralization processes. As a result, large potential energy of HCI is being deposited into a nanometar size area [21,22]. An highly charged ion has large potential energy defined as summation of ionization energies to produce an ion with charge state q from the neutral atom [23]. The transfer of potential energy from the HCI to the surface starts long before the HCI reaches the surface and within a short-time interval <100 fs, the ion is neutralized by capturing surface electrons [24]. For insulator or semiconductor surfaces, the electron response time is much longer, which may lead to a local buildup of charge or electron-hole pairs within 100 fs by transfer of electrons from the surrounding solid, the intense Coulomb repulsion can cause an explosion, ejecting ions and atoms from a localized region and thus forming a crater on the surface [19]. Over 50 % of the initial Coulomb energy is deposited to the substrate. Part of the energy converts to the shock wave propagation, but the larger portion of energy is stored in the amorphous pocket that is developed around the crater, during the nano-explosion [23]. Electronic stopping phenomena are dominating collisions of keV multiply charged ions with fullerenes [25]. Projectile velocity, atomic number and charge state provide a vast array of parameters to tune the electronic as well as vibrational excitation.

The purpose of present work is to observe structural modification in fullerene thin films after irradiation by different multiple charged ions. We wanted to see whether multiple charged nitrogen, boron and iron ion bombardment would induce the formation of C-N, B-C, Fe-C bonds in used specimens. Irradiation by highly charged ions modifies fullerene films in amorphous diamond like structure phase possesing high content of sp^3 bonds. New C-N, B-C bonds could be identified by FTIR spectroscopy. On the contrary, any Fe-C bonds could not be observed in bombarded films. Atomic force microscope observations reveal morphological changes in fullerene films after bombardment. The obtained values of optical band gap are similar to those obtained for CN films and BC films, respectively.

2. Experimental Procedure

The fullerene thin films were deposited on silicon wafers by resistive heating in high vacuum at 700 K. The silicon substrate was placed 60 mm above the crucible with fullerene mixture powder which was composed of 90% of C_{60} and 10% of C_{70} molecules. The thickness of deposited films measured by profilometry were in the range of 100 to 500 nm.

Fullerene thin films were irradiated at the low energy experimental channel (L3A) of the TESLA accelerator installation, at the Vinča Institute of Nuclear Sciences, in Belgrade [26]. The L3A experimental facility for the surface modification of materials is connected to the electron cyclotron resonance (ECR) ion sources (mVINIS) that can produce and accelerate a wide range of multiple charged ions from gases and solid substances [27,28]. The heavy ion beams obtained from mVINIS are separated by charge to mass ratio (q/m) and transported to target chamber for the fullerene modification.

The production of the boron and iron beams was tested by the MIVOC method [28]. Beam energies were in the range of 15 keV for helium-like boron ions (B^{3+}) to 90 keV for Ca-like iron ion and fluences were between $2x10^{14}$ and $2x10^{16}$ cm^{-2}. The well-known ferrocene powder $Fe(C_5H_5)_2$ was used as a source of iron atoms. The ECR source was optimized for the Ca-like ion (q=6+) and achieved beam current up to 60 μA.

The advantages of a new system are the following: its production is cheap (for every compound can be fabricated its own vessel), use is quite simple, construction is space-saving and convenient at high voltage side, the produced ion beams are stable in a long period, and the rate of consumption of compounds is a very low (approximately 5 mg/hour) making it possible to use expensive, rare isotopes. Some disadvantages should be also remarked: often the compounds are toxic and inflammable, plasma chamber can be contaminated with carbon, and influences of water vapor pressure can disable the ion source optimization.

A new special sampler holder with a calibrated Faraday cup designed to improve the quality of ion irradiation, enables direct measurements of implanted particles and achieves the maximal reasonable multiple charged ion doses.

Collision velocity of the primary projectile and potential energy of multiple charged ions are given in Table 1.

Table 1. Collision velocity and potential energy of different multiple charged ions.

	N^{5+} ions	B^{3+} ions	Fe^{6+} ions
Collision velocity 10^5 ($m{\cdot}s^{-1}$)	10.13	9.28	5.5
Potential energy (eV)	97.89	71.375	283.6

Raman spectra of as-deposited and implanted fullerene films were obtained by Micro Raman Chromex 2000 using 532 nm of a frequency doubled Nd:YaG laser. The spectral resolution was 1 cm^{-1}. Micro-Raman spectra were recorded at room temperature.

Fourier transform infrared (FTIR) spectra were measured for as-deposited and irradiated fullerene thin films at room temperature in the spectral range from 400 to 4000cm^{-1}, on a BOMEM spectrometer.

The morphology of as-deposited and irradiated films was investigated by means of an AutoProbe CP AFM microscope. A conventional Si_3N_4 cantilever (force constant =2.1 N/m) was used for all measurements.

For studies of the optical properties, reflectance measurements were carried out on the Fiber optics UV/Vis AVANTES spectrometer.

3. Results and Discussion

3.1.Raman Analysis

Raman spectroscopy is a useful nondestructive tool for the characterization of crystalline, nanocrystalline and amorphous carbon and carbon based materials [29,30]. The Raman spectrum is considered to depend on clustering of the sp^2 phase, bond disorder, presence of sp^2 rings or chains and sp^2/sp^3 ratio. These factors act as competing forces on the shape of the Raman spectra [31]. Ten Raman-allowed ($2A_g+8H_g$) modes of C_{60} molecule have been experimentally observed and theoretical calculated [29]. Fig. 1a depicts the Raman spectra of as-deposited and irradiated fullerene films by different ions at the highest fluences ($2x10^{16}$ cm^{-2}). As for pristine thin films only two sharp peaks at 495 and 1469 cm^{-1} (A_g modes) belong to C_{60} molecule while sharp peak at 1565 cm^{-1} and several weaker lines are due to presence of C_{70} molecules [32]. Silicon peaks appears in the spectra at 521 and 960 cm^{-1}. After boron and nitrogen implantation peaks of C_{60} and C_{70} have decreased and at fluences of $2x10^{15}$ cm^{-2}, broad asymmetric band between 1100 and 1700 cm^{-1} is observed [7,33]. During bombardment of fullerene thin films by highly charged iron ions (Fe^{6+}), amorphous structure could be observed even at the lowest fluences ($2x10^{14}$ cm^{-2}). The complete disintegration of C_{60} molecules occurred at the lowest applied fluences ($2x10^{14}$ cm^{-2}) [34]. The appearance of broad band indicated the formation of amorphous carbon. The silicon peaks disappear which is an indication of the increased opacity of the film which covers the silicon substrate. As can be seen from Fig. 1, the intensity of Raman amorphous band depends on the applied energy. After irradiation of fullerene thin films by Fe^{6+} ions (the applied energy was 90 keV), one could see a very small amorphous band.

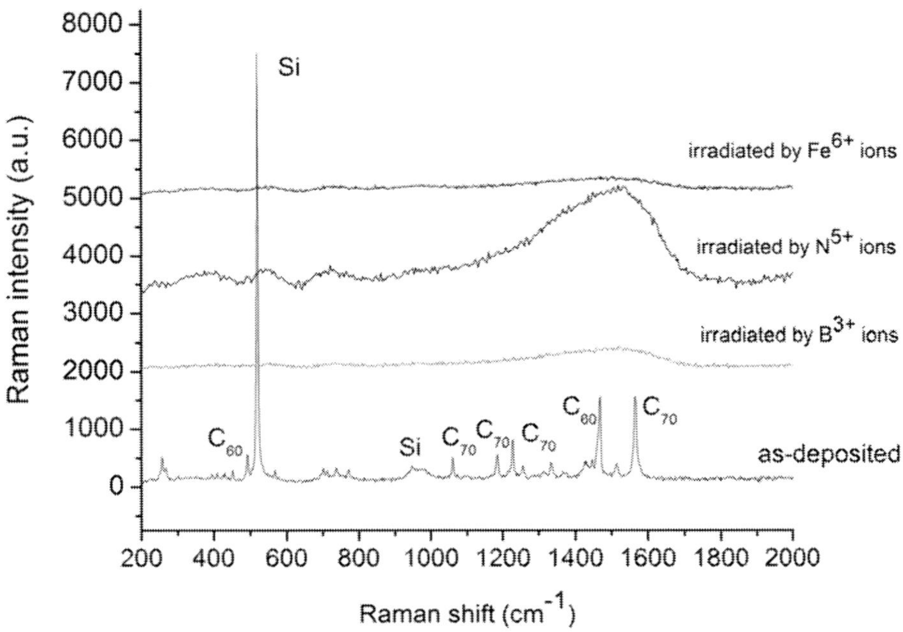

Figure 1. Raman spectra of as-deposited and fullerene thin films irradiated by different ions. The spectra have been vertically displaced for clarity but otherwise plotted on the same vertical scale.

In order to analyze the change of bonding structure in implanted thin films the deconvolution of recorded Raman spectra has been performed by using Gaussian fitting [35].

The fitting parameters are the position, width and intensity for each Gaussian peak. The main features of these deconvoluted spectra are two broad peaks (referred to the D band and G band). The D band around 1355 cm^{-1} is a breathing mode of A_{1g} symmetry involving phonons near the K zone boundary. The D mode is dispersive; it varies with phonon excitation energy, even when the G peak is not dispersive [31]. Its intensity is strictly connected to the presence of sixfold aromatic rings. The G mode of graphite at 1581 cm^{-1} has E_{2g} symmetry. This mode involves the in-plane bond-stretching motion of pairs of C sp^2 atoms. This mode does not require the presence of sixfold rings and so it occurs at all sp^2 sites, not only those in rings. It always lies in the range of 1500-1630 cm^{-1}, as it does in aromatic and olefinic molecules [31]. At lower fluences (2x10^{14} cm^{-2}), D peak dominates in the deconvoluted spectra while at higher fluences G peak increases. By increasing of fluences the distinction between the D and G peak becomes more apparent since the G peak shifts down and gets narrower while D peak becomes lower and wider [7,33]. As for deconvoluted Raman spectra of irradiated specimens by Fe^{6+} ions, the intensities of G and D peaks are almost the same at all applied fluences while D peak is wider than G peak regardless the fluences [34].

Figures 2(a,b,c) show the ratio of the two peaks I_D/I_G, G peak position and G band width as a function of fluences, respectively.

I_D/I_G ratio is decreased with fluences for all used projectiles-Fig. 2a. The largest variation of I_D/I_G ratio is obtained during bombardment by N^{5+} ions. The value of I_D/I_G during irradiation by Fe^{6+} ions is varied slightly. As for G peak position, there is a significant downshift toward lower wavenumber (from 1580 to 1530 cm^{-1})-Fig. 2b. The FWHM of Raman bands are increased with the increase of fluence-Fig. 2c.

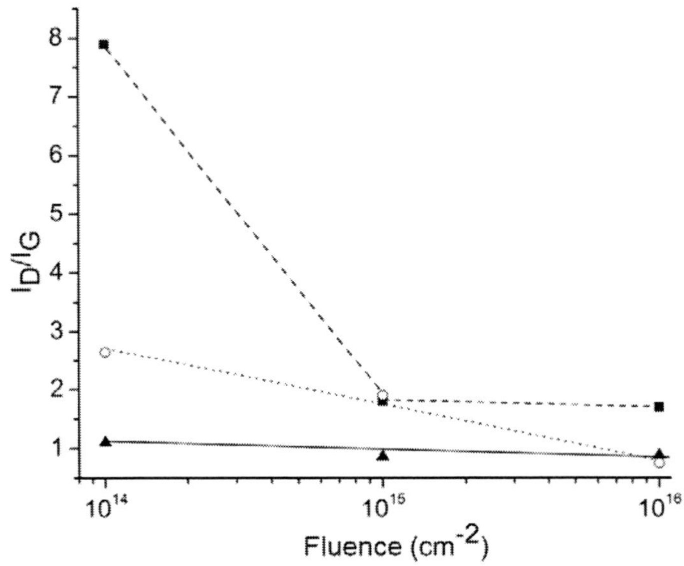

Figure 2a. The integrated-intensity ratio of D and G peaks vs. fluences:(■) irradiated by N^{5+}, (○) irradiated by B^{3+} and (▲) irradiated by Fe^{6+} ions.

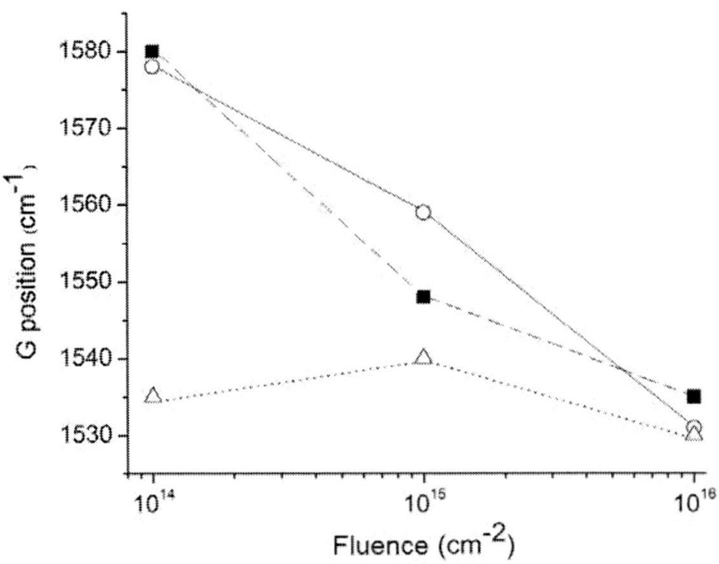

Figure 2b. The variation of the position of G peak vs. fluences: (■) irradiated by N^{5+}, (○) irradiated by B^{3+} and (Δ) irradiated by Fe^{6+} ions.

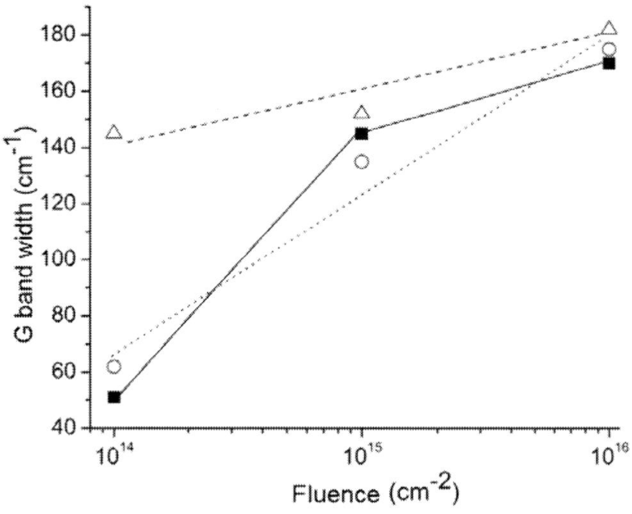

Figure 2c. The width of G peak vs. fluences: (■) irradiated by N^{5+}, (○) irradiated by B^{3+} and (Δ) irradiated by Fe^{6+} ions.

After irradiation of specimens by different multiple charged ions, the disintegration of fullerene molecules starts and the layer of amorphous carbon is formed. The downshift of G peak is associated to the decrease of sp^2-C sites probably by conversion of some sp^2 into sp^3 sites. The increase of G band width could be explained by the distortion of the sp^2 carbon bond angle, giving origin to five and seven-membered rings [17, 36]. Previous investigation showed that the effect of single nitrogen ion irradiation was the development of amorphous structure with 100 % sp^2 carbon bonds [37]. These authors observed the increase of I_D/I_G ratio and shifting of G peak upwards. We have observed the opposite effect in fullerene thin films bombarded by N^{5+} ions. The similar effect could be seen in fullerene thin films after irradiation by B^{3+} and Fe^{6+} ions. The decrease of I_D/I_G ratio with fluences is caused by the increase of sp^3 bonds.

When highly charged ions approach the surface of fullerene thin films, electrons contributing to fullerene bonding are extracted within a short time by the strong Coulomb potential of incident projectile. Electrons are instantly depleted in the ion irradiated area, and the rapid reconstruction of bonding leads to the formation of unstable structures. After irradiation of fullerene thin films by Fe^{6+} ions, nanodiamonds possessing sp^3 bonds could be formed [38]. During irradiation of fullerene thin films by N^{5+} and B^{3+} ions, amorphous carbon nitride and amorphous boron carbide were formed. The crystallites of carbon nitride and boron carbide are so small that they could not be seen by XRD technique. The thinness of the modified layer is prohibitive for many analytical methods.

3.2. FTIR Spectroscopy

FTIR measurements have been performed on pristine and irradiated specimens. As for pristine specimen, four IR peaks at 527, 576, 1183 and 1428 cm^{-1} could be identified [39]. These peaks could be observed in FTIR spectra of fullerene thin films irradiated by B^{3+} and N^{5+} at fluence of $2x10^{14}$ cm^{-2}. The complete disintegration of fullerene molecules has not been occurred. FTIR spectra (1000-2500 cm^{-1}) of fullerene thin films irradiated by B^{3+} and N^{5+} ions (ID=$2x10^{16}$ cm^{-2}) are presented in Figs. 3 (a and b). After bombardment of fullerene thin films by B^{3+} ions, a few peaks and bands which belong to B-C bonds could be identified. The band at 1100 cm^{-1} is assigned to B-C bonds in the icosahedra while the broad peak at 1280 cm^{-1} has the same origin like 1200 cm^{-1} line observed in polycrystalline boron carbide [40]. The bands between 1500-1600 cm^{-1} correspond to C=C bond (sp^2 bonding) [41]. As for

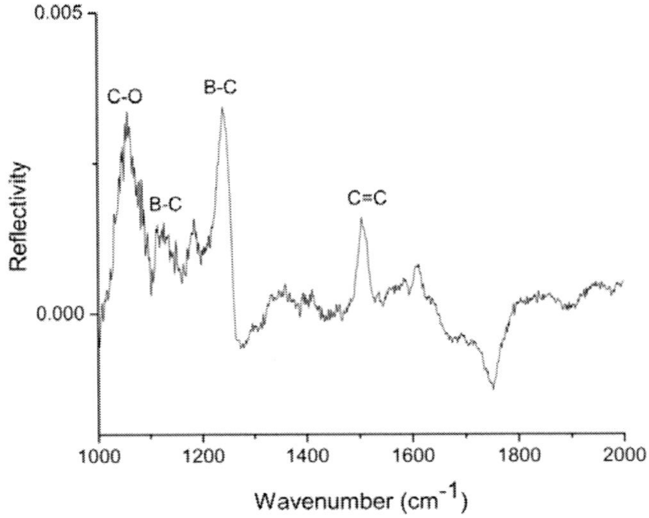

Figure 3a. The Fourier transform infrared spectra of fullerene thin films irradiated by B^{3+} ions at fluences of $2x10^{16}$ cm^{-2}. Spectra were baseline corrected.

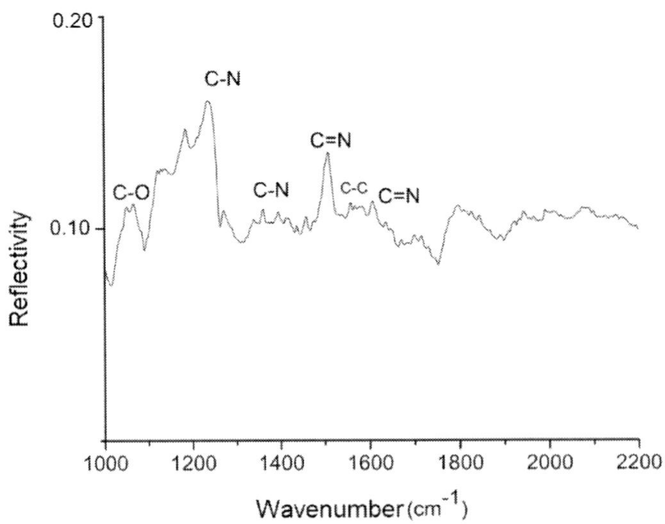

Figure 3b. The Fourier transform infrared spectra of fullerene thin films irradiated by N^{5+} ions at the fluences of $2x10^{16}$ cm^{-2}. Spectra were baseline corrected.

irradiated specimens by N^{5+} ions, a few C-N bonds could be observed in FTIR spectrum. The small peak at 1103 cm^{-1} corresponds to C-O stretching [42]. The broad peak at 1230 cm^{-1} is assigned to C-N stretching modes [43] while the broad peak at 1350 cm^{-1} is related to C-N [44]. The peak at 1570 cm^{-1} could be identified as C=C bond [45] while the peak at 1600 cm^{-1} could be assigned to both C=N [46] and sp^2 bonds in graphitic microdomains [47]. After irradiation with Fe^{6+} ions even at the lowest fluences (2x1014 cm^{-2}), neither peaks which correspond to Fe-C bonds nor the peaks of C_{60} could not be identified in FTIR spectrum.

3.3. Surface Morphology

The morphologies of as-deposited and bombarded samples were examined by atomic force microscopy (AFM). Figs. 4 (a,b,c) depict 3D view images of pristine and irradiated thin films by different ions (B^{3+}, Fe^{6+}). AFM images reveal a change in surface microstructure after bombardment by different multiple charged ions. The bombardment of fullerene thin films causes the modification on the surface of fullerene thin films. The cluster size in pristine thin films is 360 nm. In Table 2, cluster sizes in irradiated films by three different multiple charged ions at the lowest and highest fluences are given.

Table 2. Cluster size in fullerene thin films after bombardment by different ions.

Fluences (cm^{-2})	Cluster sizes in bombarded fullerene films by N^{5+} ions(nm)	Cluster sizes in bombarded fullerene films by B^{3+} ions (nm)	Cluster sizes in bombarded fullerene films by Fe^{6+} ions (nm)
$2x10^{14}$	400	250-300	400
$2x10^{16}$	300-400	150-200	350-400

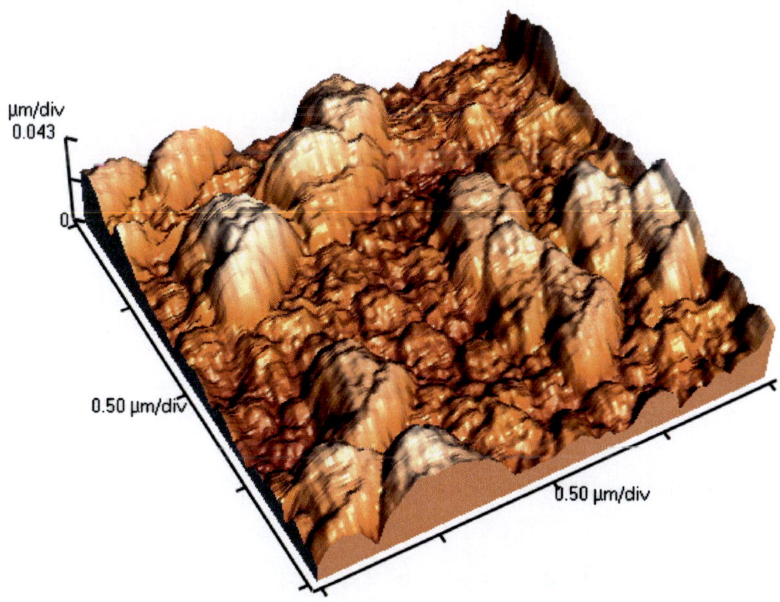

Figure 4a. AFM image of as-deposited fullerene thin films.

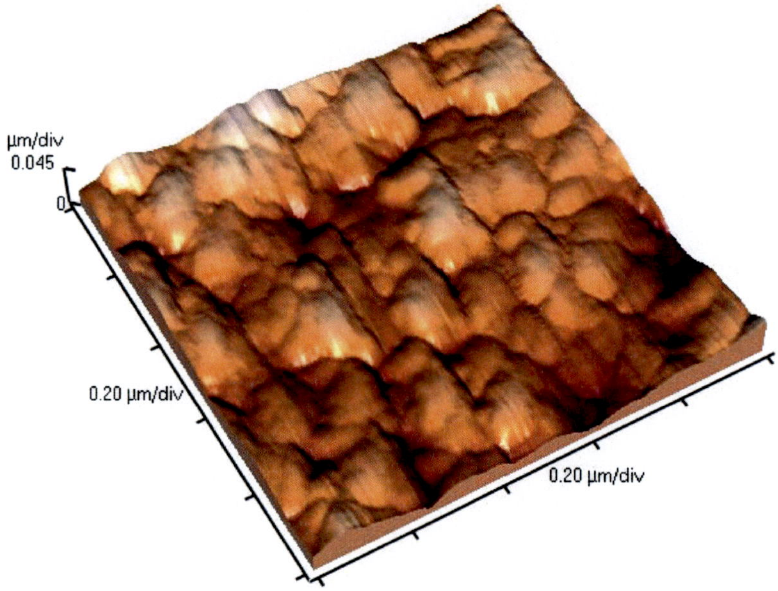

Figure 4b. AFM image of bombarded fullerene thin films by B^{3+} ions. Beam energy was 45 keV. The fluences were $2x10^{16}$ cm^{-2}.

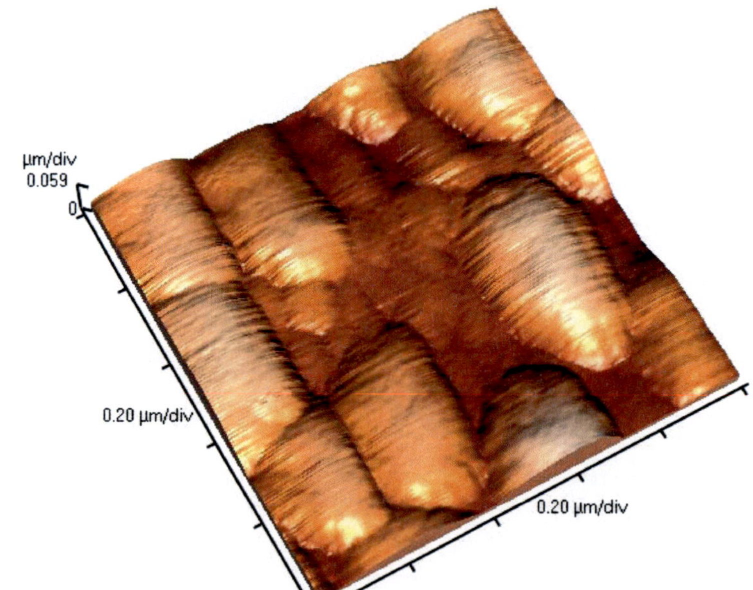

Figure 4c. AFM image of bombarded fullerene thin films by Fe^{6+} ions. Beam energy was 90 keV. The fluences were $2x10^{16}$ cm^{-2}.

The surface roughness of as-deposited and irradiated samples is differed very much. The RMS roughness of as-deposited films was 14.3 nm. After bombardment surface roughness of samples has been changed significantly. At the highest applied fluences, surface roughness of irradiated samples was decreased. In Table 3, the changes in surface roughness of bombarded fullerene thin films are given.

Table 3. RMS roughness of bombarded fullerene thin films by different ions.

Fluences (cm^{-2})	RMS roughness of bombarded samples by N^{5+} ions (nm)	RMS roughness of bombarded samples by B^{3+} ions (nm)	RMS roughness of bombarded samples by Fe^{6+} ions (nm)
$2x10^{14}$	13.1	16.67	13.435
$2x10^{15}$	12.5	15.5	12.4
$2x10^{16}$	10.6	13.9	11.5986

3.4. Optical Band Gap

The optical properties of as-deposited and irradiated fullerene thin films were examined by UV/Vis spectrometry. Tauc relationship was used to evaluate the energy gap (E_g) [48]. In Fig. 5, $(\alpha E)^{1/2}$ is presented as a function of photon energy (E). The optical band gap is obtained from the extrapolation of the linear part of the curve at the absorption coefficient α=0 using the Tauc relation. The optical band gap of fullerene is 1.7 eV. In Table 4, the values of optical band gap of irradiated fullerene thin films by different ions are given. The presented values of optical band of irradiated specimens are given for all applied fluences. The values of optical band gap of boron implanted specimens are almost identical to the values of band gaps of B_xC films produced by PECVD and sputter deposited B_xC films [49]. Compared to band gap of ion beam deposited DLC films and a:DLC films deposited by plasma decomposition of

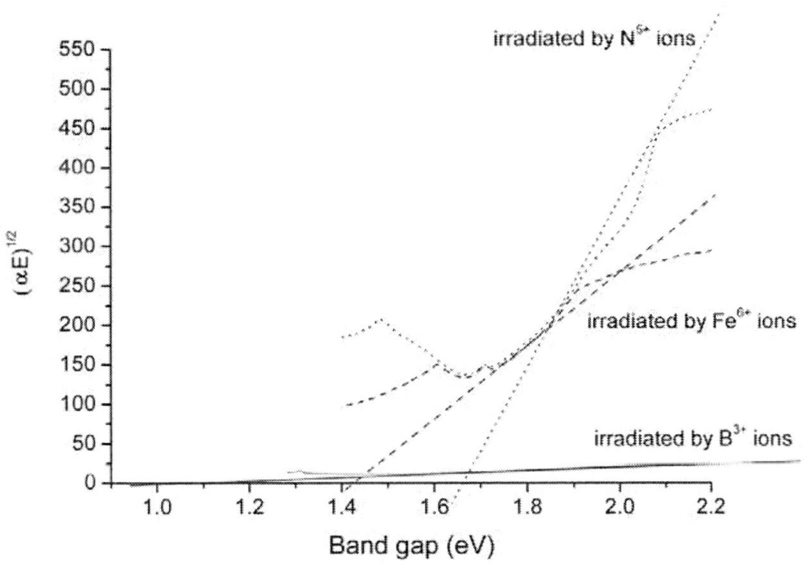

Figure 5. Optical band gap of fullerene thin films irradiated by different ions. The fluences were $2x10^{16}cm^{-2}$.

hydrocarbon gas, the obtained values of band gap of irradiated samples by Fe^{6+} ions are very similar [50-52]. As for films bombarded by N^{5+} ions optical band gap is reduced from 1.7 to 1.6 at the lowest fluences ($2x10^{14}$ cm^{-2}). By increasing the fluences, optical band is decreased to 1.3 eV. The obtained values of band gap of irradiated specimens by N^{5+} ions are in a good correspondence with the values of band gap of CN_x thin films deposited by reactive pulsed laser [53]. These values of band gap indicate that N/C ratio is low (~0.6) [54].

Table 4. Optical band gap of irradiated specimens by different ions.

Fluences (cm^{-2})	Optical band gap of irradiated films by N^{5+} ions	Optical band gap of irradiated films by B^{3+} ions	Optical band gap of irradiated films by Fe^{6+} ions
$2x10^{14}$	1.61	1.55	0.61
$2x10^{15}$	1.29	1.39	1.19
$2x10^{16}$	1.68	1.06	1.41

4. Conclusion

In this chapter, the results of structural modifiation of fullerene thin films bombarded by three different multiple charged ions have been presented. Raman analysis of irradiated specimens indicated the formation of highly disoredered structure with sp^3 bonds spread throughout sp^2 structures. Based on the results of FTIR spectroscopy we observed the formation of new C-N and B-C bonds. The optical band gap was found to be varied from 1.7 to 0.7 eV. The obtained values of band gap of boron and nitrogen implanted specimens confirmed that it was possible formation of diamond like structures.

Acknowledgements

The authors thank the support of the Ministry of Science and Environmental Protection of Republic of Serbia, via Project No. 141015b and TESLA Project No. 1247.

References

[1] Hanley L.; Sinnot S.B. *Surf. Sci.* 2002, 500, 500-522.

[2] Park J.W.; Sohn C.W.; Choi B.H. *Curr. Appl.Phys*. 2006, 6, 188-193.

[3] Gu Y.S.; Pan L.Q. *J. Mater. Sci. Lett.* 1996, 15, 1355-1357.

[4] Hoffman A.; Gouzman J.; Brener R. *Appl. Phys. Lett.* 1994, 64, 845-847.

[5] Withrow S.P.; Williams J.M.; Prawer S.; Barbara D. *J. Appl. Phys*. 1995, 78, 3060-3065.

[6] Ronning C.; Schwen D.; Eyhnsen.; Vetter U.; Hofsass H. *Surf. Coat. Technol*. 2002, 158-159, 382-387.

[7] Todorović-Marković B.; Draganić I., Vasiljević-Radović D.; Romčević N.; Romčević M.; Dramićanin M.; Marković Z. *Appl. Surf. Sci.* 2007, 253, 4029-4035.

[8] Kalish R.; Samoiloff A.; Hoffman A.; Uzan-Saguy C. Phys. Rev. B 1993,48, 18235-18238.

[9] Narayanan K.Z.; Yamaguchi M.; Dharmarasu N.; Kojima N.; Kanjilal D. *Nucl. Instr.& Meth. B.* 2001, 178, 301-304.
[10] Palmetshofer L.; Kastner *J. Nucl. Instr.& Meth. B* 1995, 96, 343-346.
[11] Prawer S.; Nugent K.W.; Biggs S.; McCulloch D.G.; Leong W.H.; Hoffman A.; Kalish R. *Phys. Rev. B.* 1995, 52, 841-849.
[12] Ren Z.M.; Long X. X.; Du Y. C.;, Ying Z.F.; Xiong X. X.; Li F. M. *Nucl. Instr.& Meth. B.* 1995, 100, 55-58.
[13] Foerster C.E.; Serbena F.C.; Lepieciski C.M.; Baptista D.Z.; Zawislak F.C. *Nucl. Instr.& Meth. B.* 1999, 148, 634-638.
[14] Cataldo F.; Baratta G.A.; Strazzulla G. *Full.Nanot.Carb.Nanostruc.*; 2002, 10, 197-206.
[15] Cataldo F.; Baratta G.A.; Ferini G.; Strazzulla G. *Full.Nanot.Carb.Nanostruc.*; 2003, 11, 191-199.
[16] Fink D.; Muller M.; Szimkowiak P.; Dobeli M.; Spieler A.; Grether M.; Richter A.; Chadderton L.T.; Zawislak F.C.; Grande P.F.P.; Behar M.; Hnatowicz V.; Vacik J. *Full. Sci&Technol.* 1998, 6, 911-962.
[17] Baptista D. L.; Zawislak F.C. *Diamond Relat. Mater.* 2004, 13, 1791-1801.
[18] Papaleo R.M.; Hallen A.; Eriksson J.; Brinkmalm G.; Demirev P.; Hakanssen P.; Sundqvist B. U. *Nucl. Instr.&.Meth.* B 1994, 91, 124-128.
[19] Arnau A.; Aumayr F.; Echenique P.M.; Grether M.; Heiland W.; Limburg J.; Morgenstern R.; Roncin P.; Schippers S.; Schuch R.; Stolterfoht N.; Varga P.; Zouros T. M. J.; Winter H. P. *Surf. Sci. Rep.* 1997, 27, 113-239.
[20] Schenkel T.; Hamza A. V.; Barnes A. V.; Schneider D. H. *Prog. Surf. Sci.* 1999, 61, 23-84.
[21] Baba Y.; Nagata K.; Takahashi S.; Nakamura N.; Yoshiyasu N.; Sakurai M.; Yamada C.; Ohtani S.; Tona M. *Surf. Sci.* 2005, 599, 248-254.
[22] Mochiji K., Yamamoto S.; Shimizu H.; Ohtani S.; Seguchi T.; Kobayashi N. *J. Appl. Phys.* 1997, 82, 6037-6040.
[23] Gebenshuber I.C.; Cernusca S.; Aumayr F.; Winter H.P. *Int. J. Mass. Spectrom.* 2003, 229, 27-34.
[24] Hedstrom M.; Cheng H. P. *J. Phys. Chem. B* 2000, 104, 4633-4611.
[25] Schlatholter T.; Hadjar O.; Hoekstra R.; Morgenstern R. *Appl. Phys. A,* 2001, 72, 281-287.
[26] Dobrosavljević A.; Milosavljević M.; Bibić N.; Efremov A.A. *Rev.Sci. Instr.* 2000, 71, 1-3.
[27] Dobrosavljević A.; Šiljegović M.; Draganić I.; Čizmić B. *Rev. Sci. Instr.* 2004, 75, 1-4.
[28] Draganić I.; Nedeljković T.; Dobrosavljević A.; Šiljegović M. *Rev. Sci. Instrum.* 2006, 77, 03A306.
[29] Dresselhaus M.S.; Dresselhaus G.; Eklund P.C. Science of Fullerenes and Carbon Nanotubes, Academic Press: New York, US, 1996, p.355.
[30] Talapatra S.; Cheng J. Y.; Chakrapani N.; Trasobares S.; Cao A.; Vajtai R.; Huang M. B.; Ajayan P.M. *Nanotechnology* 2006, 17, 305-309.
[31] Ferrari A.C.; Robertson J. *Phys. Rev. B* 2000, 61, 14095-14108.
[32] Lo K. L.; Lee M.C. C. *J.Phys.* 1993, 31, 653-661.
[33] Todorović-Marković B.; Draganić I.; Marković Z.; Romčević N.; Romčević M.; Stojanović Z.; Nikolić Z.; Mitrić M. Fullerenes, *Nanotubes&Carbon Nanostructur.* 2007, 15, 1-13.

[34] Todorović-Marković B.; Draganić I.; Vasiljević-Radović D.; Romčević N.; Blanuša J.; Dramićanin M.; Marković Z. sumitted to *Appl. Phys. A.*

[35] Wen F.; Huang N.; Sun H.; Wang J.; Leng Y.X. *Surf. &Coat. Technol.* 2004, 186, 118-124.

[36] Tallant D. R.; Parmeter J. E.; Siegal M. P.; Simpson R. L. *Diamond&Relat. Mater*. 1995, 4, 191-199.

[37] 37. Baptista D.L.; Foerster C.E.; Lepienski C.M.; Zawislak F.C. *Nucl. Instrum.&Meth. B* 2004, 218, 61-67.

[38] Meguro T.; Hida A.; Suzuki M.; Koguchi Y.; Takai H.; Yamamoto Y.; Maeda K.; Aoyagi Y. *Appl. Phys. Lett.* 2001, 79, 3866-3868.

[39] Kuzmany H.; Winkler R.; Pichler T. *J. Phys.:Condens. Matter* 1995, 7, 6601-6624.

[40] Jacobsohn L.G.; Nastase M. *Surf.Coat. Technol.* 2005, 200, 1472-1475.

[41] Podder J.; Rusop M.; Soga T.; Jimbo T. *Diamond Relat. Mater.* 2005, 14, 1799-1804.

[42] Ren Z. M.; Wang P.N.; Du Y. C.; Ying Z. F.; Li F. M. *Appl. Phys. A* 1997, 65, 407-410.

[43] Mihailescu I. N.; Gyorgy E.; Alexandrescu R.; Luches A.; Perrone A.; Ghica C.; Werckmann J.; Cojocaru I.; Chumash V. *Thin Solid Films* 1998, 323, 72-78.

[44] Gonzalez P.; Soto R.; Parada E. G.; Redondas X.; Chiussi S.; Serra J.; Pou J.; Leon B.; Perez-Amor M. *Appl. Surf. Sci.,* 1997, 109-110, 380-383.

[45] Kaufman J. H.; Metin S.; Saperstein D. D. *Phys. Rev B* 1989, 39, 13053-13060.

[46] He X. M.; Shu L.; Li W.Z.; Li H. D. *J. Mater. Res.* 1997, 12, 1595-1598.

[47] Saito K.; Koga Y. *Nucl. Inst.&Meth. B* 1997, 121, 400-403.

[48] Tauc J.; Grigorovici R.; Vancu A. *Phys. Status Solidi* 1966, 15, 627-629.

[49] Ahmad A. A.; Ianno N. J.; Hwang S. D.; Dowben P.A. *Thin Solid Films* 1998, 335, 174-177.

[50] Zhang Z.J.; Narumi K.; Naramoto H.; Wu Z.P.; Yamamoto S.; Miyashita A.; *J. Phys.:Condens.Matter* 1999, 11, L273-L277.

[51] Allon-Alaluf M.; Croitoru N. *Diamond&Relat. Mater*. 1997, 6, 555-558.

[52] Kinoshita H.; Yamashita M.; Yamaguchi T. *Jap.J. Appl. Phys.* 2006, 45, 8401-8405.

[53] Ong. C. W.; Zhao X. A.; Tsang Y. C.; Choy C. L.; Chan P. W. *Thin Solid Films* 1996, 280, 1-4.

[54] Jelinek M.; Kulisch W.; Delplancke-Ogletree M. P.; Lančok J.; Jastrabik L.; Chrostova D.; Popov C.; Bulir *J. Appl. Phys. A* 2001, 73, 167-70.

INDEX

A

B

C

D

E

F

G

H

I

J

K

L

M

N

Q

R

S

T